Oceanography: An Earth Science Perspective

Oceanography: An Earth Science Perspective

Editor: Austin Brennan

R CALLISTO REFERENCE

www.callistoreference.com

Callisto Reference,
118-35 Queens Blvd., Suite 400,
Forest Hills, NY 11375, USA

Visit us on the World Wide Web at:
www.callistoreference.com

ISBN: 978-1-64116-586-0 (Hardback)

Cataloging-in-Publication Data

Oceanography : an earth science perspective / edited by Austin Brennan.
    p. cm.
Includes bibliographical references and index.
ISBN 978-1-64116-586-0
1. Oceanography. 2. Earth sciences. 3. Geology. I. Brennan, Austin.
GC11.2 .O24 2022
551.46--dc23

Table of Contents

Preface

The main aim of this book is to educate learners and enhance their research focus by presenting diverse topics covering this vast field. This is an advanced book which compiles significant studies by distinguished experts in the area of analysis. This book addresses successive solutions to the challenges arising in the area of application, along with it; the book provides scope for future developments.

Oceanography is a fundamental study of physical and biological aspects of ocean. It is an important branch of earth science. It covers a range of topics such as ocean currents, ecosystem dynamics, waves, plate tectonics, fluxes of physical properties and chemical substances within the ocean and across its boundaries, etc. The four main branches of oceanography are biological, chemical, geological and physical oceanography. Biological oceanography deals with the investigation of the ecology of marine organisms. It involves the physical, chemical and geological characteristics of their ocean environment and the biology of individual marine organisms. Chemical oceanography studies the chemistry of ocean which includes the study and understanding of seawater properties and its changes. Geological oceanography deals with in-depth study of geology of ocean floor which also includes study of plate tectonics and paleoceanography. The study of ocean's physical attributes fall under physical oceanography, which involves the studies of temperature-salinity structure, surface waves, internal waves, etc. This book brings forth some of the most innovative concepts and elucidates the unexplored aspects of oceanography. It also traces the progress of this field and highlights some of its key concepts and applications. This book is a resource guide for experts as well as students.

It was a great honour to edit this book, though there were challenges, as it involved a lot of communication and networking between me and the editorial team. However, the end result was this all-inclusive book covering diverse themes in the field.

Finally, it is important to acknowledge the efforts of the contributors for their excellent chapters, through which a wide variety of issues have been addressed. I would also like to thank my colleagues for their valuable feedback during the making of this book.

Editor

Stable carbon isotopes of dissolved inorganic carbon for a zonal transect across the subpolar North Atlantic Ocean in summer 2014

Matthew P. Humphreys[1], Florence M. Greatrix[1], Eithne Tynan[1], Eric P. Achterberg[1,2],
Alex M. Griffiths[3], Claudia H. Fry[1], Rebecca Garley[4], Alison McDonald[5], and Adrian J. Boyce[5]

[1]Ocean and Earth Science, University of Southampton, Southampton, UK
[2]GEOMAR Helmholtz Centre for Ocean Research, Kiel, Germany
[3]Department of Earth Science and Engineering, Imperial College London, London, UK
[4]Bermuda Institute of Ocean Sciences, St George's, Bermuda
[5]Scottish Universities Environmental Research Centre, East Kilbride, UK

Correspondence to: Matthew P. Humphreys (m.p.humphreys@soton.ac.uk)

Abstract. The stable carbon isotope composition of dissolved inorganic carbon ($\delta^{13}C_{DIC}$) in seawater was measured in samples collected during June–July 2014 in the subpolar North Atlantic. Sample collection was carried out on the RRS *James Clark Ross* cruise JR302, part of the "Radiatively Active Gases from the North Atlantic Region and Climate Change" (RAGNARoCC) research programme. The observed $\delta^{13}C_{DIC}$ values for cruise JR302 fall in a range from -0.07 to $+1.95\permil$, relative to the Vienna Pee Dee Belemnite standard. From duplicate samples collected during the cruise, the 1σ precision for the 341 results is $0.08\permil$, which is similar to our previous work and other studies of this kind. We also performed a cross-over analysis using nearby historical $\delta^{13}C_{DIC}$ data, which indicated that there were no significant systematic offsets between our measurements and previously published results. We also included seawater reference material (RM) produced by A. G. Dickson (Scripps Institution of Oceanography, USA) in every batch of analysis, enabling us to improve upon the calibration and quality-control procedures from a previous study. The $\delta^{13}C_{DIC}$ is consistent within each RM batch, although its value is not certified. We report $\delta^{13}C_{DIC}$ values of $1.15 \pm 0.03\permil$ and $1.27 \pm 0.05\permil$ for batches 141 and 144 respectively. Our JR302 $\delta^{13}C_{DIC}$ data can be used – along with measurements of other biogeochemical variables – to constrain the processes that control DIC in the interior ocean, in particular the oceanic uptake of anthropogenic carbon dioxide and the biological carbon pump.

1 Introduction

The global ocean has absorbed up to half of the anthropogenic carbon dioxide (CO_2) emitted since the early 1800s (Sabine et al., 2004; Khatiwala et al., 2009, 2013) and it continues to take up about a quarter of annual CO_2 emissions at the present day (Le Quéré et al., 2009), substantially decreasing CO_2 accumulation in the atmosphere. The consequences of this uptake include a decline in pH – known as ocean acidification – with lower pH values predicted to persist for centuries longer than the atmospheric CO_2 anomaly (Caldeira and Wickett, 2003), which will have impacts on marine biogeochemistry and ecology that we are only just beginning to understand (Doney et al., 2009; Achterberg, 2014; Gaylord et al., 2015).

In order to predict the response of the oceanic CO_2 sink to the continuing rise of the atmospheric partial pressure of CO_2 (pCO_2), it is useful to first understand the existing spatial distribution of anthropogenic dissolved inorganic carbon (DIC_{anth}) in the ocean interior. Various methods have been employed to this end (Sabine and Tanhua, 2010), including back-calculation from DIC, total alkalinity (TA) and dissolved oxygen observations (Brewer, 1978; Chen and Millero, 1979; Gruber et al., 1996); inference from the oceanic distributions of other anthropogenic gases such as chlorofluorocarbons (Hall et al., 2002; Waugh et al., 2006); and multi-linear regressions using measurements from pairs of cruises in the same region, but separated in time (Friis et al., 2005; Tanhua et al., 2007). Oceanic measurements during the past few decades (Quay et al., 2007) and over longer timescales in ice cores (Rubino et al., 2013) show that the rise in $p\mathrm{CO_2}$ and DIC has been accompanied by a decline in the carbon-13 content of DIC, relative to carbon-12 (reported as $\delta^{13}\mathrm{C}$, Eqs. 1 and 2), a phenomenon called the "Suess effect" (Keeling, 1979). This is caused by the lower $\delta^{13}\mathrm{C}$ of anthropogenic $\mathrm{CO_2}$ relative to pre-industrial and present-day atmospheric $\mathrm{CO_2}$, and it provides another approach to constrain the spatial distribution and inventory of anthropogenic DIC (e.g. Quay et al., 1992, 2003, 2007; Sonnerup et al., 1999, 2007; Körtzinger et al., 2003). The Suess effect has caused significant changes in the present-day distribution of $\delta^{13}\mathrm{C_{DIC}}$ in the ocean interior (Olsen and Ninnemann, 2010). Continued observations of oceanic $\delta^{13}\mathrm{C_{DIC}}$ are essential for verification of the parameterisations of ocean carbon cycle models (Sonnerup and Quay, 2012).

Here, we present measurements of $\delta^{13}\mathrm{C_{DIC}}$ from a zonal transect across the subpolar North Atlantic Ocean in June–July 2014. The cruise, JR302 on the RRS *James Clark Ross*, was carried out as part of the "Radiatively Active Gases from the North Atlantic Region and Climate Change" (RAGNARoCC) research programme. Our observations fill important spatiotemporal gaps in the existing global data set (Schmittner et al., 2013), and will contribute towards the scientific objectives summarised above. Our analysis was carried out following the methodology presented by Humphreys et al. (2015a), but we have been able to make several improvements to the raw data processing and calibration procedures by inclusion of seawater reference material (RM) in every batch of sample analysis. This RM, produced by A. G. Dickson (Scripps Institution of Oceanography, USA), is mainly used for assessing the accuracy of non-isotopic marine carbonate chemistry measurements, and it does not have a certified $\delta^{13}\mathrm{C_{DIC}}$ value. Nevertheless, the $\delta^{13}\mathrm{C_{DIC}}$ of different RM bottles from the same RM batch should be consistent, allowing us to assess the consistency of our measurements between analysis batches. We determine $\delta^{13}\mathrm{C_{DIC}}$ values for the two RM batches that we measured (141 and 144), which could be used to check for systematic offsets between our results and those from other laboratories. We also use the RM results to carry out a statistical analysis of our measurement precision both within and between analysis batches.

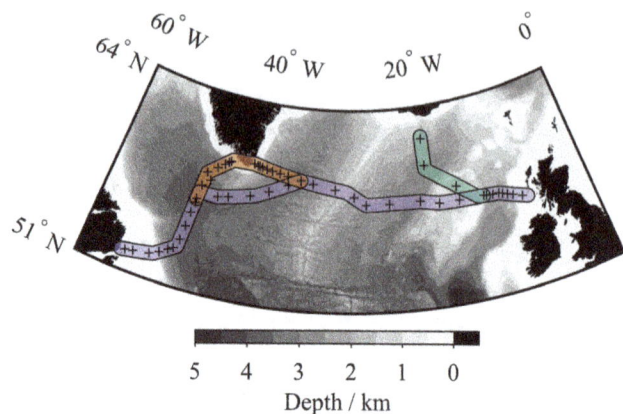

Figure 1. Bathymetric map of the subpolar North Atlantic Ocean. Black plusses show $\delta^{13}\mathrm{C_{DIC}}$ sampling locations during cruise JR302. Coloured sections indicate illustrated transects: blue for Fig. 6, orange for Fig. 7, and green for Fig. 8. Bathymetry data are from the GEBCO_2014 grid, version 20150318, http://www.gebco.net.

2 Sample collection

2.1 Cruise details

The $\delta^{13}\mathrm{C_{DIC}}$ samples were collected during RRS *James Clark Ross* cruise JR302, which was an approximately zonal transect from St John's, Newfoundland, Canada, to Immingham, UK (Fig. 1), from June to July 2014 (King and Holliday, 2015). During the crossing, several transects were sailed in towards the coast of Greenland, and in the eastern region the ship carried out a short meridional transect north towards Iceland in order to sample the Extended Ellett Line (Holliday and Cunningham, 2013).

2.2 Sample collection and storage

Prior to sample collection, the containers were thoroughly rinsed with deionised water (Milli-Q water, Millipore, $> 18.2\,\mathrm{m\Omega\,cm^{-1}}$). Samples were collected from the source (either seawater sampling bottle or underway seawater supply) via silicone tubing, following established best-practice protocols (Dickson et al., 2007; McNichol et al., 2010), as summarised here. The containers were thoroughly rinsed with excess seawater sample immediately before filling until overflowing with seawater, taking care not to generate or trap air bubbles. Two different sample containers were used: (1) 100 mL glass "bottles" with ground glass stoppers, lubricated with Apiezon® L grease and held shut with electrical tape, and (2) 50 mL glass "vials", with plastic screw-cap lids and PTFE/silicone septa. In order to sterilise each sample, 0.02 % of the sample container volume of saturated mercuric chloride solution was added before sealing. A 1 mL air headspace (i.e. 1 % of the sample volume) was also introduced to the bottles, prior to poisoning, by removing this

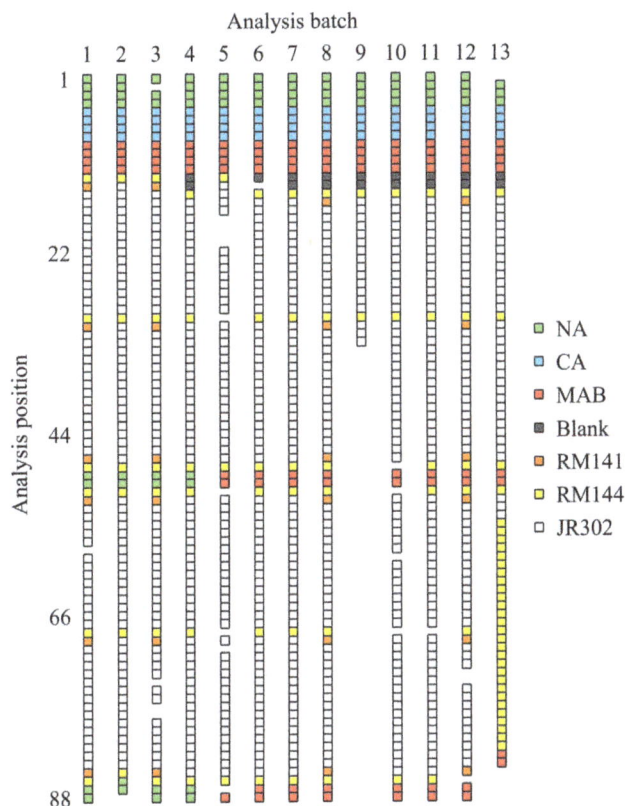

Figure 2. Schematic arrangement of calibration standards (NA, CA and MAB), blanks, RM and samples (JR302) within each analysis batch. Each square represents a separate measurement (i.e. a set of 10 technical replicates). Gaps are left where results have failed quality control.

volume of seawater via pipette. This prevents thermal expansion/contraction of the seawater from breaking the airtight seal. However, the flexible septa on the vials allowed them to be sealed when full of seawater. All samples were stored in the dark until analysis.

3 Sample analysis

All of the $\delta^{13}C_{DIC}$ samples were analysed at the Scottish Universities Environmental Research Centre Isotope Community Support Facility (SUERC-ICSF) in East Kilbride (UK), in June–July 2015. We describe the analysis procedure here only in brief, as it was identical to that of Humphreys et al. (2015a).

The samples were analysed in 13 batches. Each analysis batch consisted of up to 88 measurements, 16 of which were of calibration standards, and the remainder were of seawater samples, "blanks" or RM (Fig. 2). Each batch underwent a three-step process of overgassing, equilibration and measurement. For the overgassing step, the air in each of the measurement vials (12 mL Exetainer®) was flushed out and replaced with helium by a PAL system (CTC Analytics). For

equilibration, the standards, samples and RM were reacted with phosphoric acid to convert all DIC to CO_2. Finally, the gaseous headspace in each measurement vial was then sampled by the PAL system and transferred to a Thermo Scientific Delta V mass spectrometer via a Thermo Scientific GasBench II, and was measured 10 times (called technical replicates).

For seawater samples and RM, four drops of concentrated phosphoric acid were added to the Exetainer analysis vials prior to overgassing, and 1 mL of liquid sample was then injected into each vial for equilibration. For the standards, only the solid powder standard was added to the analysis vials prior to overgassing, and 1 mL of dilute (10 % by volume) phosphoric acid was added to each vial for equilibration. During batches 4 and 6–13, "blanks" were prepared in the same way as the standards, except that the Exetainer analysis vials were completely empty for overgassing, and had the same dilute acid added as the standards in that batch.

4 Measurement processing

We were able to make improvements to the processing of the raw measurements from our previous study (Humphreys et al., 2015a), in part by using the results from the RM that were included in every batch. The processing sequence used in this study is described in full here, including parts that are the same as in Humphreys et al. (2015a); differences between the two approaches are then discussed later. All processing was carried out using MATLAB® (MathWorks, USA).

4.1 Definitions

The relative abundance of ^{13}C to ^{12}C in a sample X is given by Eq. (1). The R_X is then normalised to a reference standard – i.e. Vienna Pee Dee Belemnite (V-PDB) (Coplen, 1995) – using Eq. (2).

$$R_X = \frac{[^{13}C]_X}{[^{12}C]_X},\tag{1}$$

where $[^{13}C]_X$ and $[^{12}C]_X$ are the concentrations of ^{13}C and ^{12}C respectively in X.

$$\delta^{13}C = \frac{R_{sample} - R_{V-PDB}}{R_{V-PDB}} \times 1000\text{‰}\tag{2}$$

4.2 General procedure

4.2.1 Anomalous measurement removal

Anomalous $\delta^{13}C$ measurements were first removed from the sets of technical replicates. These typically occurred when the CO_2 concentration in a replicate was too low, causing the peak area to fall outside the calibrated range (i.e. the range of peak areas covered by the standards). There were no

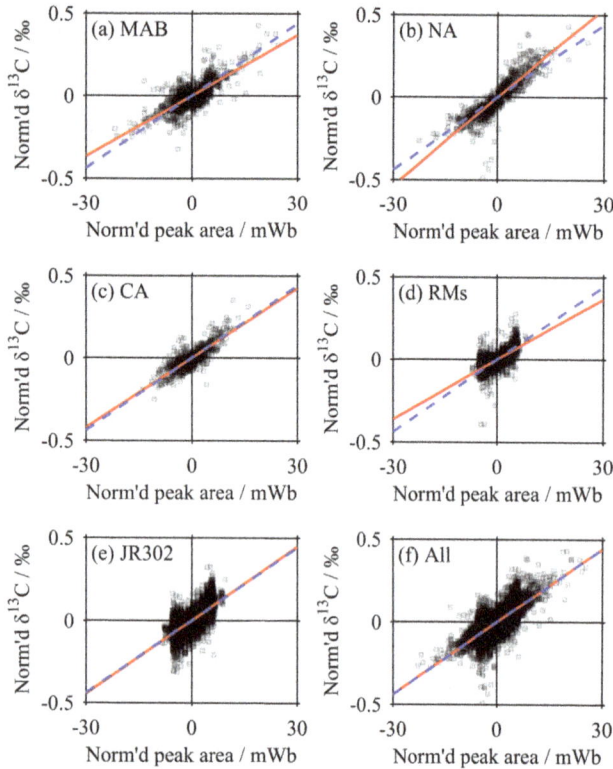

Figure 3. The peak area correction relationship. All technical replicates are plotted, normalised such that the mean peak area and $\delta^{13}C$ for each set of technical replicates are both 0 (‰ or mWb respectively). The red line is the same in each plot, showing the mean relationship used for all of the peak area corrections, while each dashed blue line shows the equivalent relationship determined only from the data scattered in each panel. Individual data points are semi-transparent.

sample measurements with peak areas greater than the calibrated range. Thus all measurements with a peak area less than 5 mWb were judged to be anomalous and discarded, and if this applied to 5 or more of the original 10 technical replicates for a given sample, the entire sample was discarded.

4.2.2 Peak area (linearity) correction

Virtually all samples, standards and RM showed a consistent decline in both peak area and raw $\delta^{13}C$ through each set of technical replicates, called "linearity" (Fig. 3). To correct for this, we first "normalised" the peak area and raw $\delta^{13}C$ of each set of technical replicates by subtracting the mean peak area and $\delta^{13}C$ respectively from each replicate; every set of technical replicates thus had a mean normalised peak area and $\delta^{13}C$ of 0 (mWb or ‰ respectively). We then performed an ordinary least-squares linear regression of normalised $\delta^{13}C$ against normalised peak area using all technical replicates from all of the samples, standards and RM. The regression was forced through the origin and had a gradient

(L) of 0.0147‰ mWb^{-1}. This was used to make a "peak area correction" to all technical replicates:

$$\delta_{\text{lin}} = \delta_{\text{raw}} - L(a - A_{\text{lin}}), \tag{3}$$

where δ_{lin} is the linearity-corrected $\delta^{13}C$, δ_{raw} is the raw $\delta^{13}C$ measurement, L is the correction gradient (i.e. 0.0147‰ mWb^{-1}), a is the peak area for the technical replicate in mWb, and A_{lin} is 20 mWb – the peak area that the correction is made to. The value of A_{lin} was chosen because it is the mean peak area for all of the seawater samples, thus minimising the magnitude of this correction.

4.2.3 Averaging

After the peak area correction had been applied, the mean $\delta^{13}C$ of each set of technical replicates was calculated. These mean values were then used for the remainder of the data processing.

4.2.4 Blank correction

A "blank correction" was then applied to the standards only (MAB, NA and CA). This was necessary because phosphoric was added to these after the overgassing step, so any CO_2 dissolved in the acid would be included in the measurement. It was not necessary for the seawater samples and RM, because here the acid was added prior to overgassing. The different procedures were necessary because of the different states of the standards and samples (solid and liquid respectively).

A pair of "blank" measurements were included during analysis batches 4 and 6–13. These were clean, empty Exetainer analysis vials that were otherwise treated in the same way as the standards: acid had been added after overgassing. The mean $\pm$ standard deviation (SD) peak area and linearity-corrected $\delta^{13}C$ of all of these blanks were 0.277 ± 0.024 mWb and 19.42 ± 2.47‰ respectively. These mean values were then used to make the blank correction to all standards:

$$\delta_{\text{blank}} = \frac{a\delta_{\text{lin}} - A_{\text{blank}}D_{\text{blank}}}{a - A_{\text{blank}}}, \tag{4}$$

where δ_{blank} is the blank-corrected $\delta^{13}C$, and A_{blank} and D_{blank} are the mean blank peak area and $\delta^{13}C$ (i.e. 0.277 ± 0.024 mWb and 19.42 ± 2.47‰ respectively).

Even after this blank correction, there remained unexplained relationships between peak area and $\delta^{13}C$ for the standards (Fig. 4). We therefore performed an ordinary least-squares regression between peak area and blank-corrected $\delta^{13}C$ for all measurements in all analysis batches of each standard, and took the value of the regression line at a peak area of 20 mWb as the $\delta^{13}C$ value for that standard in order to generate the V-PDB calibration curve (Table 1).

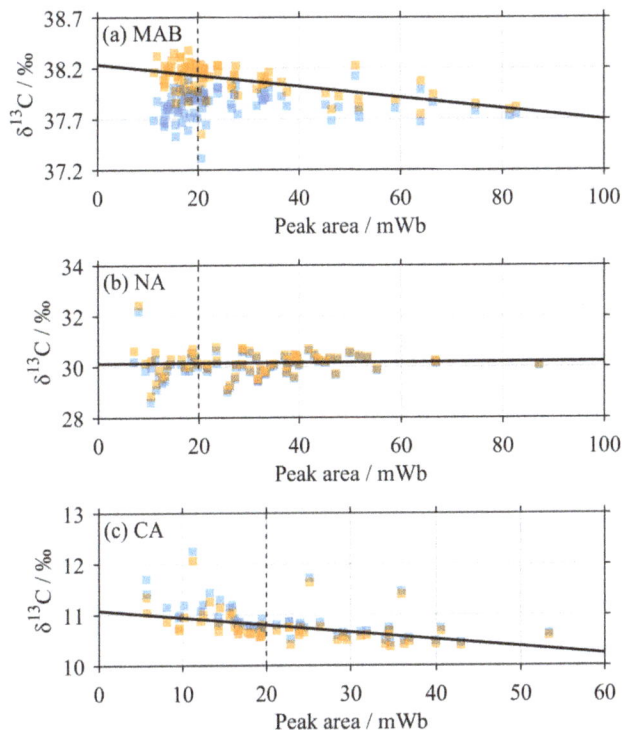

Figure 4. The blank correction for the standards. Each panel shows relationships between peak area and $\delta^{13}C$ for **(a)** MAB, **(b)** NA and **(c)** CA before (blue) and after (orange) blank correction. The solid black lines are linear least-squares regressions for the data after blank correction. Their intercepts with the dashed black lines at 20 mWb were used as the mean values for each standard to generate the V-PDB calibration curve (Fig. 5). Note that the panels have different axes resolutions.

4.2.5　Calibration to V-PDB

A second-order polynomial fit was generated to determine the certified $\delta^{13}C$ values for the standards (Table 1) from their blank-corrected mean values (Fig. 5):

$$\delta_{cert} = c\delta_{blank}^2 + s\delta_{blank} + f, \qquad (5)$$

where δ_{cert} is the certified $\delta^{13}C$ value (Table 1) and c, s and f are coefficients describing the curvature, stretch and offset respectively of the calibration fit, taking the following values: $c = -4.321 \times 10^{-3}\,\text{‰}^{-1}$, $s = 1.189$, and $f = -36.563\,\text{‰}$. Equation (5) was then used to calibrate all of the sample and RM measurements, by inputting the linearity-corrected $\delta^{13}C$ values as δ_{blank}; the output (δ_{cert}) gives the final, calibrated $\delta^{13}C_{DIC}$, relative to the V-PDB international standard (Coplen, 1995).

4.3　Quality control

4.3.1　Practical problems

We have excluded $\delta^{13}C_{DIC}$ results from our data set where practical problems were encountered and noted during sam-

Table 1. The SUERC-ICSF calibration standards.

Name	Chemical composition	Certified $\delta^{13}C$ (V-PDB)/‰	Blank-corrected $\delta^{13}C$ at 20 mWb/‰
MAB	$CaCO_3$	+2.48	+38.13
NA	$NaHCO_3$	−4.67	+30.13
CA	$CaCO_3$	−24.23	+10.80

ple analysis that discredited specific measurements. For example, during analysis batch 9 the automated needle became detached part way through the overgassing step, resulting in the loss of measurements after this point. The measurements that have been excluded in this way can be identified as gaps in Fig. 2.

4.4　Cross-over analysis

A cross-over analysis was performed using XOVER v1.0.0.1 (Humphreys, 2015) in order to evaluate the consistency of this study's results with "historical" measurements from the Schmittner et al. (2013) $\delta^{13}C_{DIC}$ compilation and our previous study (Humphreys et al., 2014a, b, 2015a). This compilation is probably the best available source for oceanic $\delta^{13}C_{DIC}$ measurements at present, although it may soon be superseded by ongoing efforts to merge and quality-control marine $\delta^{13}C_{DIC}$ data sets (e.g. Becker et al., 2016). The XOVER program follows a similar procedure to the secondary quality control toolbox of Lauvset and Tanhua (2015). Firstly, all historical sampling stations within 150 km of a JR302 CTD station were selected. The 150 km distance is the best compromise for minimising the spatial offset between the JR302 and historical observations while still capturing enough historical data to perform an effective cross-over analysis. At each of these historical stations, a piecewise cubic Hermite interpolating polynomial (PCHIP) fit was generated to predict $\delta^{13}C_{DIC}$ from depth. Values of $\delta^{13}C_{DIC}$ at depths which were equivalent to the JR302 observations were then interpolated using these PCHIP fits. Only JR302 data from deeper than 200 m were used, in order to limit the effect of seasonal variability in $\delta^{13}C_{DIC}$, which is relatively high near the ocean surface due to biological processes. The differences between the JR302 and historical $\delta^{13}C_{DIC}$ values were calculated and combined into a mean $\pm$ SD value for each cruise in the historical data sets.

4.5　Precision from duplicates

The SD obtained if one sample was measured many times (i.e. 1σ precision, 68.3 % confidence interval) can also be estimated from many duplicate measurements of different samples: it is equal to the mean of the absolute differences between the duplicate pairs divided by $2/\sqrt{\pi}$ (Thompson and Howarth, 1973; Humphreys et al., 2015a), as follows. For this purpose, each pair of duplicate measurements of a sample i is considered to be two values ($d_{i,1}$ and $d_{i,2}$) that have been randomly selected from a normal distribution with a SD equal to the 1σ measurement precision (P) and a mean equal to the "true" $\delta^{13}C_{DIC}$ value for that sample. The "duplicate pair difference" (D) for the sample i is then calculated by subtracting the $\delta^{13}C_{DIC}$ of the first duplicate from that of the second (i.e. $D_i = d_{i,2} - d_{i,1}$). As P is the same for every sample, D is normally distributed with a mean of 0 and a SD of $P\sqrt{2}$. The 1σ precision P can thus be estimated from the SD of all D by dividing it by $\sqrt{2}$. Alternatively, the absolute values of D follow a half-normal distribution, which has a mean value $2P/\sqrt{\pi}$; thus, P can also be estimated from the mean of the duplicate pair absolute differences (i.e. the mean of all $|D|$) by dividing the latter by $2/\sqrt{\pi}$. This last calculation (Eq. 6) was carried out for all of the analytical duplicate pairs and separately for the sampling duplicates to determine their respective 1σ confidence intervals:

$$P = \frac{\sqrt{\pi}}{2N} \sum_{i=1}^{N} |D_i|, \tag{6}$$

where N is the total number of duplicate pairs.

5 Results and discussion

5.1 Results in context

5.1.1 Interior $\delta^{13}C_{DIC}$ distribution

Our final $\delta^{13}C_{DIC}$ results are presented in Figs. 6–8. To first order, $\delta^{13}C_{DIC}$ is highest in surface waters (shallower than ca. 40 m) taking values up to 2‰. Then, $\delta^{13}C_{DIC}$ decreases with depth to minima just below 0‰ at about 500 m, before increasing again to intermediate values of around 1‰ in deeper waters. This pattern is in general agreement with previous studies (Schmittner et al., 2013).

5.1.2 Cross-over analysis

The cross-over analysis compared the results from this study with three nearby historical cruises: OACES93, 58GS20030922 and D379 (Table 2, Fig. 9). The mean $\delta^{13}C_{DIC}$ residual was significantly different from 0 for all of these cruises ($p < 0.01$). Despite this, the mean ($\pm$SD) $\delta^{13}C_{DIC}$ residuals for OACES93 and D379 (-0.08 ± 0.14‰ and $+0.08 \pm 0.16$‰ respectively) were no larger than our reported measurement precision of 0.08‰. Although the mean ($\pm$SD) residual for 58GS20030922 was greater (-0.19 ± 0.16‰), it must firstly be considered that this depends on only three matching $\delta^{13}C_{DIC}$ measurements, and secondly that it is still within the range of the accuracy of 0.1–0.2‰ reported for its parent data set (Schmittner et al., 2013). These cruises and JR302 span a time interval of just over 20 years, so invasion of anthropogenic CO_2 could have modified the $\delta^{13}C_{DIC}$ through the Suess effect (Keeling, 1979), potentially inhibiting the use of cross-over anal-

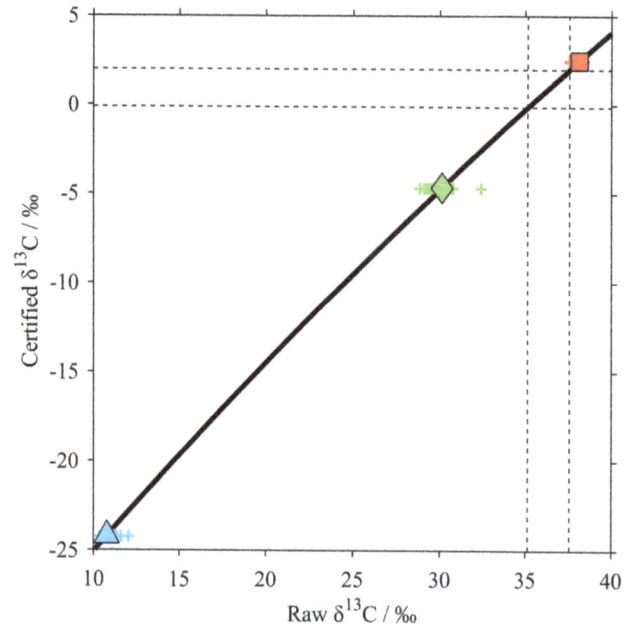

Figure 5. The V-PDB calibration. Mean values for each standard: red square for MAB, green diamond for NA, and blue triangle for CA; results of individual measurements of standards shown as plusses in the same colours as the means. Thick black line shows the calibration curve (Eq. 5); dashed black lines enclose the range of $\delta^{13}C$ values for the seawater samples and RM.

ysis. However, in the deeper part of the water column in this region, $\delta^{13}C_{DIC}$ has been observed to change at a rate of less than -0.01‰ yr^{-1} in recent decades (Humphreys et al., 2016). Thus the total change in $\delta^{13}C_{DIC}$ over the period between OACES93 (the earliest historical cruise) and JR302 should be no greater than the stated accuracy of the historical data set, and therefore our cross-over is valid. For future studies this will need to be reconsidered, as it might not hold true over longer timescales. We therefore conclude that any systematic bias between the results of this study and existing $\delta^{13}C_{DIC}$ data sets is negligible relative to the uncertainties of the measurements themselves.

5.2 Measurement uncertainty

5.2.1 Sample container types

In tests of duplicate samples collected in these two different container types, Humphreys et al. (2015a) were unable to find evidence of any systematic offset between $\delta^{13}C_{DIC}$ measurements from the same two sample container types that were used in this study. Here, five pairs of sampling duplicates were collected with one sample in each container type. The mean $\pm$SD difference in $\delta^{13}C_{DIC}$ for these duplicate pairs was -0.01 ± 0.04‰, with the difference always calculated as the $\delta^{13}C_{DIC}$ value measured in the sample collected in a 50 mL vial subtracted from that collected in a

Table 2. Results of the cross-over analysis. Sources: OACES93 on R/V *Malcolm Baldrige*, carbon PIs F. Millero, R. Feely and P. Quay, data from Schmittner et al. (2013); 58GS20030922 on G. O. *Sars*, carbon PIs A. Olsen and T. Johannessen, data from Schmittner et al. (2013); D379 on RRS *Discovery*, carbon PI A. M. Griffiths, data from Humphreys et al. (2015a).

Cross-over cruise	Sampling date	Mean of $\delta^{13}C_{DIC}$ residuals/‰	SD of $\delta^{13}C_{DIC}$ residuals/‰	Number of residuals
OACES93	Aug 1993	−0.08	0.14	19
58GS20030922	Oct 2003	−0.19	0.16	3
D379	Aug 2012	0.08	0.16	253

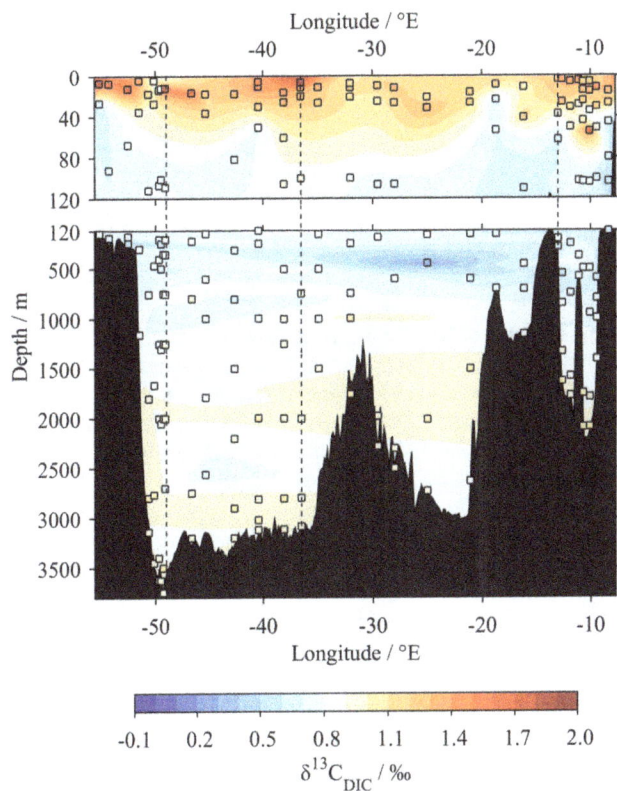

Figure 6. Zonal section of $\delta^{13}C_{DIC}$ results from this study, from west to east across the subpolar North Atlantic Ocean (blue in Fig. 1). Coloured squares show actual sample locations and $\delta^{13}C_{DIC}$ values. Vertical dashed lines indicate locations of joints with edges of Figs. 7 and 8. Bathymetry data are from the GEBCO_2014 grid, version 20150318, http://www.gebco.net.

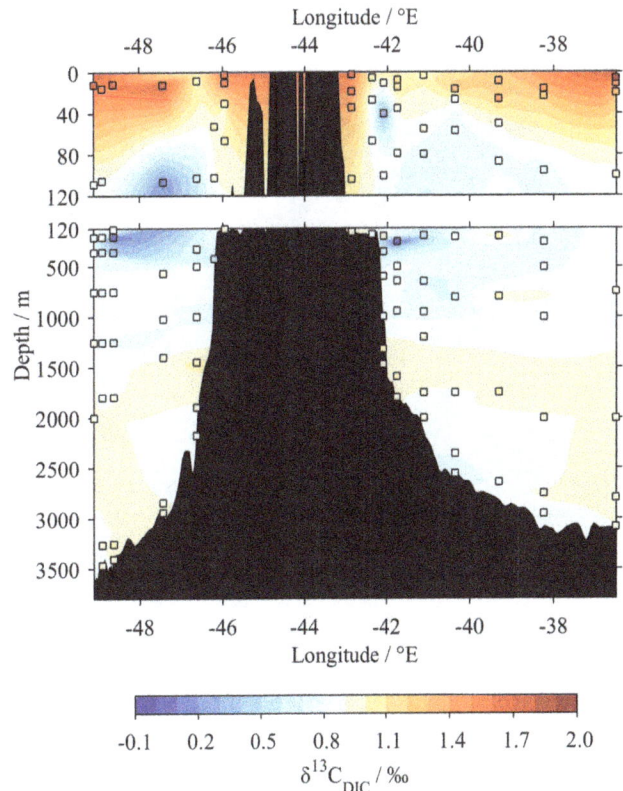

Figure 7. Zonal section of $\delta^{13}C_{DIC}$ results from this study, from west to east near southern Greenland (orange in Fig. 1). Coloured squares show actual sample locations and $\delta^{13}C_{DIC}$ values. Bathymetry data are from the GEBCO_2014 grid, version 20150318, http://www.gebco.net.

250 mL bottle. A one-sample t test could not reject the null hypothesis that this mean difference in $\delta^{13}C_{DIC}$ was equal to 0 ($p = 0.63$). We therefore conclude that the container type does not cause a systematic offset in the $\delta^{13}C_{DIC}$ measurement, in agreement with Humphreys et al. (2015a).

5.2.2 Seawater samples

The typical precision for seawater $\delta^{13}C_{DIC}$ measurements is in the range from about 0.03 to 0.23‰ (Olsen et al.,

2006; Quay et al., 2007; McNichol et al., 2010; Griffith et al., 2012), and we previously reported a value of 0.10‰ based on sampling duplicates (Humphreys et al., 2015a). In this study, we again determined the precision of the seawater sample measurements from both analytical and sampling duplicates. There were 341 analytical duplicate pairs, which had a mean absolute difference of 0.075‰ and therefore a 1σ precision of 0.067‰, and 36 sampling duplicate pairs, with a mean absolute difference of 0.090‰ and therefore a 1σ precision of 0.080‰. Although the latter is slightly greater, indicating that the sample collection and storage pro-

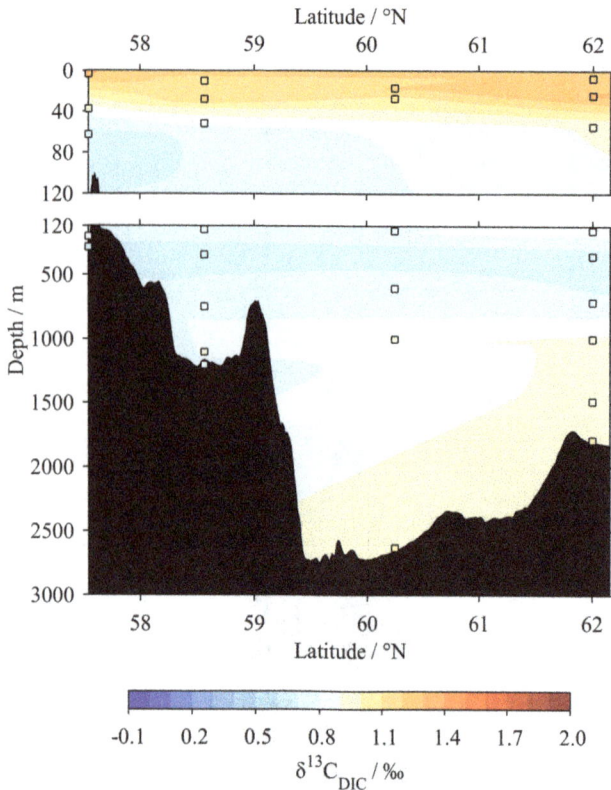

Figure 8. Meridional section of $\delta^{13}C_{DIC}$ results from this study, from south to north in the eastern subpolar North Atlantic Ocean (green in Fig. 1). Coloured squares show actual sample locations and $\delta^{13}C_{DIC}$ values. Bathymetry data are from the GEBCO_2014 grid, version 20150318, http://www.gebco.net.

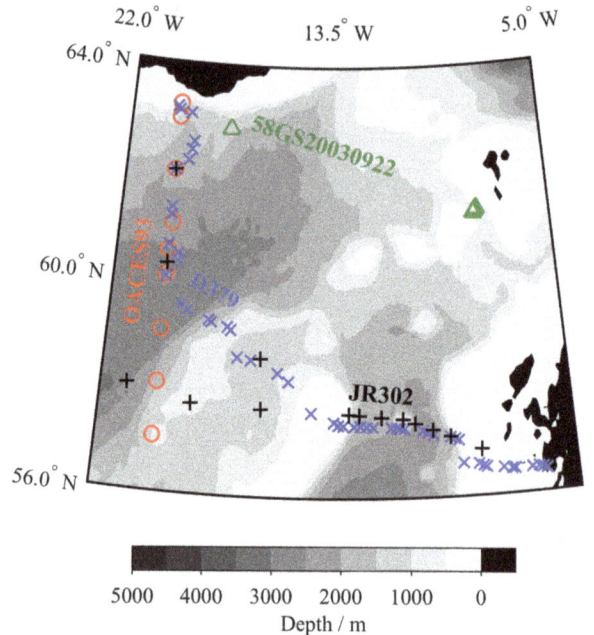

Figure 9. Map of JR302 and historical sampling stations used in the cross-over analysis (Table 2). There were no stations suitable for cross-over analysis outside of the area shown on this map. Black plusses show JR302, blue crosses are D379, green triangles are 58GS20030922, and red circles are OACES93. Only historical stations within 150 km of a JR302 station were used, as described in Sect. 4.3.2.

cedures might have adversely affected the measurement precision, Levene's test (Levene, 1960) carried out on the (non-absolute) duplicate differences could not reject the null hypothesis that the analytical and sampling precisions are in fact the same ($p = 0.33$). We therefore report the higher value of 0.08 ‰ as the 1σ precision for this data set; it falls within the range of other studies of this kind. It is important to note that this value is based on consecutively analysed samples, and so might not reflect additional uncertainty engendered by samples being measured non-consecutively or in different analysis batches. However, it is shown in the following section that any such additional uncertainty was negligible.

5.2.3 Seawater reference material

Two "batches" of RM were measured in this study: 141 and 144 (http://cdiac.ornl.gov/oceans/Dickson_CRM/batches.html). Each batch consists of multiple bottles of virtually identical seawater. Bottles from within each batch will henceforth be referred to as RM141 and RM144 respectively. The RM are primarily intended for assessment of the accu-

racy of marine carbonate chemistry measurements, in particular DIC and TA (Dickson et al., 2003); they are sterilised and sealed in airtight bottles such that the DIC and TA are consistent in all RM bottles within each RM batch, and so these variables are stable and do not change on timescales of up to a few years. Although the $\delta^{13}C_{DIC}$ value is unknown for these RM, the nature of the preparation and storage process means that we can assume that it is also consistent within each RM batch (A. G. Dickson, personal communication, 18 June 2015), thus allowing us to assess the relative accuracy of our $\delta^{13}C_{DIC}$ measurements between different analysis batches. Unpublished past measurements of $\delta^{13}C_{DIC}$ in similar RM (batches 17, 18 and 19) have supported this assumption, with the $\delta^{13}C_{DIC}$ in multiple (i.e. 3–9) bottles from the same RM batch found to have a SD of about 0.01 ‰ (A. G. Dickson, personal communication, 18 June 2015).

Typically, we made six measurements of each RM bottle, all within the same analysis batch. These were also spread throughout the analysis batch, and hence not consecutive (Fig. 2). The SD of these results for each bottle therefore represents a longer-term precision estimate than that which we get from the analytical duplicates, which were always analysed one immediately after the other. This approach therefore indicates the reproducibility of measurements carried out anywhere within a single analysis batch (rather than just consecutively). The 1σ precision from the analytical duplicates was 0.067‰, while the average SD of the measurements within each RM bottle was slightly smaller (0.058‰). This indicates that the position of samples within each analysis batch did not influence the $\delta^{13}C_{DIC}$ measurement; the relative accuracy of two consecutive measurements is no better than that between measurements from opposite ends of an analysis batch.

The next step is to verify that the difference in $\delta^{13}C_{DIC}$ between different RM bottles of the same batch is negligible. During analysis batch 13, we measured six different RM144 bottles, each up to six times (Table 3). The mean of the six measurements for each RM bottle was taken as its $\delta^{13}C_{DIC}$ value. The SD of these six mean values was 0.028‰. Next, we compared this to measurements of different RM bottles across different analysis batches. One RM144 bottle was measured during each of analysis batches 1–12. The mean value was calculated for each analysis batch, and the SD of these 12 mean values was 0.056‰. Although larger than the SD for the six RM bottles within batch 13, this value is still smaller than the overall measurement precision based on samples within the same batch. In addition, we used Levene's test (Levene, 1960) for the null hypothesis that the SD of the 6 RM bottles in analysis batch 13 was the same as the SD of the 12 RM bottles in analysis batches 1–12; the resulting p value of 0.14 was too great to confidently reject the null hypothesis, so we cannot be certain that there truly is greater variance between analysis batches than within them.

Our final $\delta^{13}C_{DIC}$ values are +1.15‰ for RM141 and +1.27‰ for RM144 (Table 3).

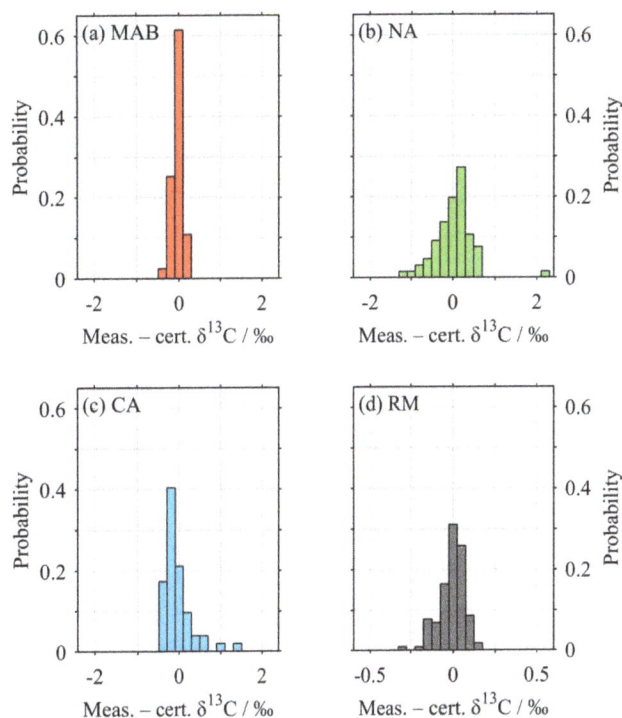

Figure 10. Histograms of offset of all standards and RM from their certified values (Table 1). Mean ± SD are **(a)** MAB, -0.03 ± 0.13‰; **(b)** NA, $+0.02 \pm 0.46$‰; **(c)** CA, -0.04 ± 0.35‰; and **(d)** all RM, -0.01 ± 0.08‰. "Certified" values for RM are our final values, 1.15‰ for RM141 and 1.27‰ for RM144 (Table 3). Note the increased horizontal resolution in **(d)**.

5.2.4 Calibration standards

The precision of measurements of the calibration standards was greater (i.e. worse) than that of the samples and RM; all calibrated MAB, NA and CA measurements had SDs of 0.13, 0.46, and 0.35‰ respectively, compared with about 0.08‰ for the RM (Fig. 10). We suggest that this is a result of the necessarily different practical treatment of the liquid seawater samples and RM compared with the powdered solid standards. The former were added to concentrated acid that had been overgassed with helium, while the latter were themselves overgassed prior to addition of dilute acid that may have contained some CO_2. The blank correction should have corrected for the influence of this CO_2, but there remained an unexplained relationship between peak area and raw $\delta^{13}C$ for the standards even after its application (Fig. 4). The poor precision for the standards might be associated with the very small quantities (0.1–1.1 mg) that were measured out into the Exetainer analysis vials, in contrast to the 1 mL of seawater sample or RM that was used each time – the former would be more susceptible to contamination. This result provides a strong incentive to develop seawater RM with a certified $\delta^{13}C_{DIC}$ value, which can be analysed following the

exact same method as the samples during future studies of this kind.

5.3 Changes from our previous study

There were three main changes to the data processing from our previous study (Humphreys et al., 2015a): the peak area correction, which previously was carried out using a different relationship for the seawater samples and for each standard; the V-PDB calibration, which was previously carried out separately for each analysis batch; and the drift correction, which is absent in the current study.

5.3.1 Peak area (linearity) correction

The linearity correction in this study was different from that in our previous work (Humphreys et al., 2015a), because we did not find the same relationships between peak area and $\delta^{13}C$. This was due to hardware changes to the mass spectrometer in the intervening time between the studies. We therefore believe that the linearity correction used in each study was appropriate, and would recommend determining the best way to apply this correction on a case-by-case ba-

Table 3. Results of the RM measurements. The RM mean and standard deviation (SD) columns contain the mean and SD of the replicate values in each row, except for the rows marked "All", which contain the mean and SD of all the RM mean values for each RM batch.

RM batch	RM bottle	Analysis batch	Replicate RM $\delta^{13}C_{DIC}$ measurements/‰						RM mean/‰	RM SD/‰
141	0585	1	1.03	1.10	1.09	1.10	1.21	1.13	1.11	0.06
141	0764	3	1.21	1.17	1.11	1.16	0.98	1.20	1.14	0.09
141	0455	8	1.21	1.13	1.20	1.22	1.19	1.14	1.18	0.04
141	0526	12	1.17	1.19	1.25	1.17	1.15	1.14	1.18	0.04
141	All	All							1.15	0.03
144	0030	1	1.20	1.14	1.25	1.19	1.28	1.21	1.21	0.05
144	1079	2	1.20	1.22	1.32	1.28	1.27	1.12	1.24	0.07
144	1141	3	1.36	1.24	1.30	1.32	1.23	1.29	1.29	0.05
144	1024	4	1.27	1.30	1.34	1.36	1.24	1.23	1.29	0.05
144	0461	5	1.29	1.24	1.18				1.24	0.05
144	0516	6	1.19	1.29	1.28	1.19	1.25	1.28	1.25	0.04
144	0399	7	1.44	1.26	1.35	1.35	1.27	1.31	1.33	0.06
144	1017	8	1.12	1.16	1.10	1.07	1.18	1.10	1.12	0.04
144	0950	9	1.31	1.29					1.30	0.01
144	0090	10	1.27	1.33	1.36				1.32	0.04
144	0881	11	1.21	1.25	1.32	1.4	1.26		1.29	0.07
144	0822	12	1.31	1.28	1.13	1.32	1.30		1.27	0.08
144	0151	13	1.36	1.34	1.36	1.34	1.23		1.33	0.05
144	0339	13	1.12	1.30	1.25	1.29	1.36	1.30	1.27	0.08
144	0575	13	1.29	1.27	1.23	1.36	1.27		1.28	0.05
144	0636	13	1.24	1.22	1.33	1.34	1.26		1.28	0.05
144	0703	13	1.17	1.29	1.25	1.33	1.30		1.27	0.06
144	0745	13	1.30	1.33	1.26	1.21	1.12		1.24	0.08
144	All	All							1.27	0.05

sis for different data sets. Another result of these hardware changes was a reduction in the mean peak area for the seawater samples from about 35 mWb to about 20 mWb. There is no evidence of any adverse (or particularly beneficial) effects on the quality of the results of either study as a result of these modifications.

5.3.2 The V-PDB calibration

The V-PDB calibration in this study was a single equation determined from all of the measurements of all of the calibration standards in every analysis batch, where previously we determined a separate equation for each batch (Humphreys et al., 2015a). This new approach delivered significantly better RM results between batches, as a result of the relatively high uncertainty in the measurements of the calibration standards; the apparent differences in calibration equations between analysis batches were in fact an artefact of these uncertainties. The consequence of this for our previous study is a decrease in precision, but it does not constitute a systematic error. If we apply a different calibration to each batch, as in our previous study, we find mean ± SD across all analysis batches of the mean $\delta^{13}C_{DIC}$ of each RM within each batch for RM141 (4 RM bottles across 4 batches) and

RM144 (18 RM bottles across 13 batches) of 1.19 ± 0.08 and 1.30 ± 0.11‰ respectively; by way of comparison, our new approach in this study yields 1.15 ± 0.03 and 1.26 ± 0.05‰ respectively. To determine the significance of these apparent differences, we took the mean RM141 and RM144 results for each analysis batch and used them to test two different null hypotheses, separately for each RM and calibration method. Firstly, we used Welch's unequal variances t test (Welch, 1947) for the null hypothesis that the mean $\delta^{13}C_{DIC}$ across all batches was the same regardless of the V-PDB calibration method. For both RM141 and RM144, the null hypothesis could not be rejected at the 5 % significance level, with p values of 0.47 and 0.28 respectively. Secondly, we used Levene's test (Levene, 1960) for the null hypothesis that the variance of these batch mean results was the same regardless of the V-PDB calibration method. For both RM141 and RM144, the null hypothesis was rejected at the 5 % significance level, with p values of 0.01 and 0.03 respectively. Thus we conclude that the change to the V-PDB calibration method – using a single equation across all analysis batches, instead of a separate one for each – results in an improvement in the precision of results from different analysis batches (i.e. reduced SD), and that it does not cause a systematic bias in these results (i.e. no change in the mean).

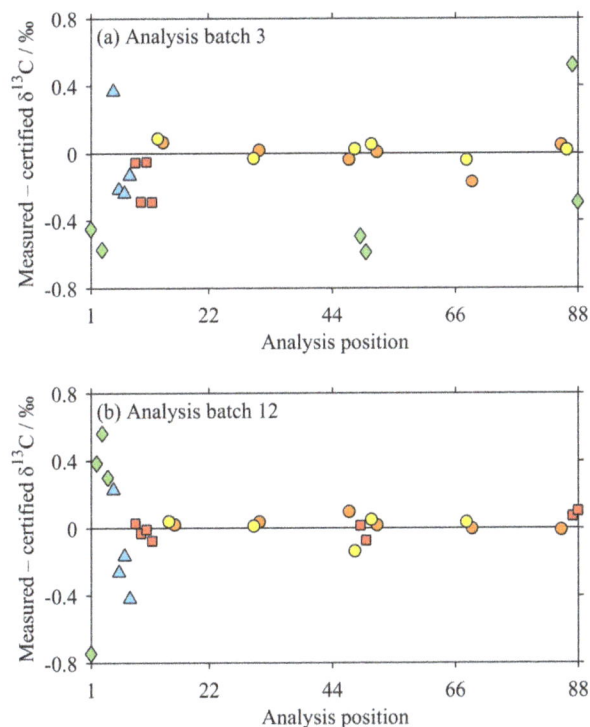

Figure 11. Examples of results of standards and RM within batches (a) 3 and (b) 12. Red squares for MAB, green diamonds for NA, blue triangles for CA, orange circles for RM141, and yellow circles for RM144.

This result cannot necessarily be applied to recalibrate the results of our previous study, as no RM measurements were carried out then. In this study, all measurements were carried out on consecutive days over a 5-week period. However, in the previous study, there were gaps of several weeks between some analysis batches, and there is no way to objectively assess the consistency of the calibration over these breaks retrospectively. Therefore, although the uncertainty estimate for the previous study was probably too generous (i.e. the reported uncertainty was lower than it should have been) and should be approximately doubled, there is no evidence of a systematic bias.

Additionally, the form of the equation used to carry out the V-PDB calibration (Eq. 5) is a second-order polynomial in this study, which differs from the circle used by Humphreys et al. (2015a). This change makes virtually no difference to the final $\delta^{13}C_{DIC}$ results, but it provides a calibration equation that only gives one possible corrected value for each input $\delta^{13}C$. The calibration equation used in this study is also much easier to interpret, with coefficients directly corresponding to the curvature (c), stretch (s) and translational offset (f) of the curve.

5.3.3   Drift correction

No drift correction was performed during this study, while Humphreys et al. (2015a) used the measurements of pairs

standards at the middle and end of each analysis batch to correct for instrumental drift. However, the RM measurements spaced throughout each analysis batch in the present study indicated that no drift correction was required, sometimes in disagreement with the calibration standards (Fig. 11). We suggest that this apparent conflict is again a result of the significantly greater uncertainty in individual measurements of the calibration standards relative to those of seawater samples and RM.

As for the V-PDB calibration, this difference does not cause an important systematic offset to the results from our previous study, but rather an increase in the variance. To support this claim, we applied a drift correction following Humphreys et al. (2015a) to our RM144 results in analysis batches 1–8 and 10–12. Analysis batches 9 and 13 were excluded due to the lack of RM data and the different arrangement of RM respectively (Fig. 2). The mean $\pm$ SD of all individual RM144 measurements (i.e. up to 6 per RM bottle) was $1.26 \pm 0.08\,\%o$ with no drift correction, but $1.22 \pm 0.17\,\%o$ when a drift correction was applied. We used Levene's test (Levene, 1960) to confidently reject the null hypothesis that the SDs were equal with and without the drift correction ($p = 0.0001$), and thus the decline in precision caused by applying the drift correction was significant. We also used Welch's unequal variances t test (Welch, 1947) for the null hypothesis that the mean value of these RM144 measurements were the same with and without drift correction; although the null hypothesis could be tentatively rejected at the 5 % significance level ($p = 0.048$) the actual magnitude of the difference in the mean values (ca. 0.04 ‰) is smaller than the measurement precision of either study, and can therefore be considered negligible.

However, for the same reasons as given regarding the V-PDB calibration, it would not be appropriate to recommend retrospective changes to the results of our previous study in the absence of any RM measurements therein.

6   Conclusions

We successfully measured $\delta^{13}C_{DIC}$ in 341 samples collected from June to July 2014 during RRS *James Clark Ross* cruise JR302 in the subpolar North Atlantic Ocean. The $\delta^{13}C_{DIC}$ values were in the range from -0.07 to $+1.95\,\%o$ relative to V-PDB and had a 1σ uncertainty of 0.08 ‰. Our results are internally consistent, with no systematic offsets between or within analysis batches, and a cross-over analysis revealed no systematic bias relative to nearby historical data in deep waters. We have also established $\delta^{13}C_{DIC}$ values for batches 141 ($+1.15\,\%o$) and 144 ($+1.27\,\%o$) of seawater RM obtained from A. G. Dickson (Scripps Institute of Oceanography, USA), and demonstrated that RM bottles within the same batch have consistent $\delta^{13}C_{DIC}$ values. These RMs greatly enhanced our ability to quantitatively assess and improve our data processing approach, and lead us to conclude that the development of an internationally available seawater RM with a certified $\delta^{13}C_{DIC}$ value would be a valuable boon to future measurements of this kind.

7   Data set availability

The $\delta^{13}C_{DIC}$ measurements described in this study are publicly available, free of charge, from the British Oceanographic Data Centre, with doi:10.5285/22235f1a-b7f3-687f-e053-6c86abc0c8a6 (Humphreys et al., 2015b). The data will also be submitted to the Carbon Dioxide Information Analysis Centre (CDIAC, Oak Ridge National Laboratory, USA) along with other carbonate chemistry and macronutrient metadata from cruise JR302 once those become available.

Author contributions. E. Tynan determined the sampling strategy, and E. Tynan, A. M. Griffiths, C. H. Fry and R. Garley collected the samples. F. M. Greatrix, A. McDonald and M. P. Humphreys carried out the measurements and data processing. M. P. Humphreys wrote the manuscript with contributions from all co-authors.

Acknowledgements. We acknowledge funding from the Natural Environment Research Council (UK) for the carbon isotope analyses (IP-1449-0514) and the RAGNARoCC award to the University of Southampton for ship time (NE-K002546-1). We are grateful to the officers, crew and scientists on board RRS *James Clark Ross* for their hard work and support during cruise JR302. We thank Matt Donnelly (BODC) for handling the archiving of our data.

Edited by: D. Carlson

References

Achterberg, E. P.: Grand challenges in marine biogeochemistry, Front. Mar. Sci., 1, 7, doi:10.3389/fmars.2014.00007, 2014.

Becker, M., Andersen, N., Erlenkeuser, H., Humphreys, Matthew. P., Tanhua, T., and Körtzinger, A.: An Internally Consistent Dataset of δ^{13}C-DIC in the North Atlantic Ocean – NAC13v1, Earth Syst. Sci. Data Discuss., doi:10.5194/essd-2016-7, in review, 2016.

Brewer, P. G.: Direct observation of the oceanic CO_2 increase, Geophys. Res. Lett., 5, 997–1000, doi:10.1029/GL005i012p00997, 1978.

Caldeira, K. and Wickett, M. E.: Anthropogenic carbon and ocean pH, Nature, 425, 365–365, doi:10.1038/425365a, 2003.

Chen, G.-T. and Millero, F. J.: Gradual increase of oceanic CO_2, Nature, 277, 205–206, doi:10.1038/277205a0, 1979.

Coplen, T. B.: Reporting of stable hydrogen, carbon, and oxygen isotopic abundances, Geothermics, 24, 707–712, doi:10.1016/0375-6505(95)00024-0, 1995.

Dickson, A. G., Afghan, J. D., and Anderson, G. C.: Reference materials for oceanic CO_2 analysis: a method for the certification of total alkalinity, Mar. Chem., 80, 185–197, doi:10.1016/S0304-4203(02)00133-0, 2003.

Dickson, A. G., Sabine, C. L., and Christian, J. R.: Guide to best practices for ocean CO_2 measurements, PICES Special Publication 3, 191 pp., 2007.

Doney, S. C., Fabry, V. J., Feely, R. A., and Kleypas, J. A.: Ocean Acidification: The Other CO_2 Problem, Annual Review of Marine Science, 1, 169–192, doi:10.1146/annurev.marine.010908.163834, 2009.

Friis, K., Körtzinger, A., Pätsch, J., and Wallace, D. W. R.: On the temporal increase of anthropogenic CO_2 in the subpolar North Atlantic, Deep-Sea Res. Pt. I, 52, 681–698, doi:10.1016/j.dsr.2004.11.017, 2005.

Gaylord, B., Kroeker, K. J., Sunday, J. M., Anderson, K. M., Barry, J. P., Brown, N. E., Connell, S. D., Dupont, S., Fabricius, K. E., Hall-Spencer, J. M., Klinger, T., Milazzo, M., Munday, P. L., Russell, B. D., Sanford, E., Schreiber, S. J., Thiyagarajan, V., Vaughan, M. L. H., Widdicombe, S., and Harley, C. D. G.: Ocean acidification through the lens of ecological theory, Ecology, 96, 3–15, doi:10.1890/14-0802.1, 2015.

Griffith, D. R., McNichol, A. P., Xu, L., McLaughlin, F. A., Macdonald, R. W., Brown, K. A., and Eglinton, T. I.: Carbon dynamics in the western Arctic Ocean: insights from full-depth carbon isotope profiles of DIC, DOC, and POC, Biogeosciences, 9, 1217–1224, doi:10.5194/bg-9-1217-2012, 2012.

Gruber, N., Sarmiento, J. L., and Stocker, T. F.: An improved method for detecting anthropogenic CO_2 in the oceans, Global Biogeochem. Cy., 10, 809–837, doi:10.1029/96GB01608, 1996.

Hall, T. M., Haine, T. W. N., and Waugh, D. W.: Inferring the concentration of anthropogenic carbon in the ocean from tracers, Global Biogeochem. Cy., 16, 1131, doi:10.1029/2001GB001835, 2002.

Holliday, N. P. and Cunningham, S.: The Extended Ellett Line: Discoveries from 65 years of marine observations west of the UK, Oceanography, 26, 156–163, doi:10.5670/oceanog.2013.17, 2013.

Humphreys, M. P.: Cross-over analysis of hydrographic variables: XOVER v1.0, Ocean and Earth Science, University of Southampton, UK, 8 pp., doi:10.13140/RG.2.1.1629.0405, 2015.

Humphreys, M. P., Achterberg, E. P., Griffiths, A. M., McDonald, A., and Boyce, A. J.: Ellett Line measurements of stable isotope composition of dissolved inorganic carbon in the Northeastern Atlantic and Nordic Seas during summer 2012, British Oceanographic Data Centre, Natural Environment Research Council, UK, doi:10/xph, 2014a.

Humphreys, M. P., Achterberg, E. P., Griffiths, A. M., McDonald, A., and Boyce, A. J.: UKOA measurements of the stable isotope composition of dissolved inorganic carbon in the Northeastern Atlantic and Nordic Seas during summer 2012, British Oceanographic Data Centre, Natural Environment Research Council, UK, doi:10/xpj, 2014b.

Humphreys, M. P., Achterberg, E. P., Griffiths, A. M., McDonald, A., and Boyce, A. J.: Measurements of the stable carbon isotope composition of dissolved inorganic carbon in the northeastern Atlantic and Nordic Seas during summer 2012, Earth Syst. Sci. Data, 7, 127–135, doi:10.5194/essd-7-127-2015, 2015a.

Humphreys, M. P., Greatrix, F. M., Tynan, E., Griffiths, A. M., Fry, C. H., Garley, R., Achterberg, E. P., McDonald, A., and Boyce, A. J.: Stable carbon isotopes of dissolved inorganic carbon for RRS *James Clark Ross* cruise JR302 in the subpolar North Atlantic Ocean from June to July 2014, British Oceanographic Data Centre, Natural Environment Research Council, UK, doi:10.5285/22235f1a-b7f3-687f-e053-6c86abc0c8a6, 2015b.

Humphreys, M. P., Griffiths, A. M., Achterberg, E. P., Holliday, N. P., Rérolle, V. M. C., Menzel Barraqueta, J.-L., Couldrey, M. P., Oliver, K. I. C., Hartman, S. E., Esposito, M., and Boyce, A. J.: Multidecadal accumulation of anthropogenic and remineralized dissolved inorganic carbon along the Extended Ellett Line in the northeast Atlantic Ocean, Global Biogeochem. Cy., 30, 2015GB005246, doi:10.1002/2015GB005246, 2016.

Keeling, C. D.: The Suess effect: ^{13}Carbon-^{14}Carbon interrelations, Environ. Int., 2, 229–300, doi:10.1016/0160-4120(79)90005-9, 1979.

Khatiwala, S., Primeau, F., and Hall, T.: Reconstruction of the history of anthropogenic CO_2 concentrations in the ocean, Nature, 462, 346–349, doi:10.1038/nature08526, 2009.

Khatiwala, S., Tanhua, T., Mikaloff Fletcher, S., Gerber, M., Doney, S. C., Graven, H. D., Gruber, N., McKinley, G. A., Murata, A., Ríos, A. F., and Sabine, C. L.: Global ocean storage of anthropogenic carbon, Biogeosciences, 10, 2169–2191, doi:10.5194/bg-10-2169-2013, 2013.

King, B. A. and Holliday, N. P.: JR302 Cruise Report, The 2014 RAGNARoCC, OSNAP and Extended Ellett Line cruise, National Oceanography Centre, Southampton, UK, 76 pp., 2015.

Körtzinger, A., Quay, P. D., and Sonnerup, R. E.: Relationship between anthropogenic CO_2 and the ^{13}C Suess effect in the North Atlantic Ocean, Global Biogeochem. Cy., 17, 5-1–5-20, doi:10.1029/2001GB001427, 2003.

Lauvset, S. K. and Tanhua, T.: A toolbox for secondary quality control on ocean chemistry and hydrographic data, Limnol. Oceanogr.-Meth., doi:10.1002/lom3.10050, 2015.

Le Quéré, C., Raupach, M. R., Canadell, J. G., Marland, G., Bopp, L., Ciais, P., Conway, T. J., Doney, S. C., Feely, R. A., Foster, P., Friedlingstein, P., Gurney, K., Houghton, R. A., House, J. I., Huntingford, C., Levy, P. E., Lomas, M. R., Majkut, J., Metzl, N., Ometto, J. P., Peters, G. P., Prentice, I. C., Randerson, J. T., Running, S. W., Sarmiento, J. L., Schuster, U., Sitch, S., Takahashi, T., Viovy, N., van der Werf, G. R., and Woodward, F. I.: Trends in the sources and sinks of carbon dioxide, Nat. Geosci., 2, 831–836, doi:10.1038/ngeo689, 2009.

Levene, H.: Robust tests for equality of variances, in Contributions to Probability and Statistics: Essays in Honor of Harold Hotelling, 278–292, Stanford University Press, USA, 1960.

McNichol, A., Quay, P. D., Gagnon, A. R., and Burton, J. R.: Collection and measurement of carbon isotopes in seawater DIC, in The GO-SHIP Repeat Hydrography Manual: A Collection of Expert Reports and Guidelines, IOCCP Report No. 14, ICPO Publication Series No. 134, Version 1, 2010.

Olsen, A. and Ninnemann, U.: Large δ^{13}C Gradients in the Preindustrial North Atlantic Revealed, Science, 330, 658–659, doi:10.1126/science.1193769, 2010.

Olsen, A., Omar, A. M., Bellerby, R. G. J., Johannessen, T., Ninnemann, U., Brown, K. R., Olsson, K. A., Olafsson, J., Nondal, G., Kivimäe, C., Kringstad, S., Neill, C., and Olafsdottir, S.: Magnitude and origin of the anthropogenic CO_2 increase and ^{13}C Suess effect in the Nordic seas since 1981, Global Biogeochem. Cy., 20, GB3027, doi:10.1029/2005GB002669, 2006.

Quay, P., Sonnerup, R., Westby, T., Stutsman, J., and McNichol, A.: Changes in the ^{13}C/^{12}C of dissolved inorganic carbon in the ocean as a tracer of anthropogenic CO_2 uptake, Global Biogeochem. Cy., 17, GB1004, doi:10.1029/2001GB001817, 2003.

Quay, P. D., Tilbrook, B., and Wong, C. S.: Oceanic Uptake of Fossil Fuel CO_2: Carbon-13 Evidence, Science, 256, 74–79, doi:10.1126/science.256.5053.74, 1992.

Quay, P. D., Sonnerup, R., Stutsman, J., Maurer, J., Körtzinger, A., Padin, X. A., and Robinson, C.: Anthropogenic CO_2 accumulation rates in the North Atlantic Ocean from changes in the ^{13}C/^{12}C of dissolved inorganic carbon, Global Biogeochem. Cy., 21, GB1009, doi:10.1029/2006GB002761, 2007.

Rubino, M., Etheridge, D. M., Trudinger, C. M., Allison, C. E., Battle, M. O., Langenfelds, R. L., Steele, L. P., Curran, M., Bender, M., White, J. W. C., Jenk, T. M., Blunier, T., and Francey, R. J.: A revised 1000? year atmospheric δ^{13}C-CO_2 record from Law Dome and South Pole, Antarctica, J. Geophys. Res. Atmos., 118, 8482–8499, doi:10.1002/jgrd.50668, 2013.

Sabine, C. L. and Tanhua, T.: Estimation of Anthropogenic CO_2 Inventories in the Ocean, Annual Review of Marine Science, 2, 175–198, doi:10.1146/annurev-marine-120308-080947, 2010.

Sabine, C. L., Feely, R. A., Gruber, N., Key, R. M., Lee, K., Bullister, J. L., Wanninkhof, R., Wong, C. S., Wallace, D. W. R., Tilbrook, B., Millero, F. J., Peng, T.-H., Kozyr, A., Ono, T., and Rios, A. F.: The Oceanic Sink for Anthropogenic CO_2, Science, 305, 367–371, doi:10.1126/science.1097403, 2004.

Schmittner, A., Gruber, N., Mix, A. C., Key, R. M., Tagliabue, A., and Westberry, T. K.: Biology and air–sea gas exchange controls on the distribution of carbon isotope ratios (δ^{13}C) in the ocean, Biogeosciences, 10, 5793–5816, doi:10.5194/bg-10-5793-2013, 2013.

Sonnerup, R. E. and Quay, P. D.: ^{13}C constraints on ocean carbon cycle models, Global Biogeochem. Cy., 26, GB2014, doi:10.1029/2010GB003980, 2012.

Sonnerup, R. E., Quay, P. D., McNichol, A. P., Bullister, J. L., Westby, T. A., and Anderson, H. L.: Reconstructing the oceanic ^{13}C Suess Effect, Global Biogeochem. Cy., 13, 857–872, doi:10.1029/1999GB900027, 1999.

Sonnerup, R. E., McNichol, A. P., Quay, P. D., Gammon, R. H., Bullister, J. L., Sabine, C. L., and Slater, R. D.: Anthropogenic δ^{13}C changes in the North Pacific Ocean reconstructed using a multiparameter mixing approach (MIX), Tellus B, 59, 303–317, doi:10.1111/j.1600-0889.2007.00250.x, 2007.

Tanhua, T., Körtzinger, A., Friis, K., Waugh, D. W., and Wallace, D. W. R.: An estimate of anthropogenic CO_2 inventory from decadal changes in oceanic carbon content, Proc. Natl. Acad. Sci. USA, 104, 3037–3042, doi:10.1073/pnas.0606574104, 2007.

Thompson, M. and Howarth, R. J.: The rapid estimation and control of precision by duplicate determinations, Analyst, 98, 153–160, doi:10.1039/AN9739800153, 1973.

Waugh, D. W., Hall, T. M., McNeil, B. I., Key, R., and Matear, R. J.: Anthropogenic CO_2 in the oceans estimated using transit time distributions, Tellus B, 58, 376–389, doi:10.1111/j.1600-0889.2006.00222.x, 2006.

Welch, B. L.: The Generalization of Student's Problem When Several Different Population Variances Are Involved, Biometrika, 34, 28–35, doi:10.1093/biomet/34.1-2.28, 1947.

2

Hydrography and biogeochemistry dedicated to the Mediterranean BGC-Argo network during a cruise with RV *Tethys 2* in May 2015

Vincent Taillandier[1], Thibaut Wagener[2], Fabrizio D'Ortenzio[1], Nicolas Mayot[1,a], Hervé Legoff[3], Joséphine Ras[1], Laurent Coppola[1], Orens Pasqueron de Fommervault[1,b], Catherine Schmechtig[4], Emilie Diamond[1], Henry Bittig[1], Dominique Lefevre[2], Edouard Leymarie[1], Antoine Poteau[1], and Louis Prieur[1]

[1]Sorbonne Universités, UPMC Université Paris 06, CNRS, LOV, Villefranche-sur-Mer, 06230, France
[2]Aix-Marseille Universiteì, CNRS/INSU, Universiteì de Toulon, IRD,
Mediterranean Institute of Oceanography (MIO), UM 110, Marseille, 13288, France
[3]Sorbonne Universités, UPMC Univ Paris 06, CNRS, IRD, MNHN, LOCEAN, Paris, France
[4]Sorbonne Universités, UPMC Univ Paris 06, CNRS, UMS 3455, OSU Ecce-Terra, Paris CEDEX 5, France
[a]now at: Bigelow Laboratory for Ocean Sciences, Maine, East Boothbay, USA
[b]now at: Laboratorio de Oceanografià Fisicà, CICESE, Ensenada, B.C., Mexico

Correspondence: Vincent Taillandier (taillandier@obs-vlfr.fr)

Abstract. We report on data from an oceanographic cruise, covering western, central and eastern parts of the Mediterranean Sea, on the French research vessel *Tethys 2* in May 2015. This cruise was fully dedicated to the maintenance and the metrological verification of a biogeochemical observing system based on a fleet of BGC-Argo floats. During the cruise, a comprehensive data set of parameters sensed by the autonomous network was collected. The measurements include ocean currents, seawater salinity and temperature, and concentrations of inorganic nutrients, dissolved oxygen and chlorophyll pigments. The analytical protocols and data processing methods are detailed, together with a first assessment of the calibration state for all the sensors deployed during the cruise.

1 Introduction

1.1 Context of the cruise

The biogeochemical functioning of the Mediterranean Sea is typical of temperate oceanic regions. Seasonal dynamics of phytoplankton follow an increase of biomass in spring even if primary production remains low during the whole year (Marty et al., 2002). The biomass distribution in the Mediterranean Sea is marked by a pronounced east–west gradient (Bosc et al., 2004). This pattern is confirmed by phenology of the underlying phytoplankton dynamics that varies from ultra-oligotrophic regimes in the eastern basin to

bloom regimes in the northwestern basin (D'Ortenzio et al., 2009). An extended study on the geographical distribution of these regimes – related to the Mediterranean bio-regions – has revealed significant changes at regional scales during the last decades (Mayot et al., 2016). Indeed, the seasonal cycle of biomass concentration turns out to be a reliable indicator of the response of pelagic ecosystems to external perturbations (Siokou-Frangou et al., 2010). Facing increasing anthropogenic effects and considered to be a regional hotspot where climate change impacts will be the largest (Giorgi and Lionello, 2008), it would appear to be essential to characterize this indicator in the Mediterranean Sea basin under a large panel of possible trophic regimes and various physical and chemical environments (Durrieu de Madron et al., 2011).

The seasonal cycles of biomass concentration have mainly been observed from satellite images of ocean color, thanks to their synoptic coverage of the area. Although limited to surface characterization, the link between biomass structuration in the water column and the underlying physical–chemical state over a seasonal scale has only been found at few ocean observation sites (Marty and Chiaverini, 2010). The emergence of BGC-Argo floats, which are autonomous profiling platforms equipped with biogeochemical sensors and programmed at weekly cycles up to 1000 m depth (Leymarie et al., 2013), now allows us to collect oceanographic profiles concomitantly for physical and biogeochemical properties (temperature, salinity, concentration of dissolved oxygen, chlorophyll *a*, nitrate). These open new perspectives for the description and comprehension of the biogeochemical functioning of the Mediterranean Sea. For example, the occurrence of phytoplankton blooms can be directly related to the availability of nutrients (D'Ortenzio et al., 2014).

Such technological advances have driven the development of a dedicated observing system over the Mediterranean Sea with a fleet of a dozen BGC-Argo floats in operation. This emerging network has been promoted and sustained by French programs such as Equipex-NAOS and the Mermex experiment, as well as at the European level through Euro-Argo infrastructure. However, sensors for biogeochemical properties, even with recent factory calibration, are subject to substantial systematic errors when deployed on BGC-Argo floats, as reported by Bittig et al. (2012) for oxygen measurements or by Pasqueron de Fommervault et al. (2015) for nitrate measurements. As a consequence, even if a BGC-Argo float is supposed to be completely autonomous after deployment, reference data for quality assessment of most of its sensors need to be collected by ship (D'Ortenzio et al., 2014; Johnson et al., 2017). Automatic quality controls are rapidly advancing for the Argo program (Schmechtig et al., 2015), although most of the methods and protocols are still under assessment. In this context, dedicated and precise efforts were necessary to ensure data quality of the Mediterranean observing system composed of BGC-Argo floats.

1.2 Objectives and achievements of the cruise

The data set presented in this paper was collected during an oceanographic cruise carried out in spring 2015 over the Mediterranean Sea. To our knowledge, it was the first cruise fully dedicated to the maintenance and the metrological verification of an autonomous observing system based on BGC-Argo floats. The objectives of the cruise were twofold: (1) to continue the time series of profile collection in operation since 2012 in the Mediterranean Sea, by deploying new BGC-Argo floats and recovering old ones and (2) to harmonize this collection with the systematic verification of calibration states for all biogeochemical sensors active in the

network, using shipboard measurements as reference standards.

The choice of a dedicated cruise instead of ships of opportunity was driven by applying the same protocol of metrological verification for all the floats, using the same instruments and methods of reference. Another crucial point remains the required flexibility to choose the location of the oceanographic stations, which mainly depended on the state of the network (i.e., the position of the different floats) at the time of the cruise.

The survey covered large parts of the western, central and eastern basins of the Mediterranean Sea with a total route of about 3000 nm (Fig. 1). The cruise started in Nice (France) on 12 May 2015 and ended up in Nice on 1 June 2015, on board the *Tethys 2*, a 24 m long research vessel of the French National Center for Scientific Research (CNRS), comprised of a crew of seven and five scientists. The cruise was divided into four legs of about 4 days each with three port calls were programmed: 18–19 May in Heraklion (Crete, Greece), 24–25 May in Heraklion and 28–29 May in Lipari (Sicily, Italy). The initial cruise planning settled upon seven oceanographic stations, which represents about two stations for each 10 h leg. Transects between stations were crossed at 8–11 kn depending on sea conditions.

Work on board during the transects was dedicated to the surface sampling, together with seawater sample analyses and data processing. During stations, a CTD carousel composed of 11 Niskin bottles was deployed and discrete samples were collected for one shallow cast (0–500 dbar) and one deep cast (0–bottom). Standard levels were chosen for the deep cast (bottom, 2000, 1500, 1250, 1000, 750, 500 dbar; salinity maximum, 200 dbar; chlorophyll maximum, 10 dbar). The shallow cast was composed of six standard levels (500, 200, 150, 50, 10, 5 dbar) and five levels dedicated to the sampling of the deep chlorophyll maximum. This sampling strategy was reduced to a single cast (0–1000 dbar) in case of rough sea conditions, or extended with another cast (0–1000 dbar) for calibration purposes. The number of casts and samples are summarized in Table 1, with a total of 60 pigment samples, 148 oxygen samples and 154 nutrient samples.

The cruise was prepared in coordination with the Euro-Argo infrastructure so that series of Argo and BGC-Argo floats were provided by different European institutes (BSH Germany, OGS Italy, LOV France). Herein, only the BGC-Argo component is considered. At the time of the cruise, there were 12 active floats; 4 of these floats were recovered and 10 new floats were deployed during the cruise. The standard method consisted of deploying BGC-Argo floats at the end of every station, as listed in Table 2. Calibration exercises were created assuming that the CTD casts and the first float profiles could be considered co-located in time and space. That is why the floats were programmed to profile everyday at noon at the beginning of their mission. The first deep profile (0–1000 m) acquired by the floats occurred on the day of

Table 1. Station summary. For bottom depth, values with asterisk indicate that the measurement was obtained from the vessel's echo-sounder rather than the altimeter interfaced to the CTD unit.

Station	Cast	Date/time UTC DD/MM/YY	Latitude	Longitude	Profile depth (m)	Bottom depth (m)	No. of samples, pigments	No. of samples, oxygen	No. of samples, nutrients
Ligurian	1	12/05/15 20:10	43°33.52′ N	7°27.78′ E	1662	1684	5	11	11
North Ionian	2	16/05/15 03:41	38°10.44′ N	18°30.12′ E	500	3038	8	5	11
	3	16/05/15 05:35	38°10.96′ N	18°30.16′ E	2990	3038	0	11	11
Central Levantine	4	21/05/15 12:21	33°33.90′ N	28°27.99′ E	500	2959*	8	11	11
	5	21/05/15 14:14	33°33.76′ N	28°28.50′ E	1240	2959*	0	11	11
West Levantine	6	22/05/15 10:33	34°13.89′ N	24°49.84′ E	1000	2244*	7	11	11
	7	22/05/15 15:02	34°12.61′ N	24°50.76′ E	500	2886	8	11	11
	8	22/05/15 17:04	34°12.66′ N	24°50.56′ E	2871	2886	0	11	11
East Ionian	9	26/05/15 12:51	36°41.84′ N	20°07.32′ E	500	3175	8	11	11
	10	26/05/15 14:44	36°41.57′ N	20°07.21′ E	3165	3175	0	11	11
South Tyrrhenian	11	30/05/15 10:05	39°10.43′ N	10°53.47′ E	500	2812	8	11	11
	12	30/05/15 13:36	39°11.44′ N	10°52.37′ E	2803	2812	0	11	11
Central Tyrrhenian	13	31/05/15 05:21	40°45.22′ N	11°30.28′ E	500	2466	8	11	11
	14	31/05/15 07:14	40°45.87′ N	11°30.66′ E	2456	2466	0	11	11

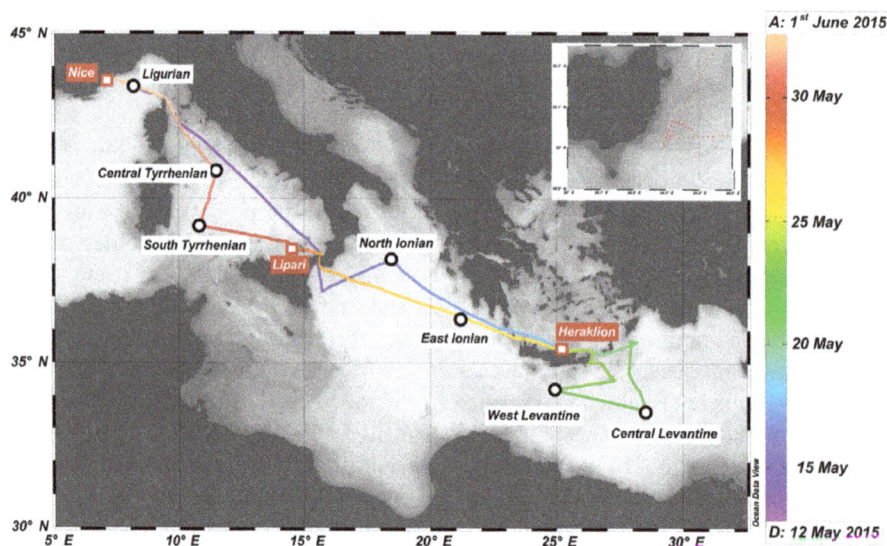

Figure 1. Cruise track plotted on a time line (color bar). Port calls are marked by red squares, and stations are marked by black circles. Detail of the L-shape track in the eastern coast of Crete.

the station if deployed early in the morning, or the day after if deployed later, as reported in Table 2.

This protocol of deployment is effective if working clearance in the area of the station was obtained in order to perform CTD casts. Unfortunately, this was not the case in the eastern Levantine Basin where the definitions of maritime exclusive economic zones are still vague. As a consequence, one BGC-Argo float was deployed without any reference CTD cast in the eastern Levantine (out of the list reported in Table 2). Two other floats were deployed in the same area some days after in the same conditions; however, the calibration exercise was performed at the west Levantine station by

clamping the floats onto the frame of the CTD carousel and acquiring a profile (identified as BCN in Table 2) concomitant with the reference CTD profile and discrete samples.

The aims of this paper are to describe the collected data set. The sensing means and the underlying processing tools for data acquired from the ship and from BGC-Argo floats are detailed in the next section. The description and the instructions on accessing the quality-controlled data set are provided. Finally, a discussion follows about the various methodological strategies to update the BGC-Argo network in the Mediterranean Sea and to provide in situ calibration of the sensors.

Table 2. BGC-Argo float summary. For every BGC-Argo float deployed with a CTD cast of reference, the distance and duration with the first profile of the float are indicated. The results of metrological verification by parameter are reported. SD stands for standard deviation. PSAL: practical salinity.

Station	ARGO WMO	First profile ID	Inter-distance (km)	Inter-duration (h)	Temp offset (°C)	PSAL offset	Optode slope	Optode offset ($\mu mol\,kg^{-1}$)	Fluo N	Fluo R^2	Fluo offset ($mg\,m^{-3}$)	Fluo slope	SUNA slope	SUNA offset ($\mu mol\,L^{-1}$)
West Lev.*	6901764	BCN	0	0	0.0059	0.0150	0.9796	11.56	7	0.98	0.01	0.67	1.00	3.20
West Lev.	6901765	000	1	19	0.0003	0.0031	1.0660	3.26	8	0.77	0.04	0.62	1.00	4.00
West Lev.*	6901766	BCN	0	0	0.0053	0.0081	1.0275	6.30	7	0.98	−0.02	0.65	1.11	0.80
Central Tyr.	6901767	000	3	7					8	0.86	0.03	0.49	1.00	−2.80
Central Tyr.	6901767	001	3	29	0.0021	−0.0009	1.1045	−3.59					1.00	−2.70
North Ion.	6901768	001	12	31	0.0052	0.0009	1.0235	6.66	8	0.89	0.04	0.63	1.00	2.10
South Tyr.	6901769	000	2	26	0.0214	0.0050	1.1626	−14.87	8	0.82	0.03	0.58	1.00	3.90
East Ion.	6901771	000	2	21	0.0085	0.0042	1.0658	0.51	8	0.93	0.02	0.55	1.17	0.10
Central Lev.	6901773	000	3	22	0.0067	0.0070	1.0923	−2.40	8	0.99	0.01	0.51		
Average			3	17	0.0069	0.0053	1.0652	0.93			0.02	0.59	1.04	1.08
SD			4	12	0.0064	0.0049	0.0564	8.12			0.02	0.07	0.07	2.74

* Metrological verification exercise: deployed at another location than the station.

2 Methods for sensing, processing and quality control

The method employed for measurement (sensor technology, analytical protocol), the method used to process the collected data, and the operated quality control on the final data set is then presented per parameter (or family of parameters).

2.1 Ocean currents

2.1.1 Presentation of the different measurements

Ocean currents were measured with acoustic Doppler current profilers (ADCP), along the ship track and at every station using two dedicated instruments.

The vessel has been equipped since January 2015 with an Ocean Surveyor 75 kHz interfaced with a GPS and a gyrocompass. For the cruise, the ship ADCP (hereafter SADCP) was programmed in broadband single-ping profile mode, over 70 bins of 8 m and a blanking distance of 8 m. The maximum range obtained was 500 m; it was reduced to 250 m in the ultra-oligotrophic waters of the eastern basin.

The CTD carousel was equipped with a lowered ADCP (hereafter LADCP) system. It was composed of two RDI Workhorse Monitors 300 kHz, one uplooker was clamped onto the upper part of the frame that removed one over 12 Niskin bottles, and one downlooker clamped onto the lower frame. The two sensors were synchronized by a command WM15. The system was supplied by an external battery box installed in the lower frame. The LADCP was programmed in narrowband mode with a sampling rate of 1 Hz and 20 bins of 8 m and a blanking distance of null, Earth coordinate which tilts the three-beam solution, and bin mapping.

2.1.2 Data processing

Data flow from SADCP was archived on board and pre-processed using the manufacturer's software VMDAS, providing 2 min averaged velocity profiles. At least once per day, the data collection was uploaded and processed using the software Cascade V6.2 (Le Bot et al., 2011): ocean currents were generated by correcting raw velocity profiles from the ship navigation and attitude. Bottom detections were masked using GEBCO 1' bathymetry, corrections of ocean tides were not applied. Two data sets were assembled: one set with a time resolution of 2 min for ocean current profiles acquired during stations, one set with a spatial resolution of 1 km for ocean current profiles acquired during transits.

Data flow from LADCP system was processed using the software LDEO IX (Thurnherr, 2014). The architecture of this software allows us to replay the processing chain with different parameterizations: depth computation either from bottom track or by using the concomitant CTD profile, the threshold of percentage of good values, the assimilation of SADCP data and the weight of this constraint, either time resolution (1 s nominal) or vertical resolution (5 m bins), adjustment of the variation of magnetic declination.

LADCP data were processed with different levels of complexity. Right after each cast, a first screening of measurements was performed in order to validate the functioning of the system and assess the percentage of good values. When CTD profiles were available, a first ocean current profile was computed with refined depth constraint. In a final step, the misfit with a mean SADCP profile during station was attempted to be minimized by iteratively processing LADCP data with this new constraint.

2.1.3 Data quality control

An in situ calibration of SADCP sensor was undertaken during the cruise. An L shape of 10 nmi length was crossed back and forth by the ship in calm seas and at moderate speeds over a shallow area off the eastern coast of Crete (see Fig. 1). Bottom track was acquired all the time which allowed comparison of ocean currents during the way in and the way back, supposedly steady over the 2 h duration of the exercise. The two transects were significantly different in amplitude and azimuth. Corrections on misalignment angle (1.1°), amplitude factor (1.004) and pitch thresholds (1 and 1.5°) for the SADCP were proposed in order to reduce the misfits between transects. Note that this calibration exercise would not have been necessary using CODAS software, which allows computation of the SADCP misalignment angle. Quality-controlled data sets of ocean currents along the ship track were post-processed thanks to these corrections.

This post-processed SADCP data set was also examined during stations in order to assess and improve the quality of the 14 LADCP profiles. As reported in Table 3, all the profiles except at casts 3 and 10 are characterized by low velocity errors and acceptable misfits with SADCP profiles. The median value of these uncertainties over the 12 acceptable casts using 1 s resolution profiles (approximately 800 ensembles) was evaluated to -0.94 ± 3.1 cm s^{-1} in module and $5.4 \pm 38°$ in azimuth without the SADCP constraint. Under SADCP constraint the median value reaches 0.17 ± 1.1 cm s^{-1} in module and $-0.02 \pm 23°$ in azimuth. It is shown that the SADCP constraint does not significantly improve the ocean current estimate in module, but does at azimuth. The quality-controlled data set of ocean currents collected at the stations was processed with SADCP constraint and binned at 5 m resolution.

2.2 Seawater temperature and practical salinity

2.2.1 Presentation of the different measurements

Temperature and practical salinity properties of seawater were continuously measured at surface along the ship track by the underway system of the vessel, and at depth by the underwater unit or by the BGC-Argo floats during the seven stations.

A SeaCAT thermosalinograph (SBE21, serial no. 3146), hereafter TSG, was mounted in the underway system of the vessel. This instrument is composed of a conductivity cell and a local temperature probe in order to derive practical salinity. A remote temperature probe (SBE38, serial no. 0528) interfaced with the TSG was located at the inlet of the underway flow to minimize thermal contamination. Factory calibration of the TSG system was performed within the year preceding the cruise (29 July 2014). The acquisition started on 13 May 00:00 UTC, and it was halted during port calls.

The underwater unit was equipped with a CTD (SBE911+, serial no. 0329), which contained an internal pressure sensor, an external temperature probe (SBE3plus, serial no. 2473) and an external conductivity cell (SBE4C, serial no. 1313). A factory calibration of the two sensors was performed within

the month preceding the cruise (16 April 2015). The GO-SHIP guidelines (Hood et al., 2010) were followed for the preparation, maintenance and deployment procedure of this instrument package.

The BGC-Argo floats were equipped with factory-calibrated CTD modules (SBE41CPs). These modules are designed as for mooring sensors to guarantee long-term stability of temperature, conductivity and pressure measurements. The probes were plumbed in a U-shaped seawater circuit with pump entrainment and taped with anti-foulant devices.

2.2.2 Data processing

The TSG data flow of 15 s resolution was archived on board together with GPS data flow as unmodifiable hexadecimal encoded files. At least once per day, the data collection was processed to feed a single time series of 5 min resolution for UTC time, geolocation, temperature and practical salinity.

During stations, seawater properties were sampled at 24 Hz with the CTD unit and transmitted on board through an electro-mechanical sea cable and slip-ring-equipped winch. At-sea processing of the archive was run after each CTD cast following GO-SHIP guidelines (Hood et al., 2010).

Data from BGC-Argo floats were transmitted to land via satellite Iridium communication and disseminated by a dedicated server. Continuous acquisition at 0.5 Hz was performed during the ascent phase of the float; pressure, temperature and practical salinity were then processed before transmission following user specifications: in the pressure range 0–10 dbar, the nominal resolution is kept; in the pressure range 10–250 dbar, averages of 2 dbar slices were computed; in the pressure range of 250–1000 dbar, averages of 10 dbar slices were computed.

2.2.3 Data quality control

The pressure measured from the CTD unit was compared on the vessel's deck with a barometer reading during port calls. No significant shift was observed that would require a post-cruise adjustment of this sensor.

There were no independent samples (such as salinity bottles) or double probes in the CTD unit that would have allowed the assessment of the temperature and conductivity sensors' stability. Thus, the quality of CTD data relies on frequent factory calibrations operated on the sensors: a pre-cruise bath was performed in April 2015 (less than 1 month before the cruise), and a post-cruise bath performed in March 2016 (less than 1 year after the cruise). The static drift of the temperature sensor between baths was 0.00008 °C which is 1 order of magnitude lower than the theoretical stability of the probe. The static conductivity ratio between baths was 1.0000321 which represents a drift of about 0.0015 mS cm^{-1}, 1 order of magnitude lower than the theoretical stability of the probe. Given the reproducibility

Table 3. Summary of ocean current profiles collected at the stations. Depth and bottom track (BT) distances, when available, are indicated. Error velocities were computed for three sets of profiles: LADCP (L) data only, SADCP (S) data only and L data processed under the constraint of S data. Final process parameters were chosen as a function that leads to the misfits between L (with final process parameters) and S currents.

Cast	Depth (m)	BT distance (m)	Error velocity (cm s^{-1})			LDEO final parameters	Misfits L ms S (cm s^{-1})	Comments
			L without S constraint	L with S constraint	S			
1	1721	26	2.5	2.5	5.5	L + S + BT	1.8	
2	498		3.4	3.4	5.7	L + S	3.8	
3	2990	53	20.3	18.9	6.5	L + S + BT	19.2	rough sea, high tilt
4	501		3.1	2.3	6.9	L + S	3.1	
5	1243		2.9	3.4	6.4	L + S	6.9	
6	996		2.5	2.6	5	L + S	2.0	
7	496		2.5	2.4	4.6	L + S	2.2	
8	2871	16	2.9	4.8	4.7	L + S + BT	6.1	
9	502		2.2	3.1	4.5	downlooker + S	2.6	uplooker failed, low battery
10	3165	17	20.7	50.4	5.2	downlooker + S + BT	36.0	
11	497		3	3	5.9	L + S	4.5	
12	2805	7	5.4	4.2	6.1	L + S + BT	4.4	
13	505		2.6	2.5	5	L + S	2.5	
14	2456	12	2.8	2.8	5.3	L + S + BT	3.6	

of the processing method, the uncertainties of measurement provided by the CTD unit should have stayed within the accuracy of the sensors, which is 0.001 °C and 0.003 mS cm^{-1} out of lowered dynamic accuracy cases (such as in sharp temperature gradients).

The data collection of temperature and practical salinity profiles at every station is thus used as reference to assess the two other sensing systems: the TSG and the BGC-Argo floats. Systematic comparisons between the profiles from the CTD unit and the neighboring data were made at every cast.

Considering TSG data set, the median value of temperature and practical salinity over a time window of 1 h around the profile date was extracted from the 5 min resolution time series. The comparison with the surface value from profiles showed a spread distribution of misfits for temperature, with an average 0.009 °C, and a narrower distribution of misfits for practical salinity with an average of 0.007. Given the nominal accuracy expected by the TSG system and in absence of systematic marked shift in the comparison, no post-cruise adjustment was performed. The uncertainty of measurement in the TSG data set should have stayed under the 0.01 °C in temperature and 0.01 in practical salinity.

Considering BGC-Argo floats, the comparison with CTD profiles was performed over the 750–1000 dbar layer, where water mass characteristics remained stable enough to ascribe misfits as instrumental calibration shifts rather than natural variability. The misfits between temperature measurements and practical salinity measurements at geopotential horizons were computed and median values provided for every BGC-

Argo float. The median offsets are reported in Table 2. Their amplitudes remained within 0.01 °C in temperature or 0.01 in practical salinity except in two cases. A large temperature offset occurred for WMO 6901769. A large practical salinity offset was reported for WMO 6901765, despite being deployed exactly concomitant with the CTD profile.

2.3 Oxygen concentration

2.3.1 Presentation of the different measurements

Concentration of dissolved dioxygen (O_2) in seawater, hereafter referred to as oxygen, was measured with three techniques: the classical iodometric Winkler method, an electrochemical oxygen sensor and optical oxygen sensors.

Oxygen concentration was measured following the Winkler method (Winkler, 1888) with potentiometric endpoint detection (Oudot et al., 1988) on discrete samples collected with Niskin bottles. For sampling, reagent preparation and analysis, the recommendations of Langdon (2010) were carefully followed.

Oxygen concentrations were measured by a Sea-Bird SBE43 (serial no. 0587) electrochemical sensor interfaced with the CTD unit. This sensor was plumbed in the pumped circuit following the GO-SHIP guidelines (Hood et al., 2010).

Oxygen optical measurements (also called optode measurements) were collected by two types of sensors. One Rinko III dissolved oxygen sensor from JFE Advanctech Co., Ltd., Japan (serial no. 171), was interfaced with the CTD unit using the analog output voltage. Aanderaa 4330 optodes were mounted on every BGC-Argo float.

2.3.2   Data processing

The titration volumes were converted to oxygen concentrations in $\mu mol\,kg^{-1}$ by following the calculation procedure proposed in Langdon (2010). The precision of the Winkler measurements was estimated by reproducibility tests based on five or six replicates for samples withdrawn from the same Niskin bottles. The standard deviation of the replicate measurements was less than $0.4\,\mu mol\,kg^{-1}$.

The sensor signal of the SBE43 was aligned to temperature and pressure scans with an advanced offset considering a unique plumbing configuration for the cruise of 3 s. The raw signal was then converted to an oxygen concentration with 13 calibration coefficients. The method is based on the Owens and Millard (1985) algorithm that has been slightly adapted by Sea-Bird in the data processing software using a hysteresis correction (Sea-Bird Scientific, 2014). A new set of calibration coefficients for this sensor was determined after the cruise; it was used to post-process the whole data set. Only 3 (the oxygen signal slope, the voltage at zero oxygen signal and the pressure correction factor) of the 13 coefficients determined by the pre-cruise factory calibration of the sensor were adjusted with the following procedure. The oxygen concentrations measured by Winkler were matched with the signal measured by the sensor at the closing of the Niskin bottles. The three values were fitted by minimizing the sum of the square of the difference between Winkler oxygen and oxygen derived from sensor signal. Outliers were discarded when the residuals exceeded 2.8 standard deviation of the residuals until no more outliers remain.

The Rinko optode provided continuous voltage output at 24 Hz, which has been directly converted to an oxygen concentration with the MATLAB code developed by the manufacturer. The original calibration coefficients were used. To process the results, the temperature measured from the CTD unit was preferred to the built-in temperature of the sensor.

The Aanderaa optodes 4330 output signal is a C1 raw phase (phase from the blue light excitation), a C2 raw phase (phase from the red light excitation) and the optode temperature. The calculation of oxygen concentrations from the optode signal follows the recommendations of Thierry et al. (2016). The calibrated phase estimated from the C1 and C2 raw phases is converted in oxygen concentration by the Stern–Volmer equation proposed by Uchida et al. (2008) using seven calibration coefficients (the so-called Stern–Volmer–Uchida coefficients). The oxygen concentration is then corrected for salinity and pressure effects. The pressure compensation is estimated following Bittig et al. (2015) with a step of phase adjustment. Finally, concentrations are expressed in $\mu mol\,kg^{-1}$ by using the potential density derived from the CTD measurements of BGC-Argo floats.

2.3.3   Data quality control

Winkler measurements of discrete samples collected during upcasts were considered as the reference oxygen value because they rely on a reference material (KIO_3 standard) given with a precision of replicate measurements lower than $0.4\,\mu mol\,kg^{-1}$. The reference Winkler measurements were used to adjust the calibration coefficients of the CTD oxygen sensor (SBE43), as described below. The corrected oxygen profiles during downcasts from the SBE43 at stations were considered as the reference profile for optode measurements from BGC-Argo floats. This quality control was based on the downcast profiles at 1 dbar resolution collected either by the electrochemical sensor SBE43 or the optode RINKO.

Residuals with Winkler measurements were expressed as the difference in an isobaric horizon between the sensor oxygen and the Winkler oxygen. A sensor error was estimated as the root mean square error of the residuals. Results are reported in Fig. 2, where the residuals over the entire cruise are plotted as a function of time and depth. Residuals appear higher and more variable in the upper part of the water column, most probably due to enhanced oxygen gradients and changes on isobaric horizons between downcasts and upcasts. For electrochemical measurements, no significant offsets or drifts were observed; the sensor error over the entire cruise is $2.4\,\mu mol\,kg^{-1}$. For RINKO optode measurements, the sensor error over the entire cruise was $6.0\,\mu mol\,kg^{-1}$ and a systematic offset of $4.8\,\mu mol\,kg^{-1}$ was observed. Moreover, a significant increase of the residuals with depth ($0.0022\,\mu mol\,kg^{-1}\,dbar^{-1}$) was observed below 200 dbar. Thus, the SBE43 data were used rather than the RINKO data in the final record.

Considering BGC-Argo floats, it has been reported that a systematic shift in the optode calibration coefficients can occur during storage and shipment of the sensors (Bittig et al., 2012). In order to compensate for this potential shift, float oxygen measurements were corrected based on a reference profile as in Takeshita et al. (2013). A slope and offset value were determined for every optode deployed in order to adjust a posteriori the calculated oxygen values from the raw signals. The adjustment of optode values was performed using a linear model, below the first 50 dbar to avoid strong variability in the surface layer, and above the last 50 dbar to get rid of possible hooks at the bottom of profiles. The results, reported in Table 2, show a consistent correlation between the two sensors and offsets ranging from -14 to $11\,\mu mol\,kg^{-1}$.

2.4   Chlorophyll a concentration

2.4.1   Presentation of the different measurements

The chlorophyll a concentration (Chl a; sum of chlorophyll a, divinyl chlorophyll a and chlorophyllide a) in seawater was measured with two methods: high-performance liquid chromatography (HPLC) and fluorescence.

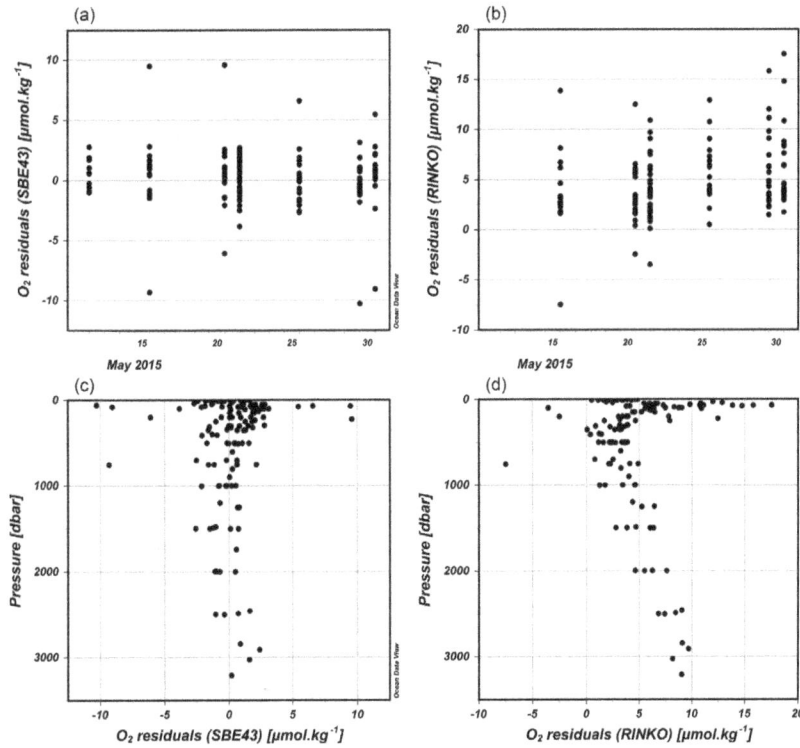

Figure 2. Oxygen residuals between sensor and Winkler measurements, plotted as a function of time (**a, b**) and as a function of depth (**c, d**). The residuals for the electrochemical sensor are plotted in (**a, c**), and those for the optode in (**b, d**).

The HPLC method is used to estimate the Chl a in discrete seawater samples collected from the TSG system or withdrawn from Niskin bottles. For this, 2.27 L of the seawater samples was filtered onto glass fiber filters (GF/F Whatman 25 mm), and all filters were then stored in liquid nitrogen at $-80\,°C$ until further laboratory analysis. The chlorophyll a and other accessory phytoplankton pigments were then extracted from the filters in 100 % methanol, disrupted by sonification and clarified by filtration (GF/F Whatman 0.7 µm) after 2 h. Extracts were injected (within 24 h of beginning of the extraction) on a reversed-phase C8 column, and 24 pigments were separated, identified and quantified according to the HPLC analytical protocol described by Ras et al. (2008).

Fluorometers provide continuous detection of chlorophyll a. Three kinds of sensors were used during the cruise: Chelsea Aqua Tracka III fluorometer (serial no. 088193) interfaced with CTD unit, ECO WetLabs fluorometers that equipped every BGC-Argo float and a Turner fluorometer (serial no. 6241) plumbed in the TSG system of the vessel. The sensing mean is based on the fluorescence concept: irradiated by blue light, chlorophyll a absorbs and re-emits in the red part of the spectrum, and the re-emitted signal (i.e., the fluorescence) is considered proportional to the Chl a (Lorenzen, 1966). However, to retrieve the exact Chl a through the raw fluorescence signal, a calibration of the signal is necessary.

Note that fluorescence is affected by non-photochemical quenching, the protection mechanism employed by phytoplankton against the effects of high light intensity. As a result, amplitude of signal is reduced for an identical Chl a when the measurement is performed under sunlight exposure in the sea surface layer.

2.4.2 Data processing

The Chl a is derived from raw fluorescence signal by a linear model using two calibration coefficients: an offset that corresponds to the value of the signal in the absence of Chl a and a scaling factor to align the signal on the exact in situ Chl a. These calibration coefficients are generally provided by the manufacturer, but an adjustment using in situ measurements of Chl a is recommended. The calibration method was based on the alignment of the fluorescence signal to exact in situ discrete measurements of Chl a provided by the HPLC method. For this, a least square linear regression was used with simultaneous measurements of Chl a from fluorescence at the time, location and depth of collected seawater samples analyzed by HPLC. The statistics associated with the linear regression were used as a quality control of the calibration.

Fluorometer-derived Chl a profiles at CTD casts were processed as follows. As a pre-processing step, the raw fluorescence measurements were corrected for possible non-photochemical quenching following the procedure of Xing

Table 4. List of parameters in the pigment data set, variable names and units; for each pigment, the detection wavelengths and the associated limits of detection in nanograms per injection (ng/inj).

Pigment	Variable name	Units	Detection wavelength (nm)	Limit of detection (ng/inj)	Limit of detection for 2 L filtered (in $mg\,m^{-3}$)
Chlorophyll c3	CHLC3	$mg\,m^{-3}$	450	0.015	0.0002
Chlorophyll c1 + c2	CHLC2	$mg\,m^{-3}$	450	0.018	0.0002
Sum chlorophyllide a	CHLDA	$mg\,m^{-3}$	667	0.016	0.0002
Peridinin	PERI	$mg\,m^{-3}$	450	0.007	0.0001
Sum phaeophorbid a	PHDA	$mg\,m^{-3}$	667	0.009	0.0001
19'-Butanoyloxyfucoxanthin	BUT	$mg\,m^{-3}$	450	0.009	0.0001
Fucoxanthin	FUCO	$mg\,m^{-3}$	450	0.009	0.0001
Neoxanthin	NEO	$mg\,m^{-3}$	450	0.009	0.0001
Prasinoxanthin	PRAS	$mg\,m^{-3}$	450	0.009	0.0001
Violaxanthin	VIOLA	$mg\,m^{-3}$	450	0.012	0.0001
19'-Hexanoyloxyfucoxanthin	HEX	$mg\,m^{-3}$	450	0.009	0.0001
Diadinoxanthin	DIADINO	$mg\,m^{-3}$	450	0.014	0.0002
Alloxanthin	ALLO	$mg\,m^{-3}$	450	0.015	0.0002
Diatoxanthin	DIATO	$mg\,m^{-3}$	450	0.015	0.0002
Zeaxanthin	ZEA	$mg\,m^{-3}$	450	0.014	0.0002
Lutein	LUT	$mg\,m^{-3}$	450	0.014	0.0002
Bacteriochlorophyll a	BCHLA	$mg\,m^{-3}$	770	0.010	0.0001
Divinyl chlorophyll b	DVCHLB	$mg\,m^{-3}$	450	0.004	0.0001
Chlorophyll b	CHLB	$mg\,m^{-3}$	450	0.004	0.0001
Total chlorophyll b	TCHLB	$mg\,m^{-3}$	450	0.004	0.0001
Divinyl chlorophyll a	DVCHLA	$mg\,m^{-3}$	667	0.011	0.0001
Chlorophyll a	CHLA	$mg\,m^{-3}$	667	0.011	0.0001
Total chlorophyll a	TCHLA	$mg\,m^{-3}$	667	0.011	0.0001
Sum phaeophytin a	PHYTNA	$mg\,m^{-3}$	667	0.007	0.0001
Sum carotenes	TCAR	$mg\,m^{-3}$	450	0.013	0.0002

et al. (2012). The linear regression was done with 61 simultaneous measurements of Chl a determined by HPLC and the fluorometer. The resulting coefficients were an offset of $0.168\,mg\,m^{-3}$ and a slope of 4.016 with a coefficient of determination equal to 0.96. An alternative estimation of the offset was performed by computing the median value of raw fluorescence profiles in the last 50 m of every profile. Indeed, when the water column is stratified (it was always the case here), the availability of light is not enough to allow the presence of active phytoplankton cells; thus the fluorescence signal should be null. This estimation considering all the fluorescence profiles provides an offset of $0.160 \pm 0.004\,mg\,m^{-3}$.

As for CTD casts, the raw fluorescence measurements from BGC-Argo floats were corrected for possible non-photochemical quenching, and offsets were determined as median values of raw fluorescence in the last 50 m of the profiles. The estimated offset values are reported in Table 2. Once offsets were adjusted, the linear regressions were performed with seven or eight simultaneous measurements of Chl a obtained by HPLC at the float deployment. The estimated slopes are reported in Table 2. On average from all

the calibration conducted, slopes range from 0.49 to 0.67 with an average value of 0.58; offsets range from -0.02 to $0.04\,mg\,m^{-3}$ with an average value of $0.02\,mg\,m^{-3}$.

Considering fluorometer-derived Chl a along the ship track, a post-cruise estimation of the calibration coefficients for the Turner fluorometer was undertaken. The linear regression was done with nine discrete seawater samples collected at night (between 19:00 and 05:00 UTC) to avoid the non-photochemical quenching. The calibration coefficients obtained were an offset of $0.059\,mg\,m^{-3}$ and a slope of 4.831 with a coefficient of determination equal to 0.70. The raw fluorescence measurements were included in the TSG data flow of 15 s resolution. The TSG data processing followed the same steps as for ship-track temperature and salinity, to provide average time series in 5 min bins.

2.4.3 Data quality control

In the Table 4, the list of quantified pigments and their limits of detections (calculated in nanograms per injection and as the concentrations corresponding to a signal-to-noise ratio of 3) is provided. Different quality control steps were applied during HPLC analysis, data processing and to the final data set. During HPLC analysis, parameters such as the stability of the baseline, the injection precision and the pressure were monitored regularly in order to detect potential anomalies in the analytical process. During data processing, chromatographic parameters were checked, including critical pair resolution, baseline noise and peak width or retention time precision. Spectral data for the different peaks were verified and used for identification purposes and peak purity assessment. The final pigment database underwent a visual verification step for each pigment of every vertical profile and quality flags were assigned for each value. The visual check confirms that the identification and quantification of all the samples did not present any issues, such as coelution problems or baseline noise leading to potential uncertainties.

Considering fluorescence measurements collected on CTD casts, the high coefficient of determination ($r^2 = 0.96$) for the linear model denotes a very good regression with HPLC data. The pair of calibration coefficients were applied in the post-processing of fluorescence data at every cast.

Considering fluorescence measurements collected by BGC-Argo floats, good alignment with in situ data was reached with coefficients of determination higher than 0.75 (see Table 2). Moreover, the homogeneity of slopes among the series of new sensors (thus recently factory calibrated) gives insight into the gain (between 1.8 and 2) to be applied afterwards to fluorescence data (Roesler et al., 2017).

Considering fluorescence measurements collected on the TSG system, its range along the ship track appears very narrow (from 0.035 to 0.112 mg m^{-3}). In addition, a low number of simultaneous HPLC measurements is available (only nine samples), and the coefficient of determination of the linear regression is lower than 0.70. Thus, the calibration effort performed is certainly not enough to provide full confidence in the adjusted coefficients, although they have been applied to the TSG time series.

2.5 Nitrate (and other nutrient) concentrations

2.5.1 Presentation of the different measurements

Concentrations of nitrate (NO$_3-$) ions in seawater were measured with two techniques: the classical colorimetric method in conjunction with nitrite, phosphate and silicate concentrations, and with an optical nitrate sensor.

Nutrient samples were collected and conserved following the recommendations of Kirkwood (1992). All nutrient samples were analyzed by a standard automated colorimetric system set up following Aminot and Kerouel (2007), using a Seal Analytical continuous flow AutoAnalyzer III (AA3).

Optical sensor measurements were performed on BGC-Argo floats. Sensors using miniaturized ultraviolet spectrophotometers allow for continuous measurement of absorbance spectra and estimations of nitrate concentrations (Johnson and Coletti, 2002). The BGC-Argo floats deployed during this cruise were equipped with the Satlantic SUNA-V2 (Submersible Ultraviolet Nitrate Analyzer) sensors.

2.5.2 Data processing

Nitrate concentrations are derived from absorbance spectra using the TCSS (temperature compensated, salinity subtracted) algorithm developed by Sakamoto et al. (2009). In the Mediterranean Sea, because of specific conditions of low nitrate concentrations and high salinity (thus high bromide concentrations), optical measurements of nitrate were extremely delicate (D'Ortenzio et al., 2014). This drove the development of a specific algorithm adapted from TCSS that substantially improved the estimation of nitrate concentrations in this area (Pasqueron de Fommervault et al., 2015).

The BGC-Argo floats deployed during the cruise transmitted the raw data of the SUNA (i.e., absorbance spectrum from 217 to 250 nm), which allowed for post-processing with the algorithm of Pasqueron de Fommervault et al. (2015). A spike test was applied in addition to a test for saturation based on the raw absorption spectrum. Nitrate concentration data computed from a spectrum for which more than 25 % of the channels saturate (i.e., reached the maximum value of numerical counts) were discarded. This was the case of one BGC-Argo float (WMO6901773).

2.5.3 Data quality control

The SUNA sensors also undergo offset and gain (Johnson et al., 2013) that were corrected using as reference the measurements of discrete samples. Given that surface nitrate concentrations in May and June in the Mediterranean Sea are below the limit of detection of the sensor (Pasqueron de Fommervault et al., 2015), an offset was computed as the difference between an assumed surface concentration of zero and the mean nitrate value measured from 5 to 30 m. A gain was then calculated with a match up between sensors measurements and nitrate concentrations at discrete depths. Gain correction was applied only if the misfits between sensor derived and reference concentrations below 950 dbar did not exceed 10 % of the deep reference value. The correction coefficients per BGC-Argo float are reported in Table 2. A slope of 1 was estimated for most of the cases, and the offsets ranged from -2.70 to 3.90 μmol L^{-1}.

3 Data availability

The final data set concatenates the different collections during the cruise, which are vertical profiles and bottle samples at CTD casts and along-track measurements at surface and at depth. This data set benefits from post-cruise corrections described in the previous sections. A unique convention was used to identify bad, absent or unreported data: they have been assigned the value -999.

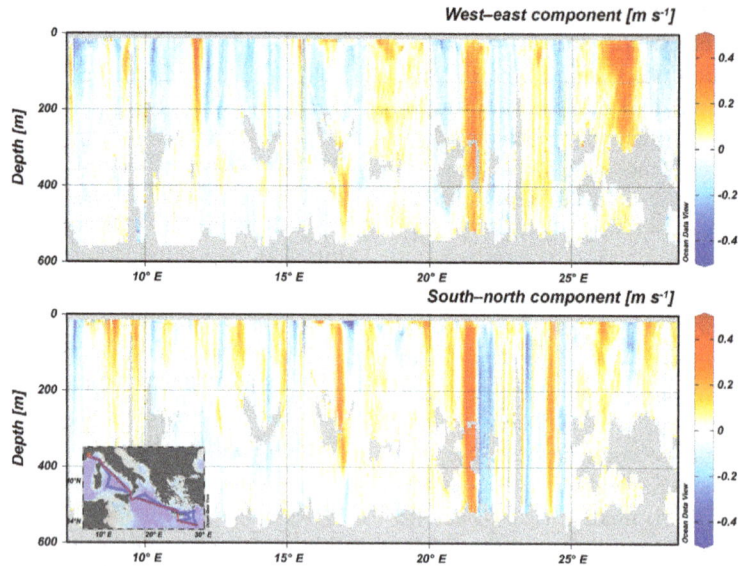

Figure 3. Velocity distribution of the upper water column along a west–east section through the Mediterranean Sea. Data are recorded by SADCP. Inner panel indicates the location of the ship track and the section. Grey areas: no data are available.

The quality control applied to discrete sample collection has been assigned with a quality flag. The quality code developed for WHP bottle parameters data was used, in particular "2: Acceptable measurement", "5: Not reported" and "9: Sample not drawn for this measurement from this bottle".

Data are published by SEANOE operated by SISMER within the framework of the information system ODATIS. Data from the stations are available at https://doi.org/10.17882/51678 (Taillandier et al., 2017a), data along the ship track are available at https://doi.org/10.17882/51691 (Taillandier et al., 2017b).

4   Discussion and conclusions

With an extension of about 25° in longitude, this cruise covered the central Mediterranean Sea and part of its northwestern and eastern basins. High-resolution ADCP data (Fig. 3) reveals some well-known patterns of the surface circulation in this area (the cyclonic gyre in the Ligurian Basin, the eastward surface flow in the Levantine) as well as ubiquitous mesoscale activity. Seven stations were chosen in this transect (one in the Ligurian, two in the Tyrrhenian, two in the Ionian, two in the Levantine) in order to provide a large-scale record on the hydrography and biogeochemistry of the Mediterranean Sea. As shown in Fig. 4a, there is a clear separation of water mass characteristics between the eastern and western basin, with a clear longitudinal gradient as deep waters and intermediate waters become saltier and warmer eastwards. Associated with this water mass distribution, biogeochemical traits clearly showed important differences among basins and relative homogeneity within basins. As shown in Fig. 4c, the oxygen minimum of the interme-

diate waters is the lowest in the western stations, and deep waters are more oxygenated in basins directly influenced by winter convection (Ligurian and Ionian). The nutrient distribution also shows the eastern depletion of nitrates in deep waters, shallower nitraclines in the western basin and the absence of nitrates in the surface layers relevant to the Mediterranean oligotrophic spring regime (see Fig. 4d). These large-scale patterns are in good agreement with observations reported by previous field surveys such as BOUM in 2009 (Moutin and Prieur, 2012) or M84/3 in 2011 (Tanhua et al., 2013). Consequently, the vertical distribution of biomass is marked by a deep chlorophyll maximum; this maximum becomes higher and shallower between the eastern to western basins (see Fig. 4b). Such spatial contrasts need to be complemented by the temporal evolution of these patterns which can be achieved thanks to the BGC-Argo floats.

The data set presented in this paper has been collected in the framework of an emerging in situ observing system in the Mediterranean Sea. In order to characterize the seasonal cycles of phytoplankton dynamics and the biogeochemical functioning of the Mediterranean Sea, this network of twelve BGC-Argo floats collects data on physical and biogeochemical properties (temperature, salinity, concentration of dissolved oxygen, chlorophyll a, nitrate) along 1000 m depth profiles at a weekly sampling rate. In spring 2015, shipboard measurements were acquired with the objective of providing a reference data set for each core parameter of the in situ observing system, verified through the inter-comparison of several in situ sensing methods. This data set allowed performing metrological verification of the deployed sensors, considering the misfits between the first profile of the float and the shipboard data. This data set can provide ancillary data

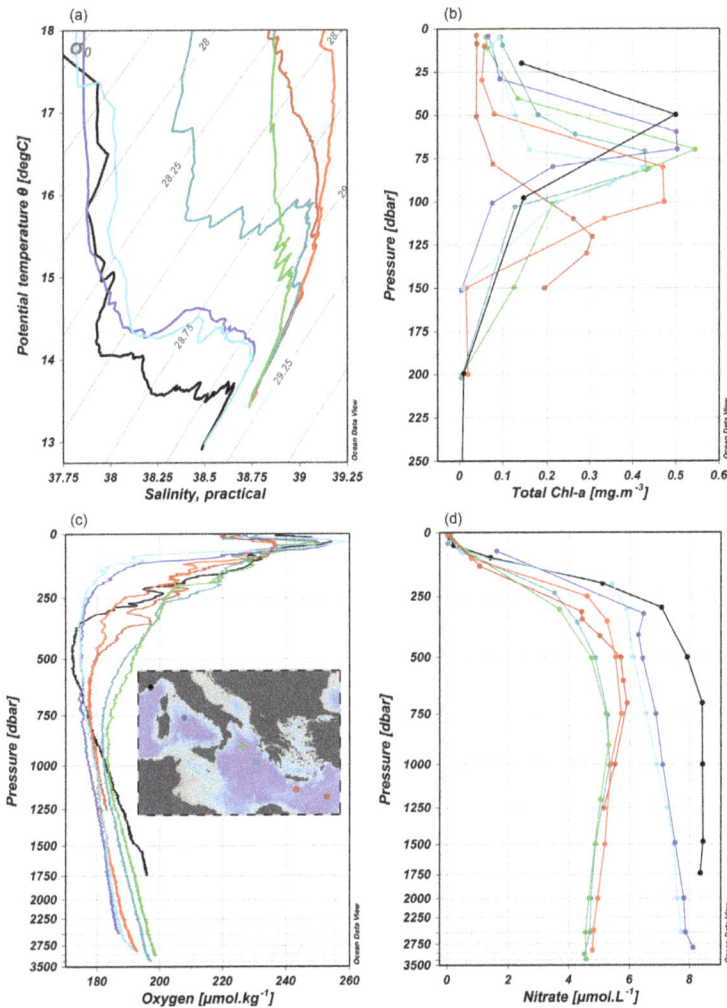

Figure 4. TS diagram comprised of CTD data (**a**); total chlorophyll *a* concentration profiles by HPLC method (**b**); dissolved oxygen concentration profiles from CTD data (**c**); nitrate concentration profiles by colorimetric method (**d**). The inner panel shows the locations of CTD stations.

for performing and distributing delayed-mode adjustments in the time series of these BGC-Argo floats to end users (e.g., Schmechtig et al., 2015).

First, the presented data set provides an in situ characterization of the environmental conditions in which the verification exercises were conducted. Thanks to ocean current and surface hydrography information collected along the ship track, a first assessment of the circulation patterns neighboring every station can be made. Complemented with satellite observations (altimetry, images of sea surface temperature or ocean color), the degree of stability of the water column can be diagnosed in order to relate (or not) the co-location in space and time of the BGC-Argo float profile with reference data.

Second, the presented data set provides material for the systematic calibration of the biogeochemical sensors active in the network. The crucial role of this operation on newly

deployed sensors has been shown (Table 2). Concerning the oxygen optode sensors, their linear response does not seem to be affected; however, offsets reaching amplitudes of $15\,\mu\,\mathrm{mol\,kg^{-1}}$ have been reported, without any systematic bias among the set of sensors. Concerning fluorometer sensors, offsets can be corrected considering dark values at depth; however, the amplitudes of the signals appeared to be overestimated by a factor between 1.5 and 2 depending on the sensor. Concerning nitrate sensors, their behavior at deployment is similar to the optodes in terms of calibration, with a sensor-dependent offset up to $4\,\mu\mathrm{mol\,L^{-1}}$ of amplitude. Overall, the biogeochemical sensors equipped on the BCG-Argo floats revealed inherent calibration shifts upon deployment. This is in agreement with recent works on ECO fluorometers (Roesler et al., 2017) and oxygen optodes (Bittig et al., 2015; Bittig and Körtzinger, 2015).

The data set presented is relevant for the robust evaluation of the calibration state of biogeochemical sensors at the beginning of their mission. In addition, if an equivalent data set is collected at the end of the mission when the BGC-Argo floats are recovered, the sensor drifts can be properly assessed from pre-mission and post-mission calibration states. This objective appears essential to allow for the harmonization between all the time series observed by the network.

This data set is a first attempt at evaluating the uncertainties that come up in the verification exercises. When measuring misfits between shipboard measurements and the first profile of the BGC-Argo floats, the natural variability of the environment can affect their complete attribution to calibration shifts. This natural variability can be inferred by diurnal cycles for biogeochemical sensors, or to a lesser extent by mesoscale effects. The expected variations depend on the type of parameter, the depth of inter-comparison and the duration or distance between profiles. Among the BGC-Argo floats deployed during the cruise, two benefited from a perfectly concomitant verification exercise, as they were clamped onto the CTD carousel. The first results show reduced dispersion as a function of depth for all the parameters. This dispersion criterion needs to be assessed more carefully with different types of match up, as a function of local environmental conditions and duration or distance from the first profile.

Preliminary conclusions stress the importance of evaluating the calibration state of the biogeochemical sensors and their possible drift over several mission years. The data set collected during the cruise of May 2015 provided relevant material for performing such metrological verification exercises, and justifies future deployments. The cruise also unintentionally showed it was possible to perform pre-deployment verification exercises some days before the beginning of the mission. The floats with newly verified sensors were deployed close to those recovered in order to continue their time series and to retrieve post-mission calibration states. If the propagation of reference data between missions is satisfactory, such a protocol could be applied to conventional oceanographic cruises as they demand one station of metrological verification with floats mounted on the CTD carousel and changes of route for float deployment and recovery operations.

Author contributions. This data set was collected by VT, TW, FDO, HLG and NM. TW analyzed the oxygen samples, JR analyzed the pigment samples, and ED analyzed the nutrient samples. Data processing and quality control were undertaken by HLG for ocean currents and TSG; by VT for seawater hydrological properties; by FDO and NM for chlorophyll *a* concentration; by TW, LC, HB and DL for oxygen concentration; and by OPF and FDO for nitrate concentration. VT, AP and EL organized the BGC-Argo float deployments and recoveries. Data management and availability were undertaken by CS. VT and TW prepared the manuscript with contributions from FDO, NM, JR, LP and OPF.

Competing interests. The authors declare that they have no conflict of interest.

Acknowledgements. We would like to thank Captain Dany Deneuve and the crew of RV *Tethys 2*. These observational efforts were supported the project Equipex-NAOS, the Euro-Argo infrastructure, the program MerMex, and the project BAMA funded by LEFE/GMMC. We gratefully acknowledge their support.

Edited by: Giuseppe M. R. Manzella

References

Aminot, A. and Kerouel, R.: Dosage automatique des nutriments dans les eaux marines methodes en flux continu, in: Methodes d'analyse en milieu marin, Ifremer, Editions 25 Quae, 188 pp., 2007.

Bittig, H. C. and Körtzinger A.: Tackling Oxygen Optode Drift: Near-Surface and In-Air Oxygen Optode Measurements on a Float Provide an Accurate in Situ Reference, J. Atmos. Ocean. Tech., 32, 1536–1543, https://doi.org/10.1175/JTECH-D-14-00162.1, 2015.

Bittig, H. C., Fiedler, B., Steinhoff, T., and Körtzinger A.: A novel electrochemical calibration setup for oxygen sensors and its use for the stability assessment of Aanderaa optodes, Limnol. Oceanogr.-Meth., 10, 921–933, https://doi.org/10.4319/lom.2012.10.921, 2012.

Bittig, H. C., Fiedler, B., Fietzek, P., and Körtzinger, A.: Pressure response of Aanderaa and Sea-Bird oxygen optodes, J. Atmos. Ocean. Tech., 32, 2305–2317, https://doi.org/10.1175/JTECH-D-15-0108.1, 2015.

Bosc, E., Bricaud, A., and Antoine, D.: Seasonal and interannual variability in algal biomass and primary production in the Mediterranean Sea, as derived from four years of SeaWiFS observations, Global Biogeochem. Cy., 18, GB1005, https://doi.org/10.1029/2003GB002034, 2004.

D'Ortenzio, F. and Ribera d'Alcalà, M.: On the trophic regimes of the Mediterranean Sea: a satellite analysis, Biogeosciences, 6, 139–148, https://doi.org/10.5194/bg-6-139-2009, 2009.

D'Ortenzio, F., Lavigne, H., Besson, F., Claustre, H., Coppola, L., Garcia, N., and Morin, P.: Observing mixed layer depth, nitrate and chlorophyll concentrations in the northwestern Mediterranean: A combined satellite and NO3 profiling floats experiment, Geophys. Res. Lett., 41, 6443–6451, 2014.

Durrieu de Madron, X., and the MerMex group: Marine ecosystems' responses to climatic and anthropogenic forcings in the Mediterranean, Prog. Oceanogr., 91, 97–116, 2011.

Giorgi, F. and Lionello, P.: Climate change projections for the Mediterranean region, Global Planet. Change, 63, 90–104, 2008.

Hood, E. M., Sabine, C. L., and Sloyan B. M.: The GO-SHIP repeat hydrography manual: A collection of expert reports and guidelines, IOCCP Report No.14, ICPO Publication Series No. 134, Version 1, 2010.

Johnson, K. S. and Coletti, L. J.: In situ ultraviolet spectrophotometry for high resolution and long-term monitoring of nitrate, bromide and bisulfide in the ocean, Deep-Sea Res. Pt. I, 49, 1291–1305, 2002.

Johnson, K. S., Coletti, L. J., Jannasch, H. W., Sakamoto, C. M., Swift, D. D., and Riser, S. C.: Long-term nitrate measurements in the ocean using the In Situ Ultraviolet Spectrophotometer: sensor integration into the Apex profiling float, J. Atmos. Ocean. Tech., 30, 1854–1866, 2013.

Johnson, K. S., Plant, J. N., Coletti L. J., Jannasch, H. W., Sakamoto C. M., Riser, S. C., Swift, D. D., Williams, N. L., Boss, E., Lynne, N. H., Talley, D., and Sarmiento, J. L.: Biogeochemical sensor performance in the SOCCOM profiling float array, J. Geophys. Res.-Oceans, 122, 6416–6436, https://doi.org/10.1002/2017JC012838, 2017.

Kirkwood, D. S.: Stability of solutions of nutrient salts during storage, Mar. Chem., 38, 151–164, 1992.

Langdon, C.: Determination of Dissolved Oxygen in Seawater by Winkler Titration Using the Amperometric Technique In The GO-SHIP Repeat Hydrography Manual, in: A Collection of Expert Reports and Guidelines, edited by: Hood, E. M., Sabine, C. L., and Sloyan, B. M., IOCCP Report Number 14, ICPO Publication Series Number 134, available at: http://www.go-ship.org/HydroMan.html (last access: 22 March 2018), 2010.

Le Bot, P., Kermabon, C., Lherminier, P., and Gaillard, F.: Cascade V6.1: Logiciel de validation et de visualization des mesures ADCP de coque, document utilisateur et maintenance, Report OPS/LPO 11-01, 2011.

Leymarie, E., Poteau, A., André, X., Besson, F., Brault, P., Claustre, H., David, A., D'Ortenzio, F., Dufour, A., Lavigne, H., Reste, S. L., Le Traon, P. Y., Migon, C., Nogre, D., Obolensky, G., Penkerc'h, C., Sagot, J., Schaeffer, C., Schmechtig, C., and Taillandier, V.: Development and validation of the new ProvBioII float, Mercator Ocean Quarterly Newsletter, available at: https://www.mercator-ocean.fr/wp-content/uploads/2015/05/Mercator-Ocean-newsletter-2013_48.pdf, 2013.

Lorenzen, C. J.: A method for the continuous measurement of in vivo chlorophyll concentration, Deep Sea Res. Oceanogr. Abstr., 13, 223–227, https://doi.org/10.1016/0011-7471(66)91102-8, 1966.

Marty, J. C. and Chiavérini, J.: Hydrological changes in the Ligurian Sea (NW Mediterranean, DYFAMED site) during 1995–2007 and biogeochemical consequences, Biogeosciences, 7, 2117–2128, https://doi.org/10.5194/bg-7-2117-2010, 2010.

Marty, J. C., Chiaverini, J., Pizay, M. D., and Avril, B.: Seasonal and interannual dynamics of nutrients and phytoplankton pigments in the western Mediterranean Sea at the DYFAMED time-series station (1991–1999), Deep-Sea Res. Pt. II, 49, 1965–1985, 2002.

Mayot, N., D'Ortenzio, F., Ribera d'Alcalà, M., Lavigne, H., and Claustre, H.: Interannual variability of the Mediterranean trophic regimes from ocean color satellites, Biogeosciences, 13, 1901–1917, https://doi.org/10.5194/bg-13-1901-2016, 2016.

Moutin, T. and Prieur, L.: Influence of anticyclonic eddies on the Biogeochemistry from the Oligotrophic to the Ultraoligotrophic Mediterranean (BOUM cruise), Biogeosciences, 9, 3827–3855, https://doi.org/10.5194/bg-9-3827-2012, 2012.

Oudot, C., Gerard, R., Morin, P., and Gningue, I.: Precise shipboard determination of dissolved-oxygen (Winkler Procedure) with a commercial system, Limnol. Oceanogr., 33, 146–150, 1988.

Owens, W. B. and Millard Jr., R. C.: A new algorithm for CTD oxygen calibration, J. Phys. Oceanogr., 15, 621–631, 1985.

Pasqueron de Fommervault, O., D'Ortenzio, F., Mangin, A., Serra, R., Migon, C., Claustre, H., Lavigne, H. Ribera d'Alcalà, M., Prieur, L., Taillandier, V., Schmechtig, C., Poteau, A., Leymarie, E., Besson, F., and Obolensky, G.: Seasonal variability of nutrient concentrations in the Mediterranean Sea: Contribution of Bio-Argo floats, J. Geophys. Res.-Oceans, 120, https://doi.org/10.1002/2015JC011103, 2015.

Ras, J., Claustre, H., and Uitz, J.: Spatial variability of phytoplankton pigment distributions in the Subtropical South Pacific Ocean: comparison between in situ and predicted data, Biogeosciences, 5, 353–369, https://doi.org/10.5194/bg-5-353-2008, 2008.

Roesler, C., Uitz, J., Claustre, H., Boss, E., Xing, X., Organelli, E., Briggs, N., Bricaud, A., Schmechtig, C., Poteau, A., D'Ortenzio, F., Ras, J., Drapeau, S., Haëntjens, N., and Barbieux, M.: Recommendations for obtaining unbiased chlorophyll estimates from in-situ chlorophyll fluorometers: A global analysis of WET Labs ECO sensors, Limnol. Oceanogr.-Meth., 15, 572–585, https://doi.org/10.1002/lom3.10185, 2017.

Sakamoto, C. M., Johnson, K. S., and Coletti, L. J.: Improved algorithm for the computation of nitrate concentrations in seawater using an in situ ultraviolet spectrophotometer, Limnol. Oceanogr.-Meth., 7, 132–143, 2009.

Schmechtig, C., Poteau, A., Claustre, H., D'Ortenzio, F., and Boss, E.: Processing bio-Argo chlorophyll-a concentration at the DAC level, Argo data management, https://doi.org/10.13155/39468, 2015.

Sea-Bird Scientific: Application Note 64-3: SBE 43 dissolved oxygen (DO) sensor – hysteresis corrections, available at: http://www.seabird.com/document/an64-3-sbe-43-dissolved-oxygen-do-sensor-hysteresis-corrections (last access: 22 March 2018), 2014.

Siokou-Frangou, I., Christaki, U., Mazzocchi, M. G., Montresor, M., Ribera d'Alcalà, M., Vaqué, D., and Zingone, A.: Plankton in the open Mediterranean Sea: a review, Biogeosciences, 7, 1543–1586, https://doi.org/10.5194/bg-7-1543-2010, 2010.

Taillandier, V., Wagener, T., D'Ortenzio, F., Mayot, N., Legoff, H., Ras, J., Coppola, L., Pasqueron de Fommervault, O., Bittig, H., Lefevre, D., Leymarie, E., Schmechtig, C., and Poteau, A.: Oceanographic dataset in the Mediterranean Sea collected during the cruise BioArgoMed 2015, https://doi.org/10.17882/51678, 2017a.

Taillandier, V., Wagener, T., D'Ortenzio, F., Mayot, N., Legoff, H., Ras, J., Coppola, L., Pasqueron de Fommervault, O., Bittig, H., Lefevre, D., Leymarie, E., Schmechtig, C., and Poteau, A.: Ship-track continuous dataset in the Mediterranean Sea collected during the cruise BioArgoMed 2015, https://doi.org/10.17882/51691, 2017b.

Takeshita, Y., Martz, T. R., Johnson, K. S., Plant, J. N., Gilbert, D., Riser, S. C., Neill, C., and Tilbrook, B.: A climatology-based quality control procedure for profiling float oxygen data, J. Geophys. Res.-Oceans, 118, 5640–5650, https://doi.org/10.1002/jgrc.20399, 2013.

Tanhua, T., Hainbucher, D., Schroeder, K., Cardin, V., Álvarez, M., and Civitarese, G.: The Mediterranean Sea system: a review and an introduction to the special issue, Ocean Sci., 9, 789–803, https://doi.org/10.5194/os-9-789-2013, 2013.

Thierry V., Gilbert D., Kobayashi T., Schmid C., and Kanako S.: Processing Argo oxygen data at the DAC level cookbook, Argo data management, https://doi.org/10.13155/39795, 2016.

Thurnherr, A. M.: How to process LADCP data with the LDEO Software (Versions IX.7-IX.10), Internal report, March 2014.

Uchida, H., Kawano, T., Kaneko, I., and Fukasawa, M.: In situ calibration of optode-based oxygen sensors, J. Atmos. Ocean. Tech., 25, 2271–2281, https://doi.org/10.1175/2008JTECHO549.1, 2008.

Winkler, L. W.: Die Bestimmung des im Wasser gelosten Sauerstoffes, Ber. Dtsch. Chem. Ges, 21, 2843–2853, 1888.

Xing, X., Claustre, H., Blain, S., D'Ortenzio, F., Antoine, D., Ras, J., and Guinet, C.: Quenching correction for in vivo chlorophyll fluorescence acquired by autonomous platforms: A case study with instrumented elephant seals in the Kerguelen region (Southern Ocean), Limnol. Oceanogr.-Meth., 10, 483–495, 2012.

3

Over 10 million seawater temperature records for the United Kingdom Continental Shelf between 1880 and 2014 from 17 Cefas (United Kingdom government) marine data systems

David J. Morris, John K. Pinnegar, David L. Maxwell, Stephen R. Dye, Liam J. Fernand,
Stephen Flatman, Oliver J. Williams, and Stuart I. Rogers

Cefas, Lowestoft Laboratory, Pakefield, Lowestoft, Suffolk, NR33 0HT, UK

Correspondence: David J. Morris (david.morris@cefas.co.uk)

Abstract. The datasets described here bring together quality-controlled seawater temperature measurements from over 130 years of departmental government-funded marine science investigations in the UK (United Kingdom). Since before the foundation of a Marine Biological Association fisheries laboratory in 1902 and through subsequent evolutions as the Directorate of Fisheries Research and the current Centre for Environment Fisheries & Aquaculture Science, UK government marine scientists and observers have been collecting seawater temperature data as part of oceanographic, chemical, biological, radiological, and other policy-driven research and observation programmes in UK waters. These datasets start with a few tens of records per year, rise to hundreds from the early 1900s, thousands by 1959, and hundreds of thousands by the 1980s, peaking with > 1 million for some years from 2000 onwards. The data source systems vary from time series at coastal monitoring stations or offshore platforms (buoys), through repeated research cruises or opportunistic sampling from ferry routes, to temperature extracts from CTD (conductivity, temperature, depth) profiles, oceanographic, fishery and plankton tows, and data collected from recreational scuba divers or electronic devices attached to marine animals. The datasets described have not been included in previous seawater temperature collation exercises (e.g. International Comprehensive Ocean–Atmosphere Data Set, Met Office Hadley Centre sea surface temperature data set, the centennial in situ observation-based estimates of sea surface temperatures), although some summary data reside in the British Oceanographic Data Centre (BODC) archive, the Marine Environment Monitoring and Assessment National (MERMAN) database and the International Council for the Exploration of the Sea (ICES) data centre. We envisage the data primarily providing a biologically and ecosystem-relevant context for regional assessments of changing hydrological conditions around the British Isles, although cross-matching with satellite-derived data for surface temperatures at specific times and in specific areas is another area in which the data could be of value (see e.g. Smit et al., 2013). Maps are provided indicating geographical coverage, which is generally within and around the UK Continental Shelf area, but occasionally extends north from Labrador and Greenland to east of Svalbard and southward to the Bay of Biscay. Example potential uses of the data are described using plots of data in four selected groups of four ICES rectangles covering areas of particular fisheries interest. The full dataset enables extensive data synthesis, for example in the southern North Sea where issues of spatial and numerical bias from a data source are explored. The full dataset also facilitates the construction of long-term temperature time series and an examination of changes in the phenology (seasonal timing) of ecosystem processes. This is done for a wide geographic area with an exploration of the limitations of data coverage over long periods. Throughout, we highlight and explore potential issues around the simple combination of data from the diverse and disparate sources collated here. The datasets are available on the Cefas Data Hub (https://www.cefas.co.uk/cefas-data-hub/). The referenced data sources are listed in Sect. 5.

1 Introduction

The measurement of surface and subsurface seawater temperature has been a standard activity for a significant proportion of marine researchers for the past 200 years. From the physical oceanographer to the marine chemist to the marine biologist, the original purposes for such measurements range from a desire to determine the physical properties and movements of seawater to understanding how temperature influences the distribution of marine species, their migration, growth, and reproduction, and, as a dominant feature of the collected works herein, the impacts of and upon commercial activities such as fishing. Furthermore, accurate sea temperature data are necessary for a wide range of applications, from providing boundary conditions for numerical hydrodynamic models and weather prediction systems, to assessing the performance of long-term climate modelling and understanding the drivers of observed changes in marine ecosystems. The importance of sea surface temperature (SST) to climate science is reflected in its designation as an "essential climate variable" of the Global Climate Observing System (Bojinski et al., 2014).

The Marine Biological Association (MBA) of the United Kingdom was established in 1884 in order "to foster the study of marine life, both for its scientific interest and because of the need to know more about the life histories and habitats of food fishes". In 1902 a dedicated fisheries laboratory was established in the Port of Lowestoft by the MBA together with the UK Board of Trade. This was the UK's primary contribution to the newly founded International Council for the Exploration of the Sea (ICES). From its inception, the laboratory in Lowestoft has collected information on fish stocks surrounding the British Isles, but also water temperatures at the surface and near the seabed. Much of the information collected by the Lowestoft laboratory over the past 115 years has never been made publicly available, but these datasets are now the subject of legacy data rescue (Wyborn et al., 2015) as part of a drive for "open data" within the UK government. This paper is one result of that ongoing effort. In their Preamble, Griffin and the CODATA DAR-TG (2015) describe the unglamourous reality of legacy data rescue and the reasons why heritage data are not as readily accessible as the term "archive" might imply. The approach taken here is to turn, in their terminology, old data into new data and to present, explore, and explain the new data so that they can be used within a context that includes the diverse and disparate reasons for which the old data were collected and the differences and limitations of the acquisition and measurement techniques of the day.

The methods of measuring seawater temperature range from the simple thermometer to the ubiquitous presence on a modern marine research vessel of a conductivity, temperature, and depth (CTD) instrument of some kind. Such activ-

Figure 1. RV *Huxley* 1902–1909.

ities have, for well over 100 years, formed a routine part of the sea-going and observational work of the MBA Lowestoft substation and its successors. In 1910 the Lowestoft laboratory transferred to the Board of Agriculture and Fisheries where it then became a Fisheries laboratory under MAFF (Ministry of Agriculture, Fisheries and Food) in 1920. From 1955 it was known as the DFR (Directorate of Fisheries Research); see Lee (1992) and Graham (1953). It now continues as Cefas (Centre for Environment, Fisheries & Aquaculture Science) under Defra (Department of Environment Food and Rural Affairs), with a remit focusing on the UK Continental Shelf and occasional forays into more distant waters for projects supporting UK government priorities.

Data holdings within this institution extend back beyond 1902 although these form only a very small part of the collated temperature dataset described here. The historic focus of our marine research has been biological, specifically fisheries related, but this has changed as both government policy needs and interests have widened. Figure 1 shows the RV *Huxley*, which was deployed between 1902 and 1909, with Fig. 2 highlighting the differences between the adapted trawlers of early years and the current bespoke research vessel, the RV *Cefas Endeavour*, which started service with Cefas in 2003. A wider, historic, institutional context for the 17 data sources described here is available in Cefas (2014).

Figure 2. RV *Cefas Endeavour* 2003 to present.

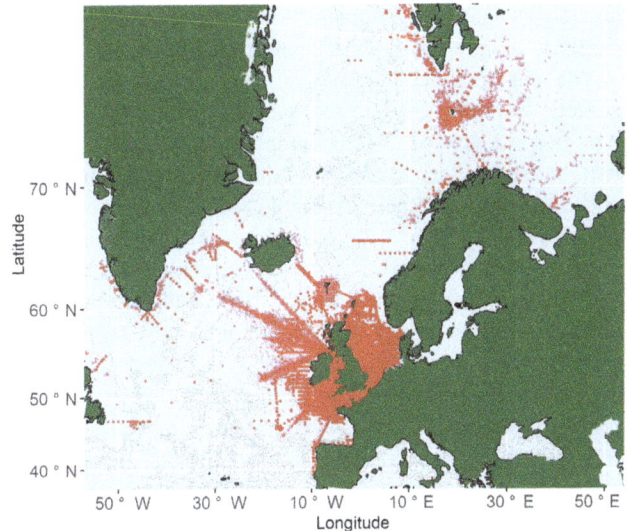

Figure 3. Overview of the locations of Cefas seawater temperature measurements with plotted point intensity reflecting data density.

The methods of measuring seawater temperature have ranged from simple mercury thermometers deployed in buckets of seawater, to pumped seawater systems on research vessels (see Kent and Taylor, 2006, for an exploration of these methods of measurement), to the ubiquitous presence on most modern research vessels of CTD instruments or, more recently, autonomous surveillance buoys, gliders, profilers, and electronic devices attached to animals. Much has been written about difficulties in calibrating information from these various data sources; see, for example, Matthews (2013) and Kennedy et al. (2011a, b). Subtle differences in the methodologies for calibrating such disparate measurements have been found to greatly impact reconstructions of time series of global climate warming (Karl et al., 2015). Both issues with ship data sources have been specifically identified, including the change from bucket samples to engine intake thermometers, and more relevant here, the increase in data density with time as buoy-mounted observation systems were deployed as sources of time-dependant bias in the global SST record. We explore such possible data bias in general terms along with examinations of the effects of data source, time dependencies, location, and numerical bias.

Many different data portals and data syntheses now exist housing collated maritime temperature records, the most notable including the International Comprehensive Ocean–Atmosphere Data Set (ICOADS; Freeman et al., 2017), the NOAA Extended Reconstruction Sea Surface Temperature (ERSST; https://www.esrl.noaa.gov/psd/data/gridded/data.noaa.ersst.v4.html) dataset, the Hadley Centre SST gridded dataset derived from observations in ICOADS (HadSST3; Kennedy et al., 2011a, b), and the Japanese Meteorological Agency centennial observation-based estimates of SSTs (COBE-SST; http://ds.data.jma.go.jp/tcc/tcc/products/elnino/cobesst/cobe-sst.html). All of these are composite SST series that ingest data from multiple different instrument platforms (ships, buoys, and some satellite data in the case of COBE-SST) and from different measurement methods to create consistent long-term time series (see Hausfather et

al., 2017). Analysis of these long-term historic datasets show that the sea surface temperatures around the British Isles have warmed at rates up to 6 times greater than the global average (Dye et al., 2013). Indeed, this region has been identified as one of 20 "hot-spots" of marine climate change globally based on an analysis of trends in ocean temperature (Hobday and Pecl, 2014).

Numerically, the data presented here start with tens of observations per year, rising to hundreds from the early 1900s, to thousands by 1959, to hundreds of thousands by the 1980s, peaking with > 1 million for some years from 2000. The majority of the data included in this paper originate from modern research and monitoring programmes executed by scientists using appropriate QA–QC (quality assurance and quality control) processes for their designated purposes, which did not include the extensive sharing and repurposing of the current day.

In this paper, 17 separate data systems are described, comprised of more than 10 million individual temperature measurements. Most are from the seas around the British Isles (ICES areas IV, VI, and VII) but there are some additional measurements in the Bay of Biscay (ICES area VIII), off Labrador and southern Greenland (ICES area XIV) and in the Norwegian and Barents seas (ICES areas I and II); see Fig. 3 (ICES, International Council for the Exploration of the Sea).

Dann et al. (2015) specifically recognise the challenges of using "data available from different surveys [that] have been collected for different purposes, using different gears and different sampling strategies over time". They were working on fish and their aim was "to provide a broad view of regional, depth related ... and temporal patterns ... by integrating as much information as possible". This paper collates and makes readily accessible data that can contribute significantly to such integrations of seawater temperature.

The data collection programmes that act here as data sources were designed to measure temperature for a specific purpose (physical oceanographic measurements and as part of Cefas SmartBuoy programmes focusing on nutrient levels or as a directly relevant contextual measurement, e.g. WaveNet and RV *Cefas Endeavour* FerryBox). Other datasets arise from research for which temperature data are collected for general context and interpretation. Two data sources are from citizen science, although the Coastal Temperature Network (CTN), which was established in the mid-1960s (with individual datasets going back over 100 years), preceded the term whilst also relying on volunteers. The majority of these temperature datasets have been previously analysed and integrated into a myriad of diverse and disparate reports and scientific papers, often in the form of summary tables and figures or as contributions to understanding the environment of fish and other biota. Most of the recent data now reside in numerous operational database systems, whilst a significant proportion of the rest now exist in organised and documented electronic forms thanks to recent legacy data rescue efforts by Cefas; all are available through the published discovery metadata Cefas Data Hub (http://data.cefas.co.uk), the UK Government Metadata Portal (https://data.gov.uk/data/search), and the MEDIN Metadata Portal (http://portal.oceannet.org/search/full).

The Cefas Data Hub extends the search for discovery metadata to include direct access to data. It provides direct access to extracts from Cefas operational databases to facilitate data reuse beyond the original purpose. This paper takes an additional step and makes comprehensive, quality-assured extracts for this key physical parameter readily available and easily accessible in simple text files of seawater temperature data, with each record standing alone and not associated with bespoke and specialist data formats. Throughout, we highlight and explore potential issues around the simple combination of data from the diverse and disparate sources collated here.

This paper focuses on seawater temperature data but we recognise the value of assembling and publishing co-located data, such as salinity and the presence of species (in the case of the plankton dataset), amongst other parameters. The Cefas Data Hub currently holds published data in source formats with the intention of making these and other datasets more accessible by using transformations similar to those executed here.

1.1 Overview of the basic characteristics of the seas covered by the dataset

Most are from the seas around the British Isles (ICES areas IV, VI, and VII) but there are some additional measurements in the Bay of Biscay (ICES area VIII), off Labrador and southern Greenland (ICES area XIV) and in the Norwe-

gian and Barents seas (ICES areas I and II); see Fig. 3. The International Council for the Exploration of the Sea (ICES) produces an annual report on the marine climate of the North Atlantic (the ICES Report on Ocean Climate). This gives a broad description of the oceanography of this region and documents the year-by-year variations using a set of hydrographic stations collected by the international community (Larsen et al., 2016). They describe the variation in the northern North Atlantic and sub-Arctic seas where the North Atlantic Current provides a source of heat and salt along the eastern margin into the Barents Sea and entry to the Arctic Ocean. Along the western margin, the Arctic influence of cold and fresh conditions extends from the Fram Strait to Cape Farewell. At the southern part of the region covered by the Cefas temperature data from the western channel down to Iberia, the influence of subtropical waters is more evident. The combination of gyres and the North Atlantic Current places the UK shelf waters at the boundary between temperate and subpolar waters exerting a heavy influence on the variability of conditions in the Greater North Sea and Celtic Seas.

1.1.1 The Greater North Sea

The temperature of the Greater North Sea is controlled by the seasonal cycle of heat exchange with the atmosphere, the vertical mixing in the water column, and the circulation of waters from the North Atlantic.

The annual mean temperature generally increases from the south (in the English Channel) to the north (near Shetland), but this pattern is not representative of all seasons. During the winter the shallow waters in the southern North Sea that are furthest from the influence of the inflowing North Atlantic waters tend to be the coolest in the entire Greater North Sea.

Northern North Sea. Modified Atlantic water flows into the region via the Fair Isle current, maintaining relatively warm winter temperatures, typically 6 to 9 °C minimum with a decrease to the south as water from the Atlantic is cooled by atmosphere and depth shallows. Summer temperatures are typically 12 to 14 °C near the surface with a cooling influence evident from the North Atlantic inflow, and it generally stratifies.

Southern North Sea. The southern North Sea is shallow, mostly less than 50 m in depth, and furthest from the inflows and influence of Atlantic water. Temperature minima in winter are typically 4 to 8 °C; they depend strongly on the weather in any one year and on depth (shallower → cooler). Likewise, the typical summer maxima of 16 to 19 °C depend on the weather and strongly on depth (shallower → warmer)

English Channel. From depths of less than 50 m near the coast and the Dover Strait, the channel deepens westwards to 100 m. The influence of Atlantic water also increases towards the west and only some parts in the very west stratify in the summer. Thus minimum winter temperatures, typically 5 to 8 °C, are strongly dependent on the weather in any one year and on depth. Summer maximum temperatures are typically 16 to 19 °C.

The Greater North Sea near-bottom temperatures differ from SST due to stratification, which takes place only during the summer. Where the region does stratify (in the northern North Sea and at the very western part of the English Channel), summer temperatures near the bottom remain cool until the breakdown of stratification in the autumn.

1.1.2 The Celtic Seas

The various temperature and salinity characteristics of the Celtic Seas are reflective of the inhomogeneity of the region, from enclosed shallow-shelf sea with large river catchments all to deep oceanic waters and across a wide range of latitudes. Surface temperature is controlled by a balance of seasonal heating, vertical mixing, and the circulation of Atlantic water, with the relative importance depending on local depth, tides, wind, and exposure to the ocean.

Celtic Sea. Sea temperatures are strongly related to the weather in any one year and to water depth. The climate being strongly maritime, typical winter minima are 8 to 11 °C and summer maxima are 14 to 18 °C. The seasonal cycle of near-bed temperature in this part of the region is controlled by the vertical mixing. When well mixed vertically in the winter, its temperature is similar to that at the surface. During the summer the area stratifies and near-bed temperatures do not reach the temperature maxima of the surface; the maximum annual temperature here is typically reached in October when the heat of surface waters is fully mixed down.

Irish Sea. Temperatures depend strongly on the weather in any one year and on water depth. Typical winter minima are 4 to 8 °C and summer maxima are 14 to 18 °C. As elsewhere, temperatures also depend on whether the area stratifies. The area is well mixed vertically in winter and typical winter minima match the SST at 4 to 8 °C. In the areas that stay well mixed throughout the year, summer maxima of 14 to 18 °C are typical, while areas that stratify in the summer reach their annual maximum of 13 to 15 °C in autumn when the heat of surface waters is fully mixed down.

Minches and western Scotland. There is some influence of (modified) Atlantic water arriving from the west. Resulting typical winter minimum temperatures are 6 to 8 °C and summer maxima are 13 to 15 °C in well mixed areas or 11 to 13 °C where stratified. Typically, there is summer stratification in the deep waters away from islands and north of the Islay front (west of Islay to Ireland).

Scottish Continental Shelf. Except for shallow areas near coasts, there is summer stratification. Temperature minima in winter are typically 9 to 10 °C at the shelf edge but 6 to 9 °C elsewhere; they depend on the weather in any one year, on depth, and on travel time for any Atlantic water arriving from the shelf edge. Summer maxima are typically 12 to 14 °C for surface water.

2 Data sources

The 17 source systems are the following.

1. The Cefas Coastal Temperature Network (CTN) is comprised of time series of measurements from a number of long-term recording stations throughout the coast of England and Wales, with measurements provided by volunteers and external suppliers who have agreed that their data can be published as part of the network (Jones, 1981). See also Joyce (2006), Jones and Jeffs (1991), Ellett and Jones (1994), and Norris (2001). In Joyce (2006), Appendix A, Table 8, and the associated figures show data at Brancaster that result in a yearly anomaly from a base period of 4–5 °C. These data have been excluded from this compilation.

2. The Cefas Fishing Survey System (FSS) is a purpose-built database used to hold and maintain Cefas fish survey data, primarily from government-mandated surveys.

3. The Cefas Oceanographic Archive (OA) is a system for managing data from a CTD system deployed during traditional oceanographic water-column profiling.

4. The Cefas Plankton Analysis System contains data from the sampling of plankton which has been carried out by Cefas since the 1940s. In recent decades, sampling has mainly been concentrated on fish eggs and larvae and other zooplankton. Pre-egg survey temperature data are profiles from stations. Egg survey temperature data are from a sensor attached to the net. Plankton samples were collected using high-speed towed nets that capture plankton from the surface to near the seabed. At each sampling position the sampler was deployed in an oblique tow from the surface to within approximately 2 m of the seabed. Veering and hauling speeds were manually adjusted with the aim of sampling each depth band equally. Since the early 1980s CTD sensor packages have been fitted to the plankton samplers to continuously monitor temperature and salinity throughout each deployment, with positions interpolated from start and end times and positions.

5. The Cefas Fisheries Ecology Research Programme covers several activities, and in this case the temperature data come from a study entitled "Diurnal and seasonal changes in water temperature in South Wales estuaries and saltmarshes". Data were collected in 1995 and 1996 from three estuarine locations in South Wales during a study of the thermal experience and tolerance of estuarine animals. The data are comprised of hourly records of temperature in brackish water creeks which are only inundated by the sea for part of the tidal cycle. Modelled depths are < 2 m when not inundated (see

http://data.cefas.co.uk/#/View/3236 for a fuller description of the data and required modelling of depth).

6. The Cefas SmartBuoy Monitoring Network consists of sensors, a platform, and supporting data acquisition and processing software. SmartBuoys are autonomous marine monitoring systems making high-frequency measurements of physical, chemical, and biological parameters (Greenwood et al., 2010). Measurements are made every second in a burst duration of between 5 and 10 min and an average is calculated. They have been deployed as part of the UK marine eutrophication monitoring programme.

7. The Defra Strategic Wave Monitoring System (WaveNet) supports a network for England and Wales (https://www.cefas.co.uk/cefas-data-hub/wavenet/), providing a single source of real-time wave data from wave buoys located in areas at risk from flooding and/or inundation. The Waverider buoys are also fitted with a sea surface temperature sensor with data recorded and transmitted half hourly.

8. The Historical Ferry Routes Monitoring System and research vessel surface logger systems contain data on near-surface temperature and salinity samples that were collected by ferries operating between Harwich and Rotterdam (Jones and Jeffs, 1991; see https://www.cefas.co.uk/cefas-data-hub/sea-temperature-and-salinity-trends/data-sets/ for full descriptions of sites and routes) and from Cefas research vessel surface logger systems. The surface logger data were used, stored, and processed as part of the vessel management system and were normally run during cruises.

9. The Cefas Electronic Data Storage-Tag Database supports the deployment of electronic tags that record temperature and depth. These tags were attached to or implanted into several species. The data provided here are from cod caught in the southern North Sea between 1999 and 2009 (for methods see Neat et al., 2014). Data from tags that were returned from recaptured cod were downloaded and the depth time series was used to estimate daily geographic location. This was done by matching the tidal and maximum depth data to known dates and locations as per the method described in Pedersen et al. (2008). Temperature data from each tag were binned into 10 m depth intervals and then averaged. Cod were at liberty to move at will, so the geographic and vertical sampling is not regularised to a grid or vertical stratification. The data describe the temperature data sampled by a total of 90 cod and are comprised of temperature data collected on a total of 10 446 days. Methods used to capture and tag cod are found in Righton et al. (2010) and Neat et al. (2014). Summary data are published in Neat and Righton (2007) and Righton et al. (2010).

10. Citizen Science Diver Recorded Temperatures come from a data source that differs from the others in this collection because it arises from an investigation into the potential for citizen science to contribute to assessments of the marine environment. The dataset is derived from a database containing over 7000 records of temperature data collected from temperature-compensated dive computers. The lowest temperature is recorded from the thermal sensor. This resulted in a quality-assured dataset of just over 5000 records (including freshwater and lake data). The subset of the global dataset provided covers the UK shelf. See Azzopardi and Sayer (2012) and Sayer and Azzopardi (2014) for additional information. Data accuracy for some instruments is limited to 1 °C.

11. The Cefas Lowestoft Sample Data Management System (LSDM) was the primary system used before and throughout the 1990s by Cefas (Lowestoft) to manage water sample processing and data. Its function was to provide a vehicle for the management of the ingestion, analysis, and recording of measurements on marine samples ranging from oceanographic water samples through sediments to "environmental materials" and radiological samples; see Sutton (1993) for an example of the supporting role of LSDM in relation to the usually high-level scientific measurement systems of the day and Sauer et al. (2002) for an example of its pivotal role in quality-assured processes and analyses. As the work profile for the Ministry of Agriculture, Fisheries and Food's Directorate of Fisheries Research changed followed by the creation of Cefas and then Defra, the need for a centralised system for the management of an extensive suite of physical samples decreased. LSDM was closed in 2015 with chemical data transferred to other systems. The temperature data held included the historical ferry routes and historical CTN data, both covered separately. The remainder from a variety of programmes and cruises are presented in this section.

12. The *Mnemiopsis* Ecology Modelling and Observation Project (MEMO) was part of a wider sampling programme in collaboration with Ifremer and ULCO (France), ILVO (Belgium), and Deltares (Netherlands). The data collected were used to produce models, such as an individual biological model and hydrodynamic, ecosystem, and socioeconomic models; see Collingridge et al. (2014) and van der Molen et al. (2015). These increased the understanding of the life cycle of warty comb jellyfish (*Mnemiopsis leidyi*). The project collected samples for the analysis of fish larvae and fish eggs, microzooplankton and mesozooplankton, and phytoplankton. Samples were collected

using a 200 µm mesh ring net of 0.5 m diameter (for zooplankton samples) and physical data were collected via a CTD attached to a ring net.

13. The Cefas Multibeam Acoustics Sound Velocity Profile Temperature Data comes from the RV *Cefas Endeavour*, which has been routinely deploying multibeam acoustic measurement techniques since 2005, with particular emphasis being placed on habitat mapping projects (Brown and Vanstaen, 2008). As part of the calibration of the various acoustic systems, a CTD cast is performed at relevant stations to provide temperature data for the necessary calculation of sound velocity.

14. Intensive plankton surveys off the north-east coast of England in 1976 were comprised of a series of 12 cruises carried out in 1976 by DFR staff to investigate the distribution, abundance, mortality, and main predators of planktonic fish eggs and larvae of important commercial fish species (e.g. plaice, cod; Harding and Nichols, 1987). Measurements of surface water temperature and salinity and bottom temperature were carried out at each sampling station on a planned survey grid.

15. The RV *Cefas Endeavour* FerryBox Monitoring System was installed in 2009. Unlike most FerryBox systems (http://www.ferrybox.org and specifically the systems described at http://noc.ac.uk/ocean-watch/shallow-coastal-seas/ferrybox), RV *Cefas Endeavour* runs a combination of regular (usually annual) monitoring cruises in UK shelf waters (with a focus on ICES-mandated surveys for fisheries assessments) and bespoke research cruises. This provides widespread coverage with some repeat components in time and space.

16. Cefas ScanFish was a programme that deployed a high-performance towed undulating CTD, initially to aid the understanding of the coupling between physical and biological processes (Brown et al., 1996). It was towed behind the vessel at approximately 8 kn and undulated from the near surface ($\sim$ 4 m) to within a few metres ($\sim$ 5 m) of the bed, down to water depths of 135 m. The vertical ascent rate was controlled so that each undulation covered a horizontal distance of 1 km regardless of water depth.

17. The Cefas ESM2 Profiler–mini CTD Logger is a Cefas-developed micro-logger for applications requiring a small low-power logger with integrated sensors and battery. It has standard sensors for conductivity, temperature, depth, optical backscatter, and roll and pitch. It was initially developed to be a handheld profiler that could be used from small boats and/or when a conventional large rosette could not be used. It is now used routinely in place of traditional CTD equipment (data held

in source system 3) and widely used on RV *Cefas Endeavour* research cruises to provide profiles of the water column for fisheries and plankton work (replacing or supplementing data in sources 2 and 4).

The date ranges and numbers of observations for each data source are summarised in Table 1.

3 Data components and methods

Each specialist data collection system is described in detail in the appropriate metadata. The data files have been extracted from the source to provide the following (with field names in parentheses):

1. Cefas data source reference number (Source);

2. date and time of measurement (Time);

3. position of measurement: latitude in decimal degrees (Lat);

4. position of measurement: longitude (Long);

5. sample depth in metres (Depth);

6. seawater temperature in degrees centigrade (tC);

7. type of sampling used (Sample);

8. type of measurement used (Measure);

9. additional source context, e.g. cruise (Ref1);

10. additional source context, e.g. station, location name, etc. (Ref2);

11. and unique identifier (ID).

The Ref1 and Ref2 fields were extracted from the source data files and provide an operational context (where this is appropriate and/or available) for the original source data, e.g. cruise and station. The Sample and Measure fields provide information on the acquisition of data and are included specifically to facilitate understanding and removal of sample bias and autocorrelation effects. The accuracy of the data is described in the metadata accompanying the data files. The number of decimal places provided reflects the source files and can generally be taken as a realistic indication of the accuracy of the position, depth, and temperature. Note that all data have standardised formats and trailing zeroes do not imply increased accuracy.

The methods used to measure parameters over the time span of the datasets vary widely in their resolution (the smallest change that can be measured), precision (the repeatability of the system used), and accuracy (the closeness of the measurement to the actual value). The data provided reflect our best estimates of accuracy when transforming the data from a wide variety of bespoke measuring, recording, and

Table 1. Summary metadata for the 17 seawater temperature sources.

Data source	Name	Type	Start year	End year	Number of data points
1	Coastal Temperature Network	Fixed coastal stations	1880	2015	836 179
2	Fishing Survey System	Surface measurements and net tows	1903	2014	35 764
3	Oceanographic Archive	CTD profiles	1981	2009	365 239
4	Plankton Analysis System	Surface measurements and net tows	1982	2004	2 639 842
5	Fisheries Ecology Research Programme	Fixed coastal stations	1995	1996	21 504
6	SmartBuoy Monitoring Network	Offshore monitoring buoy	2000	2014	1 268 832
7	Strategic Wave Monitoring System	Offshore monitoring buoy	2002	2014	1 784 092
8	Ferry routes and surface logger systems	Surface measurements	1906	2011	656 103
9	Electronic Data Storage-Tag Database	Devices attached to animals	1999	2010	13 856
10	Citizen science scuba divers	Devices attached to humans	1992	2012	2205
11	Lowestoft Sample Data Management System	CTD profiles	1960	2009	52 631
12	*Mnemiopsis* Observation Project	CTD profiles	2011	2012	506
13	Multibeam Acoustics Sound Velocity Profile Database	CTD profiles	2005	2008	9628
14	Intensive plankton surveys of NE England in 1976	Surface measurements and net tows	1976	1976	2064
15	FerryBox monitoring	Surface measurements	2009	2013	652 305
16	ScanFish undulating profiler	CTD profiles	1998	2003	2 129 341
17	ESM2 Profiler–mini CTD Logger	CTD profiles and net tows	2004	2014	210 349
All	Complete dataset – all sources	All types	1880	2015	10 680 440

use systems (some data were presented with decimal places beyond those implied by statements regarding accuracy of measurement or, in the case of position, than is known to have been feasible at the time of collection). QA–QC processes for the sources were, and are, appropriate for their particular requirements. The data published here have been subjected to additional checks in the form of minimum and maximum and outlier detection plus location plotting. These uncovered a variety of data quality issues, primarily around location but also showing sensor-related data issues. Best efforts have been made to ensure the data are clean, reliable, and representative of what was measured. A degree of selection bias is inherent in this data compilation exercise ranging from what was originally done and where and when, to what was reasonably accessible for compilation, what was removed on the grounds of quality control and uncertainty regarding validity, to what users select and do with the data. Such are the "statistical" perils of data reuse.

3.1 Source

This denotes which of the 17 data sources the record was extracted from. This field allows data to be integrated across data sources whilst retaining a reference to the source and originating resolution, precision, accuracy, and original purpose for each of the records. A significant numerical majority of data extracted from the data sources come from sensors and platforms that will be familiar to a reader around the time of publication. However, historical data, whilst of particular interest, comes with historical navigation, sensors, data gathering methods, and platforms. The following sec-

tions describe differences that a reuser of data should take into account.

3.2 Time

Across the data sources, dates and times have been recorded in a variety of ways. We have made the reasonable assumption that all times recorded used Zulu as the time zone, which equates to GMT and now UTC. Date and time were usually recorded for individual measurements unless the operational systems, such as point source data buoys, average the data at collection. Where times are not specifically recorded (usually old, shore- or vessel-based manual records) they are taken as standard for the particular source; daily reports are allocated as 12:00, morning as 08:00, and afternoon as 16:00 as best approximations for likely collection times. Some datasets take observations at local high tide. Some CTD profiles provide a start time only; depth and temperature measurements are allocated a time by interpolation using a standardised rate of descent ($0.25 \, \mathrm{m \, s^{-1}}$). The plankton data (source 4) required positional interpolation based on start and end times and positions.

3.3 Latitude and longitude

An informed use of the datasets requires an understanding of the changes in methods of measurement of location over time. Past practice separated the detailed recording of navigational data and associated uncertainties from the provision of positions to researchers. The former has not been specifically preserved.

The earliest research records consist of data from light-ships which, we assume, were reasonably accurately located. We think, based upon historical statements on intentions of best practice, that navigation on the early vessels engaged in research and monitoring would have generally always followed good practice at the time (Lee, 1992, p. 173). When in range, research vessels would have used coastal navigation techniques, including physical aids to navigation, wherever possible and positional accuracy would depend upon the navigational chart's hydrographic survey. In addition, accurately surveyed depth contours were used as position lines when useful and practical (Graham, 1953). Locations close to charted objects would have been more reliable, precise, and accurate.

Beyond coastal waters where astronomical navigation was used, positional accuracies might have been "of the order of one or two miles" (Captain R. Jolliffe, personal communication, 2017) with uncertainties deriving from the ability of the navigator, the feasibility of sextant observations in weather, and the accuracy of navigational tables. Star sights (taken at dawn and dusk when the horizon and astronomical bodies were both visible) would provide two fixes per day. Morning sun sights run up to noon latitude would give a total of up to three fixes per day. In a chapter on navigation errors, the Royal Navy (2008) indicates an accuracy of 2 miles for an experienced navigator. From fixes of whatever sort, dead reckoning (DR) or estimated positions (EPs) would be applied to derive a station position where no actual fix was possible. DR is a process of calculating a position using distance and direction from the start, whilst EP applied corrections for the set (direction) and drift (speed) of the prevailing current. Both were probably used depending on circumstances and needs, but no records of when and where are available. Pawsey et al. (1920) report that during investigations of Lousy Bank in 1920, taking observations for station fixes based on the sun and/or three stars was the preferred method, but if the weather was inclement and they had no other option, they used DR but "with concerns about strong currents".

Civilian Decca navigation systems (in general use from the late 1940s to ~2000) offered positional accuracies of the order of ~200 m to 3 miles depending on the distance from the base stations. The longer-range Loran systems (in general use from ~1974 to ~2010) were less accurate.

Satellite navigation began with the Transit system in the late 1970s, giving global coverage and a fix at intervals, depending upon satellite availability, of anywhere between 1 and 6 h. Continuous positional information became available in the 1990s with the advent of the US Navstar GPS system. GPS accuracy depended, in part, on the application of selective availability (SA), which degraded the accuracy of the system for civilian use to between 30 and 100 m. DFR used differential GPS services to overcome this problem from about 1992, improving accuracy to the order of tens of metres. In 2000 the US government abandoned SA, making standard GPS accurate to within about 15–20 m. RV *Cefas*

Endeavour routinely achieves positional accuracies of 5 m, improving to less than 10 cm if differential GPS services are used, e.g. on bathymetric surveys.

We make a reasonably secure assumption that the reference coordinate system used from the adoption of satellite navigation was the default of the system: WGS72 and then WGS84.

Other than the stated increase in accuracy with time from miles to hundreds to tens to single metres, we cannot be clearer on the actual positions of samples other than to note that the positions have been extracted "as is" and converted to decimal degrees where needed.

In addition to errors in measurement, positional data also suffer from potential human error, conversion errors and errors in electronic storage and display. Latitudes and longitudes are presented as a best estimate representing actual likely accuracy, e.g. 4 dp (~4–11 m depending on location) or 3 dp (~40–110 m). A position originally recorded in degrees, minutes, and integer seconds (2 dp for decimal degrees) would be accurate to ~400 m to 1 km.

The long-term electronic data storage tags for fish do not use GPS but indirect interpolations of position from depth and time.

3.4 Depth

This is the depth at which the sample (physical or direct measurement) was taken. The main measurement devices use pressure suitably corrected for temperature for a depth below the surface. "Surface" temperatures feature widely in the records and are taken as 0 m although there are clear sources of error with the position of the sensors (both depth and temperature) on the relevant instrument and/or sampling device. Again, these surface measurements can be affected by wind, wave, and tide. "Bottom" temperature is less used in the data sources but features for profiles and tows. Its meaning varies from the maximum depth of sample measurement (in the water column) to the measurement taken when the sampling gear is on the seabed (where the sensors may be of the order of 1 m plus above the seabed).

Depths are as recorded with an accuracy of rounded integers (or ±0.1 m for some profiles).

The NOAA bathymetric data used to create the maps used in this paper allow for the interrogation of "water depth" by using the R package marmap (Pante and Simon-Bouhet, 2013). This was used as part of the quality control process in which positional data alone were insufficient to ensure an appropriate location.

3.5 tC

Values are in degrees centigrade. The accuracy of the seawater temperature measurements varies and is summarised in Table 2 and detailed in the metadata for each data source.

Table 2. Summary of the estimated actual accuracy of seawater temperatures by data source.

Data source	Instrument type	Estimated actual accuracy ± ° C	Point and/or average	Comment
1	Thermometer, thermistor	0.1–0.2	Point and average	The sensors vary from thermometer in a bucket for very early data to handheld thermometers to in-line thermistors for the Port of Dover sensor. Calibration varies from "uncertain" to Cefas Laboratory (allowing for 0.1 estimated actual accuracy).
2	CTD-type sensor	0.1–0.2	Point	The sensors used vary with early data being less accurate. Calibration varies from "uncertain" to Cefas Laboratory (allowing for 0.1 estimated actual accuracy); these data were collected for biological not oceanographic purposes.
3	CTD	0.005	Point	CTD physical oceanographic profiles, resolution 0.001.
4	CTD-type sensor	0.1–0.2	Point, average, binned	The sensors used vary with early data being less accurate. Calibration varies from "uncertain" to Cefas Laboratory (allowing for 0.1 estimated actual accuracy); these data were collected for biological not oceanographic purposes.
5	Vemco Minilog	0.3	Point	Resolution 0.1
6	CTD sensor	0.1	Averaged	Resolution 0.01, Cefas Laboratory calibrations before deployments.
7	CTD sensor	0.1–0.2	Averaged	Resolution 0.01, Cefas Laboratory calibrations before deployments or data provider calibrations estimated to provide the lower accuracy.
8	Thermometer, thermistor	0.2	Point	Calibrated thermometers for ferry route data and pumped seawater for RV surface logger data (calibration status uncertain).
9	Thermistor	0.1	Point	Cefas Laboratory calibration. Resolution 0.03125° C at 12 bit setting (https://www.cefastechnology.co.uk/media/1105/g5.pdf).
10	Dive computers	0.2–1	Point (at max depth)	Unknown. Knowledge of general diver practice suggests factory calibration followed by no calibration.
11	Reversing thermometer	0.1	Point	0.01
12	CTD sensor	0.1	Point	Cefas Laboratory calibration.
13	CTD sensor	0.1	Point	Cefas Laboratory calibration.
14	CTD sensor	0.1	Point	Cefas Laboratory calibration (plus pumped seawater for surface temperatures; see data source 8).
15	CTD sensor	0.1	Point	Resolution 0.01, Cefas Laboratory calibrations before deployments.
16	CTD	0.005	Point	CTD physical oceanographic profile instrumentation, resolution 0.001.
17	CTD sensor	0.1	Binned	Resolution 0.01, Cefas Laboratory calibrations before deployments.

3.6 Sample

The Sample codes are the following:

- MPT (monitoring "point" or location);

- PMP (pumped water sample);

- PNT (point observation);

- PRO (profile);

- SAM (discrete water sample);

- STA (station);

- SYS (static, continuous monitoring system); and

- TOW (towed instrument).

The combination of MPT and SYS indicates a stationary data acquisition system that may need to be treated in a way that allows for data density bias and autocorrelation.

3.7 Measure

The Measure codes are the following:

- MAN (manual) and

- INS (instrument).

3.8 Ref1, Ref2

These fields record contextual data from the source systems with Ref1 providing a high-level aggregation and Ref2 a lower-level grouping. They allow data to be manipulated or interpreted in relation to their source and any relevant break-down in activities of the operations of the source system. They also provide ready links to other documentation and context, e.g. cruise reports and other data types that may be available. Direct reconnection to the originating data source is, of course, available through time and position. Since 2009 the terms Cruise and Survey have become interchangeable for the RV *Cefas Endeavour*, with the latter mandated at the time of writing.

3.9   Data ingestion, quality control, and bias estimation

3.9.1   Data ingestion and quality

In publishing scientific data, Cefas takes into account the 2013 Shakespeare Review of Public Sector Information (PSI) (https://www.gov.uk/government/uploads/system/uploads/attachment_data/file/198752/13-744-shakespeare-review-of-public-sector-information.pdf), which states the following: "A National Data Strategy for publishing PSI should include a twin-track policy for data release, which recognises that the perfect should not be the enemy of the good: a simultaneous 'publish early even if imperfect' imperative AND a commitment to a 'high quality core' ... get it all out and then improve".

The use of original, archived source data files means that any specialist QA–QC processes applied "upstream" during the original uses of the data are covered in general in the relevant publications, but the details of the data QA–QC processes deployed are not necessarily available. The historic nature of a lot of the archived data means that the focus was on the often highly specific measurement protocols with temperature as either a core or peripheral parameter. If it was core, for the bulk of the data, part of the physical oceanographic investigations that utilised a series of electronic measurement systems that were advanced and accurate at the time, each with bespoke acquisition and processing systems, ultimately created an archive with a reasonably consistent approach but over 10 often subtly different formats. If it was peripheral, data accuracy is reduced by dint of the sensors used and the calibrations employed. Formats again vary, from sensors of fishing trawls feeding into an operational database to sensors on plankton tows feeding into a large and diverse spreadsheet archive over 2 decades.

Data assembly, transformation, and scrutiny were as follows.

- Identification of Cefas data sources with public seawater temperature data and assembly of relevant datasets from source archives and extraction from operational databases.

- Extraction of required elements, primarily from text files and spreadsheets, including derivation of positions and time from start and end data where required and the reformatting of date and time from several different formats.

- The checking of date and time data consisted of format transformations which picked up systematic source differences and manual adjustments where, for example, sensor logging was not capable of recognising date changes during deployment and/or issues with early PCs, which had similar problems when interfacing with instruments.

- The checking of location by plotting on maps followed by the identification and, in some cases the removal, of plots that indicated errors in the often manual recoding of position. Positions on land indicated either a hemisphere recording error or omission or a manual positional recording error. Where the former were encountered and obvious, the relevant cruise reports were checked and adjustments to the extracted data were made. Where the latter were encountered, entire stations or sets of stations (probably associated with a watch) were omitted.

- Seawater temperature data included instrument and manual values indicating sensor errors, and these were screened by an initial ingestion filter of < -2.5 and $\geq 35\,°C$, followed by specific checks of temperature $> 25\,°C$ to remove erroneous values. These ranged from single, starting data points possibly arising from exposure to the air to transposition errors for which values of 30 in, e.g. winter, indicated a storage or transposition error in and from the raw data files usually associated with conductivity. Detection of such high values resulted in a reassessment of the bespoke ingestion programmes and a rerun to correct errors and maximise data ingestion. Sequential temperature difference plots were used to identify large changes in temperature over short time periods. In some cases, these apparent anomalies were artefacts of this simple analysis, with two sequential data points coming from different vessels in different hemispheres on different days. In other cases, this plot identified datasets, usually profiles, in which reasonably significant chunks of a profile were significantly different from the rest. These were removed. Plots of temperature against time and monthly average temperatures also highlighted potentially anomalous data, e.g. $4\,°C$ measurements at the surface in summer and significantly higher averages compared to surrounding data. The former were resolved by the identification of an unexplained switch in one source's recording date format from DD/MM/YYYY to MM/DD/YYYY with the days and months involved, e.g. 31/08 to 09/01 rather than the correct 01/09 not triggering date ingestion format check errors.

- Other test plots highlighted $0\,°C$ data near the surface in summer in the North Sea. These were identified as sensor, transmission, transcription, or storage errors because the value 0.0 appeared in data sequences of, for example, 10.1, 10.2, and 10.3. These were also removed.

- Early plots of what became Fig. 16 indicated unseasonally high or low temperatures (e.g. UK Continental Shelf near-surface waters with $14–15\,°C$ in February and $1–1.5\,°C$ in June) and apparent outliers. These prompted a final systematic check of the fully assembled data by plotting data by month, followed by the identification of suspect data. This was then replotted by individual source to provide a context against which to evaluate apparent outliers. Unseasonally high and low data revealed as outliers in the source dataset were removed. Other outlier data were removed where appropriate, although the majority of apparent high and low outliers (see Fig. 16) were attributed to sources and sites that included shallow and relatively isolated water bodies.

Given the wide variety of sources and, in a lot of cases, the non-physical oceanographic focus for the data-generating activities, a formal and rigid retrospective application of oceanographic data quality control procedures was not applied across the board. However, where appropriate they were applied at the source, e.g. the CTD and ScanFish data (sources 3 and 16). In both cases the relevant standard IOC methodology was applied. For the remaining sources, the descriptions above cover the intent of such standards, specifically basic checks for all data types, e.g. date and time, latitude and longitude, position (must not be on land), and other relevant checks, such as impossible speed, spike, global range, regional range, and check for duplicates.

Best efforts have been made to remove all obvious errors, but it is possible that some remain amongst the 10 million plus data points made available here. Please contact data.manager@cefas.co.uk to report any errors; these will be corrected and the source files on Cefas Data Hub and the relevant metadata will be updated on confirmation of any error. The same contact can be used if external users of the data wish to explore collaboration or need assistance with interpretation.

3.9.2 Bias estimation

The provision of these raw data is "as measured" with appropriate metadata to allow subsequent scientific trend analysis to be performed, which would usually include additional scrutiny for systematic bias. The main exercise here is to identify and facilitate access to a large source of hitherto unavailable data that is as yet unseen and unscrutinised by the broader community.

An assessment of accuracy and bias has been conducted by the data creators for some of the sources included here. For example for source 10, we referenced Wright et al. (2016) who examined whether the temperature data derived from hundreds of recreational scuba divers and many different models of dive computer were consistent with global sea temperature datasets. Similarly, temperature sensors on Cefas SmartBuoys and WaveNet platforms (sources 6 and 7) are calibrated annually at Cefas against certified platinum resistance thermometers. Data are subject to a full quality assurance procedure which assigns flags to poor-quality data (e.g. for sensor malfunction or drift; see https://www.cefas.co.uk/cefas-data-hub/dois/cefas-smartbuoy-monitoring-network/).

We note that ICOADs and other collated datasets (e.g. HadSST) tend to carry out their own systematic bias correction routines whenever new data are uploaded or admitted. Our intention is to make our data available so that they can be easily included (by other authors) in platforms such as the ones listed (ICOADS, COBE-SST, ERSST, and HadSST3). Within the text of the paper we include references to papers that discuss bias correction (e.g. Mathews, 2013; Kennedy et al., 2011a, b; Karl et al., 2015; Hausfather et al., 2017), but we leave it to those who might make use of the data to judge what procedures might be necessary for their own purposes.

4 Results – geographic and temporal coverage by source

4.1 Data summary by source

Table 1 provides summary metadata for each of the 17 source datasets, including their temporal coverage, the number of data points, and the type of measurement (e.g. fixed station, CTD profile, electronic device attached to an animal, etc.). Sources 1 and 2 provide the longest time series of measurements (each more than 100 years), but more recent data systems, e.g. sources 6, 7, and 8 (autonomous surveillance systems) and the undulating tow systems for plankton (4) and oceanography (16), contribute the bulk of the assembled observations.

Table 2 provides an overview of the estimated actual accuracy of the data by data source. Information on sensor resolution, accuracy, and precision is available in the relevant data source metadata or in any cited publications and/or associated documents. Where sensor resolution, precision, and calibration are unclear or unknown, conservative estimates are made based on local knowledge from internal records or cruise participants.

4.2 Summary of sources, geographic range, depth range, and temporal coverage used in data subsets

Example potential uses of the data and subsets are described using plots of data in four selected groups of four ICES rectangles covering areas of particular fisheries interest. The full dataset enables extensive data synthesis, for example in the southern North Sea where issues of spatial and numerical bias from a data source are explored. The full dataset also facilitates the construction of long-term temperature time series and an examination of changes in the phenology (seasonal timing) of ecosystem processes for a wide geographic area with an exploration of the limitations of data coverage over long periods.

Table 3 provides a summary of the subsetting of the data undertaken to illustrate potential uses and limitations of a simplistic approach to synthesis and analysis. Source is a key variable with, in this case, potentially significant temporal, spatial, and sensor resolution differences. The intervals used to subsample the data reflect the requirements of visualisation and plotting rather than any intrinsic temperature-related aspect. The highlighted geographic areas were selected to illustrate data coverage and any issues of numerical, spatial, and temporal bias. The depth ranges used reflect a primary interest in sea surface temperatures with 44 % of the data falling within a 0–5 m depth. The time range selections primarily reflect data availability.

4.3 Data summary by location

Figure 3 shows the location of measurements across all 17 data sources. It is clear that the majority of coverage is of the English Channel, the North, Irish, and Celtic seas, and the UK Continental Shelf area, reflecting historic work focused on fisheries, plankton, and oceanography as part of repeated survey programmes or bespoke research. The data from around Svalbard, Greenland, and Labrador reflect the historic interest in cod fisheries around the Arctic and the physical oceanography in those regions (see Townhill et al., 2015).

Figure 4 provides an overview of the relative data density in the English Channel, the North, Irish, and Celtic seas, and the UK Continental Shelf area. It highlights the numerical dominance of point source data, e.g. autonomous Smart-Buoys (source 6, primarily in the North and Irish seas), data from WaveNet (source 7, off the east and west coasts of Scotland), and the single year (2014) of near-continuous (1 min) data from the Coastal Temperature Network at the Port of Dover. Areas of scientific interest in the Celtic Sea (mainly source 4, plankton studies) and the North Sea (a combination of oceanographic studies, sources 3 and 16; vessel-mounted data from sources 8 and 15 and general purpose CTD data from source 17) provide more widespread but significant data densities. Subsequent sections explore data availability by source, time, geographic location, and depth in more detail.

4.4 Data summary by year

Figure 5 illustrates the inherent differences in the data coverage with time throughout the 134 years covered with low but increasing numbers of annual records between 1880 and 1956 and a 2 order of magnitude increase during the 1980s to around the year 2000. This is followed by a further order of magnitude increase as a result of the introduction of autonomous monitoring platforms that make measurements on an hourly or even minute-by-minute basis in some cases. These platforms were also deployed in research roles on the North Dogger Bank and Oyster Grounds.

Other seawater temperature data compilations (e.g. HadSST3) show similar data acquisition trends. There are challenges when attempting to reconstruct long-term trends in a region, as many thousands of records may derive from one particular sampling locality, with very few data points

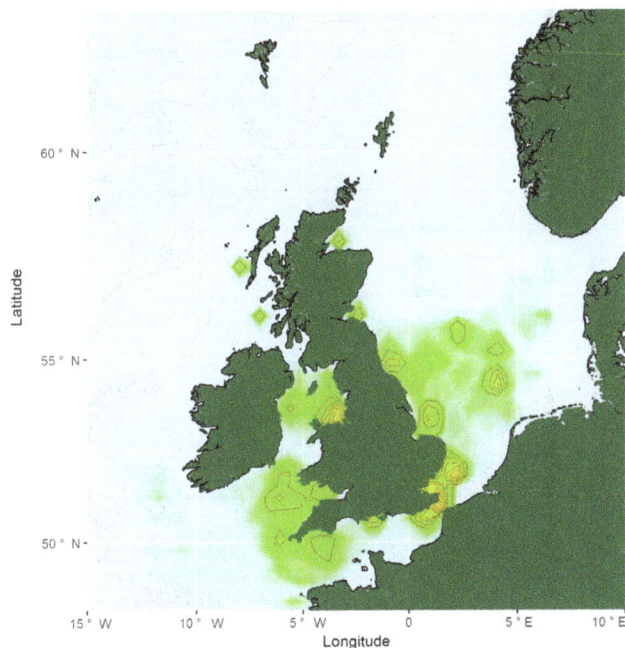

Figure 4. Overview of the relative data density in the English Channel, the North, Irish, and Celtic seas, and the UK Continental Shelf area.

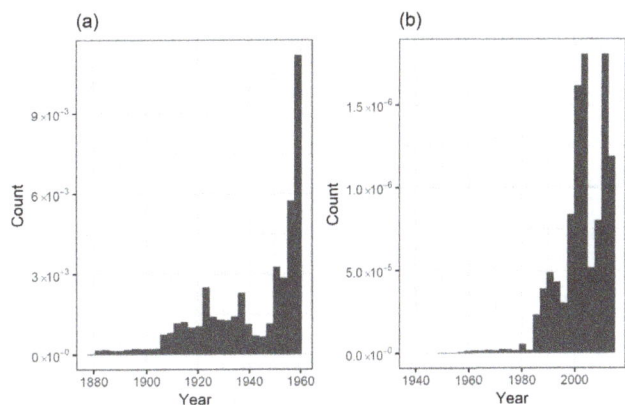

Figure 5. Illustration of data coverage with time: (a) 1880–1956 and (b) 1957–2014 (note the order of magnitude differences in counts).

elsewhere (see below and e.g. MacKenzie and Schiedek, 2007).

4.5 Data summary by depth

Figure 6 illustrates data coverage by depth. Figure 6a shows data between the surface and 10 m with high numbers (10^5 to 10^6) reflecting the preponderance of automated data collection platforms and vessel-mounted loggers. Figure 6b shows coverage between 10 and 100 m, and Fig. 6c shows data from 100 to 250 m covering the continental shelf break. Data coverage drops considerably with increasing depth as shown in

Table 3. Summary of data sources: geographic, depth, and temporal ranges for the subsetted data used in the figures.

Figure	Sources	Subset by interval	Subset by geographic area	Subset by depth range (m)	Subset by time range	Comment
3	All	1	None		None	None
4	All	5	UKCS*		None	None
5	All	1	None		None	(a) < 1960 (b) $\geq$ =1940
6	All	1	None	(a) $\leq$ 10 (b) > 10 $\leq$ 100 (c) > 100 $\leq$ 250 (d) > 250	None	
7	All	1	ICES**	$\leq$ 5	None	
8	All	1	SNS***	$\leq$ 5	None	
9	3, 4, 6, 7, 8, 11, 15, 16, 17	1	SNS	$\leq$ 5	None	
10	3, 4, 6, 7, 8, 11, 15, 16, 17	1	SNS	$\leq$ 5	None	
11	3, 4, 6, 7, 8, 11, 15, 16, 17	1 / 1	SNS	(a) $\leq$ 5 (b) < =1	$\geq$ 2000	SmartBuoy sensor $\sim$ 1 m WaveNet sensor $\sim$ 0.4 m
12	(a) 3, 4, 8, 11, 15, 16, 17 (b) 6, 7	1	SNS	(a) > 1 $\leq$ 5 (b) $\leq$ 1	$\geq$ 2000	
13	All	1	SNS	$\leq$ 5	$\geq$ 1925	Missing or limited data before 1925
14	All	All	Small "belt" around 3.2° E, 54.5° N	None	None	3, 4, 8, 9, 11, 15, 16, 17 only in selected areas
15	All	1	UKCS	(a) $\leq$ 5 (red) (b) $\geq$ 20 $\leq$ 25 (blue)	None	
16	All	1	UKCS	(a) $\leq$ 5 (red) (b) $\geq$ 20 $\leq$ 25 (blue)	1970–1984 1985–1999 2000–2015	Mid-water data sparse before 1970 Source 1 (coastal) excluded from Mid-water subset by definition
17	All	1	UKCS	(a) $\leq$ 5 (red) (b) $\geq$ 20 $\leq$ 25 (blue)	None	Monthly averages by year

* UKCS: UK Continental Shelf area. ** ICES: a selection of four groups each with four ICES rectangles; covers the Irish Sea, the Celtic Sea, the English Channel, and the Thames Estuary.
*** SNS: southern North Sea.

Fig. 6d, which illustrates data availability in the hundreds and then tens per 1 m bin for depths below 250 m.

Most of the sampling programmes involving the Lowestoft laboratory over the past 130+ years have focussed exclusively on the continental shelf, where the most productive commercial fish stocks exist and water depths rarely exceed 200 m. Only occasional forays have been made into the deeper North Atlantic, and these records are contained primarily in sources 3 and 11.

It is important to note that most of the existing data portals containing seawater temperature measurements (e.g. ERSST, HadSST3, COBE-SST) only accommodate records at the sea surface. The World Ocean Database (https://www.nodc.noaa.gov/OC5/WOD/pr_wod.html) and the Met Office EN4 database (https://www.metoffice.gov.uk/hadobs/en4/) do contain subsurface data and ICES (http://www.ices.dk/marine-data/data-portals/Pages/ocean.aspx) attempts to provide insights into near-seabed temperature conditions in certain geographical areas, but data are generally sparser than for the surface. Argo is a global array of 3800 free-drifting profiling floats that measure the temperature and salinity of the upper 2000 m of the ocean. Argo deployments began in 2000, and by November 2007, the millionth profile was collected, greatly increasing the knowledge base with regard to open-ocean and deep-water temperature conditions (see Riser et al., 2016).

The emergence of novel undulating platforms, such as ScanFish (source 16), electronic instruments attached to animals (source 9), and more recently autonomous gliders, will steadily increase the availability of measurements at depth, as will opportunistic data obtained from recreational scuba divers (source 10).

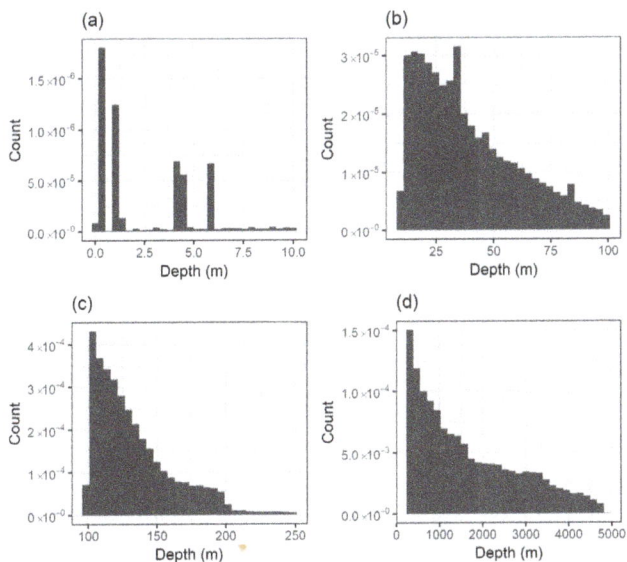

Figure 6. Illustration of data coverage with depth: **(a)** ≤ 10, **(b)** ≥ 100, **(c)** > 100 ≤ 250, **(d)** > 250 m (note the orders of magnitude differences in counts).

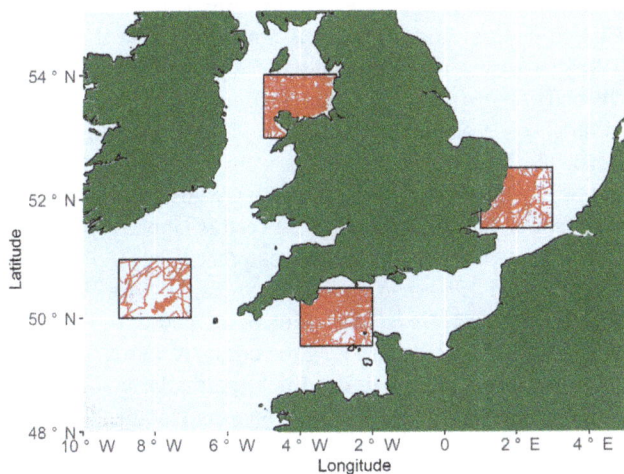

Figure 7. Illustration of "near-surface" (0–5 m) data coverage for four ICES rectangle groups: **(a)** Liverpool Bay, **(b)** Celtic Sea, **(c)** Brixham, and **(d)** the Thames area. Plotted point intensity reflects data density.

4.6 Data summary by ICES statistical rectangle group – areas of fisheries interest

To demonstrate data coverage in more detail, groups of four ICES rectangles of particular fisheries interest were selected with summary plots of the available "near-surface" data (0–5 m). This depth range specifically includes the large datasets from vessel-mounted pumped seawater systems. The four areas shown in Fig. 7 are (from N, W, S, and E)

– Liverpool Bay (Irish Sea),

– Haig Fras (Celtic Sea),

– Brixham (English Channel), and

– the Thames Estuary and the East Anglian coast (southern North Sea).

Liverpool Bay is an inshore area of langoustine (*Nephrops*), herring, and plaice fisheries but also an area characterised by major development of offshore wind farms in recent years. The ICES rectangles selected are 35E5, 35E6, 36E5, and 36E6 with a geographic bounding box of 54° N, 3° W, 53° N, and 5° W. They include extensive sampling along the North Wales coast as part of fisheries research projects and surveys centred on Red Wharf Bay in the 1960s. Figure 7 shows the intensive sampling efforts that occurred throughout the 1960s and 1970s and again after 2000 when the autonomous Liverpool Bay SmartBuoy (source 6) was installed, taking hundreds of new measurements each day. A number of long-term Coastal Temperature Network (source 1) monitoring stations have existed in this area, notably at Wylfa, Amlwch, Moelfre, and Bangor.

The ICES rectangles in the Celtic Sea (29E1, 29E2, 30E1, 30E2; geographic bounding box of 51° N, 7° W, 50° N, 9° W) were selected because this is known as an important area for cod, hake, angler fish, and megrim. The selected area includes Haig Fras, a 45 km long submarine granitic rocky outcrop which, because of the diverse fauna associated with its bedrock reef habitat, is protected as a special area of conservation (SAC). Other seawater temperature records have only been collected on an occasional basis in this region, although more surveys have been conducted in recent years associated with the designation of this feature as a new marine protected area.

Brixham is now one of the most important fishing ports in England and home to major beam-trawl fishing fleets. Important sole, plaice, and lemon sole fisheries exist inshore, and a cuttlefish fishery extends offshore. The ICES rectangles selected are 28E6, 28E7, 29E6, and 29E7 with a geographic bounding box of 50°30' N, 2° W, 49°30' N, and 4° W. Temperature sampling in this region, particularly in recent years, has generally been focussed around the annual Channel Groundfish Surveys, with a particular concentration of data measurements in quarter 1 (March) and quarter 3 (July).

The Thames Estuary and East Anglian coast are important for sea bass, sole, and elasmobranch fisheries. The ICES rectangles selected are 32F1, 32F2, 33F1, and 33F2 with a geographic bounding box of 52°30' N, 3° E, 51°30' N, and 1° E. Some of the longest-running time series exist for this region, in particular from the Coastal Temperature Network (source 1) monitoring stations that have existed at Bradwell since 1964, Leigh on Sea and Southwold since 1966, and Sizewell since 1967. Earlier temperature measurements were taken primarily during fisheries research surveys and, in addition, regular sampling was begun aboard the Harwich to Rotterdam ferry after 1970. A major intensification of sam-

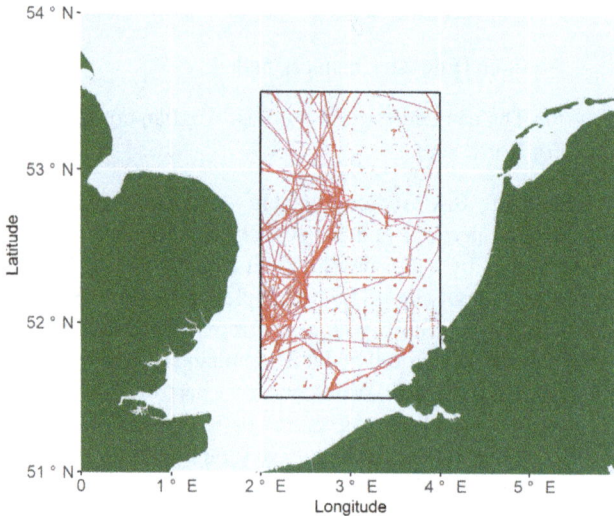

Figure 8. Illustration of near-surface (0–5 m) data coverage in the southern North Sea. Plotted point intensity reflects data density. Note that the area inside the West Frisian Islands is primarily sandbanks and reclaimed land, not sea.

pling occurred after 2000 following the installation of the autonomous Warp and Gabbard SmartBuoys (source 6).

4.7 Southern North Sea geographic data coverage – spatial, source, and numerical bias

The southern North Sea is an area of particular interest because it is one of the regional seas that is reported to have warmed the most dramatically over the 20th century (Dye et al., 2013; Hobday and Pecl, 2014). Figure 8 shows the geographic distribution of Cefas near-surface (between 0 and 5 m) seawater temperature data (specifically chosen to include data from vessel-mounted pumped systems). It also shows a clear geographical bias in terms of data coverage in the selected offshore area (geographic bounding box of 54° N, 4° E, 52° N, 2° E). This does not overlap with the Thames Estuary and East Anglian coast data plot above. The area selected specifically includes data from autonomous platforms to highlight potential issues with data density in any reuse of this data.

Figure 8 shows concentrations of measurements around major offshore fishing grounds on the North Norfolk sandbanks (e.g. Leman Ground, Smiths Knoll, Swarte Bank, Indefatigable Banks), line transects across the North Sea from ferry routes, ScanFish and the FerryBox system (sources 8, 15 and 16), and a background pattern of gridded stations from the ICES International Bottom Trawl Survey Programme (source 2).

The distribution of numbers of data points within this area led to the sources being grouped as follows:

- $> 100\,000$ data points (represented as red in Figs. 9 and 10),

- $\geq 30\,000 \leq 50\,000$ (represented as blue in Figs. 9 and 10),

- $\geq 2000 \leq 6000$ (represented as green in Figs. 9 and 10),

- and < 2000 (removed from this analysis to aid clear visualisation).

Figure 9 breaks down the temporal and numerical coverage of the data illustrated in Fig. 8, illustrating the temporal dominance of source 8 (ferry routes and surface logger systems) and the combined, post-2000 numerical dominance of the single SmartBuoy and WaveNet moored autonomous platforms (sources 6 and 7), both located in the western part of the selected area.

4.8 Southern North Sea data coverage by number and time

Figure 10a illustrates the numerical dominance of sources 6 and 7 highlighted above. Figure 10b combines plots of the selected seawater temperature records with time, using the colours from Fig. 8 to further clarify the temporal influences of major data sources. Several patterns can be discerned. Firstly, a slight upward trend is apparent across the whole 100-year time series with generally warmer temperatures at the end of the 20th century compared to the beginning. There is an absence of data from the periods of both World Wars when the DFR research vessels were requisitioned by the Admiralty for war service, mines were installed in coastal waters, and all research at the Lowestoft laboratory ceased. Several extremely cold winters are apparent, most obviously the winter of 1962–1963 (also known as the "Big Freeze"), which was one of the coldest winters on record. In February to March 1963, seawater along the coasts of Essex and Kent froze over and catches of dead fish (particularly sole) were recorded throughout much of the region (Woodhead, 1964). It is also clear that from around 2000 onwards, winter minima rarely fall below around 5 °C. It is not clear whether this is related to the beginning of the operational deployments of SmartBuoy and WaveNet stations by Cefas around this time.

In addition to the potential influences of data volumes with time on, e.g. trend interpretation, there are potential geographic and depth biases associated with source. These are illustrated in Fig. 11, which partitions the data shown in Fig. 7 by time (focusing on the period after the year 2000 identified in Fig. 10b) and by depth; 95 % of all the available data in the selected area are between 0 and 5 m, with 90 % of the 0–5 m data in the top 1 m.

Figure 11a shows the geographical distribution of data post-2000 between 1 and 5 m, whilst Fig. 11b shows data between 0 and 1 m (dominated by sources 6 and 7). The locations of the two autonomous monitoring stations are shown as orange spots in Fig. 11b; both are in the south-west quadrant of the selected area. This provides a numerical, geographical, and depth bias in the data available since 2000.

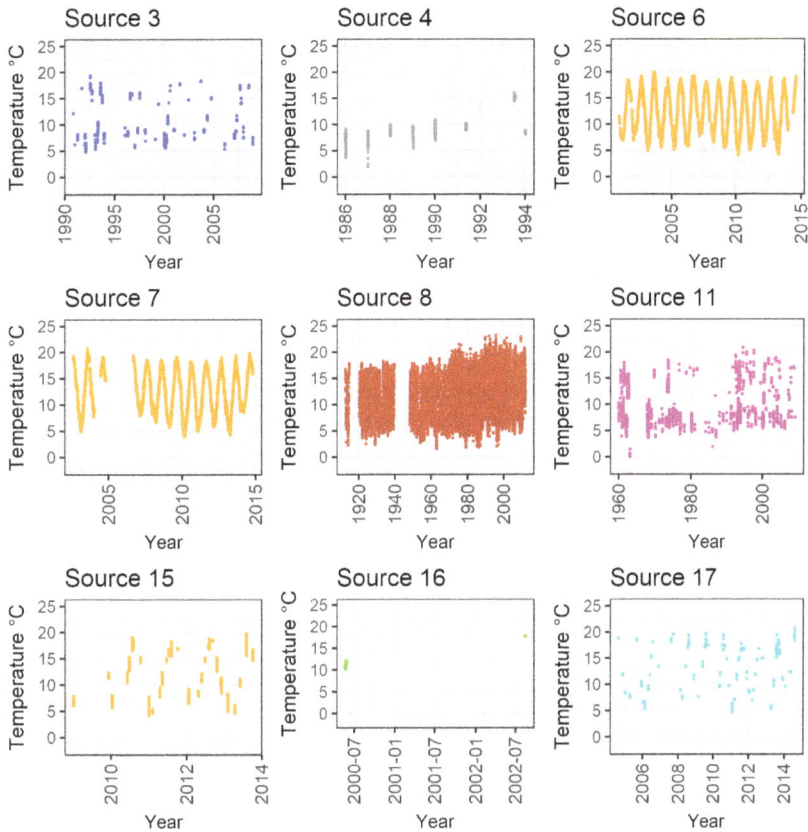

Figure 9. Near-surface (0–5 m) data coverage and temperatures in the southern North Sea by source and time. Note the different timescales for each data source.

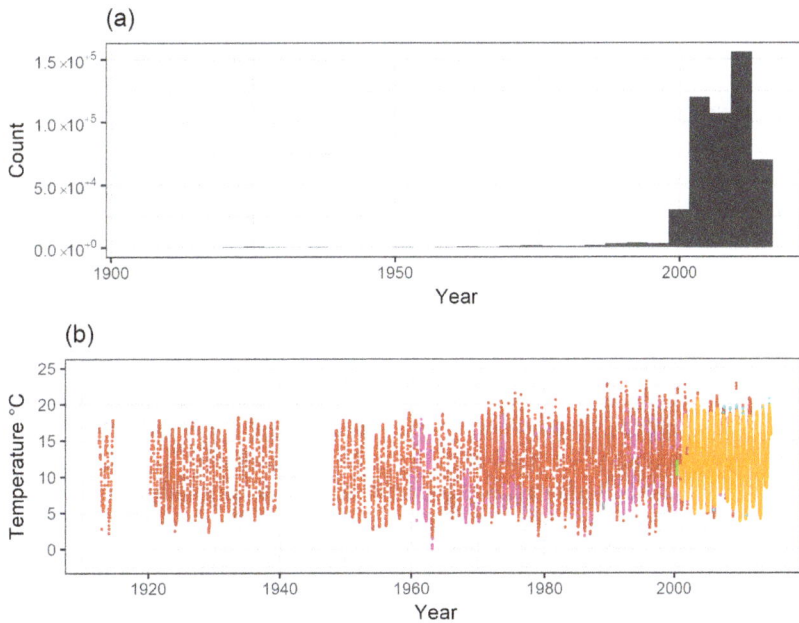

Figure 10. (a) Near-surface (0–5 m) data counts in the southern North Sea. **(b)** Seawater temperature in the southern North Sea by year and data source. Blue: sources 8 and 15 (vessel-mounted pumped systems), green: sources 3, 4, 11, 16, and 17 (other source), and red: sources 6 and 7 (autonomous platforms).

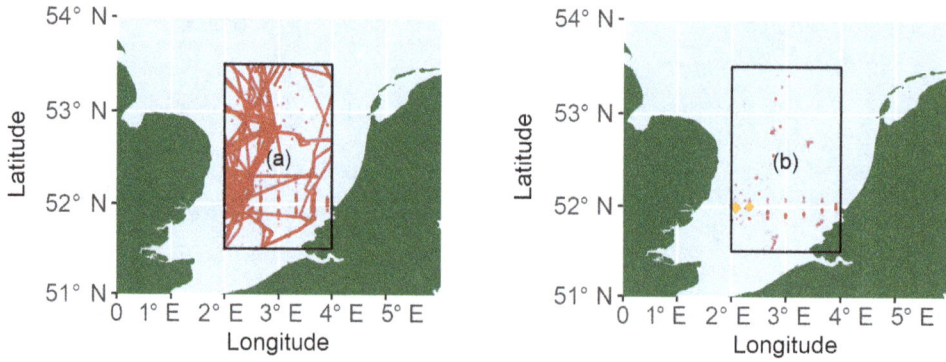

Figure 11. Illustration of potential numerical, geographical, and depth biases associated with data source in the southern North Sea from the year 2000 on: **(a)** 1–5 m and **(b)** 0–1 m (primarily autonomous platforms, sources 6 and 7). Plotted point intensity reflects data density.

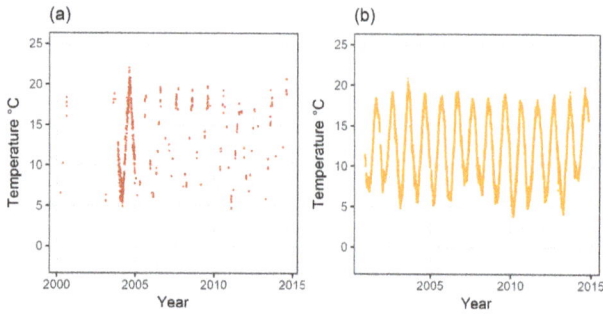

Figure 12. Plot of seawater temperature in the southern North Sea against time post-2000. **(a)** Data from sources other than autonomous platforms. **(b)** Data from the two autonomous monitoring stations in the selected area.

These factors would need to be taken into account in any investigation into the causes of the absence of minimum annual data less than around 5 °C, e.g. using models that allow for spatio-temporal trends and correlation. It is beyond the scope of this paper to construct the statistical models necessary to clarify the influences of data availability in space, time, and number; however, we do provide a further, simple examination of the potential effects of depth, location, and data number bias.

Figure 12 compares the seawater temperature records of the data in the selected area post-2000. Figure 12a shows data that do not come from the two autonomous monitoring stations, whilst Fig. 12b does. The patterns in the plots of individual data points are similar with some higher individual readings in Fig. 12a, possibly reflecting data acquired at the surface where aerial exposure during deployment is a known possible influence.

Figure 13 explores the potential influence of numerical differences in data numbers with time using all available data in the selected area of the southern North Sea to calculate annual seawater temperature statistics. It plots annual statistics (all sources, all depths) as points before 1955 when data are particularly sparse. This limited data coverage gives rise

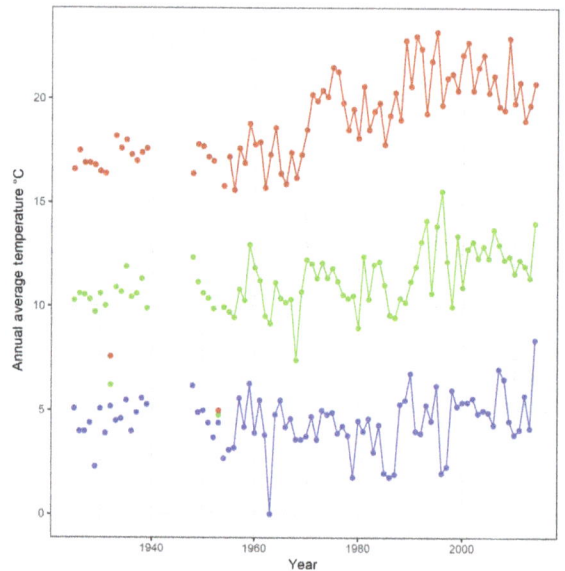

Figure 13. Average (green), minimum (blue), and maximum (red) annual temperatures for the southern North Sea including all sources and all depths.

to apparent anomalies with maximum average temperatures below 10 °C in the early 1930s and one year below 5 °C in the early 1950s. Post-1955, the increase in data volumes provides a more coherent picture (plotted as points and lines), reflecting to some degree the trend in increasing maximum and mean temperatures expected from the scientific papers cited above. The observed winter of the "Big Freeze" in the early 1960s is again very clear. However, the post-2000 absence of data below 5 °C at the surface (shown in Fig. 10b) is not reflected in the annual minimum data for all depths.

Figure 14 illustrates the depth component of the data sources in a small selected geographic "belt". Source 3 (Oceanographic Archive) is represented by a vertical CTD profile. Source 4 (Plankton Analysis System) shows temperature data gathered during a "V" profile plankton tow.

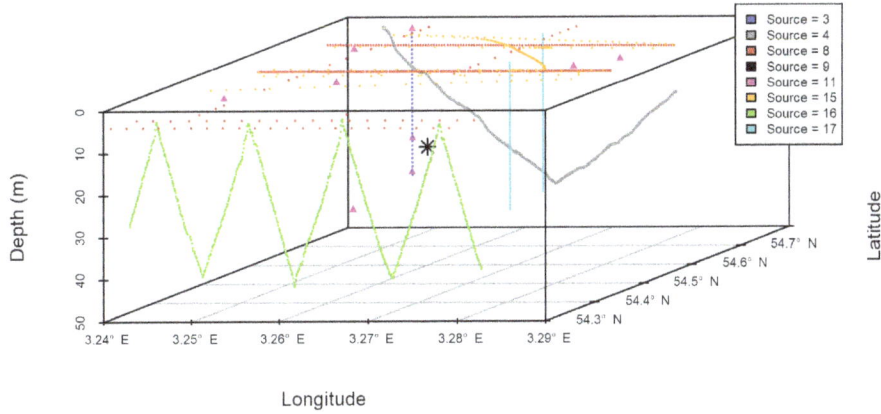

Figure 14. Selected small-scale location illustrating the diversity of data sources available and their associated depth profiles. Note that the data illustrated were not collected at the same time.

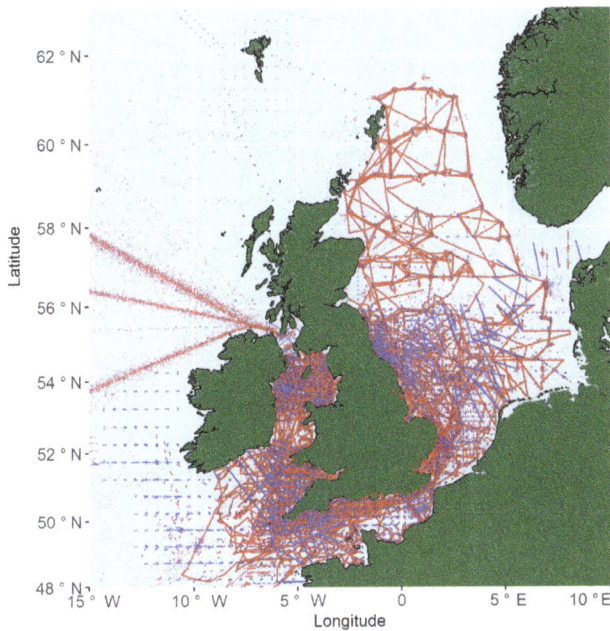

Figure 15. Illustration of the distribution of near-surface data (0–5 m, in red) and mid-water data (20–25 m, in blue) for the bulk of the UK Continental Shelf.

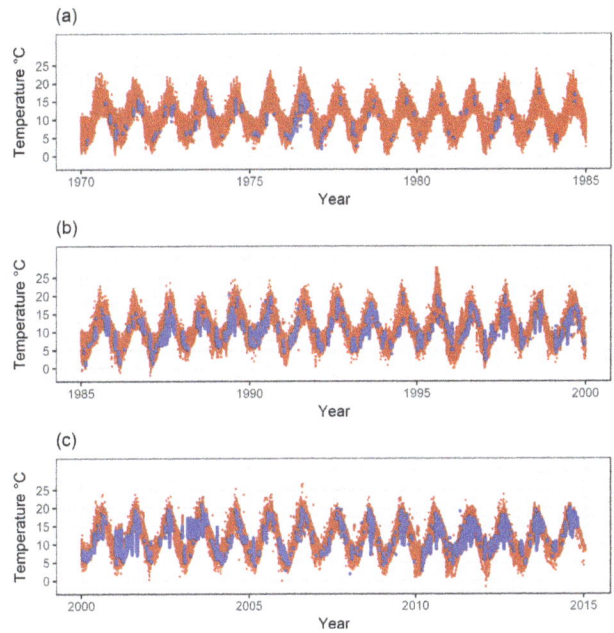

Figure 16. Seawater temperature cycles for three 14-year periods: 1970–1984 (upper chart), 1985–1999 (centre chart), and 2000–2015 (lower chart) for seas around the British Isles (area 48 to 58° N and 10° W to 10° E), separating near-surface 0–5 m (red) and 20–25 m (blue) data.

Source 8 shows data from Cefas or predecessor RV surface logging, whilst in complete contracts source 9 shows the single data point obtained from a fish tag on a cod. Source 11, the Lowestoft Sample Data Management System, shows data collected from near-surface and vertical-profile water samples. The RV *Cefas Endeavour* FerryBox System (source 15) shows research and/or transit data collection, whilst source 16 shows the data collected from the CTD mounted on the undulating ScanFish system. Source 17 shows vertical CTD profiles using the ESM2 logging system.

4.9 Distribution and patterns in seawater data for the bulk of the UK Continental Shelf area for the near surface and mid-water

This section widens the geographic coverage of the data exploration to the bulk of the assembled data (Fig. 15; see also Figs. 3 and 4 for context). We retain the near-surface 0–5 m (red) subsetting and extend it to "mid-water" at 20–25 m (blue). As already shown in the Fig. 7a (Liverpool Bay) subset, near-surface data coverage is extensive in the Irish Sea

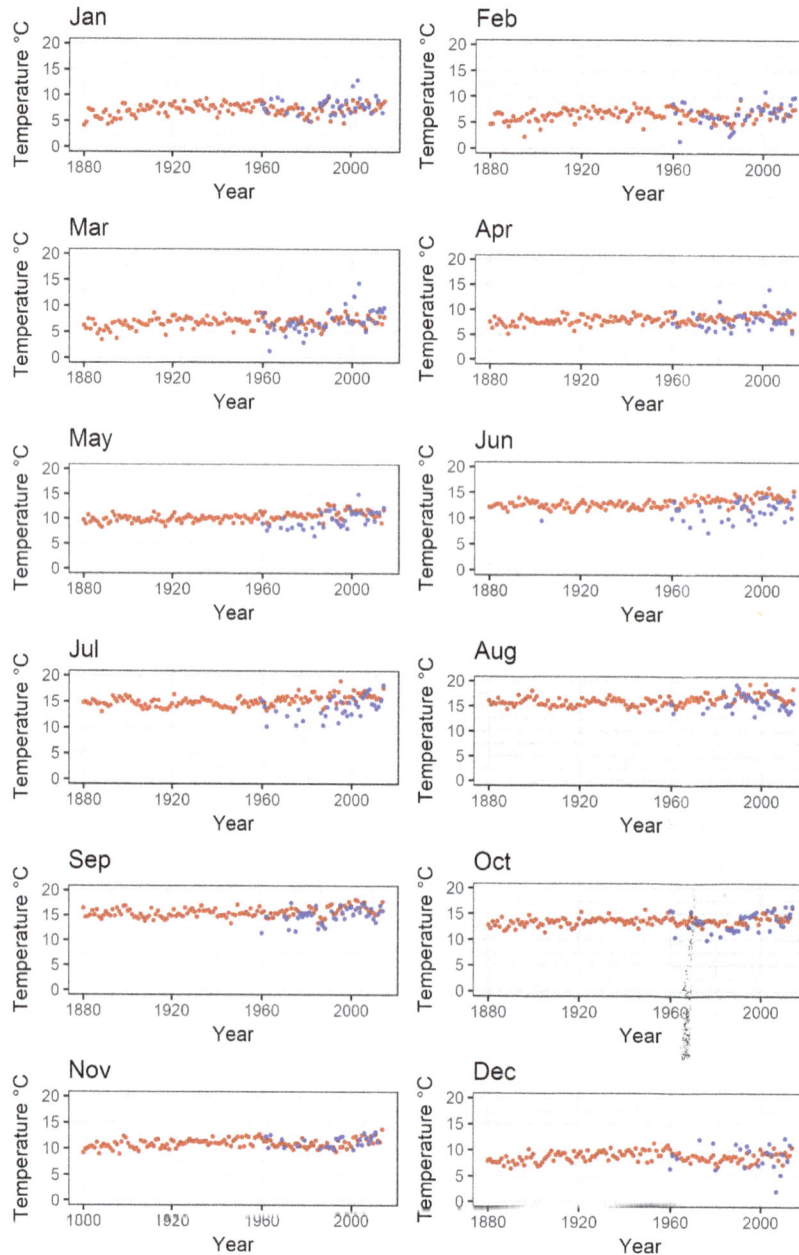

Figure 17. Average near-surface and mid-water seawater temperature around the British Isles by month from 1880 to 2015 (surface 0–5 m in red, mid-water 20–25 m in blue).

area but this is masked to some extent by the overplotting of the mid-water data distribution. This overplotting effect also applies elsewhere.

Figure 15 illustrates some of the characteristics of the data sources. Source 4, the Plankton Analysis System, provides more data at depth and this is illustrated in the south-western quadrant, an area of particular interest for plankton studies. Further north, routes to and from a series of set stations (source 8) provide data from the late 1950s to the mid-1990s. In the North Sea, the bulk of data offshore and at depth come

from an extensive series of ScanFish tows (source 16; see Brown et al., 1996).

4.10 Surface and mid-water seawater temperature around the British Isles from 1880 to 2015

The data subsets described above are comprised of surface measurements and temperatures at depth, so it is possible to extract time series with different depth bands to illustrate the breadth and depth of the data coverage with time; see Fig. 16. There are apparent artefacts in Fig. 16, e.g. high and low values that appear to be outliers (see above). High and low data points in Fig. 16 illustrate the importance of recognising the source of the data. Source 1 (Coastal Temperature Network) and source 5 (Fisheries Ecology Research Programme) both have data from relatively isolated bodies of water that have higher and lower temperatures than the surrounding sea (e.g. North Norfolk coast and South Wales inlets respectively). In addition, there are other source-affected influences on patterns and plots. In August 2001, for example, the surface data were primarily coastal in the south and in the Liverpool Bay area. The mid-water data were in the eastern North Sea, were dominated by ScanFish measurements, and were in and around the thermocline. In August 2009 the mid-water data were from CTD casts in the North Sea as far north as the Orkney Islands, whilst the surface data are coastal and in or south of the Humber Estuary. In the summer of 2012, The majority of surface data are from the RV *Cefas Endeavour* FerryBox system (source 10), which recorded tracks across the North Sea, again as far north as the Orkney Islands, whilst the mid-water data were dominated by citizen science diver data, especially on the coast of Northern Ireland.

Figure 16 clearly shows the annual cycle of seawater temperatures around the British Isles and interesting features such as the run of three cold winters (1985–1987) followed by three warm winters (1988–1990) plus warm summers (1995, 2006). The datasets are very comprehensive for the sea surface (0–5 m depth) but are sparser for deeper depths (in this case 20–25 m). Typically, and as expected, sea surface temperatures are slightly higher than temperatures at depth in this region. Dulvy et al. (2008) have shown that many fish in the North Sea have responded to rising seawater temperatures by shifting their distributions into deeper and therefore cooler waters. They suggested that the whole North Sea demersal fish assemblage deepened by ~ 3.6 m per decade in response to climate change between 1980 and 2004.

4.11 Average surface and mid-water seawater temperature around the British Isles by month from 1880 to 2015 – limitations of data density and coverage with time

Figure 17 shows the near-surface and mid-water seawater temperatures for seas around the British Isles (UKCS area 48 to 58° N and 10° W to 10° E) from 1880 to 2015 plotted by month. It shows that, for most months of the year, the sea surface temperatures around the British Isles increased throughout the 20th century, with stronger upward trends in the spring and summer months (March to August) and smaller increases in autumn and winter (September to February).

Such long-term trends have been associated with a number of observed changes in biological systems, including a clear seasonal shift to the earlier appearance of fish larvae at Helgoland Roads in the southern North Sea (Greve et al., 2005), linked to marked changes in zooplankton com-

Table 4. Rounded statistical summary of data used to calculate monthly averages.

Date range	Statistical summary of number of data points used in calculation of monthly averages			
	Minimum	Median	Mean	Maximum
Surface				
Pre-1950	4	25	27	89
1950 to 1990	50	425	535	6746
Post-1990	198	9380	14 731	89 235
Mid-water				
Pre-1950	2	2	2	2
1950 to 1990	1	14	395	7212
Post-1990	1	98	1596	67 822

position and sea surface temperature in this region (Beaugrand et al., 2002). Greve et al. (2005) suggested that in 10 cases, both the "start of season" and "end of season" (Julian date on which 15 and 85 % of all larvae were recorded respectively) were correlated with sea surface temperature. Similarly, ichthyoplankton sampling suggests that winter-breeding species in the English Channel region also spawn earlier in cooler years, while summer-spawning fish tend to spawn later (Genner et al., 2010). Phenology is the study of the timing of recurrent biological events, such as the return of migrating species or the first flowering of certain trees each year. Though most examples of phenological change in the literature have been drawn from terrestrial systems, the year-class size of marine fish is greatly influenced by the timing of spawning and the resulting match–mismatch with their prey and predators (Cushing, 1990), which are in turn greatly influenced by seawater temperatures.

The data now readily available here can contribute to further explorations of these changes although we note that the low average mid-depth seawater temperatures for the month of December in 2007 and 2009 arise from single data points forming that average. The high data point for mid-water in April 2011 comes from a diver. The following statistics (Table 4) are derived for the data used in Fig. 17. They indicate the importance of the statistical modelling outlined above, especially for earlier periods and for large areas.

5 Data availability

Data are available from the Cefas Data Hub.

- All sources: https://doi.org/10.14466/Cefasdatahub.4.

- Source 1: Coastal Temperature Network, https://doi.org/10.14466/Cefasdatahub.5.

- Source 2: Fishing Survey System, https://doi.org/10.14466/Cefasdatahub.6.

- Source 3: Oceanographic Archive, https://doi.org/10.14466/Cefasdatahub.7.

- Source 4: Plankton Analysis System, https://doi.org/10.14466/Cefasdatahub.8.

- Source 5: Fisheries Ecology Research Programme, https://doi.org/10.14466/Cefasdatahub.9.

- Source 6: SmartBuoy Monitoring Network, https://doi.org/10.14466/Cefasdatahub.10.

- Source 7: Defra Strategic Wave Monitoring System, https://doi.org/10.14466/Cefasdatahub.11.

- Source 8: Historical Ferry Routes Monitoring System and RV surface logger systems, https://doi.org/10.14466/Cefasdatahub.12.

- Source 9: Electronic Data Storage-Tag Database, https://doi.org/10.14466/Cefasdatahub.13.

- Source 10: Citizen Science Diver Recorded Temperatures, https://doi.org/10.14466/Cefasdatahub.14.

- Source 11: Lowestoft Sample Data Management System, https://doi.org/10.14466/Cefasdatahub.15.

- Source 12: *Mnemiopsis* Ecology Modelling and Observation Project, https://doi.org/10.14466/Cefasdatahub.16.

- Source 13: Multibeam Acoustics Sound Velocity Profile Temperature, https://doi.org/10.14466/Cefasdatahub.17.

- Source 14: Intensive plankton surveys off the northeast coast of England in 1976, https://doi.org/10.14466/Cefasdatahub.18.

- Source 15: FerryBox Monitoring, https://doi.org/10.14466/Cefasdatahub.19.

- Source 16: ScanFish, https://doi.org/10.14466/Cefasdatahub.20.

- Source 17: ESM2 Profiler–mini CTD Logger, https://doi.org/10.14466/Cefasdatahub.21.

The contents of the Cefas Data Hub website are provided as part of the Cefas role as a Defra agency under the Defra Open Data Strategy.

Cefas requires users to make their own decisions regarding the accuracy, reliability, and applicability of information provided. The data provided by the Cefas Data Hub are believed by Cefas to be reliable for their original purposes and are accompanied by discovery metadata that provide a copy of the information available to Cefas scientists, describing the original purposes of data collection. It is the responsibility of the data user to take this information into account when reusing

data. Regardless of any quality control processes, Cefas does not accept any liability for the use the data provided; use is at the users' own risk. Cefas does not give any warranty as to the quality or accuracy of the information or the medium on which it is provided or its suitability for any use. All implied conditions relating to the quality or suitability of the information and the medium and all liabilities arising from the supply of the information (including any liability arising from negligence) are excluded to the fullest extent permitted by law.

The use of data from the Cefas Data Hub requires that the correct and appropriate interpretation is solely the responsibility of the data users, that results, conclusions, and/or recommendations derived from the data do not imply endorsement from Cefas, that data sources must be acknowledged, preferably using a formal citation, that data users must respect all restrictions on the use of data such as for commercial purposes, and that data may only be redistributed, i.e. made available in other data collections or data portals, with the prior written consent of Cefas.

6 Conclusions

This data rescue, assembly, integration, and publication exercise stemmed from what seemed at the time to be a relatively simple plea made at an internal workshop to make all temperature datasets held within the Lowestoft laboratory available via a common data portal. What emerged was a general realisation that there were 17 separate data systems, each containing records of varying quality, on paper and stored electronically in a myriad of different formats and archaic file types, some of which could no longer be easily read without bespoke computer software. Potentially valuable information was collected for various operational reasons over the past 134 years, but every system was tailored for a specific purpose. Where temperature was specifically measured by oceanographers, some form of CTD was deployed, and in these cases semi-standardised data were often transferred to national repositories, for example the British Oceanographic Data Centre (https://www.bodc.ac.uk/data/bodc_database/ctd/) or the ICES Data Centre. However, in most cases, the data described here have never been made publicly available before, except within the context of summary outputs from the individual research projects published in peer-reviewed journal articles. The internal workshop wanted "all the temperature data in one place in the same format" so that anyone could use it. The initiating request for access without having to understand the originating formats was driven primarily by requirements for studying long-term climate change but also encompassed biological and ecological uses and work on linked data. These requirements became even more pressing given a UK government-wide drive to make publicly funded scientific datasets available. Whomersley et al. (2015) describe the reuse of data by specialists who did not need to understand the dataset and format or the associated limitations. This paper has taken a step further and decomposed the original data formats with a view to making the seawater temperature data more accessible and available, thereby widening access and reuse.

In June 2015 Defra's Secretary of State, Elizabeth Truss, announced her vision for the future of British food, farming, and the natural environment, stating that "at least 8000 datasets – will be made freely available to the public, putting Britain at the forefront of the data revolution". She stated that "vast data reserves from Defra are set to transform the world of food and farming in the single biggest government data giveaway the UK has ever seen". As a result of this initiative, Cefas has released more than 1950 individual datasets via the Cefas Data Hub (www.cefas.co.uk/cefas-data-hub/), a majority of which currently provide data in the original format.

The data presented here have not been corrected or adjusted in any way to take account of the different sampling methodologies used, as has been attempted for the most well-known data collation efforts such as ERSST, HadSST3, and COBE-SST (see Mathews, 2013; Kennedy et al., 2011a, b). Inherent biases have been partially addressed by the provision of contextual fields (Source, Sample, Measure, Ref1, Ref2), and areas for easy but potentially misleading uses of the data have been explored above. Some of the datasets described here have contributed to the ICES Report on Ocean Climate (IROC), which provides summary information on climatic conditions in the North Atlantic on an annual basis (see https://ocean.ices.dk/iroc/).

The archive of processed Coastal Temperature Network data has been widely cited (see https://scholar.google.co.uk/citations?hl=en&user=GkV5fMwAAAAJ&view_op=list_works). This paper has made the underlying data readily available (source 1). Other datasets, such as source 10 comprised of temperature and depth records obtained via a citizen science project from recreational scuba divers (see Wright et al., 2016), represent a hitherto largely untapped resource for oceanographic researchers.

Author contributions. SD, LF, and OW provided data, data processing, and deeper insights into specialist areas along with SF, who also advised on the choice of ICES rectangles to demonstrate data in areas of high fisheries interest. JP was the Cefas staff member who, during a workshop in frustration, asked "Why can't I just get access to all Cefas temperature data?". This paper provides the requested access and expands on "just". DJM took on the challenge, identified and ingested the data, decomposed the temperature data from their multiple originating formats, ran the QA–QC, and prepared the paper. DM provided statistical inputs regarding the selection and presentation of data designed to illustrate the effects of bias and spatial and temporal influence. SR (through Cefas Seedcorn) provided the internal funding (part of Cefas Seedcorn DP705 "Delivering Linked Data"), early comments on the form of the paper, the support and resources needed to build the Cefas Data Hub, and ongoing support for what turned out to be more than a year of effort.

Competing interests. The authors declare that they have no conflict of interest.

Acknowledgements. O. Andres, R. Ayers, A. Brown, E. Cappuzzo, K. Cooper, N. Greenwood, K. Hyder, S. Jennings, D. Pearce, S. Pitois, D. Righton, and N. Taylor provided access to and information on the Cefas data sources. W. Meadows assisted A. Brown in retrieving data. D. Haverson, T. Hull, and S. Wright extracted the seawater temperature data from the original complex, specialist, or bulk-originating datasets at the request of the data owners. Captain R. Jolliffe explained the realities of navigation before GPS. D. Righton also provided advice on the figures.

Most of the data-sampling programmes presented in this paper were funded by the UK Department for Environment Food and Rural Affairs (Defra) and its predecessor the Ministry of Agriculture, Food and Fisheries (MAFF). Additional support was provided by the European Union under the Interreg IVa MEMO-2 Seas Programme (source 12). Digitisation of historic datasets from paper records was funded by the Cefas Seedcorn project DP302 "Long term decrease in carrying capacity of the North Sea (2005)" and project DP332 "Trawling Through Time" (Cefas, 2014).

WaveNet data (source 7) are published with specific permission from the Agri-Food and Biosciences Institute, Defra, DONG Energy, East Anglia Offshore Wind, EDF Energy, Environment Agency, Met Office, Natural Resources Wales, Scarborough Borough Council, and the Scottish Environment Protection Agency.

Citizen Science Diver Recorded Temperatures: this work was part of the Defra-funded citizen science investigations empowering the public through the use of novel technologies to collect policy-relevant marine data (CSI); MF1230, http://randd.defra.gov.uk/Default.aspx?Module=More&Location=None&ProjectID=18553.

The authors are grateful to Sue Dale, Michelle Dann, and Suzy Angelus from JDP Management Services for their detailed work in the Cefas Legacy Data Rescue Programme and to Cefas staff (Kate Collingridge, Tiago da Silva, and Paul Dolder) who participated in a "hack" to highlight potential uses of the data and flush out positional and other data anomalies.

Brian Lockwood provided external Python programming to extract data from the multiple legacy CTD formats that emerged and evolved during the early decades of electronic oceanographic measurements.

Extensive use was made of R (R Development Core Team, 2017) and RStudio (RStudio Team, 2017) to handle and present the data. The R code used to generate the data subsets and figures is provided in a separate document.

Edited by: Giuseppe M. R. Manzella

References

Azzopardi, E. and Sayer, M.: Estimation of depth and temperature in 47 models diving decompression computer, Underwater Technol., 31, 3–12, 2012.

Beaugrand, G., Reid, P. C., and Ibañez, F.: Reorganization of North Atlantic marine copepod biodiversity and climate, Science, 296, 1692–1694, 2002.

Bojinski, S., Verstraete, M., Peterson, T. C., Richter, C., Simmons, A., and Zemp, M.: The concept of essential climate variables in support of climate research, applications, and policy, B. Am. Meteorol. Soc., 95, 1431–1443, 2014.

Brown, A. and Vanstaen, K.: The Role of multibeam sonar in Cefas habitat mapping projects, Caris 2008: 12th International User Group Conference, Bath, UK, available at: http://www.caris.com/conferences/caris2008.dsbld/proceedings/presentations/brown/BROWN-Paper.pdf (last access: 15 December 2017), 2008.

Brown, J., Brander, K. M., Fernand, L., and Hill, A. E.: Scanfish: A high performance towed undulator. A new PC controlled towed undulator is aid to understanding coupling between physical and biological processes, Sea Technol., 37, 23–27, 1996.

Collingridge, K., Van Der Molen, J., and Pitois, S.: Modelling risk areas in the North Sea for blooms of the invasive comb jelly *Mnemiopsis leidyi* A. Agassiz, 1865, Aquat. Invasions, 9, 21–36, 2014.

Cefas: Trawling Through Time: Cefas Science and Data 1902–2014, Cefas, Lowestoft, 16 pp., available at: https://www.gov.uk/government/uploads/system/uploads/attachment_data/file/364393/TTT_FINAL_11Jun14.pdf (last access: 15 December 2017), 2014.

Cushing, D. H.: Plankton production and year-class strength in fish populations: an update of the match/mismatch hypothesis, Adv. Mar. Biol., 26, 249–293, 1990.

Dann, N., Heessen, H., and Ellis, J.: Data Processing, in: Fish Atlas of the Celtic Sea, North Sea and Baltic Sea: Based on International Research-vessel Surveys, edited by: Heessen, H., Daan, N., and Ellis, J., KNNV Publishing, Chapter 5, 41–49, 2015.

Dulvy, N. K., Rogers, S. I., Jennings, S., Stelzenmüller, V., Dye, S. R., and Skjoldal, H. R.: Climate change and deepening of the North Sea fish assemblage: a biotic indicator of warming seas, J. Appl. Ecol., 45, 1029–1039, 2008.

Dye, S. R., Hughes, S. L., Tinker, J., Berry, D. I., Holliday, N. P., Kent, E. C., Kennington, K., Inall, M., Smyth, T., Nolan, G., Lyons, K., Andres, O., and Beszczynska-Möller, A.: Impacts of climate change on temperature (air and sea), MCCIP Science Review, 2013, 1–12, https://doi.org/10.14465/2013.arc01.001-012, 2013.

Ellett, D. J. and Jones, S. R.: Surface temperature and salinity time-series from the Rockall Channel, 1948–1992, Data Report, MAFF Directorate of Fisheries Research, Lowestoft, 36, 24 pp., 1994.

Freeman, E., Woodruff, S. D., Worley, S. J., Lubker, S. J., Kent, E. C., Angel, W. E., Berry, D. I., Brohan, P., Eastman, R., Gates, L., Gloeden, W., Ji, Z., Lawrimore, J., Rayner, N. A., Rosenhagen, G., and Smith, S. R.: ICOADS Release 3.0: a major update to the historical marine climate record, Int. J. Climatol., 37, 2211–2232, https://doi.org/10.1002/joc.4775, 2017.

Genner, M. J., Halliday, N. C., Simpson, S. D., Southward, A. J., Hawkins, S. J., and Sims, D. W.: Temperature-driven phenological changes within a marine larval fish assemblage, J. Plankton Res., 32, 699–708, 2010.

Graham, M.: English Fishery Research in Northern Waters, Arctic, 6, 252–259, 1953.

Greenwood, N., Parker, E. R., Fernand, L., Sivyer, D. B., Weston, K., Painting, S. J., Kröger, S., Forster, R. M., Lees, H. E., Mills, D. K., and Laane, R. W. P. M.: Detection of low bottom water oxygen concentrations in the North Sea; implications for monitoring and assessment of ecosystem health, Biogeosciences, 7, 1357–1373, https://doi.org/10.5194/bg-7-1357-2010, 2010.

Greve, W., Prinage, W. S., Zidowitz, H., Nast, J., and Reiners, F.: On the phenology of North Sea ichthyoplankton, ICES J. Mar. Sci., 62, 1216–1223, 2005.

Griffin, R. E. and the CODATA Task Group "Data At Risk" (DAR-TG): When are old data new data?, GeoResJ, 6, 92–97, https://doi.org/10.1016/j.grj.2015.02.004, 2015.

Harding, D. and Nichols, J. H.: Plankton surveys off the north-east coast of England in 1976: an introductory report and summary of the results, MAFF Directorate of Fisheries Research, Lowestoft, Fisheries Research Technical Report, 86, 56 pp., 1987.

Hausfather, Z., Cowtan, K., Clarke, D. C., Jacobs, P., Richardson, M., and Rohde, R.: Assessing recent warming using instrumentally homogeneous sea surface temperature records, Science Advances, 3, e1601207, https://doi.org/10.1126/sciadv.1601207, 2017.

Hobday, A. J. and Pecl, G. T.: Identification of global marine hotspots: sentinels for change and vanguards for adaptation action, Rev. Fish Biol. Fisher., 24, 415–425, 2014.

Jones, S. R.: Ten years of measurement of coastal sea temperatures, Weather, 36, 48–55, 1981.

Jones, S. R. and Jeffs, T. M.: Near-surface sea temperatures in coastal waters of the North Sea, English Channel and Irish Sea, Data Report, MAFF Directorate of Fisheries Research, Lowestoft, 24, 70 pp., 1991.

Joyce, A. E.: The coastal temperature network and ferry route programme: long-term temperature and salinity observations, Science Series Data Report, Cefas, Lowestoft, 43, 129 pp., 2006.

Karl, R. R., Arguez, A., Huang, B., Lawrimore, J. H., McMahon, J. R., Menne, M. J., Peterson, T. C., Vose, R. S., and Zhang, H.: Possible artefacts of data biases in the recent global surface warming hiatus, Science, 384, 1469–1472, https://doi.org/10.1126/science.aaa5632, 2015.

Kennedy, J. J., Rayner, N. A., Smith, R. O., Parker, D. E., and Saunby, M.: Reassessing biases and other uncertainties in sea surface temperature observations measured in situ since 1850: 1. Measurement and sampling uncertainties, J. Geophys. Res., 116, D14103, https://doi.org/10.1029/2010JD015218, 2011a.

Kennedy, J. J., Rayner, N. A., Smith, R. O., Parker, D. E., and Saunby, M.: Reassessing biases and other uncertainties in sea surface temperature observations measured in situ since 1850: 2. Biases and homogenization, J. Geophys. Res., 116, D14104, https://doi.org/10.1029/2010JD015220, 2011b.

Kent, E. C. and Taylor, P. K.: Toward Estimating Climatic Trends in SST. Part I: Methods of Measurement, J. Atmos. Ocean. Tech., 23, 464–475, 2006.

Larsen, K. M. H., Gonzalez-Pola, C., Fratantoni, P., Beszczynska-Möller, A., and Hughes, S. L. (Eds.): ICES Report on Ocean Climate 2015, ICES Cooperative Research Report No. 331, 79 pp., 2016.

Lee, A. J.: The Ministry of Agriculture, Fisheries and Food's Directorate of Fisheries Research: Its Origins and Development, MAFF, Lowestoft, UK, 332 pp., 1992.

MacKenzie, B. and Schiedek, D.: Daily ocean monitoring since the 1860s shows record warming of northern European seas, Glob. Change Biol., 13, 1335–1347, https://doi.org/10.1111/j.1365-2486.2007.01360.x, 2007.

Matthews, J. B. R.: Comparing historical and modern methods of sea surface temperature measurement – Part 1: Review of methods, field comparisons and dataset adjustments, Ocean Sci., 9, 683–694, https://doi.org/10.5194/os-9-683-2013, 2013.

Neat, F. and Righton, R.: Warm water occupancy by North Sea cod, P. R. Soc. B, 274, 789–798, 2007.

Neat, F., Bendall, V., Berx, B., Wright, P., O'Cuaig, M., Townhill, B., Schon, P.-J., Lee, J., and Righton, D.: Spatial dynamics of cod in UK waters, J. Anim. Ecol., 51, 1564–1574, https://doi.org/10.1111/1365-2664.12343, 2014.

Norris, S. W.: Near surface sea temperatures in coastal waters of the North Sea, English Channel and Irish Sea – Volume II, Science Series Data Report, Cefas, Lowestoft, 40, 31 pp., 2001.

Pante, E. and Simon-Bouhet, B.: marmap: a package for importing, plotting and analyzing bathymetric and topographic data in R, PLoS ONE, 8, e73051, https://doi.org/10.1371/journal.pone.0073051, 2013.

Pawsey, E. L., Atkinson, G. T., and Davis, F. M.: The exploration of Lousy Bank, Fish Investigation Series II, 4, 1–12, 1920.

Pedersen, M. W., Righton, D., Thygesen, K. H., and Madsen, H.: Geolocation of North Sea cod (Gadus morhua) using hidden Markov models and behavioural switching, Can. J. Fish. Aquat. Sci., 65, 2367–2377, 2008.

R Development Core Team: R: A language and environment for statistical computing. R Foundation for Statistical Computing, Vienna, Austria, available at: http://www.R-project.org, 15 December 2017.

Righton, D., Andersen, K. H., Neat, F., Thorsteinsson, V., Steingrund, P., Svedang, H., Michalsen, K., Hinrichsen, H. H., Bendall, V., Neuenfeldt, S., Wright, P., Jonsson, P., Huse, G., van der Kooij, J., Mosegaard, H., Hüssy, K., and Metcalfe, J.: Thermal niche of Atlantic cod (Gadus morhua): limits, tolerance and optima, Mar. Ecol. Prog. Ser., 420, 1–13, 2010.

Riser, S. C., Freeland, H. J., Roemmich, D., Wijffels, S., Troisi, A., Belbéoch, M., Gilbert, D., Xu, J., Pouliquen, S., Thresher, A., Le Traon, P.-Y., Maze, G., Klein, B., Ravichandran, M., Grant, F., Poulain, P.-M., Suga, T., Lim, B., Sterl, A., Sutton, P., Mork, K.-A., Vélez-Belchí, P. J., Ansorge, I., King, B., Turton, J., Baringer, M., and Jayne, S. R.: Fifteen years of ocean observations with the global Argo array, Nature Climate Change, 6, 145–153, https://doi.org/10.1038/nclimate2872, 2016.

Royal Navy: The Principles of Navigation, The Admiralty Manual of Navigation Volume 1, 10th Edn., The Nautical Institute, London, UK, 700 pp., 2008.

RStudio Team: RStudio: Integrated Development for R, RStudio, Inc., Boston, MA, available at: www.rstudio.com/, last access: 15 December 2017.

Sayer, M. D. J. and Azzopardi, E.: The silent witness: using dive computer records in diving fatality investigations, Diving Hyperb. Med., 44, 167–169, 2014.

Sauer, C., Van der Strict, S., Hornung-Lauxmann, L., and Tanner, V.: Verifications under the terms of Article 35 of the Euroatom Treaty, Dungeness Power Stations, Kent, United Kingdom, 6 to 10 November 2000, Art. 35 Technical Report – UK-00/2, 81 pp., 2002.

Smit, A. J., Roberts, M., Anderson, R. J., Dufois, F., Dudley, S. F. J., Bornman T. G., Olbers, J., and Bolton, J. J.: A Coastal Seawater Temperature Dataset for Biogeographical Studies: Large Biases between In Situ and Remotely-Sensed Data Sets around the Coast of South Africa, PLoS ONE, 8, e81944, https://doi.org/10.1371/journal.pone.0081944, 2013.

Sutton, G. A.: The analysis of environmental materials using gamma spectrometry. Aquatic Environment Protection: Analytical Methods, MAFF Directorate of Fisheries Research, Lowestoft, 10, 22 pp., 1993.

Townhill, B. L., Maxwell, D., Engelhard, G. H., Simpson, S. D., and Pinnegar, J. K.: Historical Arctic Logbooks Provide Insights into Past Diets and Climatic Responses of Cod, PLoS ONE, 10, e0135418, https://doi.org/10.1371/journal.pone.0135418, 2015.

van der Molen, J., van Beek, J., Augustine, S., Vansteenbrugge, L., van Walraven, L., Langenberg, V., van der Veer, H. W., Hostens, K., Pitois, S., and Robbens, J.: Modelling survival and connectivity of Mnemiopsis leidyi in the south-western North Sea and Scheldt estuaries, Ocean Sci., 11, 405–424, https://doi.org/10.5194/os-11-405-2015, 2015.

Whomersley, P., Murray, J. M., Mcilwaine, P. S. O., Stephens, D., and Stebbing, P.: More bang for your monitoring bucks: Detection and reporting of non-indigenous species, Mar. Pollut. Bull., 94, 14–18, https://doi.org/10.1016/j.marpolbul.2015.02.031, 2015.

Woodhead, P. J. M.: The death of North Sea fish during the winter of 1962/63 particularly with reference to the sole, Solea vulgaris, Helgoland. Wiss. Meer., 10, 283–300, 1964.

Wright, S., Hull, T., Sivyer, D. B., Pearce, D., Pinnegar, J. K., Sayer, M. D. J., Mogg, A. O. M., Azzopardi, E., Gontarek, S., and Hyder, K.: SCUBA divers as oceanographic samplers: The potential of dive computers to augment aquatic temperature monitoring, Sci. Rep.-UK, 6, 30164, https://doi.org/10.1038/srep30164, 2016.

Wyborn, L., Hsu, L., Lehnert, K. A., and Parsons, M. A.: Guest Editorial: Special issue Rescuing Legacy data for Future Science, GeoResJ, 6, 106–107, https://doi.org/10.1016/j.grj.2015.02.017, 2015.

4

An improved and homogeneous altimeter sea level record from the ESA Climate Change Initiative

Jean-François Legeais[1], Michaël Ablain[1], Lionel Zawadzki[1], Hao Zuo[2], Johnny A. Johannessen[3],
Martin G. Scharffenberg[4], Luciana Fenoglio-Marc[5], M. Joana Fernandes[6,7], Ole Baltazar Andersen[8],
Sergei Rudenko[9,10], Paolo Cipollini[11], Graham D. Quartly[12], Marcello Passaro[9], Anny Cazenave[13,14],
and Jérôme Benveniste[15]

[1]Collecte Localisation Satellite (CLS), 31520 Ramonville-Saint-Agne, France
[2]European Centre for Medium-Range Weather Forecasts, Reading, UK
[3]Nansen Environmental and Remote Sensing Center (NERSC), Bergen, Norway
[4]University of Hamburg, Hamburg, Germany
[5]University of Bonn, Bonn, Germany
[6]Faculdade de Ciências, Universidade do Porto, 4169-007 Porto, Portugal
[7]Centro Interdisciplinar de Investigação Marinha e Ambiental (CIIMAR), 4450-208 Matosinhos, Portugal
[8]DTU Space, 2800 Kongens Lyngby, Denmark
[9]Deutsches Geodätisches Forschungsinstitut, Technische Universität München, 80333 Munich, Germany
[10]Helmholtz Centre Potsdam – GFZ German Research Centre for Geosciences, 14473 Potsdam, Germany
[11]National Oceanography Centre, Southampton, SO14 3ZH, UK
[12]Plymouth Marine Laboratory, Plymouth, PL1 3DH, UK
[13]LEGOS, 31400 Toulouse, France
[14]ISSI, Bern, Switzerland
[15]ESA/ESRIN, 00044 Frascati, Italy

Correspondence: Jean-François Legeais (jlegeais@cls.fr)

Abstract. Sea level is a very sensitive index of climate change since it integrates the impacts of ocean warming and ice mass loss from glaciers and the ice sheets. Sea level has been listed as an essential climate variable (ECV) by the Global Climate Observing System (GCOS). During the past 25 years, the sea level ECV has been measured from space by different altimetry missions that have provided global and regional observations of sea level variations. As part of the Climate Change Initiative (CCI) program of the European Space Agency (ESA) (established in 2010), the Sea Level project (SL_cci) aimed to provide an accurate and homogeneous long-term satellite-based sea level record. At the end of the first phase of the project (2010–2013), an initial version (v1.1) of the sea level ECV was made available to users (Ablain et al., 2015). During the second phase of the project (2014–2017), improved altimeter standards were selected to produce new sea level products (called SL_cci v2.0) based on nine altimeter missions for the period 1993–2015 (https://doi.org/10.5270/esa-sea_level_cci-1993_2015-v_2.0-201612; Legeais and the ESA SL_cci team, 2016c). Corresponding orbit solutions, geophysical corrections and altimeter standards used in this v2.0 dataset are described in detail in Quartly et al. (2017). The present paper focuses on the description of the SL_cci v2.0 ECV and associated uncertainty and discusses how it has been validated. Various approaches have been used for the quality assessment such as internal validation, comparisons with sea level records from other groups and with in situ measurements, sea level budget closure analyses and comparisons with model outputs. Compared with the previous version of the sea level ECV, we show that use of improved geophysical corrections, careful bias reduction between missions and inclusion of new altimeter missions lead to improved sea level products with reduced uncertainties on different spatial and temporal scales. However, there is still room for improvement since the uncertainties remain larger than the GCOS requirements (GCOS, 2011). Perspectives on subsequent evolution are also discussed.

1 Introduction

Present-day global mean sea level (GMSL) rise primarily reflects the amount of heat added to the ocean, as well as land ice melt in response to anthropogenic global warming (e.g. IPCC, 2013; von Schuckmann et al., 2016). Accurate monitoring of sea level is required to better understand its variability and distinguish between natural and anthropogenic forcing factors as the origin of observed changes. It also allows validation of climate models developed for projecting future changes, as the models reproduce present-day and recent-past changes. Since 1993, satellite altimetry missions have delivered accurate sea level measurements, allowing the monitoring of sea level variations on different spatial and temporal scales (e.g. Pujol et al., 2016; Ablain et al., 2017; Escudier et al., 2017). About a decade ago, the Global Climate Observing System (GCOS) defined a list of key parameters of the Earth system, or "essential climate variables" (ECVs) that need to be accurately monitored in order to meet the needs of the climate change community (Bojinski et al., 2014). To respond to this need for climate-quality satellite data, the European Space Agency (ESA) developed the "Climate Change Initiative" (CCI) program. This program aims to realize the full potential of the long-term global Earth Observation archives from satellites as a significant and timely contribution to the ECV databases for climate modellers and researchers. Sea level is one of the listed ECVs of the CCI program. During the first phase (2010–2013) of the sea level CCI project (SL_cci), the first version of the ECV over the 1993–2010 time span was produced and distributed to the user community. Details of the production and validation protocol of this ECV are described in Ablain et al. (2015).

Within the second phase of the project (2014–2017), the objective was not only to extend the length of the sea level record by additional 5 years (2010–2015) but also to provide a full reprocessing of the sea level ECV during the altimetry period thanks to the development and selection of new altimeter algorithms to improve the ECV accuracy, stability and homogeneity. The details of the orbit solutions, the geophysical corrections, the altimeter standards and processing algorithms selected for the production of this v2.0 ECV are fully described in Quartly et al. (2017).

This paper describes the SL_cci v2.0 ECV and presents some validation results obtained through different approaches. After a short description of input data and altimeter standards used in the production system (Sect. 2), a presentation of the v2.0 SL_cci products is provided in Sect. 3.

The quality assessment of the ECV is described in Sect. 4. The consistency with the sea level records provided by other groups has been checked, and comparisons were also performed with in situ tide gauge measurements and combined Argo-based steric and GRACE-based barystatic sea level data. Additional validations based on a sea level budget closure approach and comparisons with the output from high-resolution ocean models are also presented (Sect. 5). The sea level errors and uncertainties are discussed in Sect. 6 with respect to the GCOS requirements (GCOS, 2011). They correspond to the error levels to be met by the sea level record on different spatial and temporal scales (e.g. long-term evolution and inter-annual and annual signals). These requirements have been considered as a reference within the CCI program and especially when assessing the quality of the SL_cci ECV. The paper finishes with the discussion of perspectives on evolution of the sea level products.

2 Input data and altimeter standards

The estimation of the altimeter-based sea level is currently based on measurements from many satellite missions (spanning more than 20 years). The input data used for the production of the first version of the SL_cci ECV v1.1 were derived from TOPEX/Poseidon, Jason-1, Jason-2, ERS-1, ERS-2, GeoSat Follow-On (GFO) and Envisat satellites. The first three missions fly along the so-called "reference orbit", sampling the ocean between 66° S and 66° N. The remaining missions have a higher orbital inclination, providing improved ocean sampling and giving near-complete coverage of the Arctic. A weakness of the v1.1 ECV is the limited number of satellite altimeters used in the production system. In particular, new altimeter missions (e.g., SARAL/AltiKa, CryoSat-2) have not been included when the temporal extensions of the dataset have been produced and the v1.1 sea level record is based on a single satellite (Jason-2) after the loss of Jason-1 in June 2013. It has affected the ECV in terms of reduced spatial coverage (no measurements north of 66° N) and in terms of variance due to a deterioration of the sampling of the ocean. These elements have been improved in the SL_cci v2.0 ECV since new altimeter missions (SARAL/AltiKa and CryoSat-2) have been additionally included in the production system, covering the period January 1993 to December 2015 (see Fig. 1 of Quartly et al., 2017).

The ESA CCI objectives put strong emphasis on developing homogeneous datasets with long-term consistency, which

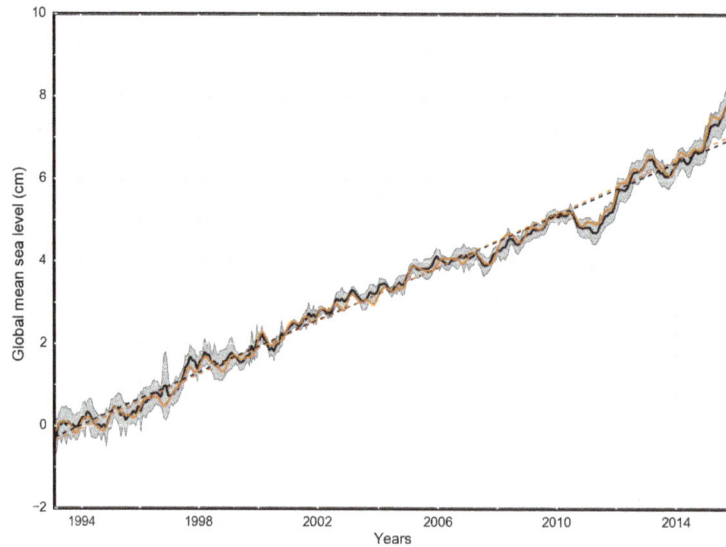

Figure 1. Comparison of the SL_cci v2.0 global MSL (solid orange line) with the associated linear trend (dashed orange line) and the ensemble mean (solid black line) of the global MSL derived from different groups (DUACS DT2014, CSIRO, Colorado University, GSFC and NOAA) with the associated linear trend (dashed black line) during the period 1993–2015. During this period, the trend of the SL_cci global MSL amounts to 3.3 ± 0.5 mm yr^{-1} with a 90 % confidence interval. The grey envelope shows 1.65 SD of the ensemble mean (90 % confidence interval). The seasonal variations have been removed and an offset has been introduced so that the mean of the 1993 data is set to zero.

necessitates not only stability for the duration of individual altimeter missions, but also great care in minimizing bias between missions. This has significantly impacted the SL_cci project since sea level estimation from altimetry requires implementation of many different algorithms to provide corrections for orbits, atmospheric delays, tides and sea surface effects (e.g. Fu and Cazenave, 2000; Ablain et al., 2017; Escudier et al., 2017). Many different solutions have been adopted over the past decades for each altimeter standard, either developed by the SL_cci consortium or provided by external projects. To select the most appropriate algorithm to ensure homogeneity and stability of the sea level product, the SL_cci project held an Algorithm Selection Meeting in November 2015, during which the latest algorithms were independently evaluated and validated according to a formal validation protocol. The associated round robin data packages (RRDPs) showing the impact of each standard and the output of the meeting are available on the SL_cci website (http://www.esa-sealevel-cci.org/PublicDocuments) (Legeais and the ESA SL_cci team, 2016c), with a synopsis of the comparisons given in Quartly et al. (2017). Many of the applied corrections have been revised, in particular modelled orbits due to time-variable gravity (Rudenko et al., 2014, 2016; Couhert et al., 2015), satellite attitude, macromodels and tropospheric correction models for DORIS observations (Rudenko et al., 2017), modifications to the wet tropospheric correction based on combined GNSS and radiometer datasets (Fernandes et al., 2015), and the latest changes in ocean tide (FES2014) and pole tide (Desai et al., 2015).

3 Description of the SL_cci v2.0 ECV

The SL_cci v2.0 products are based on multi-mission sea level measurements. They are provided as a database of different elements, referenced with the following DOI: https://doi.org/10.5270/esa-sea_level_cci-1993_2015-v_2.0-201612. The different products available for the users are the following:

1. The Fundamental Climate Data Record (FCDR) is the along-track sea level anomaly (SLA) derived from the nine altimeter missions, available at 1 Hz resolution corresponding to a ground distance of ~ 7 km. The files include a quality control indicator to remove spurious measurements and all altimeter standards applied in the SLA calculation (geophysical corrections, mean sea surface). In addition, information derived from the inter-mission sea level cross-calibration is provided in order to remove global and regional biases and to homogenize long-spatial-scale errors (e.g. due to orbit calculation). In addition to the FCDRs, gridded values of the altimeter dynamic atmospheric correction (DAC) forced by the ERA Interim reanalysis (Carrère et al., 2016) used for the production of the SL_cci products (including the SLA) are also available to the users. This may be of interest when comparing altimeter data with in situ measurements from tide gauges since both datasets should be corrected for the same atmospheric corrections.

2. The SL_cci ECV consists of monthly gridded time series (January 1993–December 2015) of multi-mission

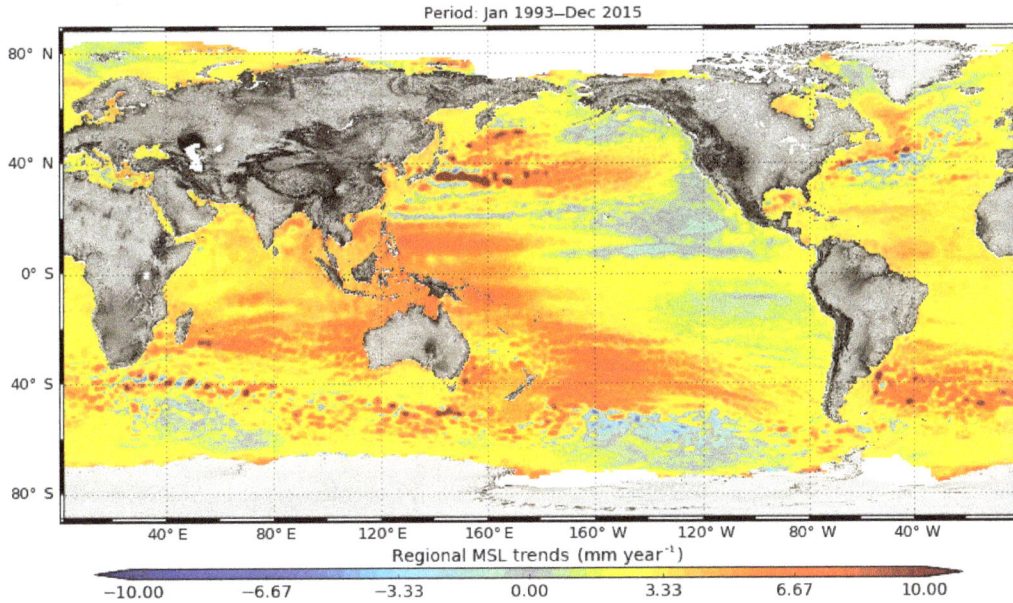

Figure 2. Regional MSL trend (from SL_cci v2.0 ECV) during the period 1993–2015.

merged SLA at a spatial resolution of 0.25°. The SL_cci v2.0 ECV has been generated thanks to the CNES/CLS DUACS production system with the same procedures as for the previous version v1.1 (Ablain et al., 2015) (except that the grids have been shifted by half a pixel in v2.0). The main processing steps (developed in Ablain and Legeais, 2014) are as follows: (i) acquire and pre-process data, (ii) perform input check and quality control, inter-calibrate and unify the multi-satellite measurements and (iii) generate along-track and gridded merged products (based on a monthly optimal interpolation). A land–sea mask derived from the LandCover_cci project has been applied to all sea level grids. The long-term stability and large-scale changes of the SL_cci v2.0 dataset are built upon the records from missions in the reference orbit (TOPEX/Poseidon, Jason-1 and Jason-2 for that period). All these satellites, called reference missions, have the same 9.92-day orbital cycle at high altitude (1336 km), making satellite trajectories less sensitive to higher-order terms of the Earth's gravity field. Data from the other missions (also called complementary missions) that contribute to improving the sampling of mesoscale processes provide the high-latitude coverage and increase the product accuracy. More details on the SL_cci ECV processing are provided in Quartly et al. (2017) and additional general information on the altimeter data processing can be found in Pujol et al. (2016).

3. Ocean indicators are derived from the SL_cci ECV: GMSL time series (Fig. 1), regional grids of sea level trends (Fig. 2), and maps of the amplitude and phase of the annual (Fig. 3) and semi-annual signals during the period available.

4. In addition to the SL_cci ECV, the along-track inter-calibrated sea level measurements of each mission (level 3 of the altimeter processing) are also available to users. The included information is the filtered and sub-sampled valid SLA, where long wavelength biases have been removed to make observed sea level measurements homogeneous and consistent between the nine altimeter missions. These data are the input measurements of the mapping procedure and can be used for data assimilation in ocean models, for instance.

5. Improving the quality of the Arctic sea level record has also been one of the key regional foci during the SL_cci project. This has led to two new Arctic sea level records available to the users. The CLS/PML product is based on improved waveform classification and retracking, applied on the Envisat and SARAL/AltiKa missions (Poisson et al., 2017). DTU/TUM proposed two versions of their Arctic sea level product, both derived from the ERS-1 and 2, Envisat, and CryoSat-2 missions: one is based on an empirical altimeter retracking and the second is based on the ALES + retracking (Passaro et al., 2017). Results of the validation and comparison of these products can be found in Carret et al. (2016).

4 Quality assessment on climate scales

The validation of the reprocessed SL_cci v2.0 ECV has been carried out over different spatial and temporal scales.

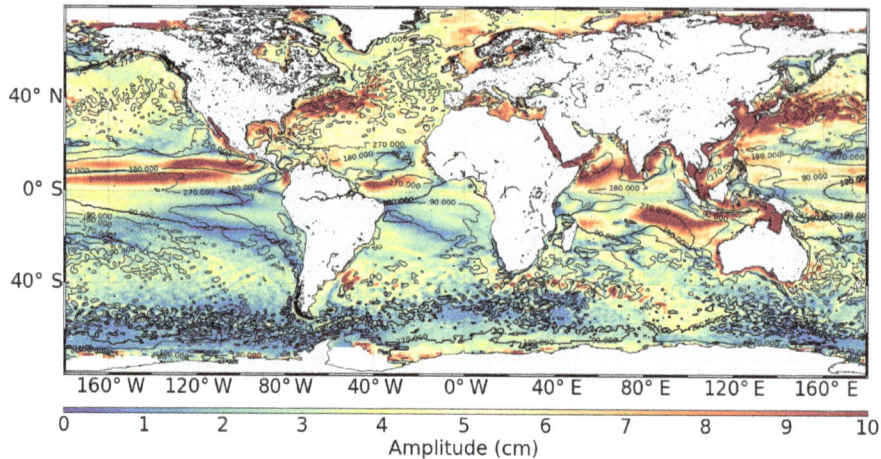

Figure 3. Global amplitude (coloured contours, between 0 and 10 cm) and phase (superimposed black isolines, between 0 and 360°) of the annual cycle of the SL_cci ECV v2.0 during 1993–2015.

4.1 Long-term GMSL evolution

The GMSL trend derived from the SL_cci ECV v2.0 during the period 1993–2015 amounts to $3.3 \pm 0.5 \, \text{mm yr}^{-1}$ with a confidence interval of 90 % (1.65 SD). The GMSL trend derived from the reprocessed dataset is the same as the one derived from the ensemble mean of GMSL from other altimeter groups (Fig. 1). When compared with the previous v1.1 ECV, no trend difference is observed during the common reduced period (1993–2014) (Fig. 4). However, over decadal timescales, the v2.0 GMSL trends are significantly different to those from v1.1 (by $-0.2 \, \text{mm yr}^{-1}$ during 1993–2003 and $+0.2 \, \text{mm yr}^{-1}$ during 2004–2014, see Fig. 4). This is mainly due to the use of the level 2 GNSS Path Delay Plus (GPD+) wet troposphere correction (Fernandes and Lázaro, 2016) for all missions in the v2.0 (except for GFO) (see Quartly et al., 2017 for more details). This is because all radiometers used in GPD+ V2.0 have been calibrated against the Special Sensor Microwave Imager and the Special Sensor Microwave Imager/Sounder, due to their known stability and independent calibration (Wentz, 2013).

For the v2.0 GMSL, the same trend of $3.3 \, \text{mm yr}^{-1}$ is found for the 1993–2003 and 2005–2015 altimetry decades, indicating a steady rise of the GMSL. However, several recent studies using different approaches suggest that an instrumental drift has affected the TOPEX-A altimeter measurements during 1993–1998 (Valladeau et al., 2012; Watson et al., 2015; Dieng et al., 2017; Beckley et al., 2017). The instrumental drift of the TOPEX-A altimeter has long been known (Hayne and Handcock, 1998), leading to the switch early 1999 to the redundant TOPEX-B altimeter. But until recently, it was considered that the TOPEX-A drift had minimal impact on the GMSL. Based on a comparison between TOPEX-A sea level and tide gauge data, Valladeau et al. (2012) challenged this conclusion but did not quantify this effect on the GMSL. More recently, three studies have

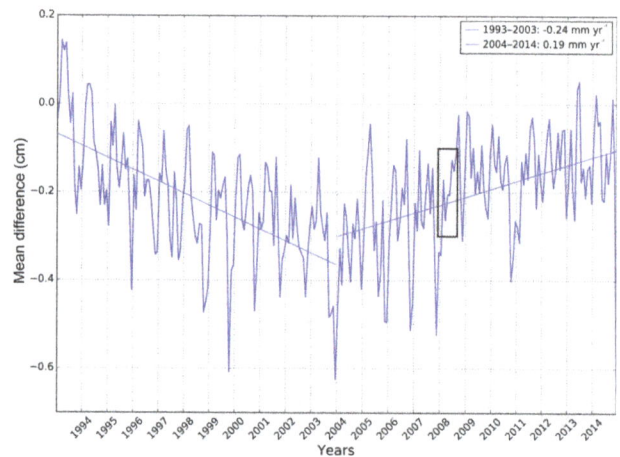

Figure 4. Global mean sea level differences between the SL_cci ECV v2.0 and v1.1. The trends are indicated for the periods 1993–2003 and 2004–2014. No trend difference is observed between ECV v1.1 and v2.0 during their common period 1993–2014 (not shown). A jump is observed in mid-2008, illustrating the anomaly of ECV v1.1 that has been corrected in ECV v2.0 (see the black box).

attempted to quantify the effect of the TOPEX-A drift on the GMSL trend over the period January 1993–December 1998. Watson et al. (2015) compared altimetry-based sea level with vertical land motion-corrected tide gauge data and estimated a TOPEX-A drift correction to the 1993–1998 GMSL trend in the range 0.9 ± 0.5 to $1.5 \pm 0.5 \, \text{mm yr}^{-1}$, with $1.5 \, \text{mm yr}^{-1}$ being the preferred value. Using a sea level budget approach, Dieng et al. (2017) also estimated the TOPEX-A drift correction to $1.5 \pm 0.5 \, \text{mm yr}^{-1}$ for 1993–1998. Another approach was followed by Beckley et al. (2017), consisting of suppressing the so-called "internal calibration-mode" range correction, included in the TOPEX-A "net instrument" correction and considered as suspect. Accounting for the TOPEX-A instrumental correction for the first 6 years of the altimetry dataset, these studies provided a revised GMSL time series that slightly reduces the average GMSL rise over the altimetry era (from 3.3 to 3.0 mm yr^{-1}), but shows clear acceleration over 1993–present. Using the corrected GMSL time series, Dieng et al. (2017) and Chen et al. (2017) found improved closure of the sea level budget compared to the uncorrected data.

In this paper, no TOPEX-A drift correction has been applied on the dataset available for the users as there is not yet consensus on the best approach to estimate it. However, ongoing work involving space agencies (National Aeronautics and Space Administration – NASA – and Centre National d' Etudes Spatiales – CNES) together with scientific teams should provide guidance and recommendations about this issue in the near future. As far as the SL_cci project is concerned, a corrected GMSL time series will be delivered to users in due time.

4.2 Inter-annual signals

The mean differences between the SL_cci v2.0 and SL_cci v1.1 are related to the different mean sea surface (MSS) used in both dataset (DTU15 and DTU10 respectively, see Quartly et al., 2017). On an inter-annual timescale, differences arise because of different mean reference periods used to compute the MSS (the period during which sea surface height measurements have been averaged). The reference period of the MSS DTU10 is 1993–2008 (15 years) whereas it is 1993–2012 (20 years) for the MSS DTU15. This is of major importance in the context of data assimilation in ocean models (Stammer and Griffies, 2017). Users interested in changing the reference period of the dataset can refer to the procedure described in annex A of Pujol et al. (2016). In addition to the reference period of the MSS, it is worth noting that a convention has been applied on the v2.0 sea level grids so that the averaged sea level during year 1993 is set to zero.

In the v1.1 SL_cci ECV, a 1 mm jump was found in the GMSL around mid-2008 (not shown). It is related to irregularities present in the Jason-1 radiometer data used to compute the GPD+ wet troposphere correction, which are enhanced near the coast using the methodology described in Brown (2010). The corresponding error was accounted for via a GMSL bias between Jason-1 and Jason-2 and was propagated to Jason-2 over the whole period. This error has been reduced in the v2.0 reprocessing and is therefore partly visible in the v2.0–v1.1 GMSL differences shown in Fig. 4.

The inter-annual variations of the SL_cci v2.0 GMSL remain in the envelope of the ensemble mean of the GMSL data from other groups (Fig. 1), which illustrates the homogeneity of the processing of the satellite measurements on these timescales. More validation details on inter-annual timescales are provided by comparison with other GMSL products (Sect. 5).

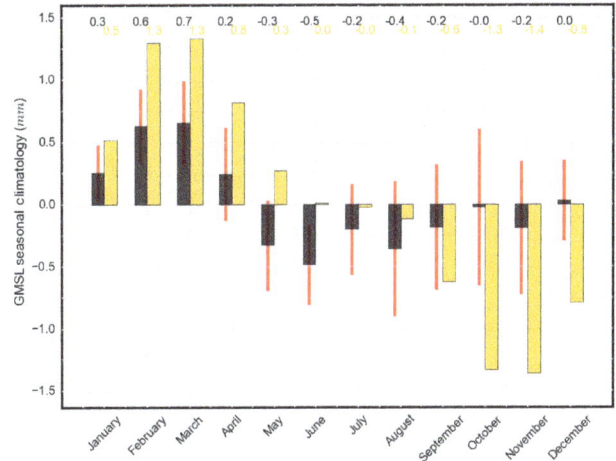

Figure 5. SL_cci v2.0 global MSL monthly climatology (yellow) compared with the ensemble mean (black) of the monthly climatology derived from different groups (DUACS DT2014, CSIRO, Colorado University, GSFC and NOAA); the red bars show the associated SD. The period considered for the monthly average is 1993–2015.

4.3 Seasonal cycle

The regional amplitude and phase of the annual cycle of the sea level ECV v2.0 are illustrated in Fig. 3. A number of factors affect the amplitude and phase of the mean seasonal cycle at a point: solar elevation, cloud cover, mixed layer depth and riverine input. The amplitude of the seasonal cycle is much greater in the Northern Hemisphere, especially for regions affected by the Gulf Stream and Kuroshio Current, whereas the deep waters associated with the Brazil–Falklands Confluence and the Agulhas Current show much weaker seasonality, with the expanse of the Antarctic Circumpolar Current showing little change. A more detailed analysis of the phase of the annual cycle (not shown) reveals that the North Atlantic and North Pacific north of 20° N generally peak in September–October, whereas the oceans south of 20° S peak in February–April. A much more complex pattern is found in the tropical belt, with our analysis confirming the work of Chen and Quartly (2005), who found October peaks for parts of the southern tropical Indian Ocean and for a region stretching from the Amazon to Africa. Chen and Quartly (2005) identified several amphidromic points in these regions, several of which are similarly located in the analysis of the longer SL_cci v2.0 dataset. The tropical Pacific remains hard to interpret because the long-term variations associated with ENSO may affect the "normal" seasonal cycle for that region. The assessment of the GMSL annual cycle from the v2.0 ECV is based on the monthly climatology during the period of the sea level record (see Fig. 5). The resulting signal displays a smoother annual cycle than the one derived from the ensemble mean of the monthly climatology derived from other products. Given the expected shape of the sea level annual cycle (Chen et al., 1998; Leg-eais et al., 2016a) and the long length of the record (that fil-ters out the potential peaks during this period), this suggests an improved estimate of the seasonal signal in the repro-cessed SL_cci ECV. Compared with the v1.1 ECV, a small difference is observed in terms of the amplitude (1 mm). It is assessed by comparison with tide gauge measurements (Val-ladeau et al., 2012). The amplitude of the annual cycle of the sea level difference computed against predominantly coastal in situ data reaches 2.4 mm with the v1.1 ECV and is reduced to 1.6 mm with the v2.0 ECV, suggesting that the annual sig-nal is better retrieved with the reprocessed dataset. This ob-served difference is related to the changes in the level 2 al-timeter standards involved in the ECV production, the main contributors being the orbit solutions (Couhert et al., 2015; Rudenko et al., 2017) and the GPD+ wet troposphere cor-rection (Fernandes and Lázaro, 2016) used for the different altimeter missions (Quartly et al., 2017). The new pole tide correction (Desai et al., 2015) also affects the amplitude of the annual cycle. As illustrated by the Taylor diagram (Tay-lor, 2001) in Fig. 6, the comparison with external indepen-dent data confirms that it leads to an improved sea level es-timation compared with the v1.1 ECV (Wahr, 1985). This figure compares the amplitude of the annual cycle of the En-visat and Jason-1 sea level computed with both pole tide cor-rections, with the sum of the steric dynamic height anomalies derived from the Argo in situ network and the GRACE ocean mass contribution (grey dot on the x axis). An increased cor-relation between both datasets and a reduced variance of the difference is obtained with the new correction.

4.4 Regional sea level trends

The regional sea level trends during 1993–2015 (Fig. 2) can deviate considerably from the global mean (typical val-ues range spatially between -5 and $+5$ mm yr^{-1} around the 3 mm yr^{-1} global estimate). Over this 23-year-long time span, this is essentially due to non-uniform thermal expan-sion (Stammer et al., 2013), in response to natural inter-nal climate variability (Meyssignac et al., 2012; Palanisamy et al., 2015a, b; Han et al., 2017). However, in some regions, like the Southern Ocean, an anthropogenically forced signal is already probably emerging. The regional sea level trends during 1993–2015 exhibit large-scale variations with regions of almost no sea level change and others with amplitudes reaching up to $+(8$–$10)$ mm yr^{-1} such as in the western trop-ical Pacific Ocean (Fig. 2). In this area, trends are mainly of thermosteric origin (Legeais et al., 2016b; Meyssignac et al., 2017) in response to increased easterly winds during the last 2 decades associated with the decreasing Interdecadal Pacific Oscillation (IPO)/Pacific Decadal Oscillation (e.g. Merrifield et al., 2012; Palanisamy et al., 2015a; Rietbroek et al., 2016).

The regional trend differences between the SL_cci v2.0 and v1.1 display values ranging between -1.5 and $+1.5$ mm yr^{-1} (Fig. 7). The large-scale differences are ex-

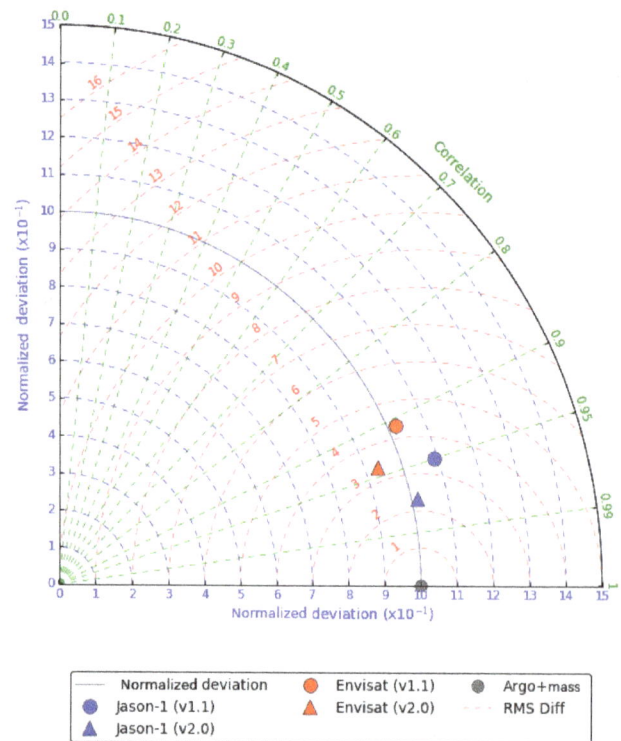

Figure 6. Taylor diagram of the annual signal of the Envisat (in red) and Jason-1 (in blue) sea level anomalies (2005–2012) calcu-lated considering the pole tide corrections derived both by Wahr (1985) (circle, used in ECV v1.1) and by Desai (2015) (triangle). They are compared with the independent sea level estimation (grey dot) derived from the in situ Argo dynamic heights anomalies (refer-enced to 900 dbar) and the GRACE ocean mass contribution (GRGS RL03v1).

plained by the differences of altimeter standards used in both versions, and the orbit solutions are the main contributor (see more details in Quartly et al., 2017). The small-scale differ-ences observed over the global ocean are related to the dif-ference in the satellite constellation between both versions of the ECV. Indeed, CryoSat-2 and SARAL/AltiKa missions are used after 2012 in v2.0 and were not included in v1.1. This means that sampling of the ocean is not the same in both datasets: the empty interleaved spaces between Jason-2 tracks in the v1.1 ECV have been sampled with CryoSat-2 and SARAL/AltiKa in the v2.0 ECV, which directly affects the trend differences, especially in the regions of high ocean variability.

4.5 Mesoscale signals

The SLA variance provides an estimate of the sea level vari-ability referenced to the mean sea surface used for the SLA calculation. The global SLA variance differences between SL_cci v2.0 and v1.1 time series are on average $+3$ cm^2 over the common period, indicating that more variability is ob-

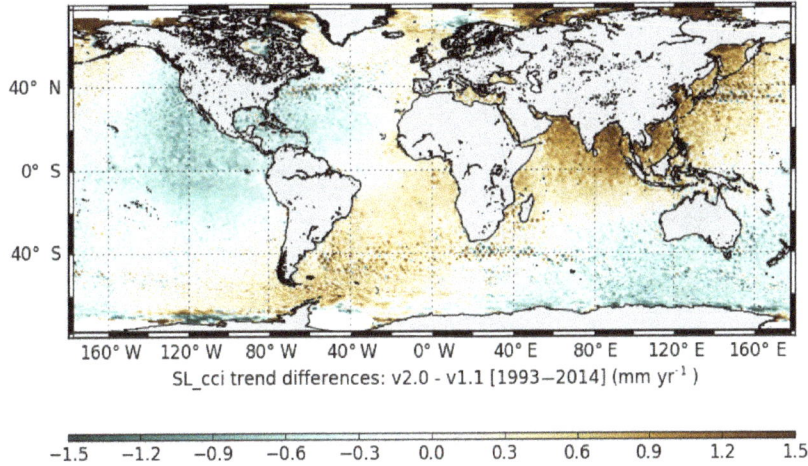

Figure 7. Regional mean sea level trend differences between the SL_cci ECV v2.0 and v1.1 during 1993–2014.

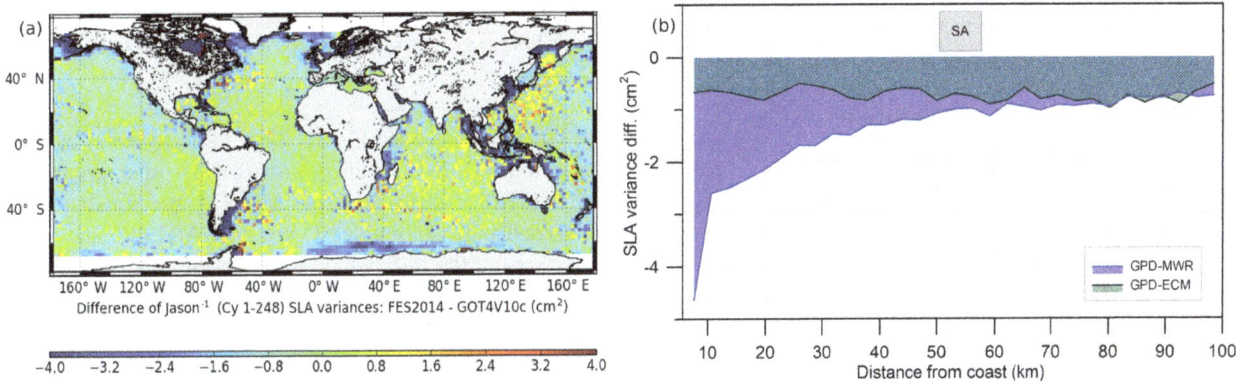

Figure 8. (a) Sea level variance differences for Jason-1 cycles 1-248 (2002–2008) using FES2014 and GOT4.10c ocean tide corrections successively. **(b)** SARAL/AltiKa sea level variance difference calculated with different wet troposphere corrections (green compares GPD+ with ECMWF operational model and purple with the initial version of the radiometer correction) as a function of the coastal distance.

served in the reprocessed ECV (even when removing the seasonal cycle). This change in the SLA variance is explained by several factors: inclusion of new missions (CryoSat-2 and SARAL/AltiKa) in the v2.0 ECV, leading to an improved mesoscale estimation and allowing better coverage of the ocean at high latitudes; use of the FES2014 ocean tide model (instead of the GOT4.8) for all altimeter missions, providing a reduced sea level variance in many coastal and shelf areas, as well as at high latitudes (Fig. 8a); and use of the GPD+ wet troposphere correction (Fernandes and Lázaro, 2016), leading to improved sea level variance estimation in coastal regions for most altimeter missions compared with the previous non-calibrated version (Fernandes et. al., 2015) and to other wet troposphere correction datasets (see Fig. 8b for the example of the SARAL/AltiKa mission). Finally, the updated sea-state bias correction used for some missions (such as Envisat) also contributes to better retrieval of mesoscale signals. The reader should refer to Quartly et al. (2017) for the details of the aforementioned corrections.

5 Sea level budget closure and comparison with model outputs

Different types of external validations of the SL_cci v2.0 products have been investigated. They are briefly described below.

5.1 Global mean sea level budget closure

Closure of the global mean sea level budget implies that

$$\Delta \text{GMSL}(t) = [\Delta M\text{Ocean}(t) + \Delta \text{SSL}(t)], \qquad (1)$$

where Δ means change of a given variable with time t; $\Delta M\text{Ocean}(t)$ and $\Delta \text{SSL}(t)$ are time-variable ocean mass and steric sea level components ($\text{SSL}(t)$ being the depth-integrated change in seawater density due to ocean temperature T and salinity S variations).

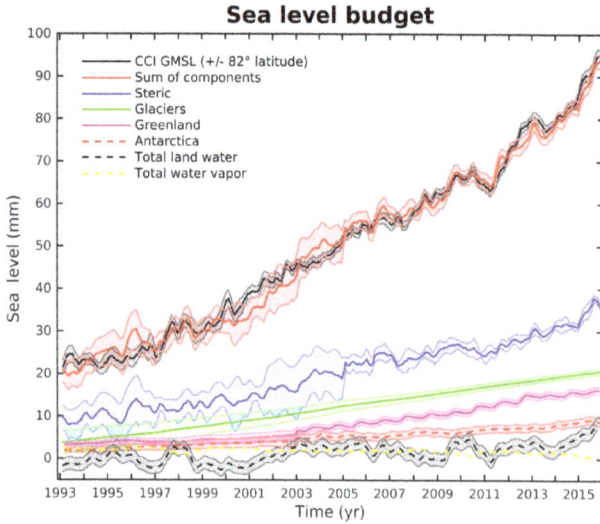

Figure 9. Sea level budget using the SL_cci v2.0 GMSL time series (adapted from Dieng et al., 2017). The $1.5\,\mathrm{mm\,yr^{-1}}$ correction supposed to represent the TOPEX-A drift has been applied (specifically in this figure) to the GMSL time series for 1993–1998 as discussed in Sect. 4.1 (see Dieng et al., 2017 for details).

Water mass conservation in the climate system implies the following:

$$\Delta M\mathrm{Ocean}(t) = -[\Delta M\mathrm{Glaciers}(t) + \Delta M\mathrm{Greenland}(t)$$
$$+ \Delta M\mathrm{Antarct.}(t) + \Delta M\mathrm{LWS}(t)$$
$$+ \Delta M\mathrm{AtmWV}(t)$$
$$+ \text{missing mass terms and errors}], \qquad (2)$$

where the $\Delta M(t)$ terms on the right-hand side refer to changes in mean glacier, Greenland and Antarctic mass balances, land water storage (LWS) and atmospheric water vapour (AtmWV).

We have investigated to what extent Eq. (1) is verified using the SL_cci v2.0 GMSL and different datasets for the ocean mass and steric components. The various contributions to the GMSL are summed to derive a "synthetic" global mean sea level. Consistency of the different products is evaluated and an error assessment study is performed. The synthetic global mean sea level is further compared with the global mean CCI sea level (both in terms of time series and trends).

We have considered the whole altimetry era (1993–2015) and used the various datasets considered in Dieng et al. (2017) to estimate the individual mass contributions (glaciers, ice sheets, stored land water, atmospheric water vapour and snow) of Eq. (1). Figure 9 shows the SL_cci v2.0 GMSL as well as the individual and sum of components over 1993–2015. The $1.5\,\mathrm{mm\,yr^{-1}}$ correction supposed to represent the TOPEX-A drift has been applied to the GMSL time series for 1993–1998 specifically in this figure as discussed in Sect. 4.1; see Dieng et al. (2017) for details.

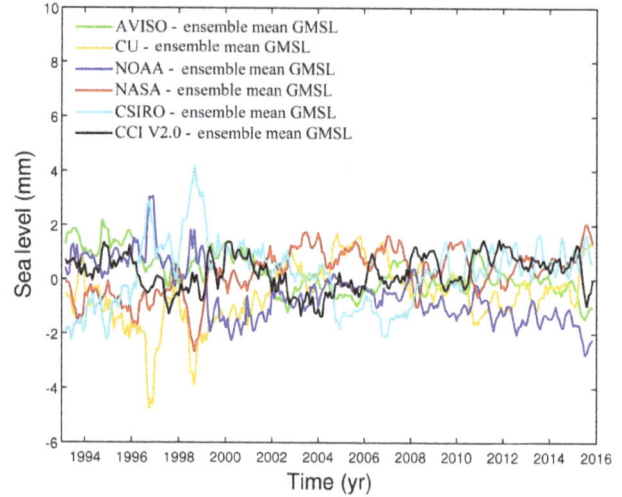

Figure 10. Differences between individual GMSL time series and ensemble mean GMSL (average of the 6 products) over January 1993–December 2015 (adapted from Dieng et al., 2017).

We have also used the SL_cci v2.0 time series as well as additional GMSL products provided by other groups (DU-ACS DT2014 distributed by CMEMS – previously AVISO – University of Colorado, NOAA, NASA/GSFC and CSIRO) to compute the differences between each individual GMSL time series and the ensemble mean GMSL over 1993–2015. Both residual trend and RMS have been estimated. These are gathered in Table 1. We note that in terms of trend, all GMSL time series are very close, residual trends ranging from -0.11 to $+0.08\,\mathrm{mm\,yr^{-1}}$ (all records used here are not corrected for the TOPEX-A instrumental drift). We conclude that looking solely at the trend does not allow us to provide significant assessment of the SL_cci v2.0 GMSL, although we note that the SL_cci v2.0 GMSL trend is one of the closest to the ensemble mean trend (difference of $0.02\,\mathrm{mm\,yr^{-1}}$). The situation looks more favourable in terms of RMS, the SL_cci v2.0 time series giving the smallest RMS of the differences (0.69 mm, Table 1 and Fig. 10). In addition, the residual GMSL obtained with the Colorado University and the CSIRO records are quite different to the ensemble mean during 1996–1999, which is not the case with the SL_cci dataset. This confirms the relatively good quality of the SL_cci v2.0 ECV on inter-annual timescales compared to other products. As the altimeter processing performed by the different groups are not the same, this inter-comparison highlights the contribution of the data processing to the GMSL uncertainty.

5.2 Comparison with ocean reanalyses

Assessment of the SL_cci v2.0 ECV has been carried out via multi-model approach by comparing with ocean reanalyses (ORA hereafter) from ECMWF. The reference ocean re-

Table 1. Trend and root mean square of the difference between individual GMSL time series and ensemble mean GMSL (mean of observed GMSL). Note that all GMSL records are not corrected for the TOPEX-A instrumental drift (see text).

	Trend (mm yr^{-1}) Jan 1993–Dec 2015	RMS (mm) Jan 1993–Dec 2015
AVISO – ensemble mean GMSL	−0.07	0.77
CU – ensemble mean GMSL	0.07	1.23
NOAA – ensemble mean GMSL	−0.11	1.13
NASA – ensemble mean GMSL	0.08	0.91
CSIRO – ensemble mean GMSL	0.02	1.22
CCI (v2.0) – ensemble mean GMSL	0.02	0.69

analysis product from ECMWF is the new ORAS5, which is closely related to the ORAP5 system (see Zuo et al., 2014 and Tietsche et al., 2015). ORAS5 is produced using the NEMO Ocean Model coupled to LIM2 sea ice model (Barnier et al., 2006). A series of observation types were assimilated in ORAS5 using NEMOVAR Ocean data assimilation system in its 3DVar FGAT (first guess at appropriate time) approach. Observations assimilated in ORAS5 include ensemble (EN4) in situ profiles, SLA from DUACS DT2014, sea surface temperature (SST) from HadSST2 and sea ice concentration (SIC) from OSTIA. It is worth noting that radar altimetry SLAs were not assimilated in ORAS5 outside of 50° S to 50° N domain, or in any coastal region with bathymetry less than 500 m. Altimeter-derived GMSL variations were also assimilated for the satellite era using a freshwater constraint in ORAS5 (see Zuo et al., 2015). A few other ORAs from ECMWF with slightly different configurations from ORAS5 (see Table 2) were also used here in order to estimate climate signal uncertainties. The gridded SLA maps (MSLA) from the SL_cci v2.0 ECV and ORAs were interpolated onto the same regular 1° × 1° grid with an optimized land–sea mask to facilitate inter-comparison. It is worth noting that the v1.1 ECV suffers from its imperfect land–sea mask, which makes the estimation of GMSL (and its seasonal cycles) non-trivial. As mentioned earlier, an adequate land–sea mask has been used in ECV v2.0.

Regional maps of MSL trends from the SL_cci v2.0 ECV were evaluated against the DUACS DT2014 and ORAs, with the results shown in Fig. 11a. Thanks to the inclusion of two additional altimetry missions (CryoSat-2 and SARAL/AltiKa), ECV v2.0 shows improved data coverage in the Arctic regions compared with ECV1.1, and a more pronounced positive trend in the Beaufort Sea. This is consistent with the ORAs and DUACS DT2014 results. This positive trend is visible both in ORAS5 and ORAS5-LW with relatively low uncertainty (± 1.5 mm yr^{-1}), suggesting a robust climate signal in the western Arctic Ocean (see Giles et al., 2012). The spatial patterns of large uncertainties are reasonably consistent between SL_cci ECVs and ORAS5 (Fig. 11b), considering that these uncertainties were estimated following very different approaches. The sea level

Table 2. Summary of the ORAs used for the SL_cci v2.0 ECV evaluation. DUACS DT2014 MSLA (Pujol et al., 2016) is now distributed by CMEMS (previously AVISO). ORAS5 is the ECMWF 0.25° resolution ocean–sea ice reanalysis; ORAS5-LW is the ORAS5-equivalent low-resolution (ORCA1 grid, with approximately 1° resolution, with meridional refinement at the Equator) reanalysis. Both ORAS5 and ORAS5-LW have five ensemble members, generated by a generic perturbation scheme (Zuo et al., 2017). SST stands for sea surface temperature, SIC for sea ice concentration and SLA for sea level anomaly.

Description	Resolution	Assimilation	Period
DUACS DT2014	0.25° × 0.25°	–	1993–2015
SL_cci ECV v1.1	0.25° × 0.25°	–	1993–2014
SL_cci ECV v2.0	0.25° × 0.25°	–	1993–2015
ORAS5	0.25° × 0.25°	SST, SIC, T, S, SLA	1993–2015
ORAS5-LW	1° × 1°	SST, SIC, T, S, SLA	1975–2015

trend uncertainty in ECMWF ORAs is due to observation representativeness errors and forcing analysis errors in the ECMWF ocean data assimilation system (Zuo et al., 2017), while sea level trend errors from SL_cci are only associated with the formal error adjustment of the trends and are thus not representative of the total regional altimeter MSL trend uncertainty (see Sect. 6 for more details). Areas with large errors are normally associated with strong mesoscale eddy activities. Moderate sea level trend uncertainties (~ 1.2 mm yr^{-1}) were also observed in the tropical Pacific and southern Indian Ocean for ECV1.1 and ECV v2.0. Compared with the SL_cci ECVs, ORAS5 is overconfident in its MSLA changes at most tropical and subtropical regions, but less confident in the Southern Ocean. An attribution study from ORAS5 reanalysis suggests that the mean sea level trend is dominated by the steric term while the mass variations are only important when considering the coastal regions (Fig. 11c). The increase in the sea level in the Beaufort Gyre is almost entirely due to halo-steric changes (steric changes due to salinity variations), which is consistent with the changes of Arctic circulation in the Beaufort Gyre (see the next paragraph) and the recent increase of freshwater there (Giles et al., 2012). The sharp front and reversing sea level trends signals in the North Atlantic suggest that the

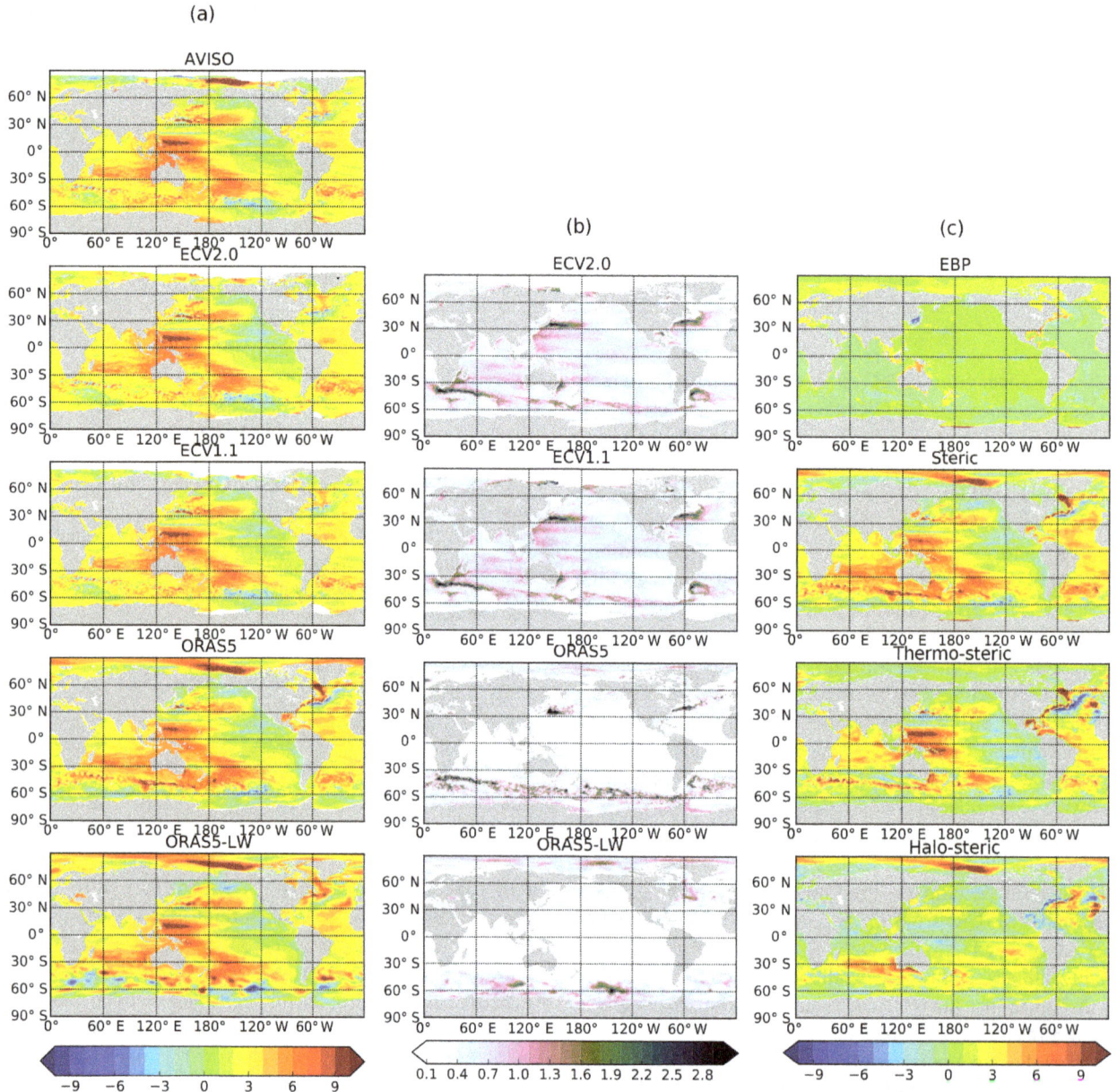

Figure 11. (a) Mean sea level trends (in $mm\,yr^{-1}$) and **(b)** uncertainties in DUACS DT2014, SL_cci ECV v2.0 and v1.1, ORAS5, and ORAS5-LW. MSL trends are computed using monthly mean sea level data from 1993–2014. Trend uncertainties have been provided as climate indicators with the ECV products and were estimated using ensemble spread from five ensemble members of ECMWF ORAs; **(c)** attributions of the MSL trends derived from ORAS5 as, from top to bottom, equivalent bottom pressure (EBP) mass variations, steric changes, and thermo-steric and halo-steric changes for the same period.

pathway of Gulf Stream extensions may be misrepresented in ORAS5, which is a common issue in ocean reanalysis.

5.3 Comparison with the TOPAZ and NorESM models in the Arctic region

The SL_cci v2.0 products for the high-latitude seas and Arctic Ocean are also compared with and assessed against complementary sea level fields derived from the TOPAZ data as-

similation system and the Norwegian Earth System Model (NorESM) for the period 1993–2016. In the Sub-Polar Gyre the models and observations show smooth seasonal variability with comparable amplitudes of around 5–7 cm (see Fig. 12a). In addition, a trend of just below $+3\,mm\,yr^{-1}$ is found for the observations and the NorESM simulation, while it is slightly less for the TOPAZ reanalysis. In the Lofoten Basin the comparison shows that the amplitude of seasonal signals is slightly larger (> 10 cm), while the trend in

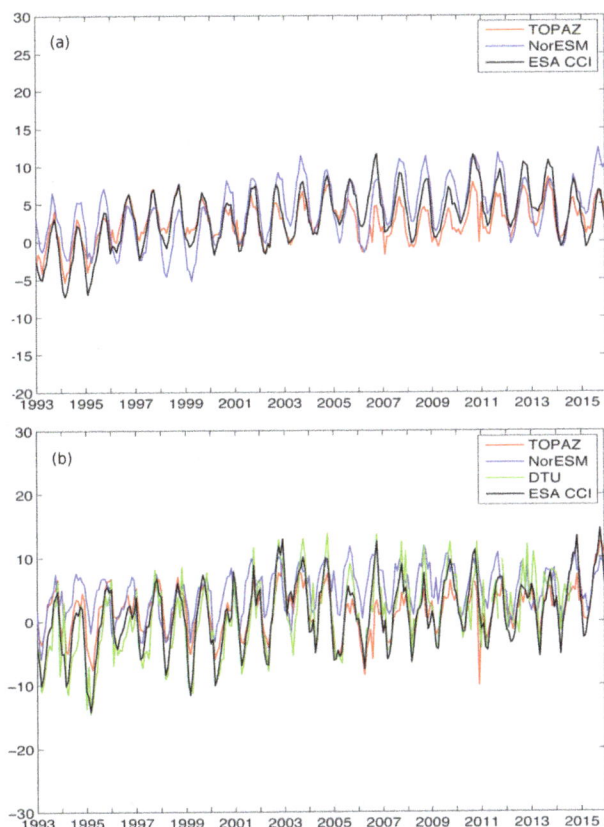

Figure 12. Seasonal to annual changes in sea level (mm yr^{-1}) for the period 1993–2016 for the Sub-Polar Gyre (a) and the Lofoten Basin (b). For the Lofoten Basin the DTU-based sea level change is also displayed for the comparison. For both plots the vertical axis is in centimetres.

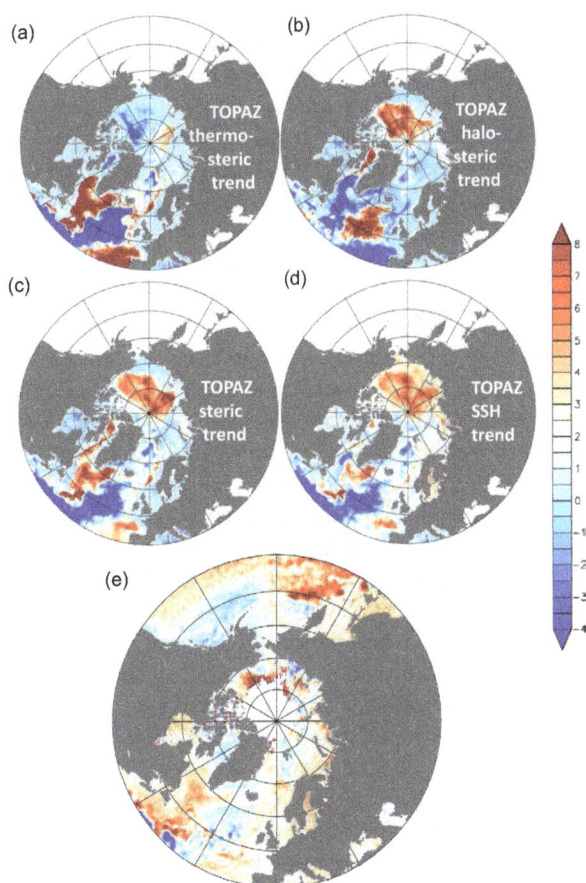

Figure 13. Contribution to the sea level trend (mm yr^{-1}) from TOPAZ4 reanalyses for the period 1993–2016 for (a) the thermosteric contribution, (b) the halosteric contribution, (c) the total steric trend and (d) the total trend. The observed trend from the ESA SL_cci v2.0 data (e).

the SL_cci v2.0 observations is about 3 mm yr^{-1} compared with about 1.5–2 mm yr^{-1} for the model fields. These differences may result from the spatial pattern of the trend in the sea level rise which is more confined in the model fields than in the observations (see also Fig. 13).

In Fig. 13 the TOPAZ4 reanalyses fields are shown for the thermosteric and halosteric trends (upper) and the steric and total trends (lower). As noted in the ECMWF comparison (Fig. 11), the thermosteric trend in TOPAZ4 has little influence on the steric and total trends in the Beaufort Gyre. In contrast the positive trend in the sea level rise appears to fully emerge from the halosteric trend except on the Siberian Shelf where it is most likely connected with the bottom pressure and hence related to water mass accumulation on the shelf. The apparent freshening of the Beaufort Gyre is consistent with findings reported by Morison et al. (2012), who proposed that this occurred as a result of persistent changes in the pathways of Arctic freshwater.

In comparison to the SL_cci v2.0 trends depicted in the Arctic Ocean, the TOPAZ4 sea surface height trends display more distinct regional structures of sea level rise and decline.

A positive trend of about 6–7 mm yr^{-1} is found in the Beaufort Gyre but appears to extend over a very large area towards the Siberian Shelf, which is not evident in the SL_cci v2.0 dataset. In the Nordic Seas and the Lofoten Basin, regions of both positive and negative trends stand out in contrast to the more gentle sea level rise expressed in the SL_cci v2.0 dataset. Moreover, the large and distinct region of strong decline in the sea level of up to 3–4 mm yr^{-1} encountered in the North Atlantic is not found in the SL_cci v2.0 dataset which only shows a very weak decline of around 1 mm yr^{-1}. Given the uncertainty of the regional sea level trends derived from altimetry and the model, this comparison will contribute to better characterize the ocean circulation changes in this region.

Looking at the Sub-Polar Gyre and the Lofoten Basin, however, one finds that the key contribution to the positive trends in steric and total signals emerge from the thermosteric trends. This is assumed to be related to an increased occupation by warm Atlantic water, which is further supported by evidence of corresponding negative halosteric trends that would be expected provided the source is the warm and saline water emerging from the northwestward extension of the Gulf Stream into the North Atlantic Current.

The steric (thermo-, halo-) trends observed in the sea level in the Sub-Polar Gyre have also been discussed by Hatun et al. (2005). They presumed that the variable dynamics of the Sub-Polar Gyre controlled the respective inflows of either cold and fresh sub-polar waters from the East Greenland Current or warm and salty subtropical waters from the Gulf Stream and its extension into the North Atlantic Current (NAC). Using salinity criteria to identify the respective sources of the water masses, they showed opposing transport variability of both source waters. Evidently, this closely mimicked a strong Sub-Polar Gyre when the cold and fresh water transport is strong and a weak Sub-Polar Gyre circulation when the warm and saline water transports dominate the inflow. This suggests a weakening of the anticlockwise circulation in the Sub-Polar Gyre during the last 20–25 years. In contrast, the distinct sea level rise encountered in the Beaufort Gyre during the same period has led to an intensification of the clockwise circulation in the gyre that may stimulate more trapping of fresh and cold Arctic surface water.

The SL_cci v2.0 products have provided new opportunities for studies of sea level changes. The assessment of these products for the high-latitude seas and the Arctic Ocean has focused on the Beaufort Gyre, the Sub-Polar Gyre and the Lofoten Basin in the Norwegian Sea. In so doing we have used the reanalyses from the TOPAZ4 operational system. The inter-comparison and assessment have documented interesting results and sometimes very good agreement and consistency between observations and models. In particular, the findings and achievements include distinct evidence of sea level rise of approximately the following values:

- 4–5 mm yr^{-1} for the inner part of the Sub-Polar Gyre explained by the thermosteric contribution together with a barotropic source;

- 6–7 mm yr^{-1} for the central Beaufort Gyre explained by the halosteric contribution and accumulation of fresh and cold Arctic water in the gyre;

- 3–4 mm yr^{-1} in the inner part of the Lofoten Basin assumed to result from the thermosteric contribution resulting from increased residence time of Atlantic Water in the basin.

Consistent with these findings, it is possible to conclude that a weakening of the anticlockwise circulation has occurred in the Sub-Polar Gyre during the last 20–25 years, while the sea level rise in Beaufort Gyre during the same period has led to an intensification of the clockwise circulation in the gyre with possible trapping of more freshwater.

5.4 Validation based on the GECCO model of the University of Hamburg

A further assessment of the quality of the different SL_cci ECVs (v1.1 and v2.0) is provided by comparison with the recent high-resolution GECCO2 ocean synthesis framework (Köhl, 2015). The GECCO2 assimilation approach uses the adjoint method to adjust uncertain model parameters to bring the model into consistency with ocean observations. In this way, the ocean state estimation ultimately leads to new estimates of the surface forcing fields that are required to simulate the observed ocean in the best possible way (given the model resolution and the model physics). The GECCO2 solution covers the period from 1948 to 2011 and had been optimized over 23 iterations. See Köhl (2015) for a description of the GECCO2 ocean state estimate and the datasets used as constraints. Starting from this already optimized state, two additional assimilation runs (G0 and G1.1) were performed as part of this study, all starting from iteration 23, carrying out 5 additional iterations. The only difference between both assimilation runs being the different SSH datasets used as constraints, G0 assimilated the AVISO SSH fields (SL0: DUACS DT2014 now distributed by CMEMS), whereas G1.1 assimilated the SSH fields from SL_cci v1.1 (SL1.1); see Scharffenberg et al. (2017).

Both daily mean GECCO2 synthesis results (G0 and G1.1) were interpolated onto the satellite tracks that matched the respective days for the respective along-track positions to be compared with DUACS DT2014 (SL0), SL_cci v1.1 (SL1.1) and SL_cci v2.0 (SL2.0) satellite datasets. In order to compensate for the scales that the GECCO2 solution is able to resolve, the satellite products SL0 and SL1.1 had to be filtered with an additional running mean filter of 9 points (f9), and SL2.0 with a running mean filter of 11 points (f11). The filter length was determined from the scales that GECCO2 manages to resolve in order to yield similar spectral characteristics of the respective signals, see Scharffenberg et al. (2017) for details. The model data comparisons have been performed separately for the ERS (ERS-1, ERS-2 and Envisat) and the TOPEX/Poseidon (TP) satellite series (TOPEX/Poseidon, Jason-1 and Jason2). A smaller difference between model and data residuals implies a better agreement between the GECCO2 model and the satellite datasets.

Figure 14 shows the ratios of RMS-difference-based skill score as defined in Scharffenberg et al. (2017), for the TP (left) and the ERS time series (right). The total improvement due to the updated satellite data SL1.1 and its assimilation into the GECCO2 synthesis can be revealed by the ratio of the differences in G0 and SL0, by using only the DUACS DT2014 dataset SL0, and of the differences in G1.1 and SL1.1 by using the previous updated SL1.1 dataset only. This ratio (G0_SL0) / (G1.1_SL1.1), as shown in Fig. 14a, highlights the reduction in the RMS differences for G1.1_SL1.1 in most regions of the world oceans, leading to an improve-

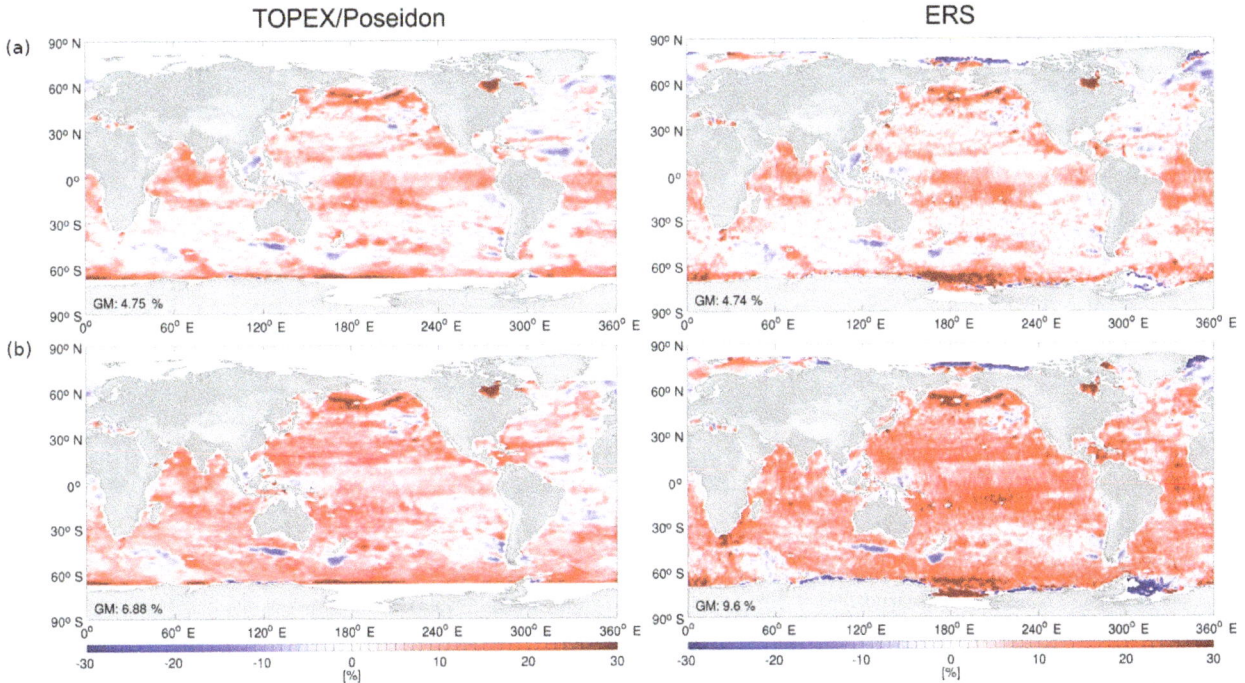

Figure 14. Ratio of RMS differences in low-pass-filtered (f9 or f11) data, for **(a)** the improvement from DUACS DT2014 (now distributed by CMEMS, previously AVISO) to SL_cci v1.1 as (G0_SL0)/(G1.1_SL1.1), and **(b)** the total improvement from DUACS DT2014 to SL_cci v2.0 as (G0_SL0)/(G1.1_SL2.0), for TP time series on the left and for ERS time series on the right.

ment (red) of more than 30 % in many regions (e.g. the equatorial regions, the Argentine shelf and parts of the ACC). Additional improvements can be seen in the northern Indian Ocean, the North Pacific, subtropical regions and large regions south of the ACC as well. Degradations of SL1.1 exist in isolated regions, where the GECCO2 synthesis adapts less well to the assimilated SL1.1 product. The regions showing a degradation (blue) match with regions of small SD (see Scharffenberg et al., 2017), implying that the assumption that the model serves as truth breaks down. The global mean improvement (between 66° N and 66° S) is 4.75 and 4.74 % for the TP and ERS datasets, respectively.

While the top panel gives the improvement from SL0 to SL1.1, the bottom panel answers the question about the total improvement from the DUACS DT2014 dataset to the latest SL_cci v2.0 ECV. Here, the ratio (G0_SL0) / (G1.1_SL2.0) compares the different assimilation runs G0 and G1.1 while calculating the RMS differences to SL0 and the latest SL_cci ECV SL2.0. The improvement of SL_cci v2.0 ECV has now a more homogenous distribution. Only isolated regions have larger RMS differences for G1.1_SL2.0, especially close to Antarctica as well as in the Arctic regions. The improvements for SL2.0 differ more between the TP and ERS datasets than was the case for SL1.1. Especially in the equatorial regions the ERS dataset has been improved. However, in most other parts of the world's ocean, both satellite datasets see a clear improvement from SL_cci v1.1

(SL1.1) to SL_cci v2.0 (SL2.0), especially in regions where SL1.1 did not improve much compared with SL0. The overall global mean improvement from DUACS DT2014 to SL_cci v2.0 ECV sums up to 6.88 % for the TP dataset and 9.60 % for the ERS dataset. As the GECCO2 synthesis had assimilated SL1.1 but not SL2.0, the GECCO2 synthesis results G1.1 are not expected to be in best agreement to SL2.0.

Furthermore, the GECCO2 synthesis itself benefits from the assimilation of the SL1.1 product as well as has been shown in Scharffenberg et al. (2017). Thereby, the SL_cci v1.1 and 2.0 ECVs, generated by the ESA SL_cci project, have been improved significantly and are now in closer agreement with the GECCO2 synthesis and the various global oceanographic datasets assimilated therein (Köhl, 2015). For a detailed description and assessment of both SL_cci ECVs, we refer to Scharffenberg et al. (2017) and to the SL_cci Climate Assessment Report (SL_cci CAR, 2017).

5.5 Regional sea level validation: agreement with ocean model outputs

The gridded SL_cci v2.0 products have also been intercompared and assessed in the Mediterranean Sea against the sea level fields derived from three regional ocean models for the period 1993–2016. Over the shorter period 2002–2014 the assessment includes the comparison of the model steric field with the steric sea level derived from the combination of altimetric and GRACE gravimetric observations.

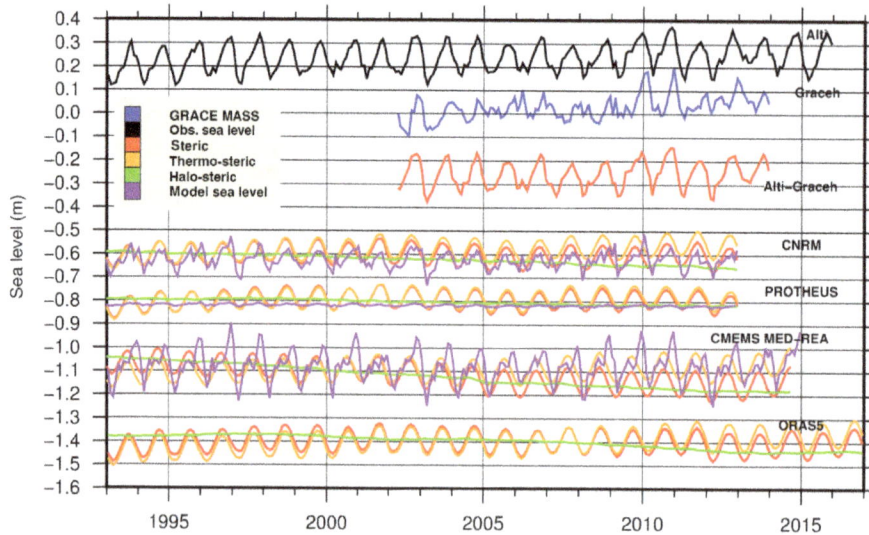

Figure 15. From top to bottom, basin average of altimeter sea level, GRACE mass change, steric derived from altimeter sea level and mass component, steric and sea level components of two simulations, and a regional model reanalysis and a global model reanalysis.

The two ocean simulations are the CNRM-RCM4 (Sevault et al., 2014) and the Protheus (Dell'Aquila et al., 2012); the ocean reanalysis is the CMEMS MEDSEA_REANALYSIS_PHIS_006_004, hereafter Med-MFC REA (http://marine.copernicus.eu). The global reanalysis ORAS5 is used as an additional comparison.

The first simulation CNRM-RCM4 is a fully coupled regional climate system model which includes a regional representation of the atmosphere, land surface, rivers and ocean. It is worth noting that the ocean NEMOMED8 model uses the "Boussinesq" approximation (Mellor and Ezer, 1995) and the relaxation of the sea surface height in the Atlantic buffer zone. The same approximation is used in the Protheus model simulation. The regional reanalysis CMEMS Med-MFC REA assimilates the DUACS DT2014 SLA (now distributed by CMEMS, previously AVISO).

The model output elevation sea surface height (SSH) and steric components are compared with the observed sea level and to its steric component. The gridded SLA maps from the SL_cci v2.0 ECV and the elevation from the models were interpolated onto the same regular 0.25° × 0.25° grid to facilitate the intercomparison. The basin averages of the observed sea level, mass change and derived steric sea level component, as well as of the model sea level and steric components, are displayed in Fig. 15. For the models, which use the Boussinesq approximation, the total sea level is the sum of the model sea level and of the steric component basin average. On basin scales we find that the observed sea level agrees at best with the sum of thermo-steric and elevation basin average components. Results are shown for the CNRM model in Fig. 16.

Regional maps of MSL trends from SL_cci v2.0 ECV were evaluated against the SSH from ocean models with the

results shown in Fig. 17 for the Protheus simulation and the CMEMS Med-MFC reanalysis. After the subtraction of the average trends of 2.5 and 0.16 mm yr^{-1} from the regional ECV and CMEMS model maps (Fig. 17a and e), the trend anomalies from observations and reanalysis show very similar spatial values (Fig. 17b and f). The regional maps of steric, thermo-steric and halo-steric trends are similar in the ocean simulations, and differences between simulation and reanalysis are higher for the halo-steric component.

Basin averages of steric sea level from each model have been compared with the difference of measured total sea level and mass from GRACE. The seasonal amplitude of the model steric component is smaller than the satellite-derived steric component and the phase is in good agreement. In Fig. 18, annual basin averages of total sea level, mass-induced sea level and steric component, grouped at the top, middle and bottom of the figure respectively, obtained from observations and models, are represented (for model datasets, the thermo-steric component is used). A significant correlation can be observed between the mass-induced sea surface heights measured by GRACE and both the observed altimeter SSH and the sum of both SSH and thermo-steric model component (see Fenoglio-Marc et al., 2012). The model and the steric sea level derived from altimeter and gravimetric observations show a similar long-term variability and some differences to be further investigated.

In summary, in the Mediterranean Sea the main differences on the one hand between the ocean model outputs and on the other hand between the ocean model outputs and the observations are related to the halo-steric component, whose trends have high negative values.

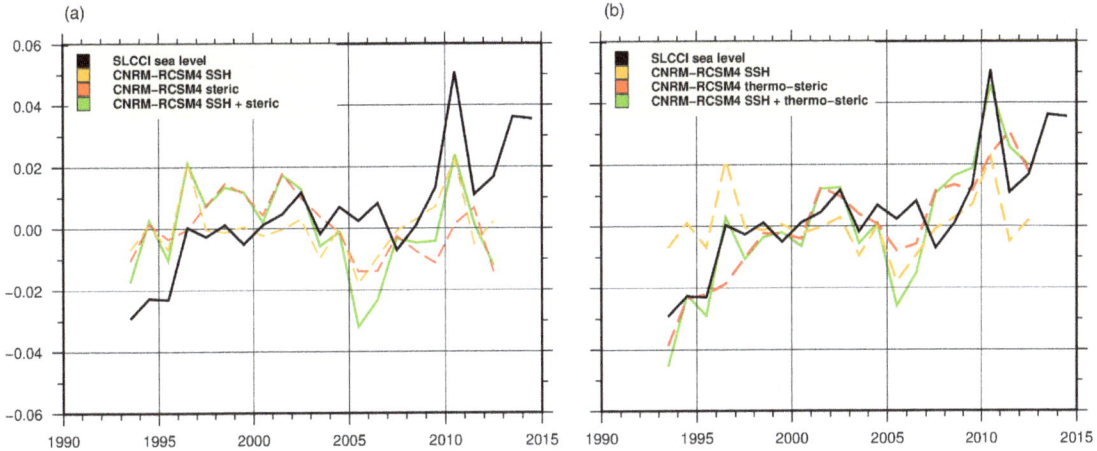

Figure 16. CNRM vs. SL_cci total sea level. **(a)** Sea surface height (green) from elevation plus steric compared with SL_cci (black), with elevation (orange) and steric (red) components. **(b)** As in **(a)** with thermo-steric instead of steric. For both plots, the vertical axis is in metres.

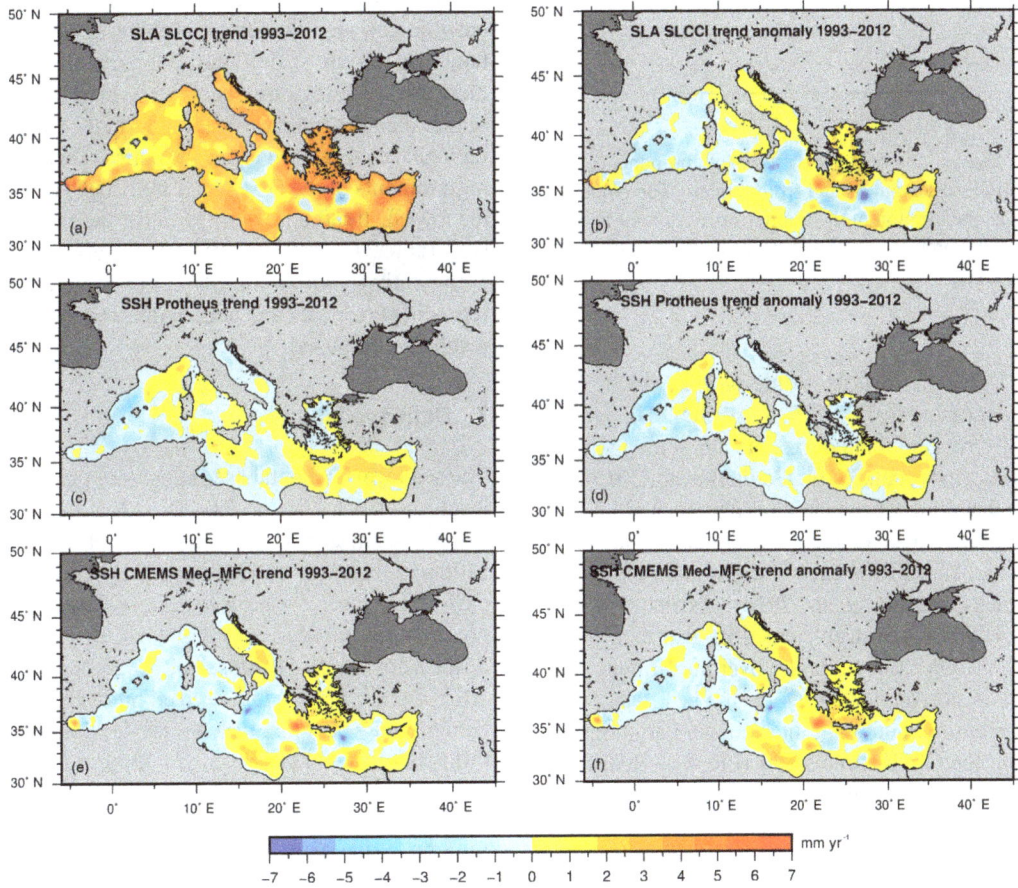

Figure 17. Trend **(a, c, e)** and trend anomalies **(b, d, f)** of SSH from SL_cci **(a, b)** and Protheus model simulation **(c, d)** and CMEMS reanalysis **(e, f)**.

6   MSL error characterization and uncertainties

Major efforts have been carried out during the past few years to provide a user-oriented error budget of the altimeter sea level estimation. Such an error budget dedicated to the main temporal scales (long-term, inter-annual and seasonal signals) has been established by Ablain et al. (2015) and is given in Table 3. The GMSL trend uncertainty has been estimated

Table 3. Mean sea level error budget for the main climate scales (Ablain et al., 2015). User requirements are from GCOS (2011).

Spatial scales	Temporal scales	Altimetry uncertainties	User requirements
Global MSL	Long-term evolution (> 10 years)	$< 0.5\,\mathrm{mm\,yr^{-1}}$	$0.3\,\mathrm{mm\,yr^{-1}}$
	Inter-annual signals (< 5 years)	$< 2\,\mathrm{mm}$ over 1 year	$0.5\,\mathrm{mm}$ over 1 year
	Annual signals	$< 1\,\mathrm{mm}$	Not defined
Regional sea level	Long-term evolution (> 10 years)	$< 3\,\mathrm{mm\,yr^{-1}}$	$1\,\mathrm{mm\,yr^{-1}}$
	Annual signals	$< 1\,\mathrm{cm}$	Not defined

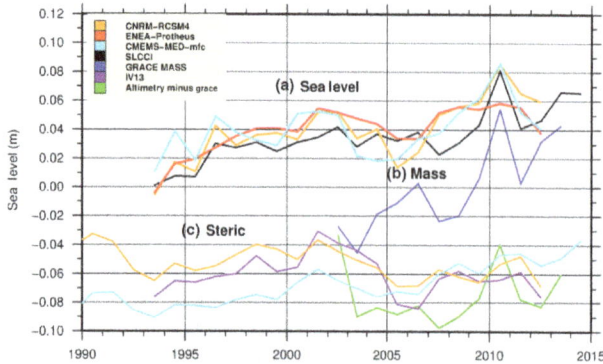

Figure 18. Sea level (a), mass (b) and steric (c) components from SL_cci and GRACE observations and from models. The colour indicates the source, the location in the figure (top, middle or bottom) indicates the type of product (sea level, mass or steric component for the model output).

as $0.5\,\mathrm{mm\,yr^{-1}}$ over the whole altimetry era (1993–2015) within a confidence interval of 90 % (1.65 SD). The associated sources of errors are related to some altimeter geophysical standards (Legeais et al., 2014; Couhert et al., 2015), the instabilities of the altimeter parameters (Ablain et al., 2012), the reference frame (Ablain et al., 2015; Couhert et al., 2015) and the issues associated with inter- intra-mission biases and inter-calibration (Zawadzki et al., 2016). As mentioned in Sect. 4.1, the TOPEX-A instrumental drift also contributes significantly to the GMSL uncertainty over the altimetry era. Note that there is not yet consensus on the best way to estimate this correction. Significant inter-annual variations are observed within the GMSL time series (Fig. 1) – mainly attributed to ENSO (Ablain et al., 2017) – and these contribute to the GMSL trend uncertainty in addition to all sources of errors described earlier (Cazenave et al., 2014). An uncertainty envelope for the GMSL has been proposed in order to better characterize inter-annual evolutions.

On a regional scale, the sea level trends vary between $\pm 5\,\mathrm{mm\,yr^{-1}}$ around the global mean of $+3\,\mathrm{mm\,yr^{-1}}$ and the associated uncertainty is of the order of 2–$3\,\mathrm{mm\,yr^{-1}}$. On a basin scale, two contributors to the altimeter trend uncertainty can be distinguished. The altimetry errors are one of the contributors. They can be related to the reduced quality of the altimeter sea level estimation in coastal areas (Cipollini

et al., 2017) and to the greater error of some geophysical altimeter corrections (ocean tide, inverse barometer and dynamic atmospheric corrections). The second contributor is related to the large internal variability of the observed ocean (and the fact that the associated trend may vary with the length of the record). The local variability is generated by regional changes in winds, pressure and ocean currents which averaged out on a global scale (e.g. Stammer et al., 2013) but this can significantly contribute to the sea level uncertainty on a basin scale. Thus, the users should be aware that there are some regions where the level of uncertainty may be higher than the trend estimations. At last, for both global and regional sea level trends, uncertainties remain higher than the requirements of the GCOS (GCOS, 2011) of $0.3\,\mathrm{mm\,yr^{-1}}$ for the GMSL trend and $1\,\mathrm{mm\,yr^{-1}}$ for the regional MSL trend (see Table 3). The MSL error characterization is still an ongoing activity within the SL_cci project, and further details regarding error bars and uncertainties will be published when results are finalized.

7 Data availability

The sea level CCI database (https://doi.org/10.5270/esa-sea_level_cci-1993_2015-v_2.0-201612; Legeais and the SL_cci team, 2016c) is freely available (upon email application to info-sealevel@esa-sealevel-cci.org). The sea level CCI website (http://www.esa-sealevel-cci.org/products) presents the gridded monthly files of sea level anomalies (https://doi.org/10.5270/esa-sea_level_cci-MSLA-1993_2015-v_2.0-201612) and derived products suitable for some climate studies: global mean sea level temporal evolution (https://doi.org/10.5270/esa-sea_level_cci-IND_MSL_MERGED-1993_2015-v_2.0-201612); regional mean sea level trends (https://doi.org/10.5270/esa-sea_level_cci-IND_MSLTR_MERGED-1993_2015-v_2.0-201612); amplitude and phase of annual cycle (https://doi.org/10.5270/esa-sea_level_cci-IND_MSLAMPH_MERGED-1993_2015-v_2.0-201612).

8 Conclusions and perspectives

The ESA Climate Change Initiative has provided the opportunity to realize the full potential of the long-term global Earth observations from satellite altimeters. This has led to the production of an improved and stable sea level record designed to answer the needs of climate modellers and researchers. The quality assessment of the SL_cci v2.0 ECV has been carried out, distinguishing different temporal and spatial wavelengths and following different approaches: comparisons with the previous version of the ECV and to altimeter products from other groups, sea level budget closure approach, and comparison with model outputs.

Compared with the previous v1.1 version of the SL_cci ECV (Ablain et al., 2015), the main observed differences are related to the updated v2.0 altimeter standards that have been selected within the SL_cci phase II project and used to calculate the altimeter sea level anomalies (see Quartly et al., 2017 for more details on these standards). One of the major differences between both versions is the increased number of altimeters available in the satellite constellation used for the SL_cci v2.0 ECV, compared with the v1.1 ECV. This has led to an improved sea level variance in the reprocessed ECV thanks to the improved sampling of the ocean at the end of the period. This highlights the importance for climate products of using a minimum of two satellites in the sea level ECV production and also to ensure that the number of such satellites remains stable in the constellation.

The different reference time periods used in both versions of the ECV are related to the different mean sea surfaces used to compute the sea level anomalies: DTU10 in v1.1 is referenced to 1993–2008 whereas DTU15 in v2.0 is referenced to 1993–2012. This has to be taken into account in the context of data assimilation for ocean models.

On a global scale, the v2.0 sea level trend is the same as in the v1.1 ECV when considering the total altimeter period (3.3 mm yr^{-1}). However, the use of the new GNSS Path delay (GPD+) wet troposphere correction (Fernandes and Lázaro, 2016) significantly affects the trend on a decadal timescale (up to 0.2 mm yr^{-1} for each altimeter decade). This is of major importance for sea level budget closure studies which usually focus on the 2005 onwards period. The SL_cci GMSL has not been corrected for the TOPEX-A instrumental drift (recently highlighted by several studies). Even if different approaches have been proposed, there is not yet consensus on the best way to estimate this correction. The recommendation of the Ocean Surface Topography Science Team is to wait for the future release of a reprocessed TOPEX dataset (currently in progress by the space agencies). On the regional scale, up to ±1 mm yr^{-1} sea level trend differences are observed compared with the previous version of the ECV and the large-scale differences are associated with the updated orbit solutions used in the v2.0 ECV (Quartly et al., 2017).

Regarding the annual cycle of the sea level, a small difference of amplitude is observed between SL_cci v1.1 and v2.0. Comparisons with the in situ measurements from tide gauges and from the combination of the dynamic heights derived from temperature and salinity profiles of Argo floats and the GRACE ocean mass contribution indicate that the SL_cci reprocessed ECV is slightly closer to the in situ reference.

The comparisons with the sea level time series from other altimetry groups and budget closure studies have demonstrated the high quality of the reprocessed SL_cci sea level record.

During the project, the altimetry measurement errors and associated uncertainties have been better estimated by separating the main temporal and spatial scales (Ablain et al., 2015). An estimation of sea level uncertainties has highlighted that in some regions, errors are greater than the signal itself. This work will significantly contribute to increasing the accuracy of climate studies. It is worth noting that in spite of the improved altimeter standards used in the product, the GCOS user requirements (GCOS, 2011) are still not reached over some specific spatial and temporal scales (e.g. 0.5 mm yr^{-1} uncertainty during the 1993–2015 period in a 90 % confidence for the GMSL trend compared with the 0.3 mm yr^{-1} requirement).

The reprocessed SL_cci v2.0 ECV is thus the state-of-the-art sea level ECV available for climate studies. Following the end of the ESA SL_cci project in 2017, the operational production of the sea level ECV has been transferred to the European Copernicus Climate Change Service (C3S), which will set up the routine and sustained production of the ECV. However, a strong need to continue research and development for the sea level record has been identified. Possibilities for evolution include the improvement of the sea level estimation in coastal areas and in ice-covered regions, the better characterization of the sea level uncertainties, and the quality improvement of the altimeter observations. This will contribute to improving the quality of the sea level ECV and achieving the GCOS requirements (GCOS, 2016).

Author contributions. Phase 2 of the Sea Level CCI project was managed by JFL, who oversaw the production and validation of the SL_cci ECV v2.0. The initial draft of the paper was written by JFL with the contributions of GQ for Sect. 2, AC for Sect. 5.1, HZ for Sect. 5.2, JJ for Sect. 5.3, MS for Sect. 5.4 and LFM for Sect. 5.5. Other authors contributed through their revision of the text.

Competing interests. The authors declare that they have no conflict of interest.

Acknowledgements. The authors acknowledge the support of ESA in the frame of the Sea Level CCI project, launched and co-ordinated by technical officer Jérôme Benveniste. It was also made possible thanks to the support of CNES for several years with the use of the DUACS altimeter processing system. We would also like to thank all contributors to this project who have participated actively in the SL_cci project, with special recognition to Sabrina Mbajon Njiche and also Americo Ambrozio and Marco Restano for their support of ESA as well as for their diligent reviewing of all the documents and datasets produced by the SL_cci team.

Edited by: Giuseppe M.R. Manzella

References

Ablain, M. and Legeais, J.-F.: Data Processing Model for the SL_cci system, Ref. CLS-DOS-NT-13-248, nomenclature SLCCI-DPM-33, available at: http://www.esa-sealevel-cci.org/webfm_send/239 (last access: 7 February 2018), 2014.

Ablain, M., Larnicol, G., Faugere, Y., Cazenave, A., Meyssignac, B., Picot, N., and Benveniste, J.: Error Characterization of Altimetry Measurements at Climate Scales, in: Proceedings of the "20 Years of Progress in Radar Altimetry" Symposium, Venice, 24–29 September 2012, Italy, edited by: Benveniste, J. and Morrow, R., ESA Special Publication SP-710, available at: http://www.congrexprojects.com/docs/12c01_docs/20ypra_abstracts_12_08_27_v9.pdf (last access: 7 February 2018), 2012.

Ablain, M., Cazenave, A., Larnicol, G., Balmaseda, M., Cipollini, P., Faugère, Y., Fernandes, M. J., Henry, O., Johannessen, J. A., Knudsen, P., Andersen, O., Legeais, J., Meyssignac, B., Picot, N., Roca, M., Rudenko, S., Scharffenberg, M. G., Stammer, D., Timms, G., and Benveniste, J.: Improved sea level record over the satellite altimetry era (1993–2010) from the Climate Change Initiative project, Ocean Sci., 11, 67–82, https://doi.org/10.5194/os-11-67-2015, 2015.

Ablain, M., Legeais, J. F., Prandi, P., Fenoglio-Marc, L., Marcos, M., Benveniste, J., and Cazenave, A.: Satellite altimetry-based sea level at global and regional scales, Surv. Geophys., 38, 9–33, https://doi.org/10.1007/s10712-016-9389-8, 2017.

Barnier, B., Madec, G., Penduff, T., Molines, J.-M., Treguier, A.-M., Le Sommer, J., Beckmann, A., Biastoch, A., Böning, C., Dengg, J., Derval, C., Durand, E., Gulev, S., Remy, E., Talandier, C., Theetten, S., Maltrud, M., McClean, J., and De Cuevas, B.: Impact of partial steps and momentum advection schemes in a global ocean circulation model at eddy-permitting resolution, Ocean Dynam., 56, 543–567, https://doi.org/10.1007/s10236-009-0180-y, 2006.

Beckley, B. D., Callahan, P. S., Hancock III, D. W., Mitchum, G. T., and Ray, R. D.: On the "cal mode" correction to TOPEX satellite altimetry and its effect on the global mean sea-level time series, J. Geophys. Res.-Oceans, 122, 8371–8384, https://doi.org/10.1002/2017JC013090, 2017.

Bojinski, S., Verstraete, M., Peterson, T. C., Richter, C., Simmons, A., and Zemp, M.: The concept of essential climate variables in support of climate. research, applications, and policy, B. Am. Meteorol. Soc., 95, 1431–1443, https://doi.org/10.1175/BAMS-D-13-00047.1, 2014.

Brown, S.: A novel near-land radiometer wet path-delay retrieval algorithm: application to the Jason-2/OSTM Advanced Microwave Radiometer, IEEE T. Geosci. Remote., 48, 1986–1992, https://doi.org/10.1109/TGRS.2009.2037220, 2010.

Carrere, L., Faugère, Y., and Ablain, M.: Major improvement of altimetry sea level estimations using pressure-derived corrections based on ERA-Interim atmospheric reanalysis, Ocean Sci., 12, 825–842, https://doi.org/10.5194/os-12-825-2016, 2016.

Carret, A., Johannessen, J., Andersen, O., Ablain, M., Prandi, P., Blazquez, A., and Cazenave, A.: Arctic sea level during the altimetry era, Surv. Geophys., 38, 251–275, https://doi.org/10.1007/s10712-016-9390-2, 2016.

Cazenave, A., Dieng, H.-B., Meyssignac, B., von Schuckmann, K., Decharme, B., and Berthier, E.: The rate of sea-level rise, Nat. Clim. Change, 4, 358–361, https://doi.org/10.1038/NCLIMATE2159, 2014.

Chen, G. and Quartly, G. D.: Annual amphidromes: a common feature in the ocean?, IEEE Geosci. Remote S., 2, 425–429, https://doi.org/10.1109/LGRS.2005.854205, 2005.

Chen, J. L., Wilson, C. R., Chambers, D. P., Nerem, R. S., and Tapley, B. D.: Seasonal global water mass budget and mean sea level variations, Geophys. Res. Lett., 25, 1944–8007, https://doi.org/10.1029/98GL02754, 1998.

Chen, X., Zhang, X., Church, J. A., Watson, C. S., King, M. A., Monselesan, D., Legresy, B., and Harig, C.: The increasing rate of global mean sea level-rise during 1993–2014, Nat. Clim. Change, 7, 492–495, https://doi.org/10.1038/NCLIMATE3325, 2017.

Cipollini, P., Birol, F., Fernandes, M. J., Obligis, E., Passaro, M., Strub, P. T., Valladeau, G., Vignudelli, S., and Wilkin, J.: Satellite altimetry in coastal regions, in: Satellite Altimetry Over Oceans and Land Surfaces, edited by: Stammer, D. and Cazenave, A., CRC Press, Taylor & Francis, Boca Raton, 2017.

Couhert, A., Cerri, L., Legeais, J. F., Ablain, M., Zelensky, N. P., Haines, B. J., Lemoine, F. G., Bertiger, W. I., Desai, S. D., and Otten, M.: Towards the $1\,mm\,yr^{-1}$ stability of the radial orbit error at regional scales, Adv. Space Res., 55, 2–23, https://doi.org/10.1016/j.asr.2014.06.041, 2015.

Dell'Aquila, A., Calmanti, S., Ruti, P., Struglia, M. V., Pisacane, G., Carillo, A., and Sannino, G.: Impacts of seasonal cycle fluctuations in an A1B scenario over the Euro–Mediterranean region, Clim. Res., 52, 135–157, https://doi.org/10.3354/cr01037, 2012.

Desai, S., Wahr, J., and Beckley, B.: Revisiting the pole tide for and from satellite altimetry, J. Geodesy, 89, 1233–1243, https://doi.org/10.1007/s00190-015-0848-7, 2015.

Dieng, H. B., Cazenave, A., Meyssignac, B., and Ablain, M.: New estimate of the current rate of sea level rise from a sea level budget approach, Geophys. Res. Lett., 44, 3744–3751, https://doi.org/10.1002/2017GL073308, 2017.

Escudier, P., Couhert, A., Mercier, F., Mallet, A., Thibaut, P., Tran, N., Amarouche, L., Picard, B., Carrère, L., Dibarboure, G., Ablain, M., Richard, J., Steunou, N., Dubois, P., Rio, M. H., and Dorandeu, J.: Satellite radar altimetry: principle, geophysical correction and orbit, accuracy and precision, in: Satellite Altimetry Over Oceans and Land Surfaces, edited by: Stammer, D. and Cazenave, A., CRC Press, Taylor & Francis, Boca Raton, 2017.

Fenoglio-Marc, L., Becker, M., Rietbroeck, R., Kusche, J., Grayek, S., and Stanev, E.: Water mass variation in Mediterranean and Black Sea, J. Geodyn., 59–60, 168–182, https://doi.org/10.1016/j.jog.2012.04.001, 2012.

Fernandes, M. J. and Lázaro, C.: GPD+ wet tropospheric corrections for CryoSat-2 and GFO altimetry missions, Remote Sens.-Basel, 8, 851–881, https://doi.org/10.3390/rs8100851, 2016.

Fernandes, M. J., Lázaro, C., Ablain, M., and Pires, N.: Improved wet path delays for all ESA and reference altimetric missions, Remote Sens. Environ., 169, 50–74, https://doi.org/10.1016/j.rse.2015.07.023, 2015.

Fu, L.-L. and Cazenave, A. (Eds.): Satellite altimetry and Earth sciences: A handbook of techniques and applications, Academic Press, San Diego, 463 pp., 2000.

GCOS: Systematic Observation Requirements for Satellite-Based Data Products for Climate (2011 Update) – Supplemental Details to the Satellite-Based Component of the "Implementation Plan for the Global Observing System for Climate in Support of the UNFCCC (2010 Update)", GCOS-154, WMO, Geneva, Switzerland, December 2011.

GCOS: Global Climate Observing System Implementation Plan 2016, GCOS-200, available at: https://library.wmo.int/opac/doc_num.php?explnum_id=3417 (last access: 7 February 2018), 2016.

Giles, K., Laxon, S., Ridout, A., Wingham, D., and Bacon, S.: Western Arctic Ocean freshwater storage increased by wind-driven spin-up of the Beaufort Gyre, Nat. Geosci., 5 194–197, https://doi.org/10.1038/ngeo1379, 2012.

Han, W., Meehl, G. A., Stammer, D., Hu, A., Hamlington, B., Kenigson, J., Palanisamy, H., and Thompson, P.: Spatial patterns of sea level variability associated with natural internal climate modes, Surv. Geophys., 38, 217–250, https://doi.org/10.1007/s10712-016-9386-y, 2017.

Hatun, H., Britt Sando, A., Drange, H., Hansen, B., and Valdimarsson, H.: Influence of the Atlantic Subpolar Gyre on the Thermohaline Circulation, Science, 309, 1841–1844, https://doi.org/10.1126/science.1114777, 2005.

Hayne, G. S. and Handcock, D. W.: Proceedings of the TOPEX/Poseidon/Jason-1 Science Working Team Meeting, October 1998, Keystone, CO, USA, 1998.

IPCC: The Physical Science Basis, in: Contribution of the Working Group I to the Fifth Assessment report of the Intergovernmental Panel on Climate Change, Cambridge University Press, Cambridge, UK, p. 1535, 2013.

Köhl, A.: Evaluation of the GECCO2 ocean synthesis: transports of volume, heat and freshwater in the Atlantic, Q. J. Roy. Meteorol. Soc., 141, 166–181, https://doi.org/10.1002/qj.2347, 2015.

Legeais, J.-F., Ablain, M., and Thao, S.: Evaluation of wet troposphere path delays from atmospheric reanalyses and radiometers and their impact on the altimeter sea level, Ocean Sci., 10, 893–905, https://doi.org/10.5194/os-10-893-2014, 2014.

Legeais, J.-F., Prandi, P., and Guinehut, S.: Analyses of altimetry errors using Argo and GRACE data, Ocean Sci., 12, 647–662, https://doi.org/10.5194/os-12-647-2016, 2016a.

Legeais, J.-F., von Schuckmann, K., Dagneaux, Q., Melet, A., Meyssignac, B., Bonaduce, A., Ablain, M., and Pérez Gomez, B.: Sea level, in: von Schuckmann, K., et al., The Copernicus Marine Environment Monitoring Service Ocean State Report, edited by: Taylor and Francis, http://www.tandfonline.com/loi/tjoo20, J. Oper. Oceanogr., 9, s235–s320, https://doi.org/10.1080/1755876X.2016.1273446, 2016b.

Legeais J.-F. and the ESA SL_cci team: the ESA SL_cci ECV v2.0, 10.5270/esa-sea_level_cci-1993_2015-v_2.0-201612 (last access: 7 February 2018), 2016c.

Mellor, G. L. and Ezer, T.: Sea level variations induced by heating and cooling: an evaluation of the Boussinesq approximation in ocean models, J. Geophys. Res., 100, 20565–20577, 1995.

Merrifield, M. A., Thompson, P. R., and Lander, M.: Multidecadal sea level anomalies and trends in the western tropical Pacific, Geophys. Res. Lett., 39, L13602, https://doi.org/10.1029/2012GL052032, 2012.

Meyssignac, B., Salas y Melia, D., Becker, M., Llovel, W., and Cazenave, A.: Tropical Pacific spatial trend patterns in observed sea level: internal variability and/or anthropogenic signature?, Clim. Past, 8, 787–802, https://doi.org/10.5194/cp-8-787-2012, 2012.

Meyssignac, B., Piecuch, C. G., Merchant, C. J., Racault, M.-F., Palanisamy, H., MacIntosh, C., Sathyendranath, S., and Brewin, R.: Causes of the regional variability in observed sea level, sea surface temperature and ocean colour over the period 1993–2011, Surv. Geophys., 38, 191–219, https://doi.org/10.1007/978-3-319-56490-6_9, 2017.

Morison, J., Kwok, R., Peralta-Ferriz, C., Alkire, M., Rigor, I., Andersen, R., and Steele, M.: Changing Arctic Ocean freshwater pathways, Nature, 481, 66–70, https://doi.org/10.1038/nature10705, 2012.

Palanisamy, H., Cazenave, A., Delcroix, T., and Meyssignac, B.: Spatial trend patterns in Pacific Ocean sea-level during the altimetry era: the contribution of thermocline depth change and internal climate variability, Ocean Dynam., 65, 341–356, https://doi.org/10.1007/s10236-014-0805-7, 2015a.

Palanisamy, H., Meyssignac, B., Cazenave, A., and Delcroix, T.: Is anthropogenic sea level fingerprint already detectable in the Pacific Ocean?, Environ. Res. Lett., 10, 084024, https://doi.org/10.1088/1748-9326/10/8/084024, 2015b.

Passaro, M., Kildegaard Rose, S., Andersen, O. B., Boergensa, E., Calafat, F. M., Dettmering, D., and Benveniste, J.: ALES+: adapting a homogenous ocean retracker for satellite altimetry to sea ice leads, coastal and inland waters, Remote Sens. Environ., in review, 2017.

Poisson, J.-C., Quartly, G., Kurekin, A., Thibaut, P., Hoang, D., and Nencioli, F.: Development of an ENVISAT altimetry processor providing sea level continuity between open ocean and Arctic leads, IEEE Geosci. Remote S., accepted, 2017.

Pujol, M.-I., Faugère, Y., Taburet, G., Dupuy, S., Pelloquin, C., Ablain, M., and Picot, N.: DUACS DT2014: the new multimission altimeter data set reprocessed over 20 years, Ocean Sci., 12, 1067–1090, https://doi.org/10.5194/os-12-1067-2016, 2016.

Quartly, G. D., Legeais, J.-F., Ablain, M., Zawadzki, L., Fernandes, M. J., Rudenko, S., Carrère, L., García, P. N., Cipollini, P., Andersen, O. B., Poisson, J.-C., Mbajon Njiche, S., Cazenave, A., and Benveniste, J.: A new phase in the production of quality-controlled sea level data, Earth Syst. Sci. Data, 9, 557–572, https://doi.org/10.5194/essd-9-557-2017, 2017.

Rietbroek, R., Brunnabend, S. E., Kusche, J., Schröter, J., and Dahle, C.: Revisiting the contemporary sea-level budget on global and regional scales, P. Natl. Acad. Sci. USA, 113, 1504–1509, https://doi.org/10.1073/pnas.1519132113, 2016.

Rudenko, S., Dettmering, D., Esselborn, S., Schöne, T., Förste, C., Lemoine, J.-M., Ablain, M., Alexandre, D., and Neumayer, K.-H.: Influence of time variable geopotential models on precise orbits of altimetry satellites, global and regional mean sea level trends, Adv. Space Res., 54, 92–118, https://doi.org/10.1016/j.asr.2014.03.010, 2014.

Rudenko, S., Dettmering, D., Esselborn, S., Fagiolini, E., and Schöne, T.: Impact of Atmospheric and Oceanic De-Aliasing Level-1B (AOD1B) products on precise orbits of altimetry satellites and altimetry results, Geophys. J. Int., 204, 1695–1702, https://doi.org/10.1093/gji/ggv545, 2016.

Rudenko, S., Neumayer, K.-H., Dettmering, D., Esselborn, S., Schöne, T., and Raimondo, J.-C.: Improvements in precise orbits of altimetry satellites and their impact on mean sea level monitoring, IEEE Geosci. Remote S., 55, 3382–3395, https://doi.org/10.1109/TGRS.2017.2670061, 2017.

Scharffenberg, M. G., Köhl, A., and Stammer, D.: Testing the quality of sea-level data using the GECCO Adjoint Assimilation Approach, Surv. Geophys., 38, 349–383, https://doi.org/10.1007/s10712-016-9401-3, 2017.

Sevault, F., Somot, S., Alias, A., Dubois, C., Lebeaupin-Brossier, C., Nabat, P., Adloff, F., Deque, M., and Decharme, B.: A fully coupled Mediterranean regional climate system model: design and evaluation of the ocean component for the 1980–2012 period, Tellus A, 66, 23967, https://doi.org/10.3402/tellusa.v66.23967, 2014.

SL_cci Climate Assessment Report (CAR): available at: http://www.esa-sealevel-cci.org/webfm_send/584, last access: 15 December 2017.

Stammer, D. and Griffies, S.: Ocean modeling and data assimilation in the context of satellite altimetry, in: Satellite Altimetry Over Oceans and Land Surfaces, edited by: Stammer, D. and Cazenave, A., CRC Press, Taylor & Francis, Boca Raton, 2017.

Stammer, D., Cazenave, A., Ponte, R. M., and Tamisiea, M. E.: Causes for contemporary regional sea-level changes, Annu. Rev. Mar. Sci., 5, 21–46, https://doi.org/10.1146/annurev-marine-121211-172406, 2013.

Taylor, K. E.: Summarizing multiple aspects of model performance in a single diagram, J. Geophys. Res., 106, 7183–7192, https://doi.org/10.1029/2000JD900719, 2001.

Tietsche, S., Balmaseda, M. A., Zuo, H., and Mogensen, K.: Arctic sea ice in the global eddy-permitting ocean reanalysis ORAP5,

Clim. Dynam., 1–15, https://doi.org/10.1007/s00382-015-2673-3, 2015.

Valladeau, G., Legeais, J. F., Ablain, M., Guinehut, S., and Picot, N.: Comparing altimetry with tide gauges and Argo profiling floats for data quality assessment and mean sea level studies, Mar. Geod., 35, 20–41, 2012.

von Schuckmann, K. and the CMEMS OSR task team: The Copernicus Marine Environment Monitoring Service Ocean State Report, J. Oper. Oceanogr., 9, s235–s320, https://doi.org/10.1080/1755876X.2016.1273446, 2016.

Wahr, J. W.: Deformation of the Earth induced by polar motion, J. Geophys. Res.-Sol. Ea., 90, 9363–9368, https://doi.org/10.1029/JB090iB11p09363, 1985.

Watson, C. S., White, N. J., Church, J. A., King, M. A., Burgette, R. J., and Legresy, B.: Unabated global mean sea level over the satellite altimeter era, Nat. Clim. Change, 5, 565–568, https://doi.org/10.1038/NCLIMATE2635, 2015.

Wentz, F. J.: SSM/I Version-7 Calibration Report, 011012, Remote Sensing Systems, Santa Rosa, CA, USA, p. 46, 11 January 2013.

Zawadzki, L. and Ablain, M.: Accuracy of the mean sea level continuous record with future altimetric missions: Jason-3 vs. Sentinel-3a, Ocean Sci., 12, 9–18, https://doi.org/10.5194/os-12-9-2016, 2016.

Zuo, H., Balmaseda, M. A., and Mogensen, K.: The ECMWF-MyOcean2 eddy-permitting ocean and sea-ice reanalysis ORAP5. Part 1: Implementation, ECMWF Tech Memo 736, ECMWF, Reading, England, 2014.

Zuo, H., Balmaseda, M. A., and Mogensen, K.: The new eddy-permitting ORAP5 ocean reanalysis: description, evaluation and uncertainties in climate signals, Clim. Dynam., 49, 791–811, https://doi.org/10.1007/s00382-015-2675-1, 2015.

Zuo, H., Balmaseda, M. A., Boisseson, E., Hirahara, S., Chrust, M., and de Rosnay, P.: A new ensemble generation scheme for ocean analysis, ECMWF Tech Memo 795, ECMWF, Reading, England, 2017.

5

Global ocean particulate organic carbon flux merged with satellite parameters

Colleen B. Mouw[1,a], **Audrey Barnett**[1,a], **Galen A. McKinley**[2], **Lucas Gloege**[2], **and Darren Pilcher**[3]

[1]Michigan Technological University, 1400 Townsend Drive, Houghton, MI 49931, USA
[2]University of Wisconsin-Madison, 1225 W. Dayton Street, Madison, WI 53706, USA
[3]NOAA, Pacific Marine Environmental Laboratory, 7600 Sand Point Way NE, Seattle, WA 98115, USA
[a]now at: University of Rhode Island, Graduate School of Oceanography, 215 South Ferry Road, Narragansett, RI 02882, USA

Correspondence to: Colleen B. Mouw (cmouw@uri.edu)

Abstract. Particulate organic carbon (POC) flux estimated from POC concentration observations from sediment traps and ^{234}Th are compiled across the global ocean. The compilation includes six time series locations: CARIACO, K2, OSP, BATS, OFP, and HOT. Efficiency of the biological pump of carbon to the deep ocean depends largely on biologically mediated export of carbon from the surface ocean and its remineralization with depth; thus biologically related parameters able to be estimated from satellite observations were merged at the POC observation sites. Satellite parameters include net primary production, percent microplankton, sea surface temperature, photosynthetically active radiation, diffuse attenuation coefficient at 490 nm, euphotic zone depth, and climatological mixed layer depth. Of the observations across the globe, 85 % are concentrated in the Northern Hemisphere with 44 % of the data record overlapping the satellite record. Time series sites accounted for 36 % of the data, while 71 % of the data are measured at $\geq 500\,\text{m}$ with the most common deployment depths between 1000 and 1500 m. This data set is valuable for investigations of CO_2 drawdown, carbon export, remineralization, and sequestration. The compiled data can be freely accessed at doi:10.1594/PANGAEA.855600.

1 Introduction

Field estimates of particulate organic carbon (POC) flux have been made over many decades in the interest of understanding the biological pump of carbon to the deep ocean. While there have been a variety of new techniques to quantify POC flux, sediment traps have been the most extensive temporally and geographically, and ^{234}Th has improved data resolution in the upper 500 m of the water column. POC flux depends largely on the biologically mediated export of carbon from the surface ocean and its remineralization with depth, thus capturing biological variables associated with POC flux are essential to understand flux variability. Here we compile POC flux estimated from sediment traps and ^{234}Th from around the globe from public repositories and directly in the literature. We then match the POC flux observations with biological and physical parameters determined from satellite

imagery along with mixed layer depth (MLD) climatology. See Table 1 for a list of products and units.

Understanding the impact of surface processes on the export of organic carbon at depth has been an ongoing challenge in the oceanographic community since the Joint Global Ocean Flux Study (JGOFS). Continued efforts with the upcoming Export Processes in the Ocean from RemoTe Sensing (EXPORTS) program along with the Pre-Aerosol, Clouds and ocean Ecosystem (PACE) satellite mission seek to connect remotely sensed estimates of net primary production, particle size distribution, phytoplankton carbon, biomass, and community composition to water column carbon processes. To do this, existing data sources capturing water column processes need to be compiled and synthesized. Our data set provides researchers with access to a comprehensive historical data set of POC flux throughout the global

Table 1. Summary of data set parameters.

Parameter	Units	Description
Satellite parameters:		
chl_gsm	$\mathrm{mg\,m^{-3}}$	Chlorophyll a concentration
kd490	$\mathrm{m^{-1}}$	Diffuse attenuation coefficient for 490 nm
par	$\mathrm{\mu mol\,quanta\,m^{-2}\,s^{-1}}$	Photosynthetically available radiation
pp_vgpm	$\mathrm{mg\,C\,m^{-2}\,d^{-1}}$	Net primary production
sfm	%	Microplankton fraction
sst	°C	Sea surface temperature
zeu	m	Base of the euphotic zone
In situ fluxes:		
al_flux	$\mathrm{\mu g\,m^{-2}\,d^{-1}}$	Flux of particulate aluminum
ba_flux	$\mathrm{\mu g\,m^{-2}\,d^{-1}}$	Flux of barium
caco3_flux	$\mathrm{mg\,m^{-2}\,d^{-1}}$	Flux of particulate calcium carbonate
chl_flux	$\mathrm{mg\,m^{-2}\,d^{-1}}$	Flux of chl
detrital_flux	$\mathrm{mg\,m^{-2}\,d^{-1}}$	Flux of detrital particles
mass_flux	$\mathrm{mg\,m^{-2}\,d^{-1}}$	Total mass flux
mn_flux	$\mathrm{\mu g\,m^{-2}\,d^{-1}}$	Flux of manganese
pheo_flux	$\mathrm{mg\,m^{-2}\,d^{-1}}$	Flux of phaeopigments
pic_flux	$\mathrm{mg\,m^{-2}\,d^{-1}}$	Flux of particulate inorganic carbon
poc_flux	$\mathrm{mg\,m^{-2}\,d^{-1}}$	Flux of particulate organic carbon
pon_flux	$\mathrm{mg\,m^{-2}\,d^{-1}}$	Flux of particulate organic nitrogen
pop_flux	$\mathrm{mg\,m^{-2}\,d^{-1}}$	Flux of particulate organic phosphorus
si_flux	$\mathrm{mg\,m^{-2}\,d^{-1}}$	Flux of total particulate silica
sio2_flux	$\mathrm{mg\,m^{-2}\,d^{-1}}$	Flux of particulate silica, in the form of SiO_2
sio4_flux	$\mathrm{mg\,m^{-2}\,d^{-1}}$	Flux of particulate silica, in the form SiO_4
tc_flux	$\mathrm{mg\,m^{-2}\,d^{-1}}$	Flux of total particulate carbon
ti_flux	$\mathrm{\mu g\,m^{-2}\,d^{-1}}$	Flux of titanium
Mixed layer depth climatology:		
mld	m	Mixed layer depth
Bathymetry:		
bathymetry	m	Total water column depth

ocean along with matched environmental parameters derived from remote sensing sources. The community can use this resource to move further towards a mechanistic understanding of the biological pump.

2 Data and methodology

2.1 Satellite products and mixed layer depth

We provide products derived from SeaWiFS (Sea-viewing Wide Field-of-view Sensor) monthly global area coverage (level 3 mapped data, 9 km, 8-day resolution, version R2014) imagery over the mission record (September 1997–December 2010) acquired from NASA Ocean Biology Distributed Active Archive Center (OB.DAAC) (http://oceancolor.gsfc.nasa.gov/). These include chlorophyll concentration ([Chl]) (Maritorena et al., 2002), diffuse attenua-

tion coefficient at 490 nm ($K_d(490)$) (O'Reilly et al., 2000), and photosynthetically available radiation (PAR) (Frouin et al., 2002). At the time of writing, only 8 % of the publicly available POC observations were measured beyond 2008, when the MODerate resolution Imaging Spectroradiometer (MODIS) replaced the SeaWiFS record, and thus we focus our data compilation here solely on SeaWiFS. Net primary production (NPP) estimates from the Vertically Generalized Production Model (VGPM) (Behrenfeld and Falkowski, 1997) are obtained from http://www.science.oregonstate.edu/ocean.productivity/ (9 km, 8-day resolution). SeaWiFS data products and NPP are retrieved as the median of a 5×5 pixel box ($2025\,\mathrm{km}^2$) centered on each POC flux location (Bailey and Werdell, 2006). AVHRR Pathfinder Version 5 (4 km, 8-day resolution) sea surface temperature (SST) imagery was acquired from the US National Oceanographic

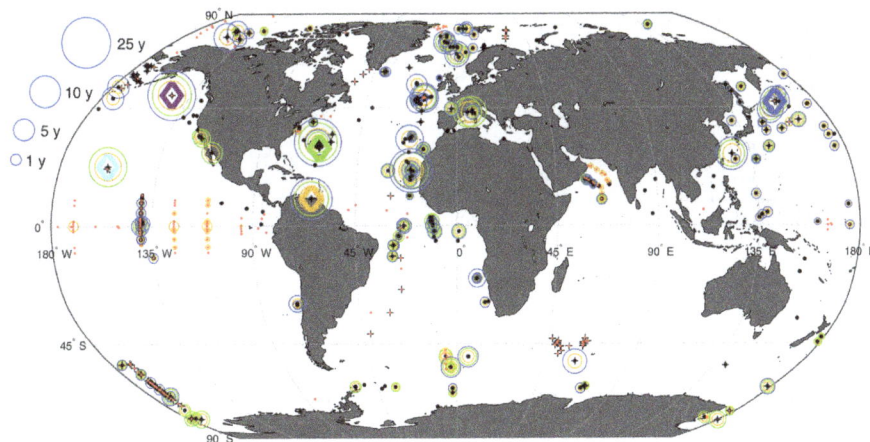

Figure 1. Geographical distribution of POC flux observations at 673 independent sites. The size of the circle indicates the length of the data record at a given site (see legend). The color of the circles indicate the depth of observation, where orange is ≤ 100 m, green is > 100 and ≤ 1000 m, and dark blue is > 1000 m. The location of sediment trap is indicated in black and the location of ^{234}Th data is in red. Plus symbols (+) indicate which observations are during the satellite era (i.e., September 1997–present). The diamonds highlight the locations of time series sites; BATS/OFP (green, 14 %), CARIACO (orange, 10 %), K2 (dark blue, 2 %), OSP (purple, 7 %), and HOT (light blue, 3 %) account for 36 % of the data record.

Data Center and GHRSST (http://pathfinder.nodc.noaa.gov) (Casey et al., 2010). To match the spatial resolution of Sea-WiFS as much as possible, SST was retrieved as the median of an 11×11 pixel box (1936 km^2) centered on each POC flux location.

The Mouw and Yoder (2010) approach is used for satellite retrieval of phytoplankton size classes from SeaWiFS imagery (9 km, monthly resolution). The imagery files were obtained from: doi:10.1594/PANGAEA.860474. The method estimates the percentage of microplankton (S_{fm}) from satellite imagery of remote sensing reflectance ($R_{rs}(\lambda)$). This is an absorption-based approach where the chlorophyll-specific absorption spectra for phytoplankton size class extremes, pico- (0.2–2 μm) and microplankton (> 20 μm), are weighted by S_{fm} (Ciotti et al., 2002; Ciotti and Bricaud, 2006). Briefly, S_{fm} is estimated from a look-up table containing simulated chlorophyll [Chl], absorption due to dissolved and detrital material at 443 nm ($a_{cdm}(443)$), $R_{rs}(\lambda)$, and S_{fm}. For a given pixel, satellite-estimated [Chl] and $a_{cdm}(443)$ (Maritorena et al., 2002) are used to narrow the search space within the look-up table. Of the remaining options, the closest simulated $R_{rs}(\lambda)$ to the satellite-observed $R_{rs}(\lambda)$ is selected and the associated S_{fm} is assigned. S_{fm} is retrieved on a monthly timescale as the median of a 5×5 pixel box (2025 km^2) centered on each POC flux location.

Export depth is often chosen as either the base of the euphotic zone or MLD (Lutz et al., 2007; Lam et al., 2011); thus both are compiled here. The depth of the euphotic zone was determined from $K_d(490)$ (O'Reilly et al., 2000) as $4.6/K_d(490)$ (Morel and Berthon, 1989) from 8-day SeaWiFS data products. MLD estimates are obtained from the IFREMER/LOS Mixed Layer Depth Climatol-ogy group (http://www.ifremer.fr/cerweb/deboyer/mld) from density profiles using a variable density threshold equivalent to 0.2 °C, which accounts for both changes in temperature and salinity (level 3, monthly climatology, 1° resolution; de Boyer Montégut et al., 2004, 2007; Mignot et al., 2007). We retrieve monthly MLD climatology for each pixel containing a POC flux location (1° resolution).

2.2 POC flux data

POC sediment trap data are acquired from public repositories and published literature (Table 2; Fig. 1). Estimates from ^{234}Th measurements are also acquired to improve the resolution of observations in the upper 500 m of the water column (Dunne et al., 2005; Henson et al., 2012; Guidi et al., 2015). These represent 4 % of the total data set. Collected field estimates of POC flux derived from ^{234}Th maintain the original authors' analysis, where POC flux is retrieved based on ^{234}Th activity in the water column accounting for the ratio of POC to ^{234}Th concentration (Buesseler and Boyd, 2009). Both sediment traps and ^{234}Th methodologies have documented challenges associated with accurately retrieving POC flux and characterizing uncertainty. Sediment traps have possible bias associated with the interaction of hydrodynamics with trap design, the capture of zooplankton ("swimmers"), and incomplete preservation of material. ^{234}Th-based measurements have associated biases accounting for local advection, quantifying particulate adsorption and with variability in the ratio of POC : ^{234}Th. See the discussions of Buesseler (1991), Buesseler et al. (2000), Lee et al. (1992), Murray et al. (1996), Quay (1997), and van der Loeff et al. (2006), for in-depth analyses of these issues.

Table 2. Summary of data sources for POC flux from sediment traps and ^{234}Th, the latter indicated in the description when applicable. Date ranges are from first deployment to last retrieval for a given data set, but do not necessarily indicate a continuous time series. Sources are listed in order of first deployment.

Latitude/longitude range	Date range yyyy-mm-dd	Description/ project	Reference
78.9° N–76.5° S All	1976-07-04 to 2005-05-09	Global collection	Lutz et al. (2007) and references therein
81.1° N–71.1° S 138.9° E–74.2° W	1982-06-07 to 2007-06-04	Atlantic Ocean Data Compilation	Torres Valdés et al. (2014)
50° N 145° W	1987-09-23 to 2006-06-04	Ocean Station Papa	Timothy et al. (2013)
60.3° N–67.8° S All	1987-06-06 to 2009-08-08	Global collection of ^{234}Th	Henson et al. (2011) and references therein
22.8° N 158° W	1988-12-01 to 2010-10-05	HOT, station ALOHA	Church and Karl (2013)
32.7–30.6° N 63.1–65.3° W	1988-12-16 to 2011-12-10	BATS	http://bats.bios.edu, last access: 27 September 2013
48–34° N 21° W	1989-04-03 to 1990-04-02	JGOFS North Atlantic Bloom Experiment	Honjo and Manganini (1995)
31.8° N 64.2° W	1989-06-09 to 2010-11-09	Ocean Flux Program	M. Conte (personal communication, 2015)
12° N–12° S 140° W	1992-01-18 to 1993-02-04	JGOFS Equatorial Pacific	Collier and Dymond (1994a, b); Honjo and Dymond (1994); Newton and Murray (1995a, b)
12° N–12° S 135–140° W	1992-02-04 to 1992-09-13	JGOFS Equatorial Pacific, ^{234}Th	Murray et al. (1996)
77–80° N 8–16° W	1992-07-19 to 1993-08-11	Northeast Water Polyna, Greenland, ^{234}Th	Cochran et al. (1995)
43.2° N 5.2° W	1993-10-16 to 2006-01-15	Mediterranean Sea	Rigual-Hernández et al. (2013)
55–88° N 34° E–176° W	1994-08-01 to 1999-07-01	Arctic Ocean, ^{234}Th	Wassmann et al. (2003)
2° N–2° S 175° E–177° W	1994-10-06 to 1996-05-01	FLUPAC and Zonal Flux Study, Western Equatorial Pacific, ^{234}Th	Dunne et al. (2000)
17.7–10.0° N 57.8–65.0° E	1994-11-11 to 1995-12-24	JGOFS Arabian Sea	Honjo (1999)
10.3° N 64.4° W	1995-11-08 to 2012-12-10	CARIACO	Thurnell (2013)
61.5–22.0° N 160° E–170° W	1996-05-15 to 2005-08-15	Review of ^{234}Th measurements	Buesseler and Boyd (2009) and references therein
73.6–76.5° S 176.9° E–178.0° W	1996-06-12 to 1999-07-25	Ross Sea	Collier et al. (2000)
73.6–76.5° S 177° E–178.0° W	1996-10-18 to 1997-04-30	Ross Sea, ^{234}Th	Cochran et al. (2000)
53.0–76.5° S Circumpolar	1996-11-28 to 1998-01-27	JGOFS Southern Ocean	Honjo and Dymond (2002)
53.0–70° S Circumpolar	1997-10-23 to 1998-03-13	JGOFS Southern Ocean, ^{234}Th	Buesseler et al. (2003)
39–25° N 147–137° E	1997-11-19 to 1999-08-12	Kuroshio Extension, Pacific	Mohiuddin et al. (2002)
36.7–36.0° N 147–154.9° E	1998-08-29 to 2000-08-29	Kuroshio Extension, Pacific	Mohiuddin et al. (2004)
44° N 155.1° E	1998-11-02 to 1999-05-26	North Pacific	Honda et al. (2002)
62.6° S 178.1° W	1999-02-12 to 2001-09-17	Antarctic Polar Front	Tesi et al. (2012)

Table 2. Continued.

Latitude/longitude range	Date range yyyy-mm-dd	Description/ project	Reference
77.0–77.8° S 172.5–180° W	2001-12-22 to 2006-02-03	Ross Sea, Antarctica	Smith Jr. et al. (2011)
51–39° N 155–165° E	2002-10-16 to 2005-03-06	NW Pacific, ^{234}Th	Kawakami and Honda (2007)
43.3° N 7.7° E	2003-03-06 to 2005-04-28	MedFlux, Mediterranean Sea	Lee (2011)
33.6° N 118.4° W	2004-01-07 to 2008-06-19	Southern California Bight	Collins et al. (2011)
34.9–29.6° N 58.2–67.2° W	2004-02-22 to 2005-03-13	New Production During Winter Convective Mixing Events	Lomas et al. (2009)
47.0–22.8° N 161° E–158° W	2004-06-22 to 2005-08-10	VERTIGO, Pacific	Lamborg et al. (2008)
47–30° N 145–160° E	2005-03-21 to 2011-07-24	OceanSITES, K2 and S1, NW Pacific	Honda (2012)
44.6° N 2.8° W	2006-06-22 to 2006-06-26	Bay of Biscay	Kuhnt et al. (2013)
10.3° N 64.4° W	2007-02-28 to 2008-12-31	CARIACO	Montes et al. (2012)
62.3–55.3° N 167.9–176.8° W	2008-03-30 to 2008-07-03	Bering Sea	Moran et al. (2012)
61.1° N 26.5° W	2008-05-05 to 2008-05-19	North Atlantic Spring Bloom	Martin et al. (2011)

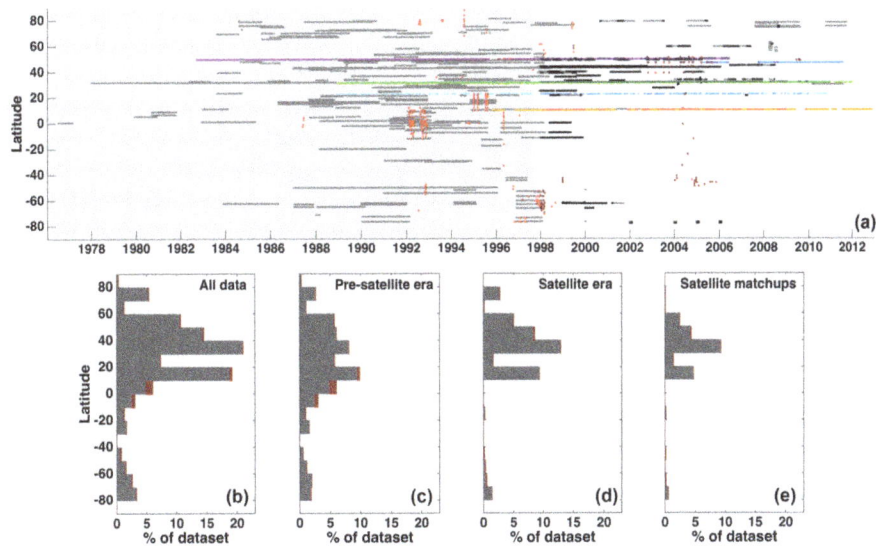

Figure 2. Latitudinal distribution of POC flux observations. **(a)** Temporal distribution showing observations prior to (shaded area) and during (right panel) the satellite era. The length of each grey bar represents a sediment trap deployment (darker bars indicate observations coincident with satellite NPP and S_{fm}); note some bars may overlap. Time series locations are denoted by color as in Fig. 1. ^{234}Th data are differentiated in all subplots (red). **(b)** The percentage of total observations binned by every 10° of latitude. **(c)** Observations prior to the continuous satellite era (before September 1997). **(d)** Observations collected during the continuous satellite era (beginning September 1997). **(e)** Observations with coincident satellite imagery within the same month of collection.

A significant number of studies occurred prior to the launch of SeaWiFS in September 1997 (see Honjo et al., 2008, and references therein). While we compiled observations across all available time frames, greater focus is placed on collecting data concurrent with the satellite record to allow corresponding imagery-based environmental parameters

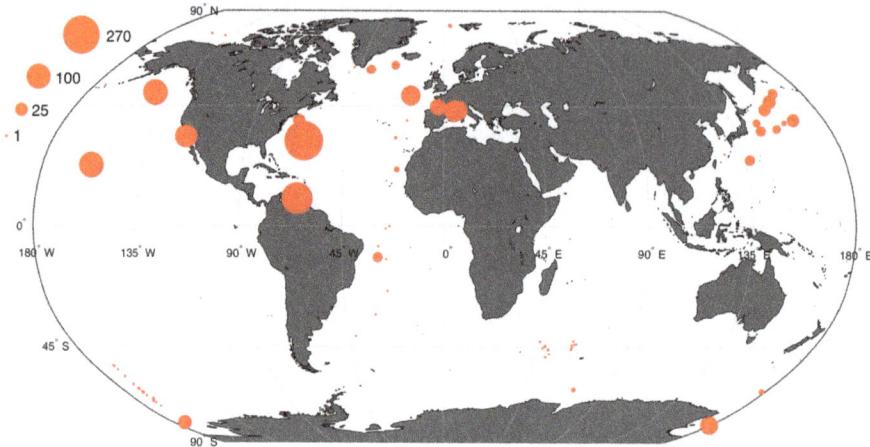

Figure 3. Spatial distribution of coincident satellite and POC flux observations. The size of the circle represents the number of coincident observations (see legend).

to be matched. Overall, the data set comprises a total of 15 792 individual measurements at 673 unique locations with 6842 (43 %) collected during the satellite record. In the interest of matching the timescale of POC flux to satellite-derived products to the greatest degree possible, we focused on collecting short-term sediment trap deployments with individual cup intervals of 30 days or less. The majority of the data set (14 555 measurements or 92 %) fell into this category with a median cup interval of 14 days and a standard deviation of 6 days. Data are skewed towards shorter deployments with 59 % of qualified measurements deployed 14 days or less and 93 % deployed 20 days or less.

2.3   Time series sites

Six long-term oceanographic time series locations are included in the compilation, providing detailed temporal resolution of POC flux export and remineralization. These were the Carbon Retention In A Colored Ocean (CARI-ACO) project site in the Cariaco Basin ($10.5° N$, $64.7° W$), K2 in the northwest Pacific ($47° N$, $160° E$), Ocean Station Papa ($50° N$, $145° W$), the Bermuda Atlantic Time Series (BATS) study site in the Sargasso Sea ($31.7° N$, $64.2° W$), the Ocean Flux Program (OFP; $31.8° N$, $64.2° W$), and the Hawaii Ocean Timeseries (HOT; $22.8° N$, $158.0° W$). Data from BATS and OFP could be combined to create a complete water column profile with BATS sediment traps deployed $\leq 300\,m$ and OFP traps deployed $\geq 500\,m$. Also, with the exception of the first deployment year, HOT only reports POC flux at a single depth.

2.4   Fluxes of other constituents, uncertainty estimates, and metadata

Where readily available, we collect concurrent flux estimates of other organic and inorganic components in addition to

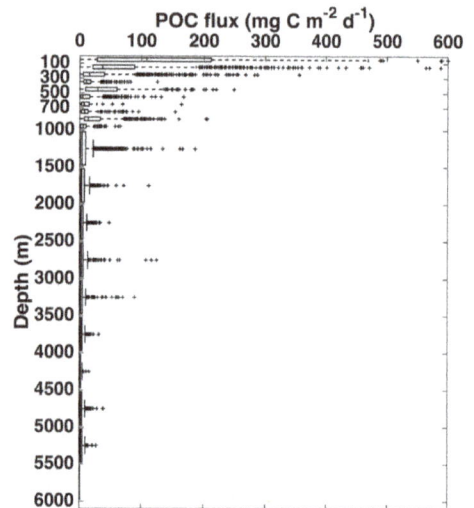

Figure 4. Global POC flux variability with depth. POC flux observations are binned every $100\,m$ in the upper $1000\,m$ and every $500\,m$ throughout the rest of the water column. Box edges enclose the 25th and 75th percentiles of data within each bin with the median shown as a vertical line. Error bars extend to the 5th and 95th percentiles and remaining outliers are indicated with $+$. There are eight data points not represented on the plot, as they were significantly higher than the majority of the data set. These values were observed $< 225\,m$ and are 620, 660, 677, 694, 830, 852, 950, and $1238\,mg\,C\,m^{-2}\,d^{-1}$.

POC flux including particulate inorganic carbon, particulate nitrogen and phosphorus, calcium carbonate, biogenic silica, trace metals, and phytoplankton pigments (Table 1). These data are included to explore relationships between POC export and remineralization and ballasting materials. Where reported by the original authors, we include uncertainty esti-

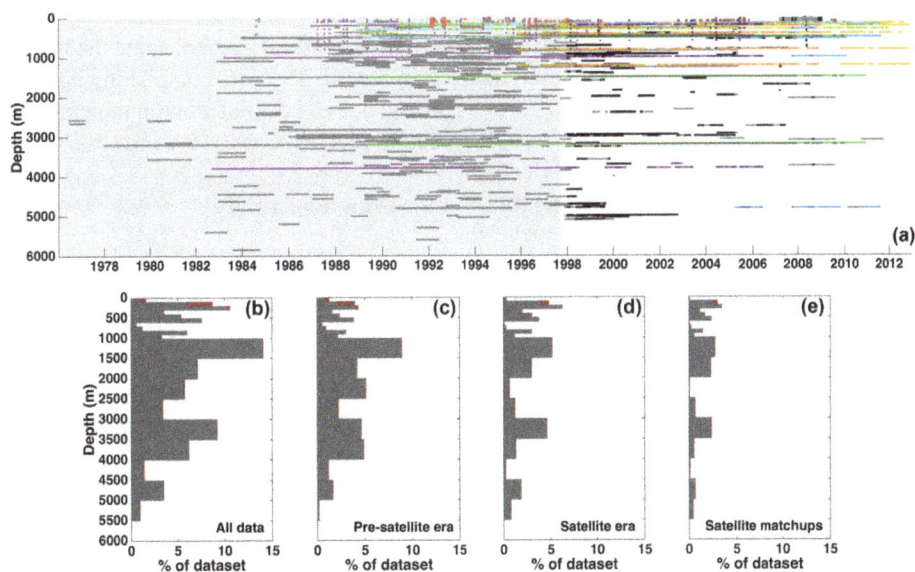

Figure 5. Depth distribution of POC flux observations. The percentage of total observations was binned every 100 m in the upper 1000 m and every 500 m throughout the rest of the water column. Coloration is the same as in Fig. 2. (a) Temporal distribution indicates observations prior to (shaded area) and during (right panel) the satellite era. (b) The percentage of depth dinned total observations. (c) Observations prior to the continuous satellite era (before September 1997). (d) Observations collected during the continuous satellite era (beginning September 1997). (e) Observations with coincident satellite imagery within the same month of observation.

mates for measured fluxes in the compilation. We also collect and include metadata as reported by the original authors. At a minimum, we require each observation be associated with latitude and longitude, deployment date, and depth to be included in the data set. Other information, such as sediment trap type and trap funnel area, is included where available. The majority of measurements (58 %) were not associated with a reported total water depth. Bathymetry was retrieved for POC flux locations from the ETOPO1 1 arcmin Global Relief Model (Amante and Eakins, 2009) from the single pixel containing the measurement location. Locations close to shore were sometimes classified as being on land by ETOPO1; bathymetry is excluded in these cases.

3 Results

The deployment, retrieval, and analysis of sediment trap and ^{234}Th samples represents a significant expenditure of both effort and resources and projects are often funded on a short-term local/regional basis (Honjo et al., 2008). This is reflected in the patchy distribution of observations across the globe in multiple dimensions: space, time, and vertical resolution (Fig. 1). Collection efforts are more prevalent in the Northern Hemisphere, with 63 % of unique station locations comprising 85 % of total observations falling north of the Equator (Fig. 2a and b). Long-term oceanographic time series locations at BATS/OFP, CARIACO, K2, OSP,

and HOT (all in the Northern Hemisphere) collectively account for 36 % of the total data set. If time series locations are removed, 77 % of remaining observations still concentrate north of the Equator. The most sampled regions in the Northern Hemisphere are at midlatitudes, with a quarter of the data set (discounting time series locations) falling between 30 and 40° N (Fig. 2b). In the Southern Hemisphere, data are concentrated at higher latitudes, with a little over half of collected measurements derived from the Southern Ocean at $\geq 60°$ S. In both hemispheres, the second-most sampled latitudes are near the Equator (10° N–10° S).

The data set spans 4 decades from 1976 to 2012 with the majority of efforts (62 %) deployed between 1990 and 2000 (Fig. 2, Table 2). In addition, 43 % of the measurements were collected after September 1997, when the SeaWiFS mission was launched. Prior to SeaWiFS, 79 % of observations are in the Northern Hemisphere (Fig. 2c). After September 1997, the latitudinal distribution becomes even more skewed with 93 % of the observations in the Northern Hemisphere concurrent with the satellite record (Fig. 2d).

While 43 % of the data were observed during the continuous satellite era, not all observations had coincidental imagery. Here we define coincident as retrieved satellite observations within the same month as sediment trap deployment or ^{234}Th measurement for a given POC flux location. We consider only the S_{fm} and NPP imagery for this purpose as they are representative of phytoplankton surface processes and the NPP product already requires SST and [Chl] imagery as inputs. This reduces the total satellite era observations from 6842 to 3722, a drop in total contribution from 43 to 24 %. These are spread over 121 unique locations (Fig. 3). Of the coincident observations, 95 % are in the Northern Hemisphere primarily between 10 and 50° N, with the majority found between 30 and 40° N (Fig. 2e). Data sets in some regions of the ocean (e.g., the equatorial Pacific and the Arabian Sea in Fig. 1) have no satellite overlap (Fig. 3).

The depth resolution of the observations is important for investigators interested in fitting export flux relationships (Martin et al., 1987; Lima et al., 2014). The greatest variability in POC flux is found in the first 500 m of the water column (Lam et al., 2011, Fig. 4). Considering all POC observations together, median POC flux rapidly diminishes from $160\,\mathrm{mg\,C\,m^{-2}\,d^{-1}}$ in the upper 100 m to $30\,\mathrm{mg\,C\,m^{-2}\,d^{-1}}$ at 500 m and $6\,\mathrm{mg\,C\,m^{-2}\,d^{-1}}$ at 1000 m. Below 1000 m, the average POC flux is $3\,\mathrm{mg\,C\,m^{-2}\,d^{-1}}$ (Fig. 4).

Overall, 70 % of the compiled data set is measured at $\geq 500\,\mathrm{m}$ (Fig. 5). Thus, the upper water column close to the depth of export is relatively underrepresented. To increase depth resolution, we consider ^{234}Th and sediment traps together (Dunne et al., 2005; Guidi et al., 2015). Guidi et al. (2015) also merged data from the underwater vision profiler (UVP), which is not included in this compilation as it has not yet been released into a public archive. Shallow observations are critical for capturing the impact of phytoplankton on POC export flux as these data are most connected to surface processes. By adding ^{234}Th measurements to the data set, 249 locations gain depths in the upper water column < 500 m. ^{234}Th data contribute 32 % of all POC flux estimates resolved at depths between 100 and 200 m (Fig. 5a). Overall, the most common deployment depths are between 1000 and 1500 m (14 %) followed by 200 to 300 m (11 %) and then 3000 to 3500 m (9 %) (Fig. 5b). The dominance of the 1000 to 1500 m observation depth is weighted to the presatellite era (Fig. 5c). During the satellite era, 200 to 300 m (6 %) became the most sampled depth, largely due to persistent time series observations at BATS and OSP, followed closely by the 1000 to 1500 and 3000 to 3500 m bins (5 % each) again the result of time series observations at CARIACO and OFP (Fig. 5d). Reasonable depth resolution is found in the observations coincident with satellite matchups (Fig. 5d).

4 Conclusions

This data set is the most comprehensive compilation of POC flux across the globe that we are aware of. By providing merged coincident satellite imagery products, the data set can immediately be used to link phytoplankton surface process with POC flux. Due to rapid remineralization within the first 500 m of the water column, shallow observations from ^{234}Th are helpful to supplement the more extensive sediment trap record. The data compilation is also insightful in terms of spatial and depth resolution to aid in decision making for future POC flux observing investments.

5 Data availability

The data set contains 15 792 individual POC flux estimates at 674 unique locations collected between 1976 and 2012. Where available, the flux of other minerals is also reported. 43 % (6842) of POC flux measurements overlap with the SeaWiFS satellite record (September 1997 to December 2010). Satellite parameters in this compilation include: chlorophyll concentration, net primary production, sea surface temperature, diffuse attenuation coefficient, euphotic depth, photosynthetically active radiation, and microplankton fraction. Estimated mixed layer depths and bathymetry are also provided. Parameters associated with observation sites are extracted as the median of a 5×5 (chlorophyll concentration, NPP, $K_{\mathrm{d}}(490)$, PAR and S_{fm}), 11×11 (SST), or 1×1 (MLD, bathymetry) pixel box. The compiled data are available on PANGAEA (https://www.pangaea.de/): doi:10.1594/PANGAEA.855600 (Mouw et al., 2016).

Author contributions. C. B. Mouw and G. A. McKinley conceived the project and acquired funding for the effort. C. B. Mouw and A. Barnett designed the data compilation. A. Barnett retrieved and processed all data and prepared figures. C. B. Mouw and A. Barnett prepared the manuscript with contributions from all co-authors.

Acknowledgements. We would like to thank the Ocean Color Processing Group at NASA GSFC for the processing and distribution of the SeaWiFS imagery and the IFREMER/LOS Mixed Layer Depth Climatology group for retrieval and distribution of MLD estimates. We would also like to thank BCO-DMO, NOAA National Centers for Environmental Information, the US JGOFS Data System at WHOI, JAMSTEC, Fisheries and Oceans Canada, and the time series efforts at BATS, OFP, OSP, CARIACO and HOT for continued collection and hosting of publicly available data sets. We would like to thank Makio C. Honda for sending data from locations in the northwest Pacific, Walker O. Smith Jr. for sharing data from the Ross Sea, Maureen Conte for sharing the OFP archive, David Timothy and Roy Hourston for providing the OSP data set, Bradley Moran for providing the arctic data set, and John Dunne for sharing the Dunne et al. (2005) data compilation. The National Aeronautics and Space Administration (NNX11AD59G and NNX13AC34G) provided financial support for this data compilation effort. This is contribution number 38 of the Great Lakes Research Center at Michigan Technological University.

Edited by: R. Key

References

Amante, C. and Eakins, B. W.: ETOPO1 1 Arc-Minute Global Relief Model: Procedures, Data Sources and Analysis, NOAA Technical Memorandum NESDIS NGDC-24, National Geophysical Data Center, NOAA, doi:10.7289/V5C8276M, 2009.

Bailey, S. and Werdell, P.: A multi-sensor approach for the on-orbit validation of ocean color satellite data products, Remote Sens. Environ., 102, 12–23, doi:10.1016/j.rse.2006.01.015, 2006.

Behrenfeld, M. and Falkowski, P.: Photosynthetic rates derived from satellite-based chlorophyll concentration, Limnol. Oceanogr., 42, 1–20, 1997.

Buesseler, K. O.: Do upper-ocean sediment traps provide an accurate record of particle flux?, Nature, 353, 420–423, 1991.

Buesseler, K. O. and Boyd, P. W.: Shedding light on processes that control particle export and flux attenuation in the twilight zone of the open ocean, Limnol. Oceanogr., 54, 1210–1232, 2009.

Buesseler, K. O., Steinberg, D. K., Michaels, A. G., Johnson, R. J., Andrews, J. E., Valdes, J. R., and Price, J. F.: A comparison of the quality and composition of material caught in a neutrally buoyant versus surface-tethered sediment trap, Deep-Sea Res. Pt. I, 47, 227–294, doi:10.1016/S0967-0637(99)00056-4, 2000.

Buesseler, K. O., Barber, R. T., Dickson, M.-L., Hiscock, M. R., Moore, J. K., and Sambrotto, R.: The effect of marginal ice-edge dynamics on production and export in the Southern Ocean along 170° W, Deep-Sea Res. Pt. II, 50, 579–603, doi:10.1016/S0967-0645(02)00585-4, 2003.

Casey, K. S., Brandon, T. B., Cornillon, P., and Evans, R.: The Past, Present, and Future of the AVHRR Pathfinder SST Program, in: Oceanography from Space, 273–287, Dordrecht: Springer, the Netherlands, doi:10.1007/978-90-481-8681-5_16, 2010.

Church, M. J. and Karl, D. M.: Primary production and sediment trap flux measurements and calculations by the Hawaii Ocean Time-series (HOT) program at Station ALOHA in the North Pacific 100 miles north of Oahu, Hawaii for Cruises HOT1-227 during 1988–2010, NOAA National Centers for Environmental Information, Accession: 0089168, available at: http://data.nodc.noaa.gov/accession/0089168 (last access: 14 May 2012), 2013.

Ciotti, A. and Bricaud, A.: Retrievals of a size parameter for phytoplankton and spectral light absorption by colored detrital matter from water-leaving radiances at SeaWiFS channels in a continental shelf region off Brazil, Limnol. Oceanogr.-Meth., 4, 237–253, 2006.

Ciotti, A., Lewis, M., and Cullen, J.: Assessment of the relationship between dominant cell size in natural phytoplankton communities and the spectral shape of the absorption coefficient, Limnol. Oceanogr., 47, 404–417, 2002.

Cochran, J. K., Barnes, C., Achman, D., and Hirshberg, D. J.: Thorium-234/Uranium-238 disequilibrium as an indicator of scavenging rates and particulate organic carbon fluxes in the Northeast Water Polynya, Greenland, J. Geophys. Res., 100, 4399–4410, doi:10.1029/94JC01954, 1995.

Cochran, J. K., Buesseler, K. O., Bacon, M. P., Wang, H. W., Hirschberg, D. J., Ball, L., Andrews, J., Crossin, G., and Fleer, A.: Short-lived isotopes (^{234}Th, ^{228}Th) as indicators of POC export and particle cycling in the Ross Sea, Southern Ocean, Deep-Sea Res. Pt. II, 47, 3451–3490, doi:10.1016/S0967-0645(00)00075-8, 2000.

Collier, R. and Dymond, J.: Sed_trap_annual, Biological and Chemical Oceanography Data System, BCO DMO, WHOI, available at: http://www.bco-dmo.org/dataset/2608 (last access: 17 September 2013), 1994a.

Collier, R. and Dymond, J.: Sed_trap_Eq_North, Biological and Chemical Oceanography Data System, BCO DMO, WHOI, available at: http://www.bco-dmo.org/dataset/2609 (last access: 17 September 2013), 1994b.

Collier, R., Dymond, J., Honjo, S., Manganini, S., Francois, R., and Dunbar, R.: The vertical flux of biogenic and lithogenic material in the Ross Sea: moored sediment trap observations 1996–1998, Deep-Sea Res. Pt. II, 47, 3491–3520, doi:10.1016/S0967-0645(00)00076-X, 2000.

Collins, L. E., Berelson, W., Hammond, D. E., Knapp, A., Schwartz, R., and Capone, D.: Particle fluxes in San Pedro Basin, California: A four-year record of sedimentation and physical forcing, Deep-Sea Res. Pt. I, 58, 898–914, doi:10.1016/j.dsr.2011.06.008, 2011.

de Boyer Montégut, C., Madec, G., Fischer, A. S., Lazar, A., and Iudicone, D.: Mixed layer depth over the global ocean: an examination of profile data and a profile-based climatology, J. Geophys. Res., 109, C12003, doi:10.1029/2004JC002378, 2004.

de Boyer Montégut, C., Mignot, J., Lazar, A., and Cravatte, S.: Control of salinity on the mixed layer depth in the world ocean: 1. General description, J. Geophys. Res., 112, C06011, doi:10.1029/2006JC003953, 2007.

Dunne, J., Armstrong, R., Gnanadesikan, A., and Sarmiento, J.: Empirical and mechanistic models for the particle export ratio, Global Biogeochem. Cy., 19, GB4026, doi:10.1029/2004GB002390, 2005.

Dunne, J. P., Murray, J. W., Rodier, M., and Hansell, D. A.: Export flux in the western and central equatorial Pacific: Zonal and temporal variability, Deep-Sea Res. Pt. I, 47, 901–936, doi:10.1016/S0967-0637(99)00089-8, 2000.

Frouin, R., Franz, B. A., and Werdell, P. J.: The SeaWiFS PAR product, in: Algorithm Updates for the Fourth SeaWiFS Data Reprocessing, edited by: Hooker, S. B. and Firestone, E. R., NASA Tech. Memo. 2003–206892, Vol. 22, NASA Goddard Space Flight Center, Greenbelt, Maryland, 46–50, 2002.

Guidi, L., Legendre, L., Reygondeau, G., Uitz, J., Stemmann, L., and Henson, S. A.: A new look at ocean carbon remineralization for estimating deepwater sequestration, Global Biogeochem. Cy., 29, 1044–1059, doi:10.1002/2014GB005063, 2015.

Henson, S. A., Sanders, R., Madsen, E., Morris, P. J., Le Moigne, F., and Quartly, G. D.: A reduced estimate of the strength of the ocean's biological carbon pump, Geophys. Res. Lett., 38, L04606, doi:10.1029/2011GL046735, 2011.

Henson, S. A., Sanders, R., and Madsen, E.: Global patterns in efficiency of particulate organic carbon export and transfer to the deep ocean, Global Biogeochem. Cy., 26, GB1028, doi:10.1029/2011GB004099, 2012.

Honda, M. C.: JAMSTEC Sediment trap data at time-series stations: K2 and S1 for Ocean SITES, JAMSTEC Environmental Biogeochemical Cycle Research Program, available at: http://ebcrpa.jamstec.go.jp/rigc/e/ebcrp/mbcrt/st_k2s1_oceansites/ (last access: 10 April 2014), 2012.

Honda, M. C., Imai, K., Norjiri, Y., Hoshi, F., Sugawara, T., and Kusakabe, M.: The biological pump in the northwestern North Pacific based on fluxes and major components of particulate matter obtained by sediment-trap experiments (1997–2000), Deep-Sea Res. Pt. II, 49, 5595–5625, doi:10.1016/S0967-0645(02)00201-1, 2002.

Honjo, S.: Sediment Properties and other data from a fixed platform from 19941102 to 19951205, NOAA National Centers for Environmental Information, Accession: 9800155, available at: http://www.nodc.noaa.gov/cgi-bin/OAS/prd/accession/9800155 (last access: 3 October 2013), 1999.

Honjo, S. and Dymond, J.: Sed_trap_Eq_South, Biological and Chemical Oceanography Data System, BCO DMO, WHOI, available at: http://www.bco-dmo.org/dataset/2618 (last access: 17 September 2013), 1994.

Honjo, S. and Dymond, J.: Deep sea sediment trap particle flux, AESOPS/Southern Ocean 1996–1997 Mooring Deployment, US JGOFS Data System, available at: http://usjgofs.whoi.edu/jg/dir/jgofs/southern/ (last access: 9 April 2014), 2002.

Honjo, S. and Maganini, S.: Sediment trap data, biogenic particle fluxes North Atlantic Bloom Experiment, US JGOFS Data System, available at: http://usjgofs.whoi.edu/jg/dir/jgofs/nabe/ (last access: 9 April 2014), 1995.

Honjo, S., Manganini, S. J., Krishfield, R. A., and Francois, R.: Particulate organic carbon fluxes to the ocean interior and factors controlling the biological pump: A synthesis of global sediment trap programs since 1983. Prog. Oceanogr., 76, 217–285, doi:10.1016/j.pocean.2007.11.003, 2008.

Kawakami, H., and Honda, M. C.: Time-series observation of POC fluxes estimated from ^{234}Th in the northwestern North Pacific, Deep-Sea Res. Pt. I, 54, 1070–1090, doi:10.1016/j.dsr.2007.04.005, 2007.

Kuhnt, T., Howa, H., Schmidt, S., Marié, L., and Scheibel, R.: Flux dynamics of planktonic foraminiferal test in the south-eastern Bay of Biscay (northeast Atlantic margin), J. Marine Syst., 109–110, S169–S181, doi:10.1016/j.jmarsys.2011.11.026, 2013.

Lam, P. J., Doney, S. C., and Bishop, J. K. B.: The dynamic ocean biological pump: Insights from a global compilation of particulate organic carbon, $CaCO_3$, and opal concentration profiles from the mesopelagic, Global Biogeochem. Cy., 25, GB3009, doi:10.1029/2010GB003868, 2011.

Lamborg, C. H., Buesseler, K. O., Valdes, J., Bertrand, C. H., Bidigare, R., Manganini, S., Pike, S., Steinberg, D., Trull, T., and Wilson, S.: The flux of bio- and lithogenic material associated with sinking particles in the mesopelagic "twilight zone" of the northwest and North Central Pacific Ocean, Deep-Sea Res. Pt. II, 55, 1540–1563, doi:10.1016/j.dsr2.2008.04.011, 2008.

Lee, C.: Sediment Trap Mass Flux, MedFlux, Biological and Chemical Oceanography Data System, BCO DMO, WHOI, available at: http://www.bco-dmo.org/dataset/3561 (last access: 22 May 2013), 2011.

Lee, C., Hedges, J. I., Wakeham, S. G., and Zhu, N.: Effectiveness of various treatments in retarding microbial activity in sediment trap material and their effects on the collection of swimmers, Limnol. Oceanogr., 37, 117–130, doi:10.4319/lo.1992.37.1.0117, 1992.

Lima, I. D., Lam, P. J., and Doney, S. C.: Dynamics of particulate organic carbon flux in a global ocean model, Biogeosciences, 11, 1177–1198, doi:10.5194/bg-11-1177-2014, 2014.

Lomas, M. W., Nelson, D. M., Lipschultz, F., Knap, A., and Bates, N.: Trap Flux, Biological and Chemical Oceanography Data System, BCO DMO, WHOI, available at: http://www.bco-dmo.org/dataset/3215 (last access: 22 May 2013), 2009.

Lomas, M. W., Roberts, N., Lipschultz, F., Krause, J. W., Nelson, D. M., and Bates, N. R.: Biogeochemical responses to late-winter storms in the Sargasso Sea. IV. Rapid succession of major phytoplankton groups, Deep-Sea Res. Pt. I, 56, 892–908, doi:10.1016/j.dsr.2009.03.004, 2009.

Lutz, M. J., Calderia, K., Dunbar, R. B., and Behrenfeld, M. J.: Seasonal rhythms of net primary production and particulate organic carbon flux to depth describe the efficiency of biological pump in the global ocean. J. Geophys. Res., 112, C100110, doi:10.1029/2006JC003706, 2007.

Maritorena, S., Siegel, D., and Peterson, A.: Optimization of a semi-analytical ocean color model for global-scale applications, Appl. Optics, 41, 2705–2714, 2002.

Martin, J. H., Knauer, G., Karl, D., and Broenkow, W.: VERTEX: Carbon cycling in the northeast Pacific, Deep-Sea Res. Pt. I, 34, 267–285, 1987.

Martin, P., Lampitt, R. S., Perry, M. J., Sanders, R., Lee, C., and D'Asaro, E.: Export and mesopelagic particle flux during a North Atlantic spring diatom bloom, Deep-Sea Res. Pt. I, 58, 338–349, doi:10.1016/j.dsr.2011.01.006, 2011.

Mignot, J., de Boyer Montégut, C., Lazar, A., and Cravatte, S.: Control of salinity on the mixed layer depth in the world ocean: 2. Tropical areas, J. Geophys. Res., 112, C10010, doi:10.1029/2006JC003954, 2007.

Mohiuddin, M. M., Nishimura, A., Tanaka, Y., and Shimamoto, A.: Regional and interannual productivity of biogenic components and planktonic foraminiferal fluxes in the northwestern Pacific Basin, Mar. Micropaleontol., 45, 57–82, doi:10.1016/S0377-8398(01)00045-7, 2002.

Mohiuddin, M. M., Nishimura, A., Tanaka, Y., and Shimamoto, A.: Seasonality of biogenic particle and planktonic foraminifera fluxes: response to hydrographic variability in the Kuroshio Extension, northwestern Pacific Ocean, Deep-Sea Res. Pt. I, 51, 1659–1683, doi:10.1016/j.dsr.2004.06.002, 2004.

Montes, E., Muller-Karger, F., Thunell, R., Hollander, D., Astor, Y., Varela, R., Soto, I., and Lorenzoni, L.: Vertical fluxes of particulate biogenic material through the euphotic and twilight zones in the Cariaco Basin, Venezuela, Deep-Sea Res. Pt. I, 67, 73–84, doi:10.1016/j.dsr.2012.05.005, 2012.

Moran, S. B., Lomas, M. W., Kelly, R. P., Gradinger, R., Iken, K., and Mathis, J. T.: Seasonal succession of net primary productivity, particulate organic carbon export, and autrophic community composition in the eastern Bering Sea, Deep-Sea Res. Pt. II, 65–70, 84–97, doi:10.1016/j.dsr2.2012.02.011, 2012.

Morel, A. and Berthon, J.-F.: Surface pigments, algal biomass, and potential production of the euphotic layer: relationships reinvestigation in view of remote-sensing applications, Limnol. Oceanogr., 8, 1545–1562, 1989.

Mouw, C. and Yoder, J.: Optical determination of phytoplankton size composition from global SeaWiFS imagery, J. Geophys. Res., 115, C12018, doi:10.1029/2010JC006337, 2010.

Mouw, C. B., Barnett, A., McKinley, G., Gloege, L., and Pilcher, D.: Global Ocean Particulate Organic Carbon flux merged with satellite parameters, PANGAEA, doi:10.1594/PANGAEA.855600, 2016.

Murray, J. W., Young, J., Newton, J., Dunne, J., Chapin, T., Paul, B., and McCarthy, J. J.: Export flux of particulate organic carbon from the central equatorial Pacific determined using a combined drifting trap-234Th approach, Deep-Sea Res. Pt. II, 43, 1095–1132, doi:10.1016/0967-0645(96)00036-7, 1996.

Newton, J. and Murray, J. W.: Poc_pn_trap, Biological and Chemical Oceanography Data System, BCO DMO, WHOI, available at: http://www.bco-dmo.org/dataset-deployment/450839 (last access: 18 September 2013), 1995a.

Newton, J. and Murray, J. W.: Poc_pn_trap, Biological and Chemical Oceanography Data System, BCO DMO, WHOI, available at: http://www.bco-dmo.org/dataset-deployment/450925 (last access: 18 September 2013), 1995b.

O'Reilly, J. E., Maritorena, S., O'Brien, M. C., Siegel, D. A., Toole, D., Menzies, D., Smith, R. C., Muller, J. L., Mitchell, B. G., Kahru, M., Chavez, F. P., Strutton, P., Cota, G. F., Hooker, S. B., McClain, C. R., Carder, K. L., Muller-Karger, F., Harding, L., Magnuson, A., Phinny, D., Moore, G. F., Aiken, J., Arrigo, K. R., Letelier, R., and Culver, M.: SeaWiFS Postlaunch Calibration and Validation Analyses, Part 3, NASA Tech. Memo. 2000-206892, Vol. 11, edited by: Hooker, S. B. and Firestone, E. R., NASA Goddard Space Flight Center, 49 pp., 2000.

Quay, P.: Was a carbon balance measured in the equatorial Pacific during JGOFS?, Deep-Sea Res. Pt. II, 44, 1765–1781, doi:10.1016/S0967-0645(97)00093-3, 1997.

Rigual-Hernández, A., Bárcena, M. A., Jordan, R. W., Sierro, R. J., Flores, J. A., Meier, K. J. S., Beaufort, L., and Heussner, S.: Diatom fluxes in the NW Mediterranean: evidence from a 12-year sediment trap record and surficial sediments, J. Plankton Res., 35, 1109–1225, doi:10.1093/plankt/fbt055, 2013.

Smith Jr., W. O., Shields, A. R., Dreyer, J. C., Peloquin, J. A., and Asper, V.: Interannual variability in vertical export in the Ross Sea: Magnitude, composition and environmental correlates, Deep-Sea Res. Pt. I, 58, 147–159, doi:10.1016/j.dsr.2010.11.007, 2011.

Tesi, T., Ravaioli, L. M., Giglio, F., and Capotondi, L.: Particulate export and lateral advection in the Antarctic Polar Front (Southern Pacific Ocean): One-year mooring deployment, J. Marine Syst., 105–108, 70–81, doi:10.1016/j.jmarsys.2012.06.002, 2012.

Thurnell, R. C.: Sediment Trap Data, CARIACO Ocean Time Series Data, available at: http://imars.marine.usf.edu/CAR/index.html (last access: 13 September 2013), 2013.

Timothy, D. A., Wong, C. S., Barwell-Clarke, J. E., Page, J. S., White, L. A., and Macdonald, R. W.: Climatology of sediment flux and composition in the subarctic Northeast Pacific Ocean with biogeochemical implications. Prog. Oceanogr., 116, 95–129, doi:10.1016/j.pocean.2013.06.017, 2013.

Torres Valdés, S., Painter, S. C., Martin, A. P., Sanders, R., and Felden, J.: Data compilation of fluxes of sedimenting material from sediment traps in the Atlantic Ocean, Earth Syst. Sci. Data, 6, 123–145, doi:10.5194/essd-6-123-2014, 2014.

van der Loeff, M. R., Sarin, M. M., Baskaran, M., Benitez-Nelson, C., Buesseler, K. O., Charette, M., Dai, M., Gustafsson, O., Masque, P., Morris, P. J., Orlandini, K., y Baena, A. R., Savoye, N., Schmidt, S., Turnewitsch, R., Vöge, I., and Waples, J. T.: A review of present techniques and methodological advances in analyzing ^{234}Th in aquatic systems, Mar. Chem., 100, 190–212, doi:10.1016/j.marchem.2005.10.012, 2006.

Wassmann, P., Bauerfind, E., Fortier, M., Fukuchi, M., Hargrave, B., Moran, S. B., Noji, T., Nöthig, E.-M., Olli, K., Peinert, R., Sasaki, H., and Shevchenko, V. P.: Particulate organic carbon flux to the Arctic Ocean sea floor, in: The Organic Carbon Cycle in the Arctic Ocean, edited by: Stein, R. and Macdonald, R. W., Springer-Verlag, Heidelberg-Berlin-New York, 102–138, 2003.

6

Sea surface salinity and temperature in the southern Atlantic Ocean from South African icebreakers

Giuseppe Aulicino[1,2], **Yuri Cotroneo**[2], **Isabelle Ansorge**[3], **Marcel van den Berg**[4], **Cinzia Cesarano**[5], **Maria Belmonte Rivas**[6,7], **and Estrella Olmedo Casal**[6]

[1]Department of Life and Environmental Sciences, Università Politecnica delle Marche, Ancona, 60131, Italy

[2]Department of Science and Technologies, Università degli Studi di Napoli Parthenope, Naples, 80143, Italy

[3]Marine Research Institute, Oceanography Department, University of Cape Town, Rondebosch, Cape Town, 7701, South Africa

[4]Department of Environmental Affairs, Cape Town, 8001, South Africa

[5]Progetto Terra, Gragnano, 80054, Italy

[6]Institute of Marine Sciences, ICM, Barcelona, 08003, Spain

[7]Royal Netherlands Meteorological Institute, KNMI, De Bilt, 3730, the Netherlands

Correspondence: Giuseppe Aulicino (g.aulicino@staff.univpm.it)

Abstract. We present here sea surface salinity (SSS) and temperature (SST) data collected on board the *S.A. Agulhas-I* and *S.A. Agulhas-II* research vessels, in the framework of the South African National Antarctic Programme (SANAP). Onboard Sea-Bird thermosalinographs were regularly calibrated and continuously monitored in-between cruises, and no appreciable sensor drift emerged. Water samples were taken on a daily basis and later analyzed with a Portasal salinometer; some CTD measurements collected along the cruises were used to validate the data. No systematic differences appeared after a rigorous quality control on continuous data. Results show that salinity measurement error was a few hundredths of a unit on the practical salinity scale. Quality control included several steps, among which an automatic detection of unreliable values through selected threshold criteria and an attribution of quality flags based on multiple criteria, i.e., analysis of information included in the cruise reports, detection of insufficient flow and/or presence of air bubbles and ice crystals in the seawater pipe, visual inspection of individual campaigns, and ex post check of sea ice maps for confirming ice field locations. This data processing led us to discard about 36 % of acquired observations, while reliable data showed an excellent agreement with several independent SSS products. Nevertheless, a sea ice flag has been included for identifying valid data which could have been affected by scattered sea ice contamination. In our opinion, this dataset, available through an unrestricted repository at https://doi.org/10.7289/V56M3545, contributes to improving the knowledge of surface water features in one of the most important regions for global climate. The dataset will be highly valuable for studies focusing on climate variability in the Atlantic sector of the Southern Ocean, especially across the Antarctic Circumpolar Current and its fronts. Furthermore, we expect that the collected SSS data will represent a valuable tool for the calibration and validation of recent satellite observations provided by SMOS and Aquarius missions.

1 Introduction

The salinity of the ocean is one of the key parameters identified by the Global Climate Observing System (GCOS) as being essential for climate studies (World Meteorological Organization, 2016). Many water masses are identified and traced through salinity and temperature values; in addition, the entire ocean circulation from surface to deep layers is largely conditioned by their influence on the density field (e.g., Rahmstorf, 2006; Helm et al., 2010; Sansiviero et al., 2017). Increasing efforts have been made in the past decades to provide a global synoptic monitoring of the sea surface salinity (SSS) in conjunction with the recent launch of two dedicated satellite missions, i.e., soil moisture and ocean salinity (SMOS) in 2009 (Kerr et al., 2010) and Aquarius in 2011 (Le Vine et al., 2010). The delivered remotely sensed data provided interesting insights into the upper ocean, especially when considering that the surface layer is strictly connected to (i) the physical and biogeochemical interactions between ocean and atmosphere and (ii) the observation of large-scale circulation features (i.e., fronts, currents) as well as mesoscale and small-scale structures (i.e., meanders, eddies) (e.g., in the open ocean: Cotroneo et al., 2013; Reul et al., 2014; Ansorge et al., 2015; D'Addezio and Subrahmanyam, 2016; in marginal seas: Cotroneo et al., 2015; Rivaro et al., 2017; Mangoni et al., 2017; Misic et al., 2017; Aulicino et al., 2018a). Nevertheless, original SMOS and Aquarius products showed limitations in completely retrieving reliable SSS values in some regions of the worldwide oceans, especially at latitudes higher than 45–50° (Tang et al., 2014; Kohler et al., 2015). Despite their oceanographic and biological importance, the southern sector of the Atlantic Ocean and the Southern Ocean are among such areas. It is recognized by the scientific community that further studies are needed to improve the satellite calibration, SSS retrievals algorithms, and better validate them in these regions (Lagerloef et al., 2009; Chen et al., 2014; Boutin et al., 2016). To this aim, all the available near-surface measurements (mostly in the upper 5 m) can provide a significant contribution and should be shared by the oceanographic community.

Since 2010, South Africa's Department of Environmental Affairs (DEA), the South African National Antarctic Programme (SANAP), and the University of Cape Town (UCT) have carried out annual research cruises across the Southern Ocean, as part of the South Atlantic Meridional Overturning Circulation – South Africa (SAMOC-SA) program (Ansorge et al., 2014), in order to collect multidisciplinary meteo-oceanographic in situ data.

In this paper, "thermosalinograph" (hereafter TSG) data and bottle samples used for their validation are described in Sect. 2, as well as the applied quality control (QC) methodology. Then, Sect. 3 presents the comparison between the TSG SSS dataset and the other reference datasets. Finally, data record details and conclusions are reported in Sect. 4.

2 Data and methods

We present here the dataset collected by South African icebreakers *S.A. Agulhas-I* and *S.A. Agulhas-II* during several research cruises in the southern Atlantic Ocean and in the adjacent Southern Ocean sector between December 2010 and February 2017 (Table 1; Fig. 2).

TSGs mounted on board the two research vessels provided high-resolution measurements (conductivity, temperature, salinity) along the cruise tracks. The TSG system (Fig. 1) is a continuous, underway monitoring system connected to a dedicated scientific seawater supply. A conductivity cell measures the conductivity of the seawater pumped in, from which salinity can be deducted, while a thermistor cell measures the sea surface temperature (SST). An additional temperature sensor is installed across from the hull water inlet for measuring the actual sea surface temperature (SSTH) before it is slightly modified during the seawater path to the conductivity cell. The measuring system is similar on both vessels, with the water inlet sitting at about 5.0 m below the waterline. The nominal accuracy of Sea-Bird TSG SSS is better than 0.01 on the practical salinity scale (PSS), while the resolution is close to 0.001 (www.seabird.com, last access: 29 May 2018); these values are largely sufficient to capture the surface variability (Gaillard et al., 2015). These sensors are regularly calibrated and continuously monitored in-between cruises, and no appreciable sensor drift emerged in the study period. Regular comparisons between bottle samples and continuous measurements are also carried out on board during each scientific voyage. However, several aspects could increase the nominal errors and corrupt the data acquired during part of a cruise, i.e., insufficient flow through the conductivity cell, the presence of air bubbles in the pipe, or fouling contamination. Thus an accurate QC of the collected dataset, as well as an eventual comparison with external observations, are strongly recommended; these aspects will be addressed in this section and in Sect. 3, respectively. Furthermore, it is important to remark that the *S.A. Agulhas-I* and *S.A. Agulhas-II* TSG systems are generally switched on underway; however, when sailing south of 55° S the presence of sea ice could block the scientific water supply, repeatedly, hampering data collection. For this reason, the TSG pumps are turned off before entering the ice field in order to reduce the potential damage to the TSG system and the possible acquisition of bad data.

For each cruise, the full-resolution TSG dataset has been processed and undersampled with a median filter over a 1 min interval. The dataset is available at https://doi.org/10.7289/V56M3545. We plan to provide updates as soon as further observations are collected and processed. This archive includes the following variables: time of the acquisition, latitude, longitude, conductivity, SSS, SST, SST at hull (SSTH), and sea ice flag. It is important to remark that SSS is the actual ocean salinity only if the flow rate to the conductivity cell is sufficient; otherwise, it would represent the salinity of

Table 1. List of scientific cruises between 2010 and 2017 included in the dataset.

Cruise name	Ship	Start date	End date	Latitude	Longitude
SANAE 2010	*Agulhas-I*	8 Dec 2010	10 Feb 2011	33.96–70.65° S	37.00° W–15.99° E
SANAE 2011	*Agulhas-I*	10 Dec 2011	8 Feb 2012	37.62–70.46° S	36.51° W–11.92° E
Winter 2012	*Agulhas-II*	9 Jul 2012	1 Aug 2012	33.87–57.16° S	0.00–43.07° E
Gough 2012	*Agulhas-II*	6 Sep 2012	10 Oct 2012	33.92–50.25° S	15.00° W–18.09° E
SANAE 2012	*Agulhas-II*	7 Dec 2012	19 Feb 2013	33.88–70.80° S	35.77° W–18.69° E
Marion 2013	*Agulhas-II*	10 Apr 2013	16 May 2013	33.88–47.79° S	18.22–43.23° E
Gough 2013	*Agulhas-II*	5 Sep 2013	12 Sep 2013	34.06–37.06° S	11.89° W–18.14° E
Marion 2014	*Agulhas-II*	2 Apr 2014	6 May 2014	33.88–58.75° S	18.25–38.75° E
Gough 2014	*Agulhas-II*	4 Sep 2014	7 Oct 2014	33.87–49.26° S	11.01° W–18.60° E
SANAE 2014	*Agulhas-II*	5 Dec 2014	16 Feb 2015	33.91–70.77° S	35.31° W–17.48° E
Marion 2015	*Agulhas-II*	9 Apr 2015	15 May 2015	34.44–47.75° S	18.35–39.37° E
Winter 2015	*Agulhas-II*	23 Jul 2015	14 Aug 2015	33.88–56.81° S	0.00–18.64° E
Gough 2015	*Agulhas-II*	4 Sep 2015	6 Oct 2015	33.90–47.73° S	11.72° W–18.61° E
SANAE 2015	*Agulhas-II*	5 Dec 2015	10 Feb 2016	34.44–70.78° S	35.62° W–17.72° E
Marion 2016	*Agulhas-II*	8 Apr 2016	16 May 2016	33.87–47.77° S	18.25–38.75° E
Winter 2016	*Agulhas-II*	5 Jul 2016	27 Jul 2016	33.35–55.11° S	0.00–29.26° E
SANAE 2016	*Agulhas-II*	30 Nov 2016	2 Feb 2017	34.05–70.78° S	33.80° W–17.64° E

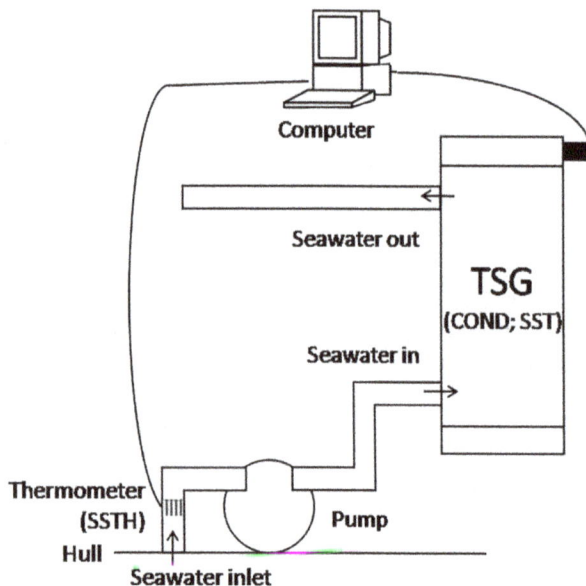

Figure 1. Schematic of the underway data collection system highlighting the seawater path to the conductivity and thermistor cells (TSG), and the temperature sensor location at the water inlet.

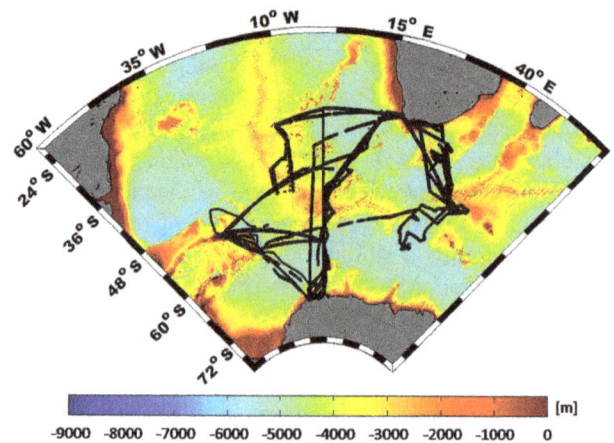

Figure 2. *S.A. Agulhas-I* and *S.A. Agulhas-II* cruise tracks (black dots) in the southern Atlantic Ocean and in the Southern Ocean between December 2010 and February 2017. Bathymetry (m) is expressed in color.

temperature difference between the seawater and the ambient temperature (Gaillard et al., 2015).

2.1 Quality control

The QC of the collected TSG measurements included three main steps, which led us to discard about 36 % of acquired observations; statistics are summarized in Table 2. Firstly, an automatic detection of unreliable values was performed using selected threshold criteria on conductivity (2.5–5.5 S m^{-1}), SSS (32.0–37.0), and SST values (varying with latitude and season) (QC-1). Then, following the World Ocean Circulation Experiment (WOCE) principles and NOAA National

the seawater trapped in the TSG. As for temperatures, please note that SST is the temperature of the water volume inside the conductivity cell (TSG) while SSTH is the temperature of the ocean at the water intake very close to the hull. Monitoring SST is necessary to ensure a precise salinity, while SSTH is essential for computing actual seawater density. These values can be slightly different because of heat exchanges along the seawater path to reach the TSG. Exchanges depend on the flow rate, the volume of water in the circuit, and on the

Table 2. Statistics of the quality control.

Total measurements	929 801	
Discarded data after QC-1	242 488	26.0 %
Discarded data after QC-2	316 391	34.0 %
Discarded data after QC-3	334 687	35.9 %
Valid SSS measurements	595 114	
Sea ice flagged data	45 063	7.5 %
Missing COND and SST values	51 275	8.6 %
Missing SSTH values	82 527	13.8 %

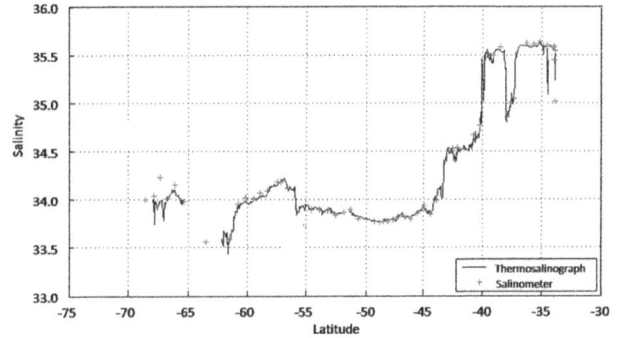

Figure 3. A comparison between TSG underway SSS (black line) collected during the southward leg of the SANAE 2012 cruise and the related bottle samples measured via the Portasal salinometer (red crosses).

Center for Environmental Information (NCEI) database requirements, quality flags (good, suspicious, bad, harbour, icefield) were attributed to the data based on the analysis of several factors, i.e., the detection of insufficient seawater flow, the contamination by air bubbles, and the presence of sea ice (QC-2). The episodic reductions in seawater flow were identified on the basis of a large and increasing difference between SST and SSTH, while SSS remains nearly constant; a threshold difference of $0.2\,°C$ was used for recognizing bad data (Gaillard et al., 2015). The analysis of the conductivity measurements pointed out the presence of episodic quick decreases to underestimated, and often unreliable, values; of course, this is also reflected in SSS, with variations that range between a few decimal places and several units on the PSS. This phenomenon is due to the presence of air bubbles in the conductivity cell usually associated with strong waves and severe sea conditions. The effect of air bubbles usually runs out in a few minutes. Measurements showing evidence of this conductivity (salinity) decrease were flagged as bad data. Furthermore, information included in the cruise reports was carefully analyzed to be aware of these events and other suspicious malfunctioning of the system. Harbor data and observations collected when sailing into ice fields were also flagged at this step (taking advantage of latitude and/or longitude information provided by cruise reports), and discarded similarly to all the other bad data.

Finally, a visual inspection of individual campaigns was carried out (QC-3), with a specific attention to data flagged as suspicious in the QC-2. An additional 2 % of bad measurements were discarded, while suspicious values passing this analysis were set to good. Nevertheless, a sea ice flag was included in the dataset for identifying valid measurements that could have been slightly affected by scattered sea ice contamination, as recorded by cruise reports, when sailing the Southern Ocean. Sea ice maps retrieved through satellite passive microwave sensors (Spreen et al., 2008; Aulicino et al., 2013, 2014); synthetic aperture radar (SAR), when available, (Wadhams et al., 2016, 2018), and thermal infrared imagery (Aulicino et al., 2018b) were used for confirming ice field locations. However, when not discarded, these flagged data do not seem to significantly affect the good agreement

between the provided TSG dataset and several independent SSS products (see Sect. 3).

It is important to note that some of the conductivity, SST, and SSTH values associated with the published SSS measurements are missing (Table 2). We inform the user that the QC-2 described above was not complete for these values, whereas an in-depth visual inspection was performed.

2.2 Validation versus bottle samples

During all *S.A. Agulhas-I* and *S.A. Agulhas-II* research voyages, ship-based scientific teams collected salinity samples from the uncontaminated underway lab supply usually at each 20–30 nautical miles. These samples were taken in 250 mL double-cap glass bottles with rubber stoppers and completely filled to minimize evaporative error. In most cases, these independent water samples were analyzed directly on board with a Portasal salinometer 8410A in order to get a potential reference for adjusting the TSG data. Triple Portasal measures of each sample (then averaging) were usually performed on bottle samples to reduce possible errors. During some cruises salinity samples were not analyzed on board but later, due to severe weather conditions. Actually, no systematic bias was found, thus no adjustment to TSG measurements was necessary. Figure 3 shows an example of the comparison between the TSG data and the bottled samples analyzed with the Portasal during the southward leg of the SANAE 2012 cruise. Only few outliers are present at the start and, mostly, at the end of this leg. The offset between TSG and Portasal salinity values is plotted in Fig. 4; the average standard deviation, correcting for the outlier, would result in a $p < 0.01$. Similar results were found for the other cruises. A compilation of these results is provided in Fig. 5 in order to better assess the quality of the whole dataset. The mean and standard deviation of the difference between TSG measurements and bottle samples per ($1°$) latitudinal $1°$ bin are reported. Although a different number of samples was processed for each latitudinal bin (much more data were

Table 3. Name and description of the main variables included in the TSG NetCDF files.

Name of variable	Unit	Description
TIME	dd/mm/yyyyhh:mm	Date and time of TSG measurement
LAT	decimal degree	Latitude of TSG measurement
LON	decimal degree	Longitude of TSG measurement
COND	Siemens per meter	TSG conductivity measurement
SSS	–	Salinity retrieved from SST and COND
SST	degrees Celsius	TSG temperature measurement
SSTH	degrees Celsius	Temperature measurement at hull
SEA ICE	0–1	Sea ice flag 0 = no ice, 1 = scatter ice

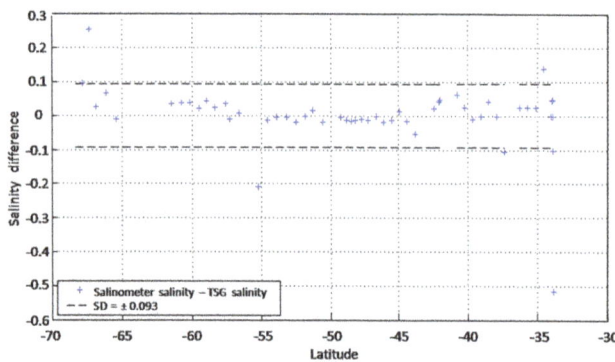

Figure 4. A diagram showing the difference in salinity between the TSG and the Portasal salinometer during the southward leg of the SANAE 2012 cruise. Standard deviation value (SD) is reported in the legend.

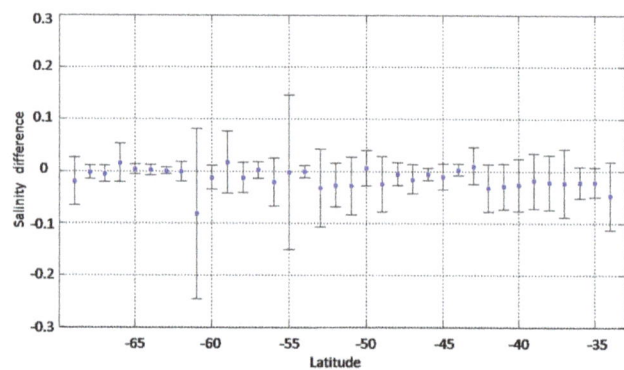

Figure 5. Mean (blue dots) and standard deviation (gray bars) values of difference in salinity between the TSG and the Portasal salinometer per latitudinal 1° bin for the 2010–2017 dataset.

collected north of 48° S), an overall general agreement was found with evident outliers identified only at 55 and 61° S. This is most likely due to careful maintenance of the TSG on board the *S.A. Agulhas-I* and *S.A. Agulhas-II* provided by the scientific teams, including cleaning of the tank, pump, and conductivity cell at the beginning of each cruise; stopping the TSG when entering the ice field; and constant monitoring of possible bio-fouling during the cruise.

3   Comparison of TSG sea surface salinities to other reference datasets

For a general assessment of our measurements, the TSG SSS dataset was compared to several global gridded reference SSS datasets over the southern Atlantic and the adjacent sectors of the Southern Ocean covered by the South African cruises. The reference datasets, which do not include South African TSG information, are the following: (i) the World Ocean Atlas 2013 (WOA13), (ii) the Global ARMOR3D L4 products, and (iii) the GLORYS ocean reanalysis. A point-to-point comparison with Argo measurements was also attempted, but the very low number of co-located observations was considered insufficient.

WOA13 is a long-term set of objectively analyzed climatologies at annual, seasonal, and monthly timescales produced by the National Oceanic and Atmospheric Administration's National Oceanographic Data Center (NOAA-NODC). We used monthly composite salinity fields on a 0.25° grid (Zweng et al., 2013) for a comparison with our TSG SSS data. All WOA13 climatological mean fields are available on the NODC website (www.nodc.noaa.gov) in NetCDF as well as other common file formats. ARMOR3D is a monthly objective reanalysis that includes salinity on a 0.25° regular grid on 33 depth levels (also at a weekly period in V4). ARMOR L4 products are obtained by assimilating satellite and in situ observations through statistical methods around a climatology. In particular, the ARMOR3D temperature–salinity (T / S) combined fields are generated using a two step procedure: synthetic fields are obtained from sea level anomalies and SST satellite information projected onto the vertical using a multiple linear regression method and the covariances deduced from historical observations; then, the synthetic fields and all available in situ T / S profiles (including Argo and CTD profiles) are combined through an optimal interpolation method (Guinehut et al., 2012). ARMOR3D data are available through the Copernicus online catalogue (http://marine.copernicus.eu/). GLORYS (V4) is a

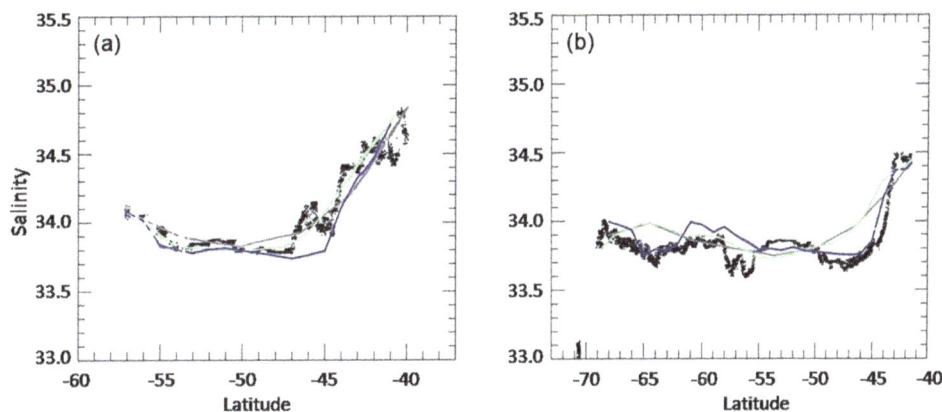

Figure 6. A comparison between TSG SSS (black dots) and monthly gridded products from WOA13 (gray), ARMOR3D (green), and GLORYS (blue) gridded salinities during July 2012 **(a)** and February 2013 **(b)** for *S.A. Agulhas-II* scientific cruises.

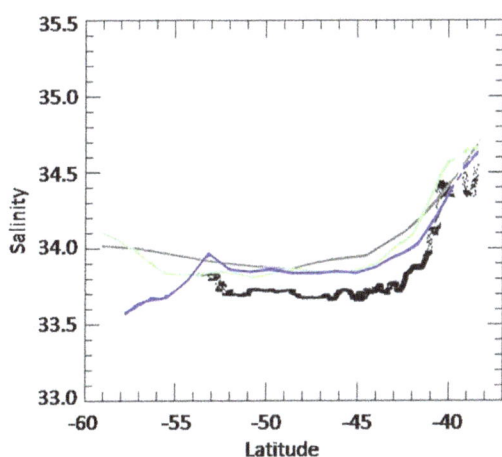

Figure 7. A comparison between TSG SSS (black) and monthly gridded products from WOA13 (gray), ARMOR3D (green), and GLORYS (blue) gridded salinities during December 2015 for *S.A. Agulhas-II* scientific cruises showing a large negative bias between transect and gridded data over Antarctic waters which is solidly supported by bottles validation.

reanalysis project carried out in the framework of the Copernicus Marine Environment Monitoring Service (CMEMS), which produces and distributes daily global ocean reanalysis on 75 levels at eddy permitting resolution (0.25°). Salinity products are generated through the assimilation (based on a reduced order Kalman filter) of in situ T/S observations (including also T/S profiles collected by sea mammals) using the NEMO dynamical ocean model in the ORCA025 configuration (Ferry et al., 2015). Data are available through the Copernicus online catalogue.

The TSG SSS dataset showed generally good agreement to the ensemble of reference datasets (WOA13, ARMOR, and GLORYS), with absolute biases generally lower than 0.1 and well within the level of spread found among the references themselves. A compilation of the statistical results is reported in Table 4. We were aware that differences due to the small-scale variability filtered out by the gridded products could emerge when comparing local measurements to larger-scale and monthly-averaged salinity fields. However, even though instances where transect data deviate from the reference ensemble were identified, overall local differences were lower than expected. The standard deviation of the local differences is between 0.1 and 0.2 and likely corresponds to mesoscale variability. The austral winter and summer examples reported in Fig. 6 show that TSG salinities agree well with gridded references regardless of the season; on the other hand, they suggest that the level of agreement changes with latitude in the study region. In the sub-Antarctic waters, TSG SSS follows very well the large-scale signature of the salinity fronts featured in the gridded products; some deviations are present, presumably related to actual salinity mesoscale and sub-mesoscale structures that the monthly maps cannot resolve. When approaching Antarctic waters and the sea ice edge, larger spreads from reference products can be found; but most of the significant deviations disappear when masking the sea ice flagged SSS values in the transect data (as in Fig. 6). These differences can be ascribed to the presence of scattered sea ice, which may influence SSS and/or could affect the TSG nominal functioning due to the presence of ice crystals (e.g., causing seawater flux to slow down or temporary interruption); but could also be due to the sparseness of in situ observations at high latitudes in the gridded products. Of particular relevance are the larger biases found between the TSG SSS and the reference datasets later in 2015 (Fig. 7), revealing a signature of surface freshening possibly associated with the low Antarctic sea ice extent anomaly of 2016, which the gridded references apparently fail to detect. In this case, the freshening signature captured in the TSG SSS is effectively supported by bottle validation, lending it further credibility.

Table 4. Compilation of the statistical results (bias and SD) obtained through the analysis of the difference between reference salinities (i.e., WOA09, ARMOR, GLORYS) and TSG salinities, for each transect of the analyzed research cruises. The comparison is limited to the end of 2015, i.e., the last available year with complete reanalysis data for the reference datasets.

Cruise	Dates year_month_dd-dd	Bias	SD	Bias	SD	Bias	SD
		WOA09		ARMOR		GLORYS	
SANAE 2010	2010_12_11-22	0.07	0.10	0.07	0.12	−0.09	0.14
SANAE 2010	2011_01_04-12	−0.11	0.12	−0.16	0.14	−0.36	0.23
SANAE 2010	2011_01_12-23	−0.11	0.09	−0.14	0.10	−0.37	0.19
SANAE 2011	2012_01_03-18	0.46	0.19	0.45	0.17	0.20	0.29
SANAE 2011	2012_01_18-25	0.26	0.30	0.25	0.27	0.05	0.26
Winter 2012	2012_07_15-23	< 0.01	0.09	< 0.01	0.08	−0.08	0.10
Winter 2012	2012_07_23-03	0.08	0.17	0.05	0.15	< 0.01	0.13
Gough 2012	2012_09_21-25	−0.04	0.20	−0.03	0.20	−0.05	0.16
Gough 2012	2012_09_25-29	−0.06	0.14	−0.08	0.20	−0.12	0.18
SANAE 2012	2012_12_11-28	0.03	0.13	0.02	0.12	−0.03	0.10
SANAE 2012	2013_01_16-25	0.12	0.18	0.08	0.18	0.05	0.21
SANAE 2012	2013_01_25-31	0.04	0.13	0.01	0.14	0.07	0.18
SANAE 2012	2013_02_01-18	0.09	0.12	0.09	0.13	0.06	0.09
Marion 2014	2014_04_07-19	0.02	0.13	−0.01	0.14	−0.01	0.12
Marion 2014	2014_04_19-05	0.07	0.13	< 0.01	0.15	0.01	0.13
SANAE 2014	2014_12_07-17	0.06	0.12	0.03	0.15	0.09	0.14
SANAE 2014	2014_12_30-15	0.22	0.17	0.20	0.17	0.09	0.18
SANAE 2014	2015_01_15-22	0.15	0.13	0.07	0.15	0.05	0.17
SANAE 2014	2015_02_09-17	0.12	0.24	0.15	0.16	0.09	0.17
Winter 2015	2015_07_29-06	0.13	0.08	0.08	0.09	0.02	0.08
Winter 2015	2015_08_06-10	0.16	0.09	0.15	0.12	0.09	0.05
SANAE 2015	2015_12_10-15	0.19	0.11	0.17	0.08	0.12	0.10
SANAE 2015	2015_12_27-08	0.19	0.11	0.13	0.11	0.07	0.11

Furthermore, several interesting geophysical features can be identified and analyzed through the TSG dataset. Among others, the presence of sharp gradients in proximity to the Antarctic Circumpolar Current (ACC) fronts is really evident. A coupled analysis of TSG SSS and expendable bathythermograph (XBT) observations collected along each cruise could be attempted for monitoring front locations and variability, and eventually analyze transport and variability in the ACC south of Africa (Swart et al., 2008), also in the perspective of climate change (Bard and Rickaby, 2009; Beal et al., 2011). Then, TSG data collected in proximity to the Agulhas current and its retroflection could be very useful for studying the current fluctuations and the eventual presence of pulses and rings (Swart and Speich, 2010).

4   Data availability

TSG data are available to the public in text format through an unrestricted repository at https://doi.org/10.7289/ V56M3545. Table 3 summarizes the main variables, while the metadata are included in the "readme" file provided with data on the NOAA-NCEI archive. One file is created for each research cruise. The naming convention is code_yyyy.txt, where code is a cruise type identification name depending

on the cruise objective or enquired area, i.e., SANAE, WINTER, MARION, GOUGH; and yyyy is the year when TSG acquisition started.

5   Data records and conclusions

We believe that this exceptional SSS dataset represents a valuable source of high-resolution, independent, and reliable information capable of completing data collected through the existing observing networks (i.e., drifters, ARGO floats, glider fleets), and current state-of-the-art gridded salinity products. A sea ice flag helps with its correct use south of the Antarctic polar front. The final goal is enlarging the amount of in situ ocean observations available to the scientific community for addressing several climatic issues. In particular, improving the knowledge of sea surface thermohaline features is one of the most important results to be achieved for advancing studies focusing on climate variability in the Southern Hemisphere. Although the Southern Ocean is a key place for atmosphere–ocean interactions at different spatial and temporal scales (Cerrone et al., 2017a, b; Buongiorno Nardelli et al., 2017; Fusco et al., 2018), mesoscale and submesoscale processes acting in the Atlantic sector are still poorly known because of the limited number of available in situ measurements and the coarse accuracy/resolution of the available SSS satellite observations (Boutin et al., 2016). That is particularly true of the ACC region and its fronts, which are characterized by complex dynamics and intense eddy activity (Cotroneo et al., 2013; Frenger et al., 2015). Furthermore, even though causes have not been firmly defined, several studies pointed out that recent salinity changes in the Southern Ocean are among the most prominent signals of climate change in the global Ocean (Böning et al., 2008; Haumann et al., 2016); the freshening signature captured in our TSG SSS could contribute to this debate (Boutin et al., 2013).

In this framework, even though limited in time (i.e., few months per year) and space (i.e., the Atlantic sector of the Southern Ocean), the present TSG SSS dataset represents an uncommon opportunity to partially fill this lack of information, and a valuable tool for improving the reconstruction of density fields in combination with numerical simulations (Chen et al., 2017) and the calibration and validation of SSS satellite observations recently provided by SMOS and Aquarius missions.

Author contributions. GA and YC conceived and designed the manuscript. GA, YC, IA and MvdB collected the measurements and organized the TSG dataset. GA and CC carried out the quality control analyses and the validation of the TSG measurements versus the bottle samples. MBR and EOC performed the comparison of TSG salinities to other reference dataset. All authors analyzed the achieved results, contributed to the writing, and approved the final manuscript.

Competing interests. The authors declare that they have no conflict of interest.

Acknowledgements. We acknowledge the support of the Department of Environmental Affairs (DEA), South Africa; the South African National Antarctic Programme (SANAP), South Africa; and the Italian National Antarctic Research Programme (PNRA), Italy. This study was made possible thanks to the contribution of the Southern Ocean Chokepoint: an Italian Contribution (SO-ChIC) project and the Multiplatform Observations and Modeling in a sector of the Antarctic circumpolar current (MOMA) project. Special thanks go to the captain, officers, and crew of the *S.A. Agulhas-I* and *II* as well as all the technicians, scientists, and students on board the *S.A. Agulhas-I* and *S.A. Agulhas-II* research vessels who contributed to the TSG data functioning and cleaning, and to the onboard SSS validation activities.

Edited by: Giuseppe M. R. Manzella

References

Ansorge, I., Baringer, M. O., Campos, E. J. D., Dong, S., Fine, R. A., Garzoli, S. L., Goni, G., Meinen, C. S., Perez, R. C., Piola, A. R., Roberts, M. J., Speich, S., Sprintall, J., Terre, T., and van den Berg, M. A.: Basin-wide oceanographic array bridges the South Atlantic, Eos T. Am. Geophys. Un., 95, 53–54, 2014.

Ansorge, I., Jackson, J., Reid, K., Durgadoo, J., Swart, S., and Eberenz, S.: Evidence of a southward eddy corridor in the southwest Indian Ocean, Deep-Sea Res. Pt. II, 119, 69–76, 2015.

Aulicino, G., Fusco, G., Kern, S., and Budillon, G.: 1992–2011 sea ice thickness estimation in the Ross and Weddell Seas from SSM/I brightness temperatures, European Space Agency, Special Publication ESA SP-712, 2013.

Aulicino, G., Fusco, G., Kern, S., and Budillon, G.: Estimation of sea ice thickness in Ross and Weddell Seas from SSM/I brightness temperatures, IEEE T. Geosci. Remote, 52, 4122–4140, 2014.

Aulicino, G., Cotroneo, Y., Ruiz, S., Sánchez Román, A., Pascual, A., Fusco, G., Tintoré, J., and Budillon, G.: Monitoring the Algerian Basin through glider observations, satellite altimetry and numerical simulations along a SARAL/AltiKa track, J. Marine Syst., 179, 55–71, 2018a.

Aulicino, G., Sansiviero, M., Paul, S., Cesarano, C., Fusco, G., Wadhams, P., and Budillon, G.: A new approach for monitoring the Terra Nova Bay polynya through MODIS ice surface temperature imagery and its validation during 2010 and 2011 winter seasons, Remote Sens., 10, 366, https://doi.org/10.3390/rs10030366, 2018b.

Bard, E. and Rickaby, R. E. M.: Migration of the subtropical front as a modulator of glacial climate, Nature, 460, 380–383, 2009.

Beal, L. M., De Ruijter, W. P., Biastoch, A., Zahn, R., and SCOR/WCRP/IAPSO Working Group 13: On the role of the Agulhas system in ocean circulation and climate, Nature, 472, 429–436, 2011.

Böning, C. W., Dispert, A., Visbeck, M., Rintoul, S. R., and Schwarzkopf, F. U.: The response of the Antarctic Circumpolar Current to recent climate change, Nat. Geosci., 1, 864–869, 2008.

Boutin, J., Martin, N., Reverdin, G., Yin, X., and Gaillard, F.: Sea surface freshening inferred from SMOS and ARGO salinity: impact of rain, Ocean Sci., 9, 183–192, https://doi.org/10.5194/os-9-183-2013, 2013.

Boutin, J., Chao, Y., Asher, W. E., Delcroix, T., Drucker, R., Drushka, K., Kolodziejczyk, N., Lee, T., Reul, N., Reverdin, G., Schanze, J., Soloviev, A., Yu, L., Anderson, J., Brucker, L., Dinnat, E., Santos-Garcia, A., Jones, W. L. Maes, C., Meissner, T., Tang, W., Vinogradova, N., and Ward, B.: Satellite and in situ salinity: Understanding near-surface stratification and subfootprint variability, B. Am. Meteorol. Soc., 97, 1391–1407, 2016.

Buongiorno Nardelli, B., Guinehut, S., Verbrugge, N., Cotroneo, Y., Zambianchi, E., and Iudicone, D.: Southern Ocean mixed-layer seasonal and interannual variations from combined satellite and in situ data, J. Geophys. Res., 122, 10042–10060, 2017.

Cerrone, D., Fusco, G., Simmonds, I., Aulicino, G., and Budillon, G.: Dominant covarying climate signals in the Southern Ocean and Antarctic sea ice influence during the last three decades, J. Climate, 30, 3055–3072, 2017a.

Cerrone, D., Fusco, G., Cotroneo, Y., Simmonds, I., and Budillon, G.: The Antarctic Circumpolar Wave: Its presence and interdecadal changes during the last 142 years, J. Climate, 30, 6371–6389, 2017b.

Chen, J., Zhang, R., Wang, H., Yuzhu, A., An, Y., Wang, L., and Wang, G.: An analysis on the error structure and mechanism of soil moisture and ocean salinity remotely sensed sea surface salinity products, Acta Oceanol. Sin., 33, 48–55, 2014.

Chen, J., You, X., Xiao, Y., Zhang, R., Wang, G., and Bao, S.: A performance evaluation of remotely sensed sea surface salinity products in combination with other surface measurements in reconstructing three-dimensional salinity fields, Acta Oceanol. Sin., 36, 15–31, 2017.

Cotroneo, Y., Budillon, G., Fusco, G., and Spezie, G.: Cold core eddies and fronts of the Antarctic Circumpolar Current south of New Zealand from in situ and satellite data, J. Geophys. Res.-Oceans, 118, 2653–2666, 2013.

Cotroneo, Y., Aulicino, G., Ruiz, S., Pascual, A., Budillon, G., Fusco, G., and Tintoré, J.: Glider and satellite high resolution monitoring of a mesoscale eddy in the Algerian basin: effects on the mixed layer depth and biochemistry, J. Marine Syst., 162, 73–88, 2015.

D'Addezio, J. M. and Subrahmanyam, B.: Sea surface salinity variability in the Agulhas Current region inferred from SMOS and Aquarius, Remote Sens. Environ., 180, 440–452, 2016.

Ferry, N., Parent, L., Masina, S., Storto, A., Haines, K., Valdivieso, M., Barnier, B., Molines, J. M., Zuo, H., and Balmaseda, M.: Product user manual for Global Ocean Reanalysis Products, CMEMS version scope: Version 1.0, Copernicus Marine Environment Monitoring Service (CMEMS), 2015.

Frenger, I., Muennich, M., Gruber, N., and Knutti, R.: Southern Ocean eddy phenomenology, J. Geophys. Res.-Oceans, 120, 7413–7449, 2015.

Fusco, G., Cotroneo, Y., and Aulicino, G.: Different behaviours of the Ross and Weddell seas surface heat fluxes in the period 1972–2015, Climate, 6, 17, https://doi.org/10.3390/cli6010017, 2018.

Gaillard, F., Diverres, D., Jacquin, S., Gouriou, Y., Grelet, J., Le Menn, M., Tassel, J., and Reverdin, G.: Sea surface temperature and salinity from French research vessels, 2001–2013, Scientific Data, 2, 150054, https://doi.org/10.1038/sdata.2015.54, 2015.

Guinehut, S., Dhomps, A.-L., Larnicol, G., and Le Traon, P.-Y.: High resolution 3-D temperature and salinity fields derived from in situ and satellite observations, Ocean Sci., 8, 845–857, https://doi.org/10.5194/os-8-845-2012, 2012.

Haumann, F. A., Gruber, N., Münnich, M., Frenger, I., and Kern, S.: Sea-ice transport driving Southern Ocean salinity and its recent trends, Nature, 537, 89–92, 2016.

Helm, K. P., Bindoff, N. L., and Church, J. A.: Changes in the global hydrological-cycle inferred from ocean salinity, Geophys. Res. Lett., 37, L18701, https://doi.org/10.1029/2010GL044222, 2010.

Kerr, Y., Waldteufel, P., Wigneron, J. P., Delwart, S., Cabot, F., Boutin, J., Escorihuela, M. J., Font, J., Reul, N., Gruhier, C., Juglea, S., Drinkwater, M., Hahne, A., Martin-Neira, M., and Mecklenburg, S.: The SMOS mission: A new tool for monitoring key elements of the global water cycle, Proceedings of the IEEE, 98, 666–687, 2010.

Kohler, J., Sena Martins, M., Serra, N., and Stammer, D.: Quality assessment of spaceborne sea surface salinity observations over the northern North Atlantic, J. Geophys. Res.-Oceans, 120, 94–112, 2015.

Lagerloef, G., Boutin, J.,Chao, Y., Delcroix, T., Font, J., Niiler, P., Reul, N., Riser, S., Schmitt, R., Stammer, D., and Wentz, F.: Resolving the global surface salinity field and variations by blending satellite and in situ observations, in: Proceedings of OceanObs'09: Sustained Ocean Observations and Information for Society (Vol. 2), Venice, Italy, 21–25 September 2009, edited by: Hall, J., Harrison, D. E., and Stammer, D., 11 pp., 2009.

Le Vine, D., Lagerloef, G., and Torrusio, S.: Aquarius and remote sensing of sea surface salinity from space, Proceedings of the IEEE, 98, 688–703, 2010.

Mangoni, O., Saggiomo, V., Bolinesi, F., Margiotta, F., Budillon, G., Cotroneo, Y., Misic, C., Rivaro, P., and Saggiomo, M.: Phytoplankton blooms during austral summer in the Ross Sea, Antarctica: driving factors and trophic implications, PLoS ONE, 12, e0176033, https://doi.org/10.1371/journal.pone.0176033, 2017.

Misic, C., Covazzi Harriague, A., Mangoni, O., Aulicino, G., Castagno, P., and Cotroneo, Y.: Effects of physical constraints on the lability of POM during summer in the Ross Sea, J. Marine Syst., 166, 132–143, 2017.

Rahmstorf, S.: Thermohaline Ocean Circulation, in: Encyclopedia of Quaternary Sciences, edited by: Elias, S. A., Elsevier, Amsterdam, 2006.

Reul, N., Chapron, B., Lee, T., Donlon, C., Boutin, J., and Alory, G.: Sea surface salinity structure of the meandering Gulf Stream revealed by SMOS sensor, Geophys. Res. Lett., 41, 3141–3148, 2014.

Rivaro, P., Ianni, C., Langone, L., Ori, C., Aulicino, G., Cotroneo, Y., Saggiomo, M., and Mangoni, O.: Physical and biological forcing of mesoscale variability in the carbonate system of the Ross Sea (Antarctica) during summer 2014, J. Mar. Syst., 166, 144–158, 2017.

Sansiviero, M., Morales Maqueda, M. Á., Fusco, G., Aulicino, G., Flocco, D., and Budillon, G.: Modelling sea ice formation in the Terra Nova Bay polynya, J. Marine Syst., 166, 4–25, 2017.

Spreen, G., Kaleschke, L., and Heygster, G.: Sea ice remote sensing using AMSR-E 89-GHz channels, J. Geophys. Res., 113, C02S03, https://doi.org/10.1029/2005JC003384, 2008.

Swart, S. and Speich, S.: An altimetry-based gravest empirical mode south of Africa: 2. Dynamic nature of the Antarctic Circumpolar Current fronts, J. Geophys. Res., 115, C03003, https://doi.org/10.1029/2009JC005300, 2010.

Swart, S., Speich, S., Ansorge, I., Goni, G., Gladyshev, S., and Lutjeharms, J. R. E.: Transport and variability of the Antarctic Circumpolar Current south of Africa, J. Geophys. Res., 113, C09014, https://doi.org/10.1029/2007JC004223, 2008.

Tang, W., Yueh, S. H., Fore, A. G., and Hayashi, A.: Validation of Aquarius sea surface salinity with in situ measurements from Argo floats and moored buoys, J. Geophys. Res.-Oceans, 119, 6171–6189, 2014.

Wadhams, P., Aulicino, G., Parmiggiani, F., and Pignagnoli, L.: Sea ice thickness mapping in the Beaufort Sea using wave dispersion in pancake ice – a case study with intensive ground truth, European Space Agency, Special Publication ESA SP-740, 2016.

Wadhams, P., Aulicino, G., Parmiggiani, F., Persson, P. O. G., and Holt, B.: Pancake ice thickness mapping in the Beaufort Sea from wave dispersion observed in SAR imagery, J. Geophys. Res.-Oceans, 123, 2213–2237, https://doi.org/10.1002/2017JC013003, 2018.

World Meteorological Organization: GCOS 2016 Implementation Plan, "The Global Observing System for Climate: Implementation Needs", 2016.

Zweng, M. M, Reagan, J. R., Antonov, J. I., Locarnini, R. A., Mishonov, A. V., Boyer, T. P., Garcia, H. E., Baranova, O. K., Johnson, D. R., Seidov, D., and Biddle, M. M.: World Ocean Atlas 2013, Volume 2: Salinity, edited by: Levitus, S. and Mishonov, A., NOAA Atlas NESDIS, 74 pp., 2013.

7

Wind and wave dataset for Matara, Sri Lanka

Yao Luo[1], Dongxiao Wang[1], Tilak Priyadarshana Gamage[2], Fenghua Zhou[1],
Charith Madusanka Widanage[1], and Taiwei Liu[3]

[1]State Key Laboratory of Tropical Oceanography, South China Sea Institute of Oceanology,
Chinese Academy of Sciences, Guangzhou 510301, China
[2]University of Ruhuna, Matara 810000, Sri Lanka
[3]China Harbour Engineering Company Ltd., Beijing 100027, China

Correspondence: Dongxiao Wang (dxwang@scsio.ac.cn)

Abstract. We present a continuous in situ hydro-meteorology observational dataset from a set of instruments first deployed in December 2012 in the south of Sri Lanka, facing toward the north Indian Ocean. In these waters, simultaneous records of wind and wave data are sparse due to difficulties in deploying measurement instruments, although the area hosts one of the busiest shipping lanes in the world. This study describes the survey, deployment, and measurements of wind and waves, with the aim of offering future users of the dataset the most comprehensive and as much information as possible. This dataset advances our understanding of the nearshore hydrodynamic processes and wave climate, including sea waves and swells, in the north Indian Ocean. Moreover, it is a valuable resource for ocean model parameterization and validation. The archived dataset (Table 1) is examined in detail, including wave data at two locations with water depths of 20 and 10 m comprising synchronous time series of wind, ocean astronomical tide, air pressure, etc. In addition, we use these wave observations to evaluate the ERA-Interim reanalysis product. Based on Buoy 2 data, the swells are the main component of waves year-round, although monsoons can markedly alter the proportion between swell and wind sea.

1 Background and summary

Ocean observational data are difficult to obtain but are needed for model validation and data assimilation. Specifically, wind and wave observational data are of great importance for the study of ocean surface mixing, local air–sea interactions, coastal hydrodynamic characteristics, sediment movement, monsoon studies, and ocean engineering.

In this study, wind and wave observational datasets were simultaneously collected in a nearshore area off Matara, Sri Lanka. Matara is the southernmost city of the Indian Peninsula and is bordered by the northern Indian Ocean. The northern Indian Ocean is an important area for studying swells in the global wave climate, which persistently form in the southern oceans owing to the absence of land barriers (Alves, 2006). In addition, Sri Lanka is a good observation point for the southern Asian monsoon. Many wind and wave obser-

vations have been made in the Bay of Bengal (Glejin et al., 2013a; Nayak et al., 2013) and Arabian Sea (Aboobacker et al., 2011a; Glejin et al., 2013b); however, investigations of wind and waves in the waters south of Sri Lanka, which hosts one of the busiest shipping lanes in the world, are scarce.

Although datasets from reanalysis projects (Semedo et al., 2010) and models (Sabique et al., 2012; Bhaskaran et al., 2014; Murty et al., 2014) are increasingly common, in situ observations (Rapizo et al., 2015) remain the most important reflections of the "true" ocean and atmosphere state. Therefore, most major developments in the marine and atmospheric sciences have been closely related to the development of in situ observations.

The wind and wave dataset (Table 2) presented herein was collected from an automated weather station (AWS) and wave buoys in the south of Sri Lanka (see Fig. 1 for locations). Construction of the observation system, which in-

Table 1. Dataset profile.

Dataset title	Wind and wave dataset for Matara, Sri Lanka		
Time range	Wave observations: 2013.9–2014.2 and 2013.4–2014.4 Wind and air pressure observations: 2012.12–2014.6 and 2015.12–2016.10		
Geographic scope	Sri Lanka; north Indian Ocean		
Data format	".xlsx" for observed data	Data volume	5.63 MB for the 30 or 10 min observations of wind and waves
Data service system	http://www.sciencedb.cn/dataSet/handle/447 https://doi.org/10.11922/sciencedb.447		

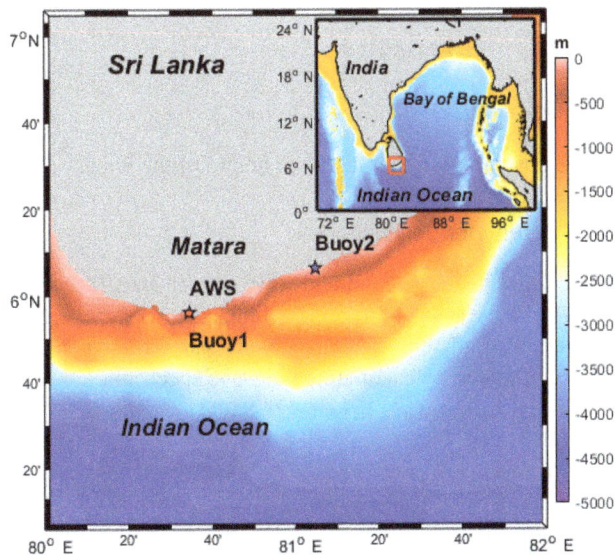

Figure 1. Observation deployment locations and topography (m). The distance between AWS and Buoy1 is less than 1000 m.

cludes a 25 m meteorological tower and an acoustic sounder, began in 2011 and is ongoing. To date, a variety of marine atmospheric parameters have been measured, including temperature, air pressure, rainfall, wind, wave, and tide. Data from this observational system will be available for monsoon, air–sea interactions, and coastal dynamics research.

2 Basic data description and quality control

The wind and wave dataset for Matara, Sri Lanka, consists of data obtained mainly from two different observation systems: the AWS and wave buoys.

Table 2 shows the details of the dataset and Fig. 1 shows the locations of the instruments. After rigorous quality control, the time series of wind, air pressure, and surface waves from the AWS, Buoy 1, and Buoy 2 were used to generate datasets with temporal resolutions of 30 min, 10 min, and 1 h, respectively. At the data collection stage, 24 elements were measured. It was necessary to rigorously control data qual-

ity for all elements. The main technical processes involved in generating the dataset are as follows:

1. calibration of all sensors,

2. transfer of a data in real time to a storage facility (to reduce or remove manual intervention),

3. error detection,

4. deviation testing and removal of records that deviate by > 3 standard deviations from the data mean.

To coincide with this publication, and in addition to the accompanying archived dataset spanning 2012–2016, unrestricted access is now available from a live online data repository located at http://www.sciencedb.cn/dataSet/handle/447. Contact details are provided to external organizations (to which the authors have no affiliation) to request additional wave spectra, tidal and wind parameters, or other elements of real-time observations after October 2016.

2.1 Data collection methods

The data collected in southern Sri Lanka were taken from an integrated observation system that included the wave buoys (Buoy 1 and Buoy 2) and the AWSs (AWS1 and AWS2) (for details, see Fig. 2 and Table 2).

Buoy 1, which was anchored at the bottom of the sea at a depth of 20 m in September 2013, was operated normally and continuously to collect water level and wave data for about 5 months. Buoy 1 measured many types of non-directional wave parameters, as well as air pressure at the sea surface and water level. It was powered with a cable and accepted wave information with a pressure sensor, which was developed at the South China Sea Institute of Oceanology. Pressure data were recorded continuously at a rate of 1 Hz over 5 min. The data for every 10 min are processed as one record. At the location of Buoy 1, a new buoy will be installed in the near future.

Buoy 2, which measured both wave and water level for 1 year from April 2013 to April 2014, consists of two instruments, a self-contained wave buoy at a depth of 10 m and

Table 2. List of datasets.

Instrument		Location		Period	Frequency	Variable
		Latitude	Longitude			
AWS	AWS1	5.936° N	80.575° E	Nov 2012–Jun 2014	30 min	Temp, rel. humidity, pressure, wind speeds and direction
	AWS2			Nov 2015–Oct 2016		Temp, rel. humidity, air pressure, wind speed, and direction
Wave buoy	Buoy 1	5.934° N	80.574° E	Sep 2013–Feb 2014	10 min	H_s, H_{max}, H_{10}, H_{avg}, T_p, air pressure, and water level
	Buoy 2	6.106° N	81.080° E	Apr 2013–Apr 2014	1 h	H_s, T_p, MWD, and P_{dir}

Figure 2. AWS1 (a) and AWS2 (b).

a water level gauge. To measure the waves, the instrument was set to a frequency of 2 Hz with a sampling period of 15 min. In addition, 15 min sampling was carried out at 1 h intervals. For water level measurements, sample recording was performed at 10 min intervals. Tide level was computed using the 10 min average of the pressure height data.

The seaside AWS (AWS1, Fig. 2) was installed in December 2012 and can measure meteorological parameters, including wind speed and direction, air temperature, air humidity, and air pressure at an elevation of 12 m. In December 2015, the AWS was upgraded and moved approximately 5 m, with all sensors installed at an elevation of 1.5 m (AWS2). The sampling frequency and parameters were based on AWS1, with the addition of a shortwave radiation sensor. AWS2 is still working and is planned to remain in operation indefinitely.

Since AWS1 and AWS2 collected wind speed at heights of 12 and 1.5 m, respectively, wind speed from the buoy (W_z) was transformed into wind speed at a 10 m elevation (W_{10}) using the Prandtl 1/7 law approximation (Streeter et al., 1998) with the equation

$$W_{10}/W_z = (10/z)^{1/7}, \tag{1}$$

where z is the wind measurement height of the buoy.

2.2 Parameter descriptions

The dataset includes wind and wave data in Excel files. For Buoy 1, wave parameters were obtained directly using the zero-crossing method. The main wave parameters for Buoy 2 were obtained by calculating the wave spectrum.

The significant wave height (H_s, SWH) is given as

$$H_s = 4\sqrt{m_0},$$

$$T_m = \frac{m_0}{m_1},$$

$$T_z = \sqrt{\frac{m_0}{m_2}}, \tag{2}$$

where m_n is the nth moment of spectral density, T_m and T_z are the mean wave period and zero-crossing wave period, respectively.

The energy-averaged mean wave direction α_m was determined as

$$\alpha_m = \frac{1}{m_0} \int S(f)\alpha(f)\mathrm{d}f, \tag{3}$$

where $S(f)$ and $\alpha(f)$ are the directional distributions.

3 Verification of reanalysis wave data

To study the impact of climate change on oceans, long-term datasets are needed; however, the lengths of continuous wave records are usually only a few to tens of years. An important use of observational wave and wind data is model validation and assimilation. The ERA-Interim is based on the

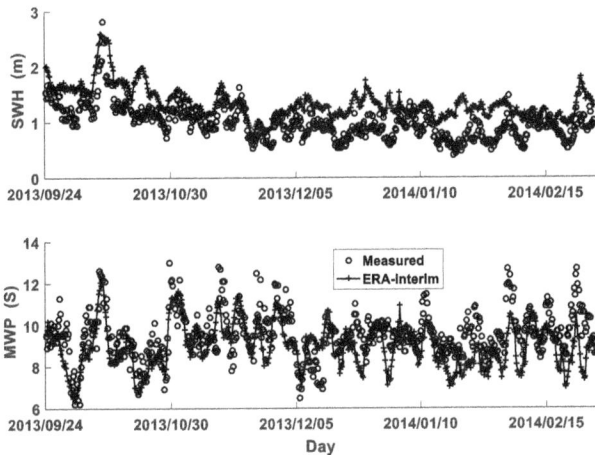

Figure 3. Time series of significant wave height (SWH) and mean wave period (MWD) for 24 September 2013–28 February 2014 (Buoy 1).

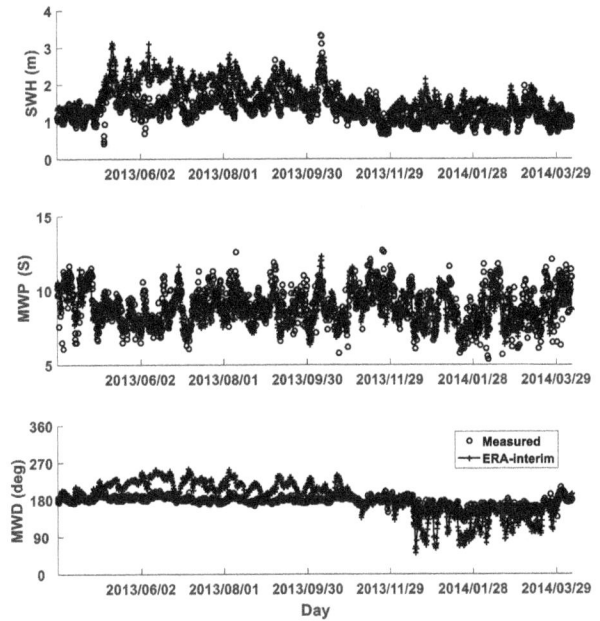

Figure 4. Time series of significant wave height (SWH), mean wave period (MWP), and mean wave direction (MWD) for 3 April 2013–9 April 2014 (Buoy 2).

global atmospheric reanalysis product of the European Centre for Medium-Range Weather Forecasts and is available from 1979 onwards (Dee et al., 2011). It comprises long-term and continuous grid data and is useful for wave extreme values, long-term variability of wave climate, and monsoon studies. Relative to the ERA-40 system, ERA-Interim incorporates many important Integrated Forecast System improvements, such as model resolution and physics changes, the use of four-dimensional variational (4D-Var) assimilation, and various other changes in the analysis methodology (Dee and Uppala, 2008, 2009). Furthermore, a reduction in the root-mean-square error (RMSE) of H_s against buoy data makes it smaller than that in ERA-40 (Bidlot et al., 2007). By comparison with observation data, the accuracy of ERA-Interim reanalysis data can be assessed, providing the basis to quantify the monthly, inter-annual, and decadal variability of wind and wave climate.

Extensive inter-comparison and evaluation of wind stress estimates from the reanalysis data set (ERA-Interim) against available in situ observations have been performed for the tropical Indian Ocean (Kumar et al., 2011). The verification of H_s in the reanalysis dataset has been performed only for the Arabian Sea and Bay of Bengal (Shanas and Sanil Kumar, 2014; Shanas and Kumar, 2015). In the present study, the nearest available ERA-Interim H_s data were compared to the data measured from situ buoy data (Figs. 3, 4 and 5). The comparison of the reanalysis and measured H_s data shows a general mean correlation (correlation coefficient = 0.747) with a general mean RMSE of 0.435 m during both years. For the mean wave period (MWP), the mean correlation coefficient and mean RMSE are 0.618 and 1.084; therefore, there are significant errors in the measurements, especially for MWD. These could be caused by the complicated near-

shore geometry and/or by the resolution of the large-scale ERA-Interim model, which is so low that it cannot describe changes in complex seabed terrain.

4   Wind data records

The wind is mainly controlled by northeast (NE) and southwest (SW) monsoons in the area. The wind speed and wind direction were measured at AWS1 and AWS2 (see Fig. 6 for the time series of wind speed). The maximum measured wind was $8.9 \, \mathrm{m \, s^{-1}}$ and a maximum gust $13.9 \, \mathrm{m \, s^{-1}}$. The wind sensors of AWS1 and AWS2 are located at heights of 12 and 1.5 m, respectively. Due to its high position, the coastal orography had no apparent impact on AWS1. However, for AWS2, the wind speed was small during NE monsoons and was clearly effected by the landform of the land to the north (see Fig. 1). During SW monsoons, the wind originated from sea and the effect was weak for AWS2. This explains the significantly lower wind speed during the winter months than the summer months (see Table 3). The monthly mean wind speed was the highest in September 2016 and lowest in December 2015 (Table 3).

5   Data availability

The data set was deposited in Science Data Bank, whose DOI is https://doi.org/10.11922/sciencedb.447. It can be downloaded publicly from http://www.sciencedb.cn/dataSet/handle/447.

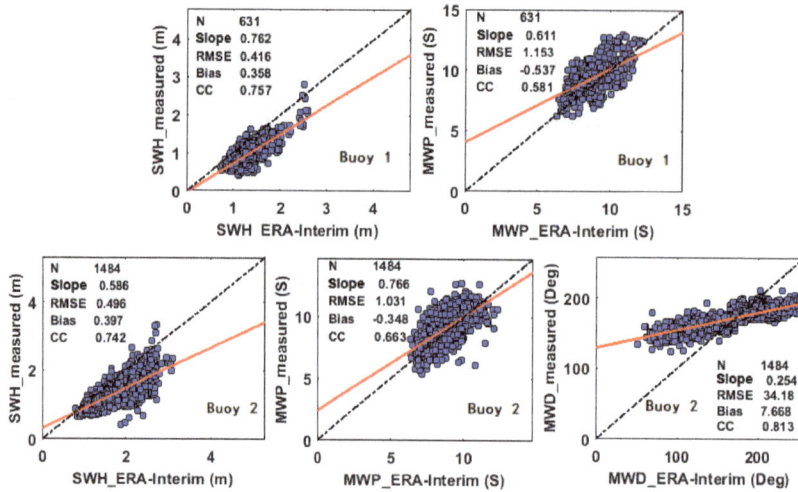

Figure 5. Scatter plots of SWH and MWP from the ERA-Interim. The displayed statistics are *N*, number of samples; Slope, slope of least-squares regression; RMSE, root-mean-square error; Bias, bias; and CC, Pearson's correlation coefficient.

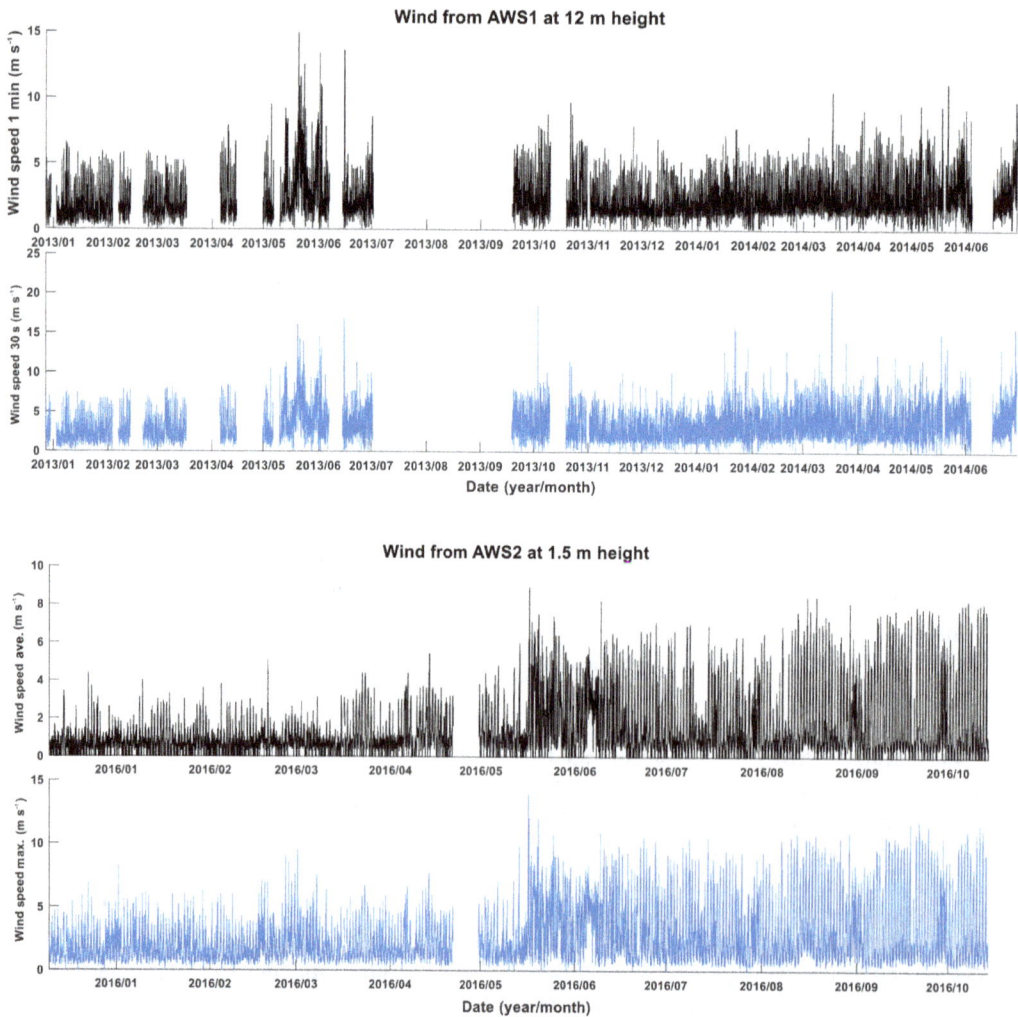

Figure 6. Wind speed measurements.

Table 3. Monthly average values for the averaged wind speed and maximum wind speed from AWS2 (WSA: averaged wind speed; WSM: maximum wind speed; num: the number of effective data points).

	WSA $\mathrm{m\,s^{-1}}$	$\mathrm{WSA_{max}}$ $\mathrm{m\,s^{-1}}$	WSM $\mathrm{m\,s^{-1}}$	$\mathrm{WSM_{max}}$ $\mathrm{m\,s^{-1}}$	num
201 512	0.69	4.40	1.65	6.9	1067
201 601	0.77	4.00	1.81	8.3	1488
201 602	0.75	5.10	1.84	9	1383
201 603	0.89	4.40	1.98	9.5	1447
201 604	1.27	5.40	2.14	7.7	1011
201 605	1.76	8.90	3.00	13.9	1456
201 606	1.89	8.20	3.40	10.9	1435
201 607	1.44	7.00	2.80	10.5	1470
201608	1.72	8.40	3.20	10.9	1484
201 609	2.17	7.90	3.67	11.8	1418
201 610	2.05	8.20	3.38	11.5	632

Table 4. Monthly percentage dominance of swell and sea.

Month	Swell (%)	Sea (%)
Jan	59.1	40.9
Feb	67.9	32.1
Mar	66.8	33.2
Apr	74.4	25.6
May	57.6	42.4
Jun	61.9	38.1
Jul	68.2	31.8
Aug	68.8	31.2
Sep	67.5	32.5
Oct	60.5	39.5
Nov	77.6	22.4
Dec	62.9	37.1

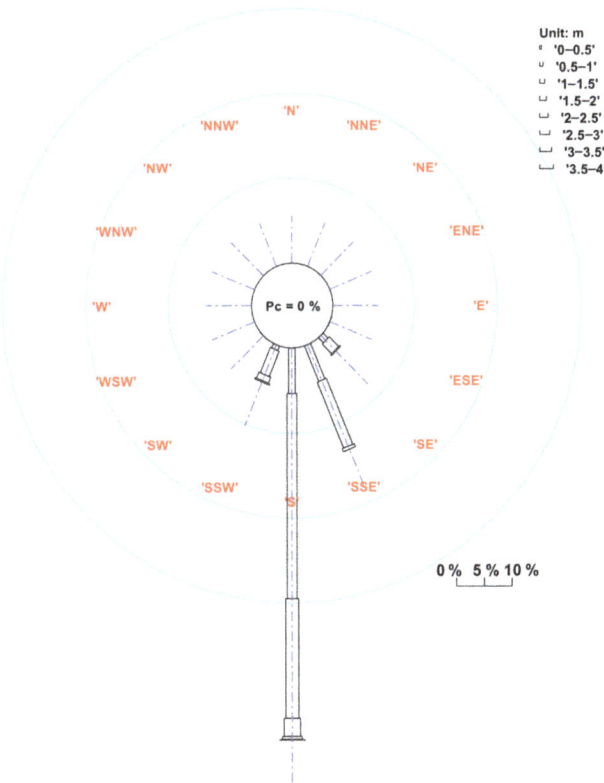

Figure 7. Wave rose of significant wave height for Buoy 2. South waves accounted for 69.12 % of the total.

6   Wave data records

The integrated rate of wave data in Buoy 2 is high (96.38 %); therefore, we investigated the characteristics of the wind sea and swell using the data from Buoy 2. The frequency range of the measured wave spectra was 0.026–0.3 Hz. The frequency associated with the peak energy is indicative of the domi-

nant wave system over the region at that time. The wind sea and swell parameters were separated by frequency. Figure 8 shows the time series H_s, wind sea H_s, and swell H_s. Waves in the Indian Ocean vary in response to prevailing wind systems according to the month, and include inter-Monsoon-1 (IM1), SW monsoon (May–September), inter-Monsoon-2 (IM2) (October–November), and NE monsoon (December–February).

Table 4 shows the monthly dominance of swell and wind sea on wave energy. Swells are predominant (>50 % in height) throughout the year. The maximum swell dominance is 77.6 %, which occurs in November. The role of wind seas is significant in January and May with a percentage dominance greater than 40 %, and the maximum wind seas dominance is 42.4 % in May. Annually, 66.1 % of the waves are dominated by swells. Even though swells are predominant year-round, wind seas contribute sufficiently to the resultant waves during the SW (35.18 %) and NE (36.7 %) monsoon seasons. During the inter-monsoon periods, wind weakening and wind sea strengthening are reduced accordingly. Waves from the south accounted for 69.12 % of all waves annually (Fig. 7), mainly driven by swells, which predominate throughout the year. As a result, the wave climate in this area is controlled mainly by long swells.

From Fig. 8b, the wind sea directions are SW and NE. In the NE monsoon period, the wind sea H_s is less than that during the SW monsoon period. Regarding swells, the wave direction throughout the year is predominantly south, although there are some small fluctuations after the SW monsoons. However, as with wind sea, the H_s is greater in the SW monsoon period. These phenomena originate from a mixture of the geographic location of Buoy 2 and prevailing wind systems. Buoy 2 is almost completely unsheltered from the south. In the SW monsoon season, except for long waves from the Southern Ocean, strong winds and sufficient fetch allow swells to grow and evolve in the northern Indian Ocean. However, because Buoy 2 is located in the nearshore zone with a short fetch, wind sea during the NE monsoon

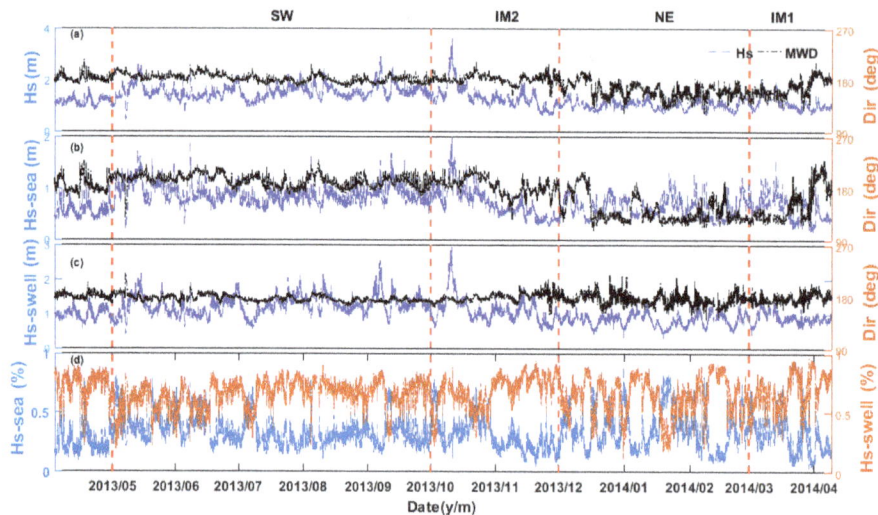

Figure 8. Time series of (a) H_s, (b) mean wave direction for H_s, (c) wind sea and swell, and (d) the percentage dominance of wind sea and swell (Buoy 2).

season is weaker. The NE monsoon has an inhibitory effect on swells from the south; therefore, there are small fluctuations in wave direction, and H_s is reduced during the NE monsoon season.

The unique geographic location of Sri Lanka to the south of the Indian Peninsula yields a different wave climate from those in the Bay of Bengal (Glejin et al., 2013a, b) and Arabian Sea (Aboobacker et al., 2011a, b), which are also located in the northern Indian Ocean and experience the same prevailing wind systems. Compared with the above two water bodies, swells have a larger impact on wind sea.

Author contributions. YL and DW designed the study and wrote the manuscript. YL, FZ, TPG, CMW, and TL performed the data collection, quality control, generation, and validation.

Competing interests. The authors declare that they have no conflict of interest.

Acknowledgements. We thank the China-Sri Lanka Joint Center for Education and Research for providing the hydro-meteorology observations. We also gratefully acknowledge the European Centre for Medium-Range Weather Forecasts Science Team for providing the ERA-Interim dataset. The work is supported by the Major State Research Development Program of China (2016YFC1402603) and the International Partnership Program of the Chinese Academy of Sciences (grant no. 131551KYSB20160002).

Edited by: Giuseppe M. R. Manzella

References

Aboobacker, V. M., Rashmi, R., Vethamony, P., and Menon, H. B.: On the dominance of pre-existing swells over wind seas along the west coast of India, Cont. Shelf Res., 31, 1701–1712, 2011a.

Aboobacker, V. M., Vethamony, P., and Rashmi, R.: "Shamal" swells in the Arabian Sea and their influence along the west coast of India, Geophys. Res. Lett., 38, 1–7, 2011b.

Alves, J. H. G. M.: Numerical modeling of ocean swell contributions to the global wind-wave climate, Ocean Model., 11, 98–122, 2006.

Bhaskaran, P. K., Gayathri, R., Murty, P. L. N., Bonthu, S. R., and Sen, D.: A numerical study of coastal inundation and its validation for thane, cyclone in the Bay of Bengal, Coast. Eng., 83, 108–118, 2014.

Bidlot, J., Janssen, P., and Abdalla, S.: A revised formulation of ocean wave dissipation and its model impact, in: Technical Report Memorandum 509, ECMWF, 2007.

Dee, D. P. and Uppala, S. M.: Variational bias correction in ERAInterim, Technical Memorandum No. 575, ECMWF: Reading, UK, available at: https://www.ecmwf.int/sites/default/files/elibrary/2008/8936-variational-bias-correction-era-interim.pdf, 2008.

Dee, D. P. and Uppala, S.: Variational bias correction of satellite radiance data in the era-interim reanalysis, Q. J. Roy. Meteor. Soc., 135, 1830–1841, 2009.

Dee, D. P., Uppala, S. M., Simmons, A. J., Berrisford, P., Poli, P., Kobayashi, S., Andrae, U., Balmaseda, M. A., Balsamo, G.,Bauer, P., Bechtold, P., Beljaars, A. C. M., van de Berg, L., Bidlot, J., Bormann, N., Delsol, C., Dragani, R., Fuentes, M., Geer, A. J., Haimberger, L., Healy, S. B., Hersbach, H., Hólm, E. V., Isaksen, L., Kållberg, P., Köhler, M., Matricardia, M., McNally, A. P., Monge-Sanz, B. M., Morcrette, J.-J., Park, B.-K., Peubey, C., de Rosnay, P., Tavolato, C., Thépaut, J.-N., and Vitart, F: The era-interim reanalysis: configuration and performance of the data assimilation system, Q. J. Roy. Meteor. Soc., 137, 553–597, 2011.

Glejin, J., Kumar, V. S., and Nair, T. M. B.: Monsoon and cyclone induced wave climate over the near shore waters off Puduchery, south western Bay of Bengal, Ocean Eng., 72, 277–286, 2013a.

Glejin, J., Sanil Kumar, V., Balakrishnan Nair, T. M., and Singh, J.: Influence of winds on temporally varying short and long period gravity waves in the near shore regions of the eastern Arabian Sea, Ocean Sci., 9, 343–353, https://doi.org/10.5194/os-9-343-2013, 2013b.

Kumar, B. P., Vialard, J., Lengaigne, M., Murty, V. S. N., and McPhaden, M. J.: TropFlux: air-sea fluxes for the global tropical oceans-description and evaluation, Clim. Dynam., 38, 1521–1543, 2011.

Luo, Y., Wang, D., Priyadarshana Gamage, T., Zhou, F., Widanage, C. M., and Liu, T.: A wind & wave dataset at Matara, Sri Lanka, Science Data Bank, https://doi.org/10.11922/sciencedb.447, 2017.

Murty, P. L. N., Sandhya, K. G., Bhaskaran, P. K., Jose, F., Gayathri, R., Nair, T. M. B., Srinivasa Kumar, T., and Shenoi, S. S. C.: A coupled hydrodynamic modeling system for PHAILIN cyclone in the Bay of Bengal, Coast. Eng., 93, 93, 71–81, 2014.

Nayak, S., Bhaskaran, P. K., Venkatesan, R., and Dasgupta, S.: Modulation of local wind-waves at Kalpakkam from remote forcing effects of southern ocean swells, Ocean Eng., 64, 23–35, 2013.

Rapizo, H., Babanin, A. V., Schulz, E., Hemer, M. A., and Durrant, T. H.: Observation of wind-waves from a moored buoy in the southern ocean, Ocean Dynam., 65, 1275–1288, 2015.

Sabique, L., Annapurnaiah, K., Nair, T. M. B., and Srinivas, K.: Contribution of southern Indian Ocean swells on the wave heights in the northern Indian Ocean – a modeling study, Ocean Eng., 43, 113–120, 2012.

Semedo, A., Sušelj, K., Rutgersson, A., and Sterl, A.: A global view on the wind sea and swell climate and variability from ERA-40, J. Climate, 24, 1461–1479, 2010.

Shanas, P. R. and Kumar, V. S.: Trends in surface wind speed and significant wave height as revealed by ERA-Interim wind wave hindcast in the Central Bay of Bengal, Int. J. Climatol., 35, 2654–2663, https://doi.org/10.1002/joc.4164, 2015.

Shanas, P. R. and Sanil Kumar, V.: Temporal variations in the wind and wave climate at a location in the eastern Arabian Sea based on ERA-Interim reanalysis data, Nat. Hazards Earth Syst. Sci., 14, 1371–1381, https://doi.org/10.5194/nhess-14-1371-2014, 2014.

Streeter, V. L., Wylie, E. B., and Bedford, K. W.: Fluid Mechanics, McGraw-Hill, Singapore, 1998.

8

Mediterranean Sea Hydrographic Atlas: towards optimal data analysis by including time-dependent statistical parameters

Athanasia Iona[1,2], Athanasios Theodorou[2], Sylvain Watelet[3], Charles Troupin[3], Jean-Marie Beckers[3], and Simona Simoncelli[4]

[1]Hellenic Centre for Marine Research, Institute of Oceanography, Hellenic National Oceanographic Data Centre, 46,7 km Athens Sounio, Mavro Lithari P.O. Box 712 19013 Anavissos, Attica, Greece

[2]University of Thessaly, Department of Ichthyology & Aquatic Environment, Laboratory of Oceanography, Fytoko Street, 38 445, Nea Ionia Magnesia, Greece

[3]University of Liège, GeoHydrodynamics and Environment Research, Quartier Agora, Allée du 6-Août, 17, Sart Tilman, 4000 Liège 1, Belgium

[4]Istituto Nazionale di Geofisica e Vulcanologia (INGV), Sezione di Bologna, Via Franceschini 31, 40128 Bologna, Italy

Correspondence: Athanasia Iona (sissy@hnodc.hcmr.gr)

Abstract. The goal of the present work is to provide the scientific community with a high-resolution atlas of temperature and salinity for the Mediterranean Sea based on the most recent datasets available and contribute to the studies of the long-term variability in the region. Data from the pan-European marine data infrastructure SeaDataNet were used, the most complete and, to our best knowledge, best quality dataset for the Mediterranean Sea as of today. The dataset is based on in situ measurements acquired between 1900 and 2015. The atlas consists of horizontal gridded fields produced by the Data-Interpolating Variational Analysis, in which unevenly spatial distributed measurements were interpolated onto a $1/8° \times 1/8°$ regular grid on 31 depth levels. Seven different types of climatological fields were prepared with different temporal integration of observations. Monthly, seasonal and annual climatological fields have been calculated for all the available years, seasonal to annual climatologies for overlapping decades and specific periods. The seasonal and decadal time frames have been chosen in accordance with the regional variability and in coherence with atmospheric indices. The decadal and specific-period analysis was not extended to monthly resolution due to the lack of data, especially for the salinity. The Data-Interpolating Variational Analysis software has been used in the Mediterranean region for the SeaDataNet and its predecessor Medar/Medatlas Climatologies. In the present study, a more advanced optimization of the analysis parameters was performed in order to produce more detailed results. The past and present states of the Mediterranean region have been extensively studied and documented in a series of publications. The purpose of this atlas is to contribute to these climatological studies and get a better understanding of the variability on timescales from months to decades and longer. Our gridded fields provide a valuable complementary source of knowledge in regions where measurements are scarce, especially in critical areas of interest such as the Marine Strategy Framework Directive (MSFD) regions and subregions.

1 Introduction

In oceanography, a climatology is defined as a set of gridded fields that describe the mean state of the ocean properties over a given time period. It is constructed by the analysis of in situ historical data sets and has many applications such as initialization of numerical models and quality control of observational data in real time and delayed mode, and it is used as a baseline for comparison to understand how the ocean is changing. The Mediterranean Sea is among the most interesting regions in the world because it influences the global thermohaline circulation and plays an important role in regulating the global climate (Lozier et al., 1995; Béthoux et al., 1998; Rahmstorf, 1998). It is therefore essential to improve our understanding of its dynamics and its process variability. It is a semi-enclosed sea divided by the Sicily Strait in two geographical basins, the western Mediterranean and the eastern Mediterranean, and is characterized by peculiar topographic deep depressions where nutrient-rich deep-water masses are stored for long time (Manca et al., 2004). It is a concentration basin, where evaporation exceeds precipitation. The general mean circulation pattern is driven by the continuous evaporation, heat flux exchanges with the atmosphere, the wind stress and the water mass exchanges between its basins and subbasins, as described in Robinson et al. (2001). In the surface layer (0–150 m) there is an inflow of warm and relatively fresh Atlantic water (AW; $S \approx 36.5$ ppt, $T \approx 15\,°C$) which is modified along its path to the eastern basin following a general cyclonic circulation. The intermediate layers (150–600 m) are dominated by the saline Levantine Intermediate Water (LIW; $S \approx 38.4$ ppt, $T \approx 13.5\,°C$), regularly formed at the Levantine basin. LIW is one of the most important Mediterranean water masses because it constitutes the higher percentage of the outflow from the Gibraltar Strait towards the Atlantic Ocean. In the deep layers, there are two main thermohaline cells. Deep waters formed via convective events in the northern regions of the western Mediterranean (WMDW – Western Mediterranean Deep Water; $S \approx 38.44$–38.46 ppt, $T \approx 12.75$–13.80 °C) in the Gulf of Lion and in the northern regions of the eastern Mediterranean (EMDW – Eastern Mediterranean Deep Waters) in the Adriatic ($S \approx 38.65$ ppt, $T \approx 13.0\,°C$) and Cretan seas ($S \approx 39$ ppt, $T \approx 14.8\,°C$). On top of this large-scale general pattern are superimposed several subbasin-scale and mesoscale cyclonic and anticyclonic motions due to topographic constraints and internal processes. The first climatological studies of the Mediterranean go back to 1966 with Ovchinnikov who carried out a geostrophic analysis to compute the surface circulation. The circulation features were describing a linear, stationary ocean. In the 1980s and 1990s, through a comprehensive series of observational studies and experiments in the western Mediterranean (La Violette, 1990; Millot, 1987, 1991, 1999), the subbasin and mesoscale patterns were discovered and the crucial role of eddies in modifying the mean climatological circulation and mixing properties inside the different subbasins which include the Tyrrhenian, Ligurian, and Alboran seas was emphasized (Bergamasco and Malanotte-Rizzoli, 2010). In 1987, Guibout constructed charts with the typical structures observed at sea, but with limited climatological characteristic as this atlas was based on the quasi-synoptic information of the selected cruise used. In 1982, an international research group was formed, called POEM (Physical Oceanography of the Eastern Mediterranean, 1984), under the auspices of the IOC/UNESCO and of CIESM (Commission Internationale pour l'Exploration Scientifique de la Méditerranée), which is focused on the description of the phenomenology of the eastern Mediterranean, by analyzing historical data and collecting new data (Malanotte-Rizzoli and Hecht, 1988). However, the focus was on the eastern Mediterranean Sea, as there was very little knowledge of this basin compared to the western Mediterranean and other world regions, and still there was no global data set to describe the eastern Mediterranean and its interaction with the western part efficiently. It is worth noting here that it was the POEM cruises that provided the observational evidence of the profound changes in the eastern Mediterranean deep circulation between the late 1980s and mid-1990s. During this phenomenon, known as Eastern Mediterranean Transient (EMT) the deep water production in the eastern Mediterranean shifted from the Adriatic to the Aegean. The bottom layers of the eastern Mediterranean were replaced by waters warmer but also much saltier and hence denser than the previous Eastern Mediterranean Deep Waters (Malanotte-Rizzoli et al., 1999; Lascaratos et al., 1999; Klein et al., 1999; Roether et al., 1996). The EMT affected not only the thermohaline characteristics of intermediate and deep water masses of eastern Mediterranean but also the western Mediterranean deep water production (Schröder et al., 2006; Schroeder et al., 2016). The EMT "signature" can be captured to the climatological references of a region and can be used to improve the near-real-time control procedures of operational and delayed mode data (Manzella et al., 2013). Hecht et al. (1988), by analyzing hydrographic measurements in the southeastern Levantine Basin, succeeded in describing the climatological water masses of the region and identifying their seasonal variations. In the western Mediterranean, Picco (1990) conducted an important climatological study and constructed a climatological atlas analyzing around 15 000 hydrological profiles from various sources for the period 1909–1987. In 1982, Levitus published the first climatological atlas of the world ocean at a 1° resolution, using temperature, salinity, and oxygen data from CTD, bottle stations, mechanical, and expendable bathythermographs of the previous 80 years. An objective analysis at standard levels was carried out to compute the climatology. Since 1994, a new version is released every 4 years. The current version that includes the Mediterranean (World Ocean Atlas 2013, V2; Locarnini et al., 2013) is a long-term set of objectively analyzed climatologies of temperature, salinity, oxygen, phosphate, silicate, and nitrate for annual, seasonal, and monthly periods for the world ocean. It also includes associated statistical fields of observed oceanographic profile data from which the climatologies were computed. In addition to increased vertical resolution, the 2013 version has both 1 and $1/4°$ horizontal resolution versions available for annual and seasonal temperature and salinity for 6 decades, as well as monthly for the decadal average. Brasseur et al. (1996) introduced a new method to reconstruct the three-dimensional fields of the properties of the Mediterranean Sea. Seasonal and monthly fields were analyzed using a variational inverse method (VIM) to generate the climatological maps, instead of the objective analysis introduced by Gandin (1966) and Bretherton et al. (1976) on the meteorology and oceanography which was widely used by then. More than 34 000 CTD and bottle data were used and integrated in the so-called MED2 historical database. Comparison of the results obtained with both methods showed that VIM was mathematically equivalent but numerically more efficient than the objective analysis. The method has been adopted by the Medar/Medatlas and its successor SeaDataNet Project.

At the beginning of the 2000s, many research projects and monitoring activities have produced large amounts of multidisciplinary in situ hydrographic and biochemical data for the whole Mediterranean Sea, but still the data were fragmented and inaccessible from the scientific community. The aim of the EU/MAST Medar/Medatlas Project, 2001 (http://www.ifremer.fr/medar/, last access: 9 July 2018) was to rescue and archive the dispersed data through a wide cooperation of countries and produce an atlas for 12 core parameters: temperature and salinity, dissolved oxygen, hydrogen sulfur, alkalinity, phosphate, ammonium, nitrite, nitrate, silicate, chlorophyll and pH. Gridded fields have been computed on a $1/4°$ grid resolution using the VIM and the DIVA tool developed by the GHER group of the University of Liège. Interannual and decadal variabilities of temperature and salinity were computed as well. The atlas is available at http://modb.oce.ulg.ac.be/backup/medar/contribution.html (last access: 9 July 2018) and on CD-Rom (MEDAR Group, 2002). The atlas contains a selection of figures and three-dimensional fields in netCDF. The EU FP7 SeaDataNet project, 2006–2011 and 2012–2016 (the successor of the EU/MAST Medar/Medatlas), has integrated historical, multidisciplinary data on a unique, standardized online data management infrastructure and provides value-added aggregated datasets and regional climatologies based on these aggregated datasets for all the European sea basins. SeaDataNet has adopted the VIM method and the DIVA software tool. Temperature and salinity monthly climatologies have been produced on a $1/8°$ grid resolution (Simoncelli et al., 2015a, 2016, https://doi.org/10.12770/90ae7a06-8b08-

4afe-83dd-ca92bc99f5c0). These climatologies are based on the V1.1 historical data collection of publicly available temperature and salinity in situ profiles (Simoncelli et al., 2014, https://doi.org/10.12770/90ae7a06-8b08-4afe-83dd-ca92bc99f5c0) spanning the time period 1900–2013 (Simoncelli et al., 2015a).

1.1 Objectives

The objective of this study is the computation of an improved atlas compared with the existing products in the Mediterranean Sea using the latest developments of the DIVA tool with the aim to contribute to the better representation of the climatological patterns and understanding of the long-term variability of the regional features of the basin. The originality of this product compared to the SeaDataNet climatology (both products use similar techniques) is the higher temporal resolution, up to decadal, and more advanced calibration of the analysis parameters for improving the results and the representation of general circulation patterns at timescales smaller than the climatic means. Besides the WOA13 decadal periods climatologies (1955–1964, 1965–1974, 1975–1984, 1985–1994, 1995–2004, and 2005–2012) the Medar/Medatlas provides interannual and decadal gridded fields and therefore it is compared with the present atlas. The present product has two major improvements: higher spatial and vertical resolutions and error fields that accompany all the analysis results, therefore allowing a more reliable assessment of the results. The advantage of this product in relation to the WOA13 climatology is the higher spatial resolution (up to $1/8° \times 1/8°$) longitude–latitude grids used and the higher temporal resolution as the analysis uses running decades instead of successive decades. Another important difference with respect to all previous existing climatologies in the Mediterranean is that additional and more recent data are used. The atlas constructed consists of climatological fields (called climatology hereafter) to depict the "mean" state of the Mediterranean region at monthly, seasonal (winter: January–March; spring: April–June; summer: July–September; fall: October–December), and annual scale. Additionally, climatological fields at seasonal and annual scale for 57 running decades have been produced to depict the interannual and decadal variability of the system and reveal the decadal trends.

One additional period from 2000–2015 was produced for those users or applications who are interested to reference to the latest data and new platform types such as Argo floats. This period is provided with the five previous successive decadal periods, e.g., 1950–1959, 1960–1969, 1970–1979, 1980–1989, 1990–1999, and 2000–2015. In summary, the following gridded products are available:

- monthly climatological gridded fields obtained by analyzing all monthly data of the whole period 1950 to 2015;

- seasonal climatological gridded fields obtained by analyzing all monthly data of the whole period from 1950
to 2015 falling within each season;

- seasonal gridded fields obtained by analyzing all data
falling within each season for each of 57 seasons over
the running decades from 1950–1959 to 2006–2015;

- seasonal gridded fields obtained by analyzing data
falling within each season for six periods: 1950–
1959, 1960–1969, 1970–1979, 1980–1989, 1990–1999,
2000–2015;

- annual climatology obtained by analyzing all data (regardless month or season) for the whole period from
1950 to 2015;

- annual gridded fields obtained by analyzing all data regardless month or season for each of the 57 seasons over
the running decades from 1950–1959 to 2006–2015;

- annual gridded fields obtained by analyzing data falling
with each season for six periods, 1950–1959, 1960–
1969, 1970–1979, 1980–1989–1990–1999, 2000–2015.
The first five periods coincide with the corresponding
decades.

The climatologies provided cannot rely on sufficient high-
frequency and high-resolution data to allow mesoscale features, which play an important role and modify the large-
scale flow fields, to be resolved (Robinson et al., 2001). We
focus thus on the seasonal and decadal variations. This time
filtering also results in a spatial filter as later shown by the
spatial correlations found in the data. The spatial scales the
data can capture are of the order of 300–350 km at the surface, much larger than the Rossby radius of deformation
scale (10–15 km) associated with mesoscale motions (30–
80 km, Robinson et al., 2001). These mesoscale features are
thus filtered out from the analysis and hence the numerical
grids we will use only need to resolve the large scales. The
same holds for the output files, where there is no reason to
save at very high resolution (much smaller than the deformation radius) as in any case the analysis provides large-scale
fields.

The atlas covers the geographical region 30–46° N,
6.25° W–36.5° E on 31 standard depth levels from 0
to 4000 m. The fields are stored in netCDF files. The
atlas is accessible in netCDF from the Zenodo platform using the following DOIs: annual climatology
(https://doi.org/10.5281/zenodo.1146976), seasonal climatology for 57 running decades from 1950–1959 to 2006–
2015 (https://doi.org/10.5281/zenodo.1146938); seasonal
climatology (https://doi.org/10.5281/zenodo.1146953), annual climatology for 57 running decades from 1950–1959 to
2006–2015 (https://doi.org/10.5281/zenodo.1146957);
seasonal climatology for six periods: 1950–1959,
1960–1969, 1970–1979, 1980–1989, 1990–1999,

2000–2015 (https://doi.org/10.5281/zenodo.1146966);
annual climatology for six periods: 1950–1959, 1960–
1969, 1970–1979, 1980–1989, 1990–1999, 2000–2015
(https://doi.org/10.5281/zenodo.1146970); monthly climatology (https://doi.org/10.5281/zenodo.1146974).

Tables 1 and 2 give the state of the art of the existing climatologies in the Mediterranean Sea.

2 Data

2.1 Data source

The SeaDataNet historical temperature and salinity data
collection V2 for the Mediterranean Sea was used
(Simoncelli et al., 2015b, http://sextant.ifremer.fr/record/
8c3bd19b-9687-429c-a232-48b10478581c/, last access: 9
July 2018). The collection includes 213 542 temperature and
138 691 salinity profiles from in situ measurements covering the 1911–2015 period, which corresponds to all open-
access data available through the European SeaDataNet Marine Data Infrastructure (www.seadatanet.org, last access: 9
July 2018). These datasets were collected from 102 data
providers, are quality controlled and archived in 32 marine
and oceanographic data centers, and distributed into the infrastructure by 27 SeaDataNet partners. The data range from
9.25° W to 37° E and include a part of the Atlantic (not included in the atlas) and the Marmara Sea. Figure 1 shows the
locations of the profiles and Table 3 the main instruments
of the collection. Users have to register at the Marine-ID
(https://users.marine-id.org, last access: 9 July 2018) to get
an account for downloading the SeaDataNet V2 data collection. Registration is done only once and thereafter users can
have access not only to SeaDataNet but also to all EMODnet
and Copernicus Marine Data services.

The profiles in the SeaDataNet collection come from
in situ observations collected with various instruments and
platforms such as CTD and bottles data by discrete water
samplers operated by research vessels or other smaller vessels, bathythermographs (mechanical - MBT - or expendable
- XBT). Table 3 summarizes the major numbers for temperature and salinity profiles and their temporal spanning as derived from the station type, the instrument, and the platform
type.

There are 21 682 vertical profiles of temperature and salinity from CTD and bottle stations with no information on the
instrument used. These figures correspond to profiles before
the quality control and the processing of data for preparation
of climatology.

The bathythermograph data were maintained in the collection despite the incorrect fall rate and the resulting warm
bias in the measurements (Wijffels et al., 2008). Yet, there is
not a XBT/MBT correction for the Mediterranean as for the
global ocean (https://www.nodc.noaa.gov/OC5/XBT_BIAS/
xbt_bias.html, last access: 9 July 2018). These data significantly improve the geographical coverage of the temperature

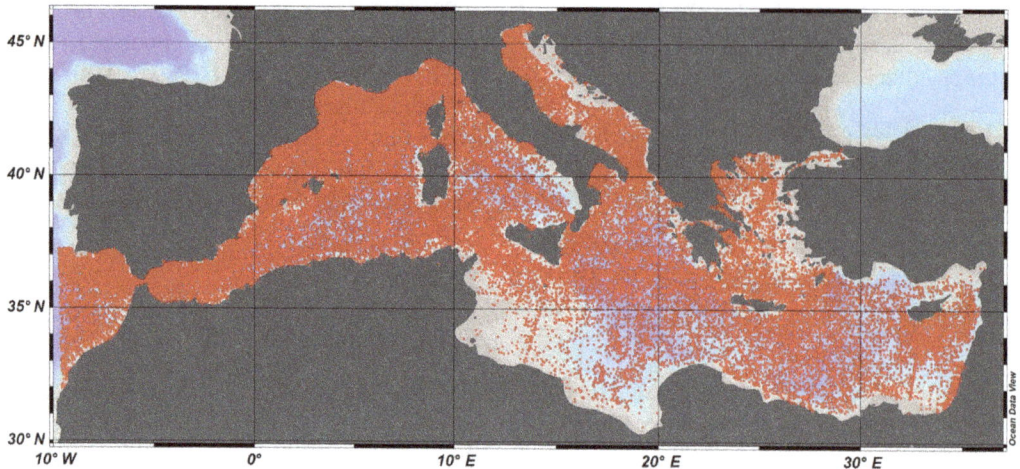

(a) Temperature

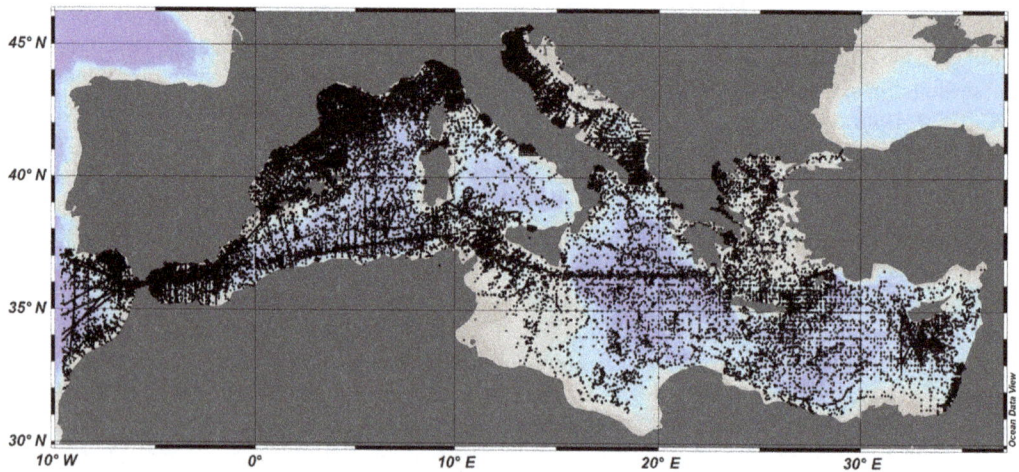

(b) Salinity

Figure 1. Geographical distribution of (**a**) temperature and (**b**) salinity profiles of the SeaDataNet V2 historical data collection (Simoncelli et al., 2015b, http://sextant.ifremer.fr/record/8c3bd19b-9687-429c-a232-48b10478581c/, last access: 9 July 2018).

records during the 1950–1985 period and reduce the resulting analysis error due to data gaps. Therefore, it was decided to keep them despite their known (small) bias.

2.1.1 Preparation for the analysis

The following operations were applied to the initial data before performing the analysis:

1. Data were extracted at the selected depth levels and temporal frames (months, seasons, decades, and periods).

2. Spatial binning was performed to improve the quality of the parameters (correlation length and signal-to-noise ratio) optimization by averaging a part of the highly correlated observations (taken by a research vessel in a re-

stricted area, for instance). Spatial binning was applied to the data only during the parameter optimization step but then the original data (and nonbinned) were used for the analysis itself.

3. Data outside the analysis domain were excluded.

4. Weights were applied to the data to reduce the influence of a large number of data located in a small area within a short period. This is particularly useful for time series data. The characteristic length of weighting was set to be equal to 0.08° and the characteristic time of weighting was set to be equal to 90 days for the seasonal analysis and 30 days for the monthly analysis. The scales were chosen according to the spatial and temporal resolution of the analysis.

Table 1. Overview of characteristics of existing T/S climatologies in Mediterranean Sea. OSD = Ocean Station Data, CTD = conductivity–temperature–depth, MBT = mechanical–digital–micro bathythermograph, XBT = expendable bathythermograph, SUR = surface, APB = autonomous pinniped bathythermograph, MRB = moored buoy data, PFL = profiling float data, DRB = drifting buoy, UOR = undulating oceanographic recorder, GLD = glider.

Climatology	World Ocean Atlas 2013	Medar/Medatlas	SeaDataNet	Present
Date	2013	2002	2015	2017
Instruments/platforms	OSD, High-resolution CTD, MBT, XBT, SUR, APB, MRB, PFL, DRB, UOR, GLD	MBT, XBT, discrete water samplers, CTD, thermistor chains	MBT, XBT, discrete water samplers, CTD, thermistor chains, thermosalinographs, DRB, PFL, MRB	MBT, XBT, discrete water samplers, CTD, thermistor chains, thermosalinographs, DRB, PFL, MRB
Horizontal extent	Global	Mediterranean, Black Sea	Mediterranean Sea	Mediterranean Sea
Parameters	Temperature, salinity, density beta version, conductivity, dissolved oxygen, percent oxygen saturation, apparent oxygen utilization, silicate, phosphate, nitrate	Temperature, salinity, alkalinity, pH, dissolved oxygen, ammonium, nitrite, nitrate, phosphate, silicates, chlorophyll-a, hydrogen sulfide	Temperature, salinity	Temperature, salinity
Horizontal resolution	$1/4° \times 1/4°$	$1/4° \times 1/4°$	$1/8° \times 1/8°$	$1/8° \times 1/8°$
Vertical extent	0–1500 m, 0–5500 m	0–4000 m	0–5500 m	0–5500 m
Vertical levels	57 levels for monthly fields, 102 levels for annual, seasonal fields	25	33	33
Temporal data coverage	1864–2013	1890–2000	1900–2013	1900–2015
Temporal resolution	Climatic, monthly, seasonal, averaged periods, decadal (1955–1964, 1965–1974, 1975–1984, 1985–1994, 1995–2004, 2005–2012 years)	Climatic, monthly, seasonal, interannual*, running 3 years*, running 5 years, running decades*	Monthly	Climatic, monthly, seasonal, running decades, averaged periods

No error fields are available for the Medar/Medatlas, interannual*, running 3-years* and running decades* computations.

Table 2. Analysis parameters for variational inverse method (VIM) implementation in the Mediterranean.

Climatology	Medar/Medatlas	SeaDataNet	Present
Method	VIM	DIVA (VIM)	DIVA (VIM)
Correlation length	Constant	Constant	Variable
Signal-to-noise ratio	Variable	Constant	Variable
Background field	Seminormed	Seminormed	Seminormed
Detrending	No	No	Yes (for background fields)
Observation weighting	No	No	Yes

2.1.2 Vertical interpolation

The observations do not have a uniform vertical distribution and a vertical interpolation of the profiles into standard depths is needed prior to the gridding. The DIVA tool has embedded the weighted parabolic interpolation method (Reiniger and Ross, 1968) into its workflow (Troupin et al., 2010). This method is widely used in oceanographic climatologies, such as World Ocean Atlas 2013 (WOA13) (Locarnini et al., 2013; Zweng et al., 2013) and Medar/Medatlas (MEDAR Group, 2002), as it creates less vertical instabilities. Only "good" data were used, e.g., data with quality con-

(a) Temperature

(b) Salinity

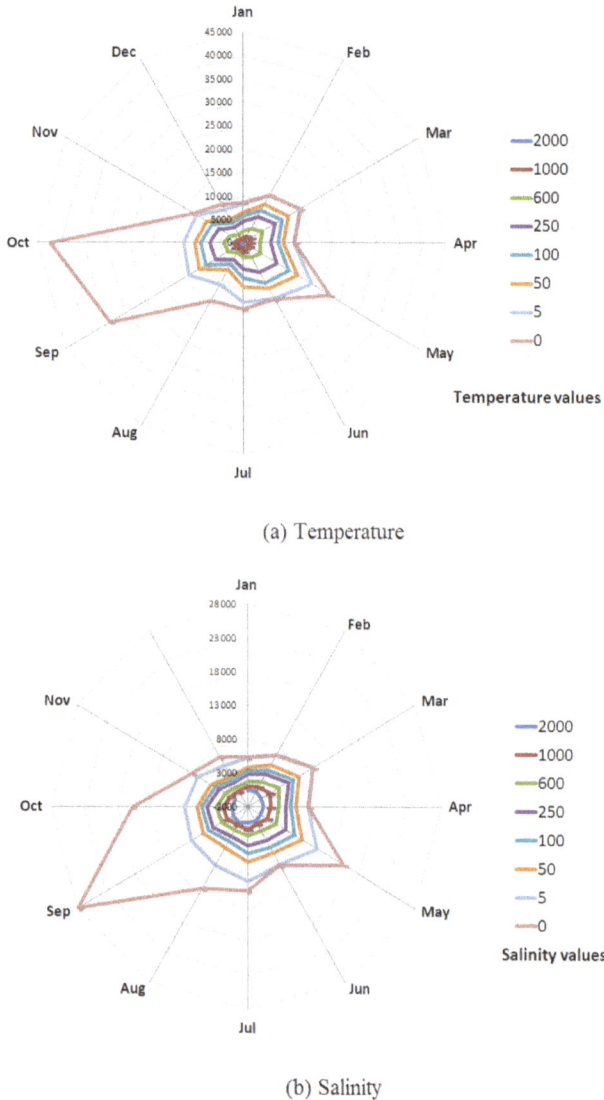

Figure 2. Number of observations per month for **(a)** temperature and **(b)** salinity.

creased. This version rather focuses on the improvement of the horizontal and temporal resolutions.

2.1.3 Quality control

Currently the SeaDataNet V2 collection is the most complete and quality-controlled data set for the Mediterranean Sea, released within SeaDataNet project. Prior to its release, it has undergone extended quality control according to the SeaDataNet standards and strategy such as elimination of duplicates, broad range checks for detecting outliers, spikes, density inversion, zero values. Regional experts performed the quality checks in close cooperation with the data originators and the responsible data centres ensuring the best result as described in Simoncelli et al. (2015b). The quality control is done with the use of the Ocean Data View tool (ODV, Schlitzer, 2002, https://odv.awi.de, last access: 9 July 2018). However, in order to avoid the influence of extremes (but not necessarily erroneous values) in the climatology, the following additional checks were applied to the data. The Mediterranean Sea was divided in 27 rectangular areas defined by Manca et al. (2004), which exhibit the typical subregional features. In each area, a mean profile was calculated at the standard depths. Values lying outside ± 3 standard deviations around the mean were considered as "outliers" and excluded from the DIVA analysis. The percentage of data excluded from the analysis was 0.8 % for the temperature and 0.9 % for the salinity values. This filter with the standard deviation was applied twice at the data sets that were used as background fields and for the optimization of the analysis parameters (the correlation length, signal-to-noise ratio and variance of the background field). In the second run of the standard deviation filter, the amount of excluded data was 0.5 and 0.7 % respectively.

2.1.4 Data temporal distributions

There is an expected seasonal distribution with more temperature and salinity profiles during summer and autumn as well as at the surface layers compared to the deeper ones (Fig. 2). However, the monthly and yearly distributions reveal a great bias for September and October and for the years 2008 and 2009. This is due to the temporary presence of 51 063 thermosalinograph data, at 2 and 3 m depth, essentially located near the Strait of Gibraltar and in the Alboran Sea. Such irregular distribution in time has to be taken into account when computing the climatological fields otherwise the spatial interpolations will be biased towards the values of these high data coverage during these 2 months. Therefore, a detrending method was applied to the data prior to the DIVA analysis to remove this effect of uneven distribution in time. The detrending tool is provided by the DIVA software. The detrending was applied to all background fields (see paragraph below) and the monthly, seasonal, and annual climatologies. It was not applied to running decades and the 6 decadal pe-

trol flag 1 and 2 according to the SeaDataNet QC flag scale (SeaDataNet Group, 2010). The interpolation was performed onto 31 International Oceanographic Data and Information Exchange (IODE) standards depths: [0, 5, 10, 20, 30, 50, 75, 100, 125, 150, 200, 250, 300, 400, 500, 600, 700, 800, 900, 1000, 1100, 1200, 1300, 1400, 1500, 1750, 2000, 2500, 3000, 3500, 4000]. These are the same depths as Simoncelli et al. (2014) used in the existing SeaDataNet V1.1 climatology, thus allowing quick visual intercomparisons. The WOA13 climatological fields offer a higher vertical resolution (87 depth levels from 0 to 4000 m), thus facilitating higher resolution models or more accurate quality control for observational data. However, the monthly WOA13 fields in the Mediterranean extend only down to 1500 m. In the next releases of the present atlas, the vertical resolution will be in-

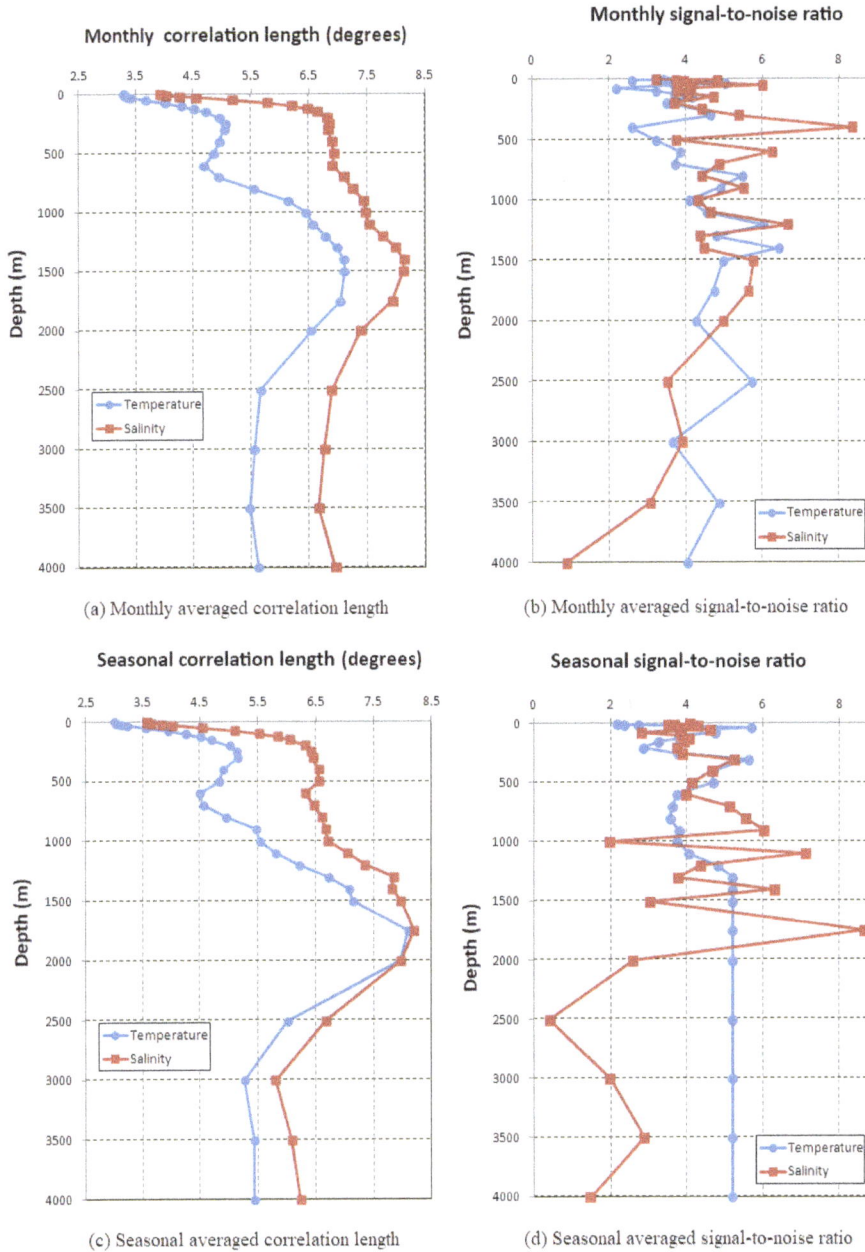

(a) Monthly averaged correlation length

(b) Monthly averaged signal-to-noise ratio

(c) Seasonal averaged correlation length

(d) Seasonal averaged signal-to-noise ratio

Figure 3. Averaged profiles of correlation length (**a, b**) and signal-to noise ratio (**c, d**) for temperature and salinity.

riod climatologies in order not to remove any of the long-term trends in temperature and salinity.

2.1.5 Data weighting

The influence of the uneven distribution in space where a large number of data points are concentrated in a very small area and within a very short period is controlled by applying different weights to each of these data points. Indeed, such points cannot be considered independent in a climatological analysis. So rather than calculating a super observation (similarly to binning), one can reduce the weight or,

in other words, increase the error attached to each individual measurement. Points which are close in time and space will undergo such a treatment. The scales below which such a weighting is done have a characteristic length of 0.08° (same unit as the data locations) and a characteristic time of 1 month (30 days). Typical examples where the weighting is particularly profitable are the cases of the thermosalinograph data (see previous paragraph) and time series data from coastal monitoring stations. Data weighting functionality is also provided by the DIVA tool. Both data weighting and de-

Table 3. Data distribution per instrument, station, and platform type.

Instrument/station type	Platform type	Number of profiles Temperature	Salinity	Period
CTD	Research vessel, moored surface buoy, fixed mooring, ships of opportunity, other type of vessels	45 914	42 420	1964–2013
Discrete water samplers	Research vessel, other type of vessels	33 449	30 423	1912–2009
Bathythermographs	Research vessel, ships of opportunity, other type of vessels	44 336		1952–2014
Thermosalinographs	Research vessel, other type of vessels	51 063	29 747	2008–2009
Salinity sensors, water temperature sensor	Drifting subsurface float	8206	8207	2005–2015
Not provided	Research vessel, ships of opportunity, other type of vessels, unknown	21 682	21 682	1914–2013

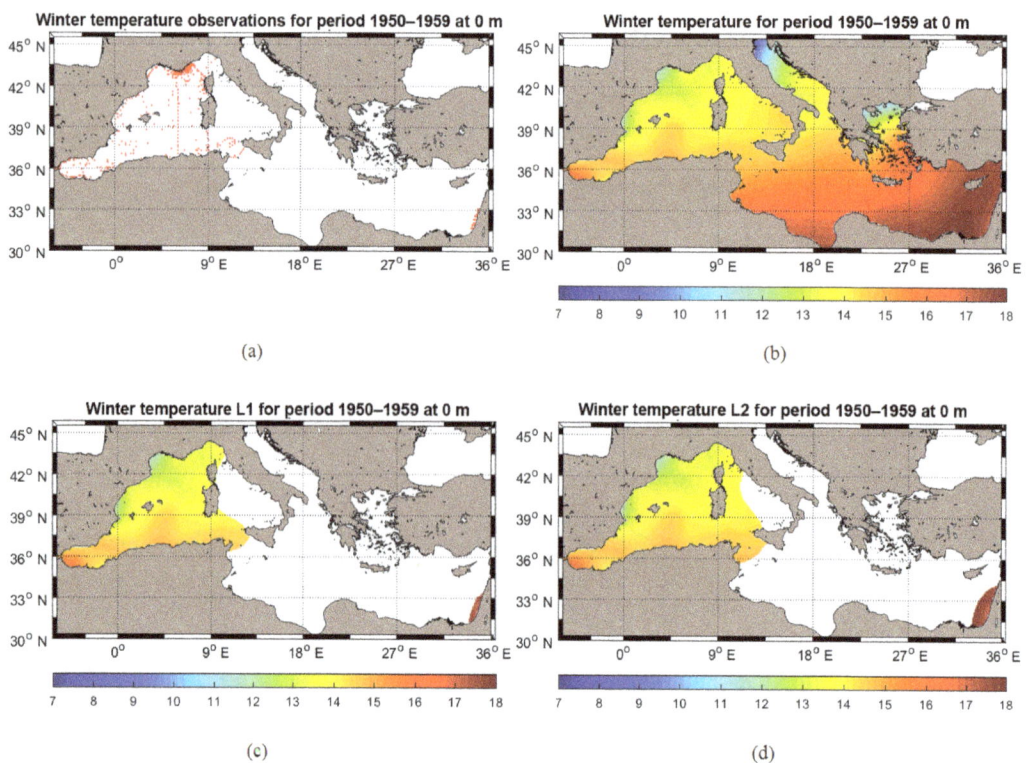

Figure 4. Example of observations distribution (a) and corresponding fields distributions depending on relative error threshold values, (b) an unmasked field, (c) a masked field with relative error threshold = 0.3, and (d) a masked field with relative error threshold = 0.5.

trending techniques have been applied for the first time in the computations for the Mediterranean Sea climatologies.

3 Method

3.1 The DIVA interpolation tool

The Data-Interpolating Variational Analysis (DIVA) is a method designed to perform spatial interpolation (analysis) of sparse and heterogeneously distributed and noisy data into a regular grid in an optimal way. The basic idea of the vari-

ational analysis is to determine a continuous field approximating data and exhibiting small spatial variations. In other words, the target of the analysis is defined as the smoothest fields that respects the consistency with the observations and a priori knowledge of the background field over the domain of interest. To do so, a cost function that takes into account the distance between the reconstructed field and the observations, and the regularity of the field is minimized. The solution of the minimization problem is obtained through a finite-element technique (Rixen et al., 2000). The main advantage is that the computational cost is independent of the number

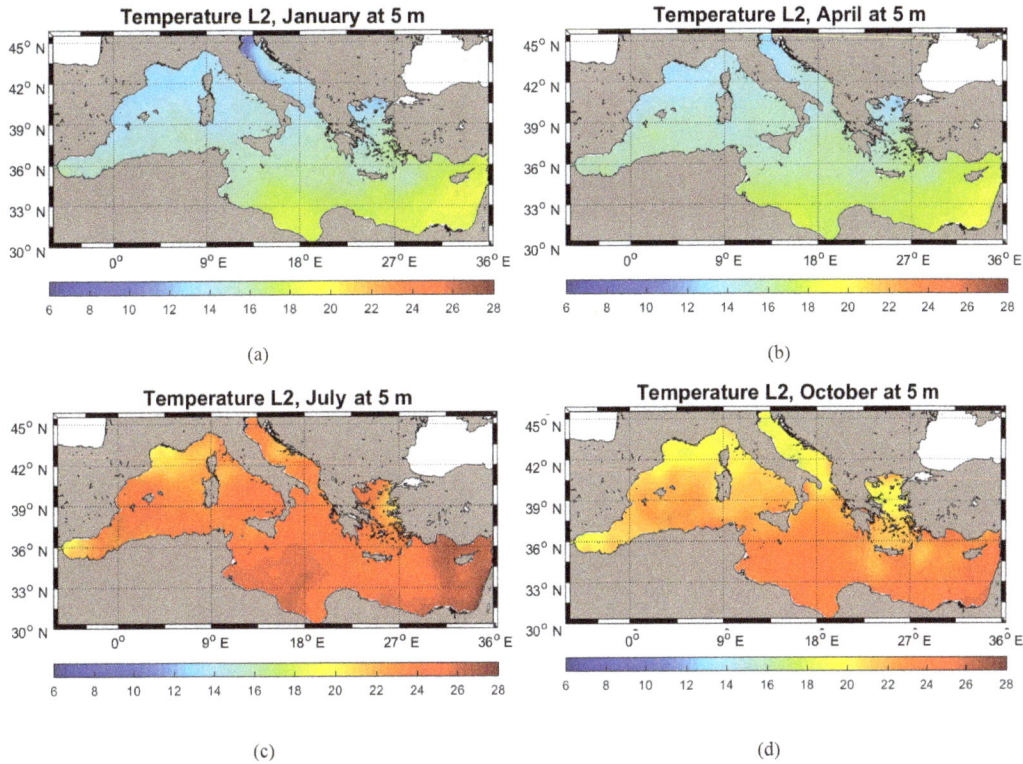

Figure 5. Surface temperature climatology at 5 m in **(a)** January, **(b)** April, **(c)** July and **(d)** October.

of data points analyzed; instead it depends on the number of degrees of freedom, i.e., on the size of the finite-element mesh. The mesh takes into account the complexity of the geometry of the domain without having to separate subbasins prior to the interpolation, and the mesh automatically prohibiting correlations across land barriers. Among other major advantages of the method, DIVA can take into account dynamic constraints allowing for anisotropic spatial correlation. Tools to generate the finite-element mesh are provided as well as tools to optimize the parameters of the analysis. The signal-to-noise ratio is optimized by a generalized cross-validation (GCV) technique (Brankart and Brasseur, 1996) while the correlation length is estimated by comparing the relation between the empirical data covariance and the distance against its theoretical counterpart (Troupin et al., 2017). Along with the analysis gridded fields, DIVA also provides error fields (Brankart and Brasseur, 1998; Rixen et al., 2000) based on the data coverage and their noise. The method computes gridded fields in two dimensions. The three-dimensional (x, y, z) and four-dimensional extensions (x, y, z, t) have been embedded into the interpolation scheme with an emphasis on creating climatologies. Detailed documentation of the method can be found in Troupin et al. (2010, 2012) and in the *Diva User Guide* (Troupin et al., 2017). The current atlas has been produced with the use of DIVA 4.6.11 (Watelet et al., 2015a) with a Linux Ubuntu 16.04.3 operating system.

3.2 Topography and coastlines

The domain where the interpolation has to be performed is covered by a triangular finite-element mesh that follows the coastline. The bathymetry used for coastline definitions and mesh generation is based on the General Bathymetric Chart of the Oceans (GEBCO) 1 min topography. Since the subsequent analyses focus on much larger scales than 1 min, the resolution of the topography was downgraded to 5 min, still fine enough to resolve topological features such as islands and straits, yet coarse enough to be coherent with the scales of interest. The step of the output grid was set to $1/8°$ for similar reasons. The geographical boundaries of the region were set to 30–46° N, 6.25° W–36.5° E. The depth contours that define the two-dimensional horizontal planes where the interpolation takes place are the 31 standard depth levels from 0 to 4000 m.

3.3 Finite-element mesh

Once the depth contour files at the specified 31 standard depth levels were prepared, the mesh was generated using an initial correlation length L_c equal to 0.5°, which means the initial size L_e of each finite triangular element is equal to 0.167° ($L_c/3$). The initial L_c scale was made as small as allowed by computing resources. This length scale is much smaller than the length scales for the analysis (typically a few degrees) and it ensures that the finite-element solution is

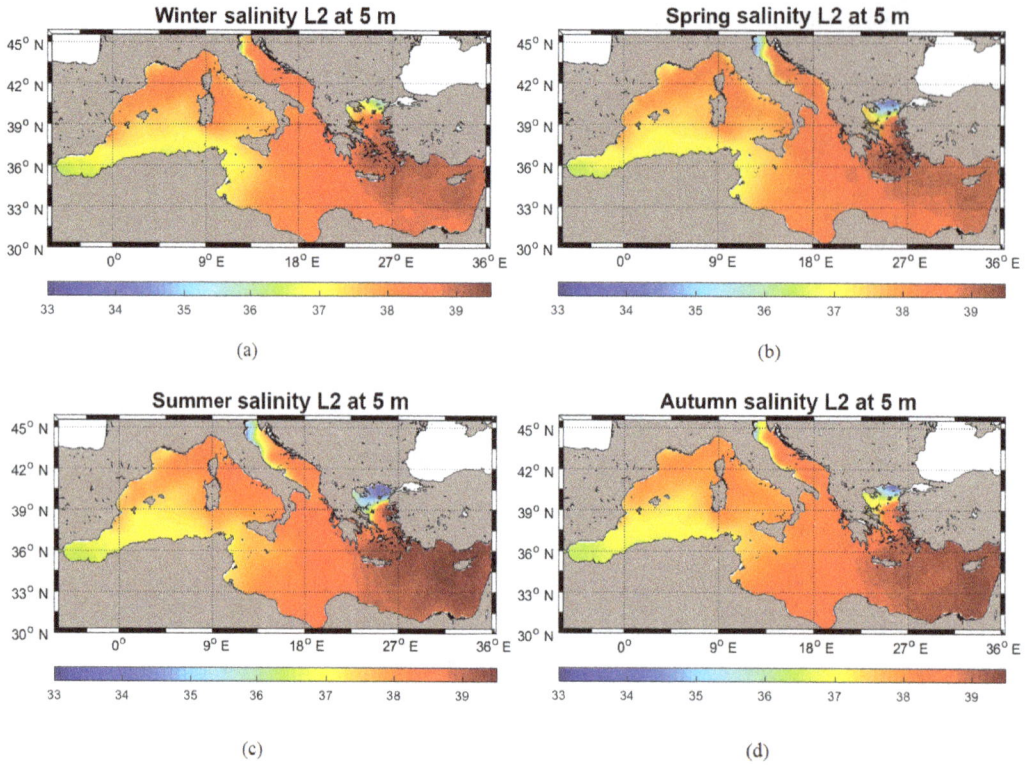

Figure 6. Surface salinity climatology at 5 m for **(a)** winter, **(b)** spring, **(c)** summer, and **(d)** autumn.

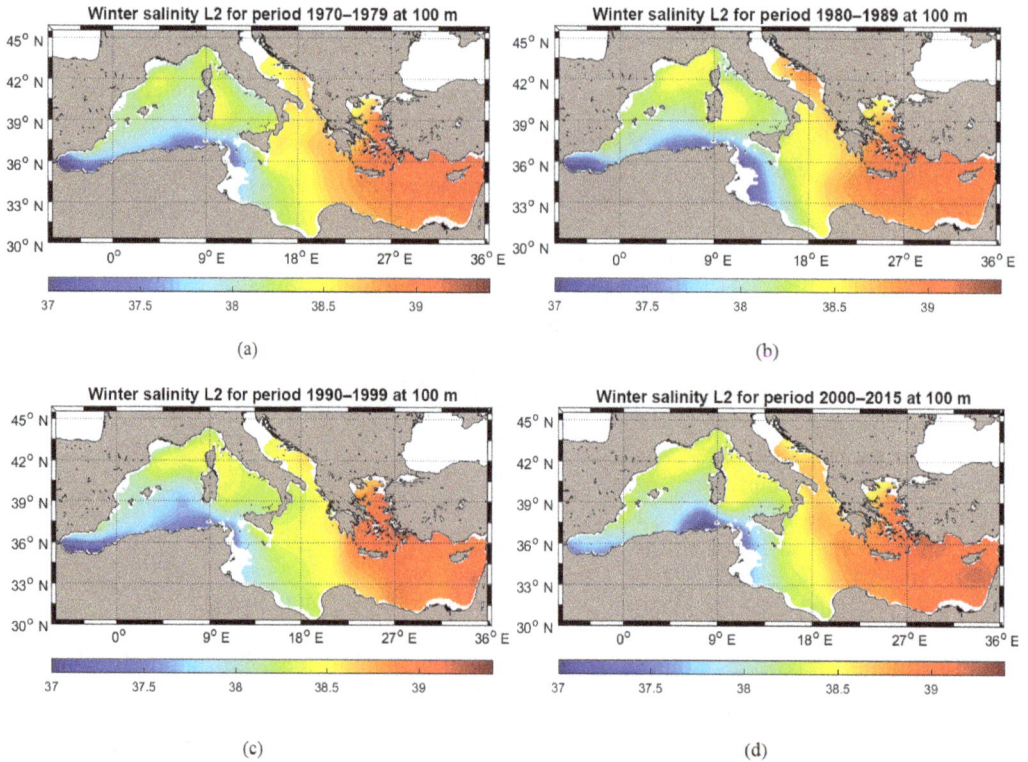

Figure 7. Winter salinity at 100 m for four periods: **(a)** 1970–1979, **(b)** 1980–1989, **(c)** 1990–1999, **(d)** 2000–2015.

solving the mathematical DIVA formulation with very high precision (it does not mean that the analysis is the truth, but that the numerical solution is actually close to the mathematical solution of the variational formulation; in other words, the discretization does not add further errors into the analysis than the analysis method and the data themselves). The mesh was generated once and used in each repetition of the analysis.

3.4 Interpolation parameters

3.4.1 Correlation length

The correlation length (L_c) is the radius of influence of data points. It can be determined objectively using a specific DIVA tool that takes into account the data distribution, or the correlation length can be provided a priori by the user according to its (subjective) experience on expected patterns occurring in the domain of interest. In the contrary to the SeaDataNet climatology where a constant value used in all depths, in this atlas, the correlation length was defined as follows. For the monthly climatology, the correlation length was calculated for every month (for all years from 1950 to 2015) and at each depth by fitting the empirical data covariance as a function of distance by its theoretical function. It was then smoothed by applying a vertical filtering, similarly to Troupin et al. (2010). Then, these monthly correlation length profiles were averaged together in order to have a smooth transition from one month to another. Indeed, the estimated correlation lengths at the surface vary between 1.8 (July) and 3.6° (December). For the seasonal and the annual climatologies, the correlation length was calculated for every season and depth (from all years between 1950 and 2015), filtered vertically and then averaged into a single profile. This approach yields slightly smaller correlation lengths than the averaging per month and in combination with the rest of the analysis parameters, the fields are not so smoothed and are closer to the data. The mean profiles for temperature and salinity are shown in Fig. 3a and c. There is a general increase from the surface to 2000 m except in a layer from 300 to 600 m where it decreases. Also there is a decrease from around 2000 to 3000 m in both cases that can be attributed to the variability of the intermediate and deep waters in the Mediterranean.

3.4.2 Signal-to-noise ratio

The signal-to-noise ratio (S / N) represents the ratio of the variance of the signal to the variance of observational errors. When measuring a variable, there is always an uncertainty on the value obtained. Noise does not only take into account instrumental errors (which are generally low), but it also includes the following:

1. the representativeness errors, meaning that what one measures is not always what one intends to analyze (e.g., skin temperature, inadequate scales);

2. the synopticity errors, occurring when the measurements are assumed to be taken at the same time (e.g., data from a cruise; Rixen et al., 2001).

Because of the multiple sources of error, a perfect fit to observations is thus not advised. In the case of climatology, the main cause of error is not the instrumental error but the representativeness errors, which cannot easily be quantified. In this atlas, unlike the SeaDataNet (Simoncelli et al., 2014, 2016) and other regional climatologies (Troupin et al., 2010), a variable S / N was defined following the same approach as with the correlation length. The mean monthly and seasonal S / N profiles for temperature and salinity were vertically filtered. The remaining extremes were further filtered out manually by replacing them with the mean of the adjacent layers. The seasonal mean profile for temperature was further filtered out to a constant value from 1300 to 4000 m. In the first 500 m (see Fig. 3b and d), both the monthly and seasonal profiles display similar behaviors with a mean value ranging from 3.5 to 4.5, with an exception of the monthly mean value for salinity at 400 m.

3.5 Background field

The background field is the first guess of the gridded field (analysis) to reconstruct. First of all, this background is subtracted from data. Then, the variational analysis is performed by DIVA using these data anomalies as an input. Lastly, the background is added to the solution so that original data and final analyses are both expressed as absolute values. In areas with very few or no data, the analysis tends to the background field since the data anomalies are close to zero (Brankart and Brasseur, 1998). In this atlas, a seminormed analysis field (roughly speaking an analysis only capturing very large scales) was chosen as the background in order to guarantee the solution is most realistic in areas void of data. Depending on the type of climatology and gridded fields, two different background fields were used: (a) a climatic seasonal field to account for seasonality from the original data, and (b) a climatic annual field to account for interannual variability. The same correlation lengths and signal-to-noise ratios were used as for the atlas computations (see previous paragraph). Analytically the following background fields were produced:

1. climatic seasonal seminormed analysis as background for the decadal seasonal fields of temperature and salinity;

2. climatic annual seminormed analysis as background for the decadal annual fields of temperature and salinity;

3. climatic seasonal seminormed analysis as background for the periodical seasonal fields of temperature and salinity;

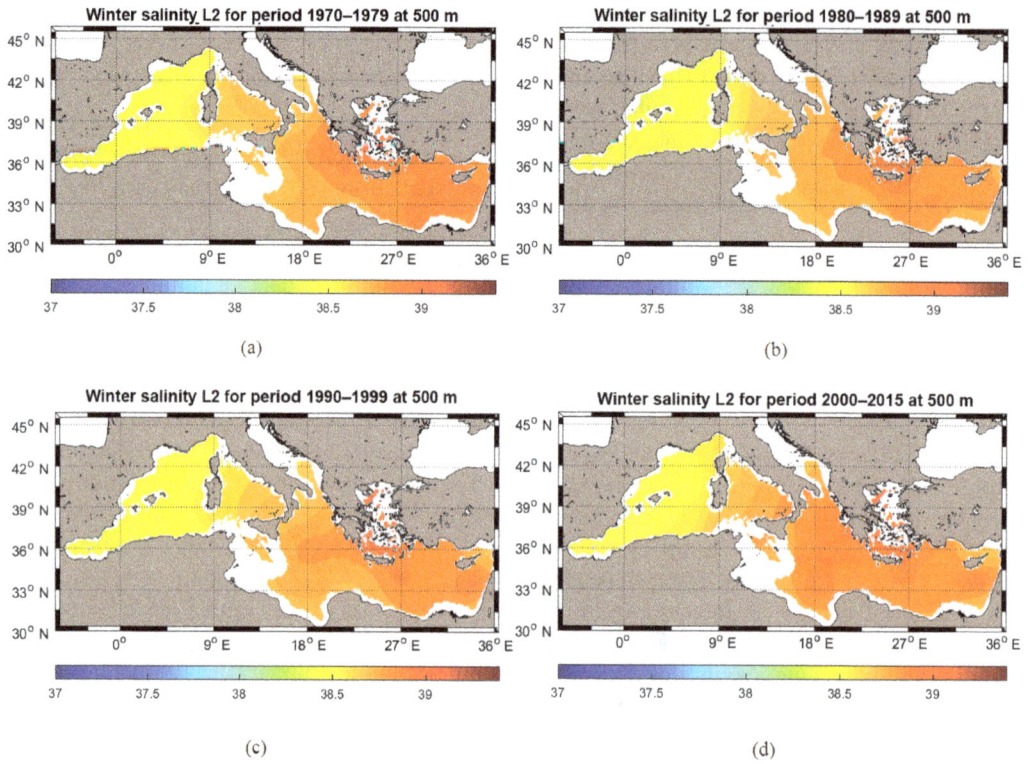

Figure 8. Winter salinity at 500 m for four periods: **(a)** 1970–1979, **(b)** 1980–1989, **(c)** 1990–1999, **(d)** 2000–2015.

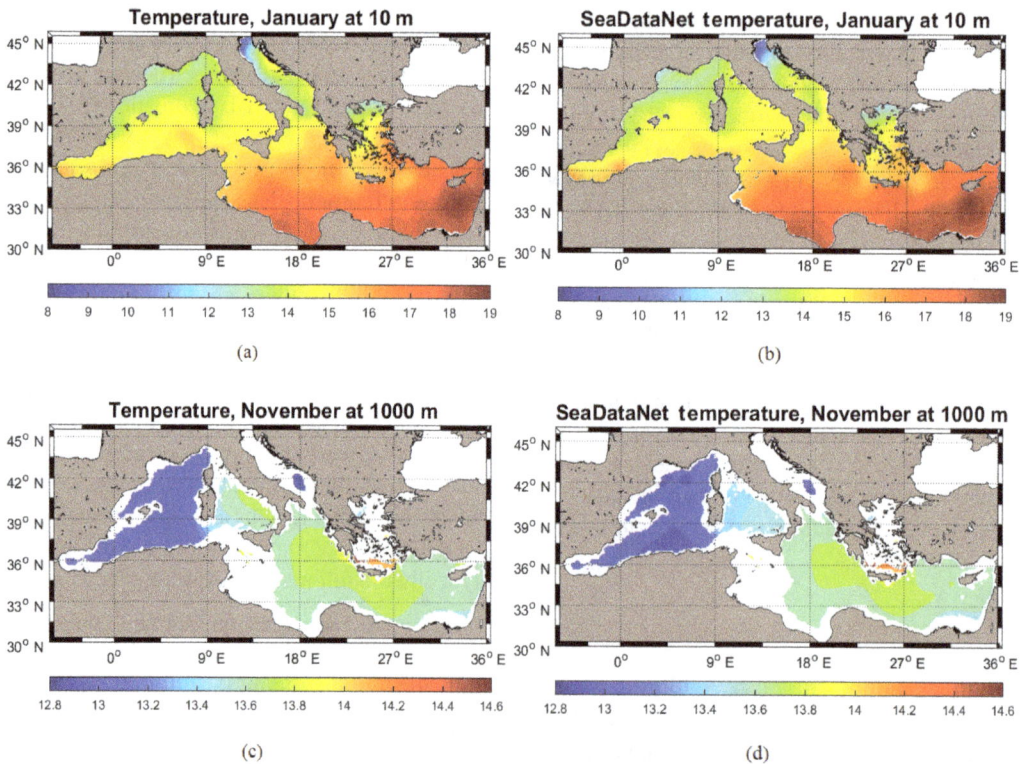

Figure 9. Comparison of atlas temperature **(a, c)** and SeaDataNet climatology **(b, d)** between January at 10 m **(a, b)** and November at 1000 m **(c, d)**.

(a) Temperature

(b) Salinity

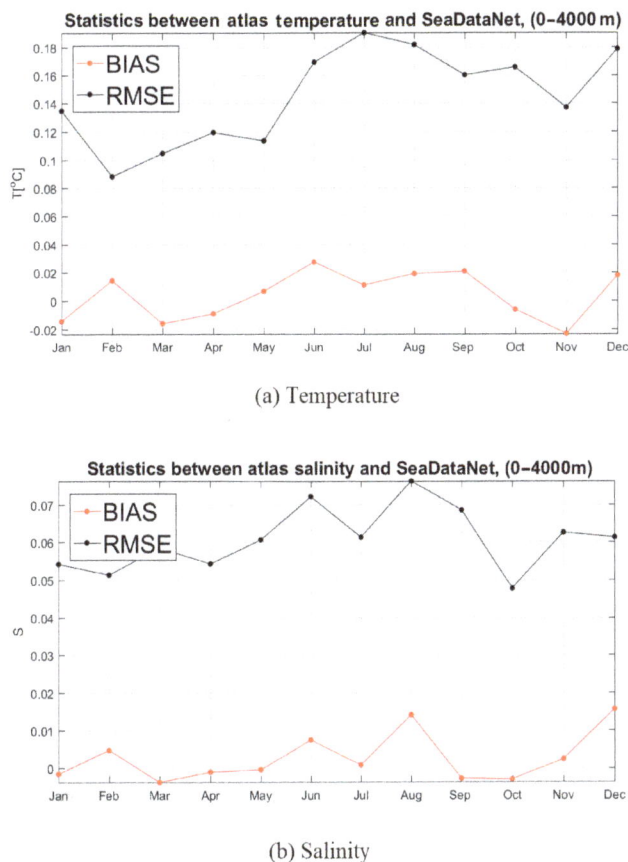

Figure 10. Basin vertical averages of BIAS and RMSE between the atlas and SeaDataNet monthly temperature (**a**), and monthly salinity (**b**) climatologies.

4. climatic annual seminormed analysis as background for the periodical annual fields of temperature and salinity;

5. climatic seasonal seminormed analysis as background for the monthly and seasonal climatology;

6. data mean value as background for the annual climatologies.

Additional monthly and seasonal climatologies were computed using the data mean value as background (the data mean value is subtracted from the data values).

The background climatic seasonal and annual fields were detrended in order to remove any seasonal and interannual trends that are due to nonuniform spatial data distribution (Capet et al., 2014). A characteristic example is the nonuniform distribution of thermosalinographs. Also, the monthly, seasonal and annual climatologies were detrended. A future goal is to apply the detrending for the estimation of the biases induced in the temperature data due to instrumental errors such as the bathythermographs.

4 Climatological gridded fields

The content of the atlas is described in Sect. 1. For each gridded error field, maps are provided that allow one to assess the reliability of the gridded fields and to objectively identify areas with poor coverage. DIVA provides two different errors, the relative (from 0 to 1) and the absolute errors (expressed in physical parameter units), which depend on the accuracy of the observations and their distribution. The relative error has been chosen in this atlas and two threshold values are used (equal to 0.3 or 0.5) for the quality assessment of the results (Fig. 4).

4.1 Description of the gridded fields

While the physical interpretation of the maps is not in the scope of this paper, it is worth noting that the main physical processes are well resolved both in space and time, such as the following:

1. the clear signature of the temperature and salinity gradient from west to east, shown in all monthly, seasonal and annual plots at all depths;

2. the temperature and its seasonal cycle;

3. the Rhodes gyre with the low temperatures is evident in all plots of temperature at surface layers regardless of month or season;

4. the Black Sea outflow with the low temperature and salinity values, which is evident in the surface plots, regardless of month or season.

Figure 5 illustrates the surface temperature at 5 m in January, April, July, and October and represents the seasonal cycle of the thermal field with a mean shift of 9 °C between January and July. The western Mediterranean exhibits smaller temperatures than the eastern basin. The differences between the northern and southern regions (meridional gradient) are also very well represented. The Rhodes Gyre, identified by minimum temperature values, is distinguished southeast of Crete island, mainly at spring and autumn periods.

Some representative seasonal distributions of salinity at 5 m are shown in Fig. 6. The signal of the fresh Atlantic water ($S \approx 36.5$ ppt) and its flow along the Algerian coast towards the eastern basin through the Sicily Strait is very evident. We can also notice the areas of low salinity due to river runoffs such as the north Adriatic and the north Aegean Sea that is influenced by the Dardanelles outflow.

Winter salinity fields at 100 and 500 m respectively are shown in Figs. 3.1 and 3.3 for the last four periods (1970–1979, 1980–1989, 1990–1999, and 2000–2015). There is a homogeneous increasing trend from west to east compared to the surface. The salinity range is between about 37 and 38.6 ppt at 100 m at the western Mediterranean, while the periods 1980–1989 and 2000–2015 are characterized by higher

(a)

(b)

(c)

(d)

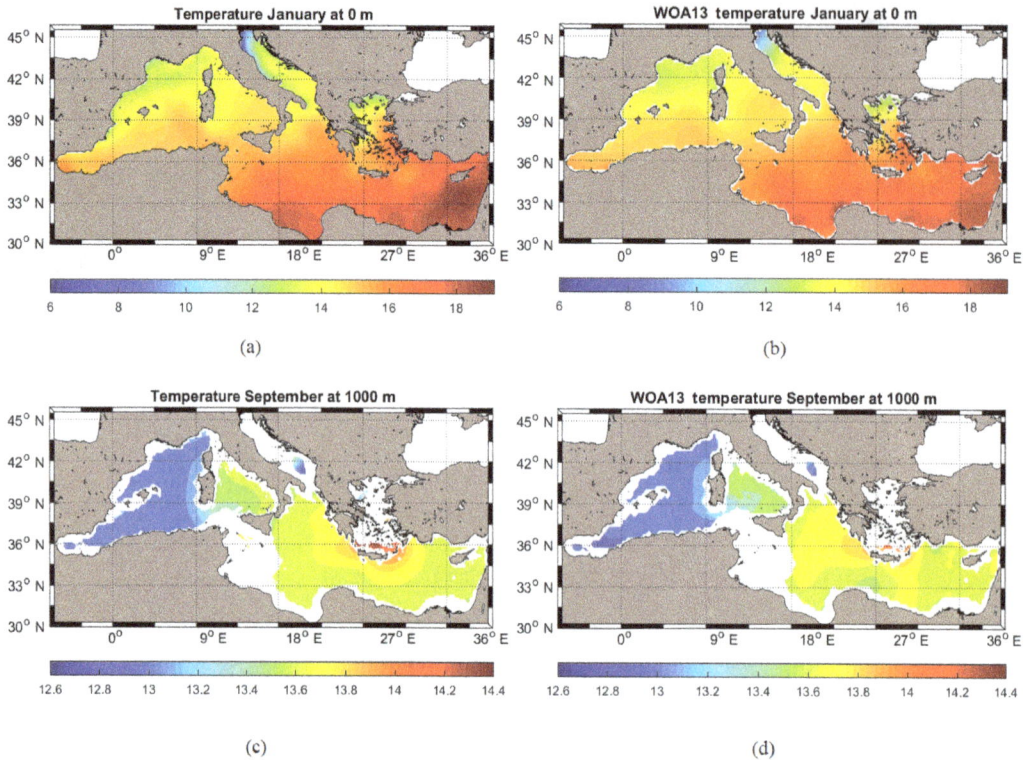

Figure 11. Comparison of atlas temperature (**a, c**) and WOA13 climatology (**b, d**) between January at 0 m (**a, b**) and September at 1000 m (**c, d**).

salinities compared to the other periods. Maximum values are found in the south Adriatic at 38.9 ppt in the period 1980–1989 and 38.8 ppt in the period 2000–2015. In the eastern Mediterranean, the salinity is between about 38.4 and 39.4 ppt at 100 m with the maxima found in the Levantine basin and Aegean Sea. The last period 2000–2015 is on average more saline than the previous ones, and the thermohaline circulation patterns (such as the Shikmona Gyre) are more intense.

In the western basin there is a characteristic increase in salinity at 500 m compared to at 100 m, where it is between about 38.4 and 38.7 ppt. This increase is due to the mixing with LIW that flows towards the Strait of Gibraltar. The Tyrrhenian Sea has higher salinities in the periods 1970–1979 and 2000–2015. In the eastern Mediterranean the salinity is between about 38.4 and 39.2 ppt, with the highest values in the north and south Aegean Sea and the period 2000–2015 showing higher salinity than the previous years.

4.2   Comparison with SeaDataNet climatology

For the current atlas, a direct comparison with Sea-DataNet V1.1 climatology (Simoncelli et al., 2015a, 2016, https://doi.org/10.12770/90ae7a06-8b08-4afe-83dd-ca92bc99f5c0) is performed using statistical indexes like bias (differences between the two climatologies) and RMSE

(root-mean-square error) implemented by Simoncelli et al. (2015a, 2016), during the SeaDataNet 2 project, as consistency analysis among different climatological products. The SeaDataNet V1.1 climatology is defined between 9.25° W–36.5° E of longitude and 30–46° N of latitude with a horizontal resolution of $1/8° \times 1/8°$ on 30 IODE vertical standard levels from 0 to 4000 m. DIVA software version 4.6.9 (Watelet et al., 2015b) has been used for both the analysis and background field computation. The salinity background field has been computed through annual seminormed analysis considering all available observations, while for temperature, 3-month seminormed background fields centered on the analysis month have been considered due to the large temperature seasonal variability. Figure 9 shows temperature distributions for atlas and SeaDataNet climatology, for January at 10 m and November at 1000 m. There is very good consistency between both products. The same good agreement exists among other months and depths (not shown here). In surface distributions (at 10 m), the data weighting applied in the atlas (Fig. 9a) has reduced the extension of the influence of Po river.

Statistical indexes like bias (differences between the two climatologies) and RMSE (root-mean-square error) show very good agreement too. Figure 10 shows the basin vertical averages of bias and RMSE between the atlas and Sea-DataNet monthly temperature (a) and monthly salinity (b) climatologies. Higher differences occur in spring and autumn, and the atlas is 0.03 °C warmer than SeaDataNet in June and less warm by 0.02 °C in November. Temperature RMSE values are ranging between 0.09 in February and 0.19 in July. Maximum differences between the two climatologies are located in the first layers (not shown here) for all months. The atlas has slightly higher salinity values than SeaDataNet salinity values in August and December and the mean bias value throughout the year is 0.003. Salinity RMSE values are ranging between 0.048 and 0.076. As for temperature, maximum differences are located in the first layers for all months.

4.3 Comparison with World Ocean Atlas (WOA13)

WOA13 provides objectively analyzed annual, seasonal, and monthly fields covering the time period 1955–2012 (average of 6 decadal means) and 6 decades: 1955–1964, 1965–1974, 1975–1984, 1985–1994, 1995–2004, 2005–2012. Each decadal climatology consists of annual (computed as 12-month averages; seasonal (winter: January–March; spring: April–June; summer: July–September; fall: October–December), computed as 3-month averages; and monthly fields (above 1500 m). Annual, seasonal, and monthly temperature and salinity fields are available on $1° \times 1°$ and $1/4° \times 1/4°$ latitude–longitude grids. All annual and seasonal fields were calculated from 0 to 5500 m depth on 102 standard levels. Monthly fields are available only above 1500 m on all grids on 57 standard levels. Figure 11 shows temperature fields for the atlas and WOA13 for January at 0 m and for September at 1000 m. At the surface, the atlas is able to represent the January mean state with more detail.

The statistical indexes applied to SeaDataNet V1.1 climatology have been calculated also for WOA2013, as in Simoncelli et al. (2015a, 2016), and they show a very good agreement between both monthly T/S climatologies. Figure 12 shows the basin vertical averages of bias and RMSE between the atlas and WOA13 monthly temperature (a) and monthly salinity (b) climatologies. Temperature bias is always positive, indicating that the atlas is slightly warmer than the WOA13 climatology. Atlas salinity is less than the WOA13 at an average of 0.007 throughout the whole year.

4.4 Comparison with Medar/Medatlas Climatology

The Medar/Medatlas climatology was computed using the variational inverse method as the current atlas. Correlation length was fixed a priori to a constant value according to the a priori knowledge of typical scales of parameters in the domain of interest. The signal-to-noise ratio was calibrated by a generalized cross-validation technique. Data closer than 15 km from the coast or in areas shallower than 50 m were rejected in order to avoid the influence of coastal features on open sea (Rixen et al., 2005).

The impact of such as rejection can be seen in Fig. 13: in the north Adriatic Sea and in the north Aegean Sea, the

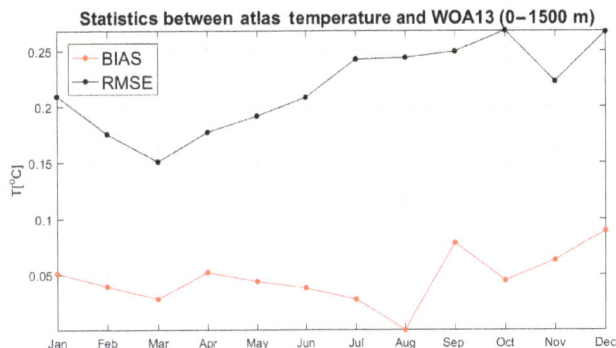

(a) Temperature

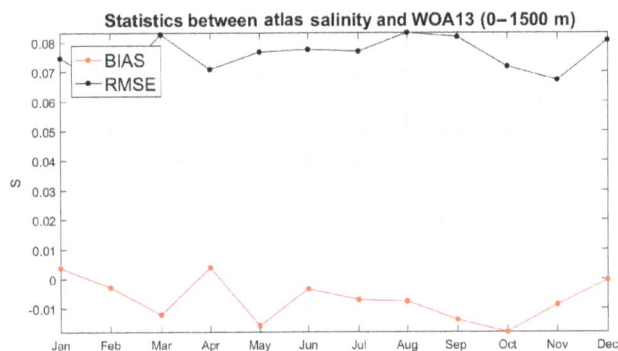

(b) Salinity

Figure 12. Basin vertical averages of BIAS and RMSE between atlas and WOA13 monthly temperature (**a**), and monthly salinity (**b**) climatologies.

influence of Po river and Black Sea outflows during winter is not captured.

The temperature at 10 m for the decade 1991–2000 is shown below for the atlas (Fig. 14a) and the Medar/Medatlas climatology (Fig. 14b). The Medar/Medatlas profiles extend until 2000, so this decade was chosen as the most complete one since there is no error field available to mask regions empty of data.

As it can be seen in Fig. 14, the current atlas describes with more detail the several features of the general circulation pattern.

5 Code and data availability

The dataset used in this work (Simoncelli et al., 2015b, http://sextant.ifremer.fr/record/8c3bd19b-9687-429c-a232-48b10478581c/) is available from the SeaDataNet catalogue at https://www.seadatanet.org/Products. Updated versions will be released periodically.

The DIVA source code is distributed via GitHub at https://github.com/gher-ulg/DIVA (last access: 9 July 2018).

(a)

(b)

(c)

(d)

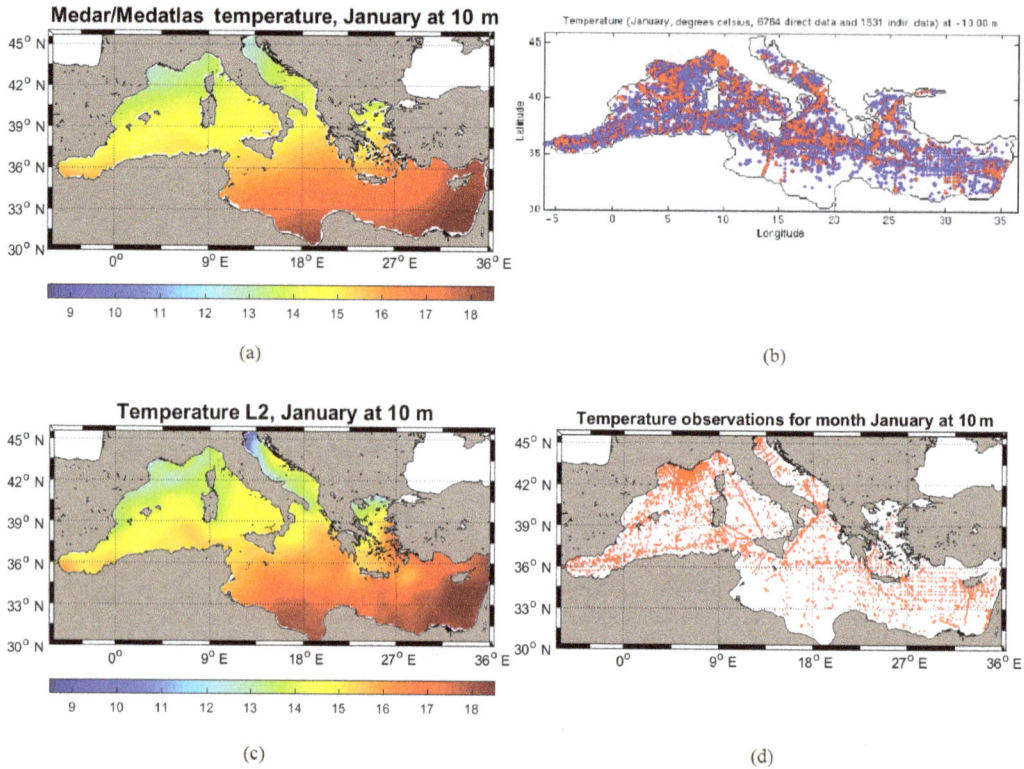

Figure 13. January temperature at 10 m: **(a)** Medar/Medatlas, coastal data removed from the analysis; **(b)** data used (imported from http://modb.oce.ulg.ac.be/backup/medar/JPGSTATIONS/medar.01.med.temp.23.3.0.stations.jpg, last access: 9 July 2018); **(c)** current climatology, coastal data included; **(d)** data used.

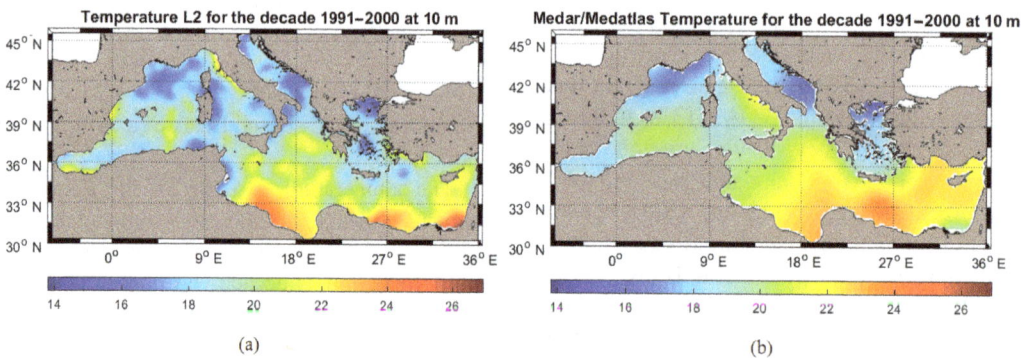

(a)

(b)

Figure 14. (a) Atlas masked temperature field for the decade 1991–2000 at 10 m, **(b)** Medar/Medatlas temperature field for the decade 1991–2000, at 10 m (no error field is available).

The different versions of the software tool are archived and referenced in Zenodo platform under the DOI: https://doi.org/10.5281/zenodo.592476.

The atlas itself is distributed through Zenodo according to the following subproducts: annual climatology: https://doi.org/10.5281/zenodo. 1146976 (Iona, 2018a); seasonal climatology for 57 running decades from 1950-1959 to 2006–2015: https://doi.org/10.5281/zenodo.1146938 (Iona, 2018b); seasonal climatology: https://doi.org/10.5281/zenodo.1146953

(Iona, 2018c). annual climatology for 57 running decades from 1950–1959 to 2006–2015: https://doi.org/10.5281/zenodo.1146957 (Iona, 2018d); seasonal climatology for six periods: 1950–1959, 1960–1969, 1970–1979, 1980–1989, 1990–1999, 2000–2015: https://doi.org/10.5281/zenodo.1146966 (Iona, 2018e); annual climatology for six periods: 1950–1959, 1960–1969, 1970–1979, 1980–1989, 1990–1999, 2000–2015: https://doi.org/10.5281/zenodo.1146970

(Iona, 2018f); monthly climatology: https://doi.org/10.5281/zenodo.1146974 (Iona, 2018g).

6 Conclusions

A new, high-resolution atlas of temperature and salinity for the Mediterranean Sea for the period 1950–2015 is presented. The analysis is based on the latest SeaDataNet V2 dataset (Simoncelli et al., 2015b, http://sextant.ifremer.fr/record/8c3bd19b-9687-429c-a232-48b10478581c/), providing the most complete, extended, and improved collection of in situ observations.

The results describe well the expected distributions of the hydrological characteristics and reveal their long-term changes. Techniques for overcoming the inhomogeneous vertical and spatial distribution of oceanographic measurements were presented. The analysis focused on the data of 1950 and onwards where more reliable instrumental observations exist. Comparisons with SeaDataNet V1.1 climatology (Simoncelli et al., 2015a, 2016, http://doi.org/10.12770/90ae7a06-8b08-4afe-83dd-ca92bc99f5c0) reveal a very good agreement. It was expected since both two climatologies use similar interpolation methodology. The atlas offers additional value-added decadal temperature and salinity distributions which already exist in the region from previous versions of Medar/Medatlas climatology, though with no error fields available. This is of particular importance.

There are still improvements to be implemented in future versions of the atlas, such as a three-dimensional analysis instead of stacking the two-dimensional horizontal levels together in order to take into account the vertical correlation of the parameters. This would reduce vertical inconsistencies that may remain in the results. In addition, the correction of the warm bias in the bathythermograph data, caused by instrumental errors, should also be addressed. However, it is anticipated that the approach followed here concerning the calibration of the analysis parameters will be followed by other groups in the future for the Mediterranean climate studies and other applications related with the long-term variability of the hydrological characteristics of the region and its climate change.

Another future improvement is the use of a dynamical constraint such as a real velocity field or an advection constraint based on topography that would allow anisotropies in the correlations to be included into the analysis. This atlas aims to contribute to existing available knowledge in the region and fill existing data gaps in space and time.

Author contributions. AI created the atlas, wrote the first version of the paper and prepared the figures. SS provided the MATLAB programs to compute the statistical indexes and plot them. JMB, SW, CT and AT reviewed the paper. CT and SW formatted the document in LaTeX.

Competing interests. The authors declare that they have no conflict of interest.

Disclaimer. It cannot be warranted that the atlas is free from errors or omissions. Correct and appropriate atlas interpretation and usage is solely the responsibility of data users.

Acknowledgements. The DIVA development has received funding from the European Union Sixth Framework Programme (FP6/2002–2006) under grant agreement no. 026212, SeaDataNet (http://www.seadatanet.org/, last access: 9 July 2018), the Seventh Framework Programme (FP7/2007–2013) under grant agreement no. 283607, SeaDataNet II, SeaDataCloud and EMODnet (http://www.emodnet.eu/, last access: 9 July 2018) (MARE/2008/03 – Lot 3 Chemistry – SI2.531432) from the http://ec.europa.eu/dgs/maritimeaffairs_fisheries/index_en.htm (last access: 9 July 2018) Directorate-General for Maritime Affairs and Fisheries.

Edited by: Giuseppe M. R. Manzella

References

Beckers, J.-M., Barth, A., Troupin, C., and Alvera-Azcárate, A.: Some approximate and efficient methods to assess error fields in spatial gridding with DIVA (Data Interpolating Variational Analysis), J. Atmos. Ocean. Tech., 31, 515–530, https://doi.org/10.1175/JTECH-D-13-00130.1, 2014.

Bergamasco, A. and Malanotte-Rizzoli, P.: The circulation of the Mediterranean Sea: a historical review of experimental investigations, Advances in Oceanography and Limnology, 1, 11–28, https://doi.org/10.1080/19475721.2010.491656, 2010.

Béthoux, J.-P., Gentili, B., and Tailliez, D.: Warming and freshwater budget change in the Mediterranean since the 1940s, their possible relation to the greenhouse effect, Geophys. Res. Lett., 25, 1023–1026, https://doi.org/10.1029/98gl00724, 1998.

Brankart, J.-M. and Brasseur, P.: Optimal analysis of In Situ Data in the Western Mediterranean Using Statistics and Cross-Validation, J. Atmos. Ocean. Tech., 13, 477–491, https://doi.org/10.1175/1520-0426(1996)013<0477:OAOISD>2.0.CO;2, 1996.

Brankart, J.-M. and Brasseur, P.: The general circulation in the Mediterranean Sea: a climatological approach, J. Marine Syst., 18, 41–70, https://doi.org/10.1016/S0924-7963(98)00005-0, 1998.

Brasseur, P., Beckers, J., Brankart, J., and Schoenauen, R.: Seasonal temperature and salinity fields in the Mediterranean Sea: Climatological analyses of a historical data set, Deep-Sea Res. Pt. I, 43, 159–192, https://doi.org/10.1016/0967-0637(96)00012-x, 1996.

Bretherton, F., Davis, R., and Fandry, C.: A technique for objective analysis and design of oceanic experiments applied to Mode-73, Deep-Sea Res., 23, 559–582, https://doi.org/10.1016/0011-7471(76)90001-2, 1976.

Capet, A., Troupin, C., Carstensen, J., Grégoire, M., and Beckers, J.-M.: Untangling spatial and temporal trends in the variability of the Black Sea Cold Intermediate Layer and mixed Layer Depth using the DIVA detrending procedure, Ocean Dynam., 64, 315–324, https://doi.org/10.1007/s10236-013-0683-4, 2014.

Gandin, L. S.: Objective analysis of meteorological fields. By L. S. Gandin. Translated from the Russian, Jerusalem (Israel Program for Scientific Translations), 1965, Q. J. Roy. Meteor. Soc., 92, 447–447, https://doi.org/10.1002/qj.49709239320, 1966.

Guibout, P.: Atlas Hydrologique de la Mediterranee, Shom, Ifremer, Paris, France, 1 Edn., 150 pp., 1987.

Hecht, A., Pinardi, N., and Robinson, A. R.: Currents, Water Masses, Eddies and Jets in the Mediterranean Levantine Basin, J. Phys. Oceanogr., 18, 1320–1353, https://doi.org/10.1175/1520-0485(1988)018<1320:CWMEAJ>2.0.CO;2, 1988.

Iona, A.: Mediterannean Sea-Temperature and Salinity Annual Climatology [Data set], Zenodo, https://doi.org/10.5281/zenodo.1146976, 2018a.

Iona, A.: Mediterannean Sea-Temperature and Salinity Seasonal Climatology for 57 running decades from 1950–1959 to 2006–2015 [Data set], Zenodo, https://doi.org/10.5281/zenodo.1146938, 2018b.

Iona, A.: Mediterannean Sea-Temperature and Salinity Seasonal Climatology [Data set], Zenodo, https://doi.org/10.5281/zenodo.1146953, 2018c.

Iona, A.: Mediterannean Sea-Temperature and Salinity Annual Climatology for 57 running decades from 1950–1959 to 2006–2015 [Data set], Zenodo, https://doi.org/10.5281/zenodo.1146957, 2018d.

Iona, A.: Mediterannean Sea-Temperature and Salinity Seasonal Climatology for six periods (1950–1959, 1960–1969, 1970–1979, 1980–1989, 1990–1999, 2000–2015) [Data set], Zenodo, https://doi.org/10.5281/zenodo.1146966, 2018e.

Iona, A.: Mediterannean Sea-Temperature and Salinity Annual Climatology for six periods (1950–1959, 1960–1969, 1970–1979, 1980–1989, 1990–1999, 2000–2015) [Data set], Zenodo, https://doi.org/10.5281/zenodo.1146970, 2018f.

Iona, A.: Mediterannean Sea-Temperature and Salinity Monthly Climatology [Data set], Zenodo, https://doi.org/10.5281/zenodo.1146974, 2018g.

Klein, B., Roether, W., Manca, B. B., Bregant, D., Beitzel, V., Kovacevic, V., and Luchetta, A.: The large deep water transient in the Eastern Mediterranean, Deep-Sea Res. Pt. I, 46, 371–414, https://doi.org/10.1016/s0967-0637(98)00075-2, 1999.

Lascaratos, A., Roether, W., Nittls, K., and Klein, B.: Recent changes in deep water formation and spreading in the eastern Mediterranean Sea: a review, Prog. Oceanogr., 44, 5–36, https://doi.org/10.1016/s0079-6611(99)00019-1, 1999.

La Violette, P. E.: The Western Mediterranean Circulation Experiment (WMCE): Introduction, J. Geophys. Res., 95, 1511, https://doi.org/10.1029/jc095ic02p01511, 1990.

Levitus, S.: Climatological Atlas of the World Ocean, Tech. rep., NOAA Professional Paper 13, US. Government Printing Office: Washington, DC, 1982.

Locarnini, R. A., Mishonov, A. V., Antonov, J. I., Boyer, T. P., Garcia, H. E., Baranova, O. K., Zweng, M. M., Paver, C. R., Reagan, J. R., Johnson, D. R., Hamilton, M., and Seido, D.: World Ocean Atlas 2013, Volume 1: Temperature, in: NOAA Atlas NESDIS, edited by: Levitus, S. and Mishonov A., 73, 40 pp., NOAA, available at: http://data.nodc.noaa.gov/woa/WOA13/DOC/woa13_vol1.pdf (last access: 9 July 2018), 2013.

Lozier, M. S., Owens, W. B., and Curry, R. G.: The climatology of the North Atlantic, Prog. Oceanogr., 36, 1–44, https://doi.org/10.1016/0079-6611(95)00013-5, 1995.

Malanotte-Rizzoli, P. and Hecht, A.: Large-scale properties of the eastern mediterranean: a rewiew, Oceanol. Acta, 11, 323–335, 1988.

Malanotte-Rizzoli, P., Manca, B. B., d'Alcala, M. R., Theocharis, A., Brenner, S., Budillon, G., and Ozsoy, E.: The Eastern Mediterranean in the 80s and in the 90s: the big transition in the intermediate and deep circulations, Dynam. Atmos. Oceans, 29, 365–395, https://doi.org/10.1016/s0377-0265(99)00011-1, 1999.

Manca, B., Burca, M., Giorgetti, A., Coatanoan, C., Garcia, M.-J., and Iona, A.: Physical and biochemical averaged vertical profiles in the Mediterranean regions: an important tool to trace the climatology of water masses and to validate incoming data from operational oceanography, J. Marine Syst., 48, 83–116, https://doi.org/10.1016/j.jmarsys.2003.11.025, 2004.

Manzella, G. M. R. and Gambetta, M.: Implementation of real-time quality control procedures by means of a probabilistic estimate of seawater temperature and its temporal evolution, J. Atmos. Ocean. Tech., 30, 609–625, https://doi.org/10.1175/JTECH-D-11-00218.1, 2013.

MEDAR Group: MEDATLAS/2002 database, Mediterranean and Black Sea database of temperature salinity and bio-chemical parameters, Climatological Atlas, Tech. rep., Ifremer, 4 Cdroms, 2002.

Millot, C.: Circulation in the Western Mediterranean Sea, Oceanol. Acta, 10, 143–148, 1987.

Millot, C.: Mesoscale and seasonal variabilities of the circulation in the western Mediterranean, Dynam. Atmos. Oceans, 15, 179–214, https://doi.org/10.1016/0377-0265(91)90020-G, 1991.

Millot, C.: Circulation in the Western Mediterranean Sea, J. Marine Syst., 20, 423–442, https://doi.org/10.1016/S0924-7963(98)00078-5, 1999.

Ovchinnikov, I.: Circulation in the surface and intermediate layers of the mediterranean, Oceanology, 6, 49–59, 1966.

Picco, P.: Climatological Atlas of the Western Mediterranean, Tech. rep., ENEA Santa Teresa Centre, La Spezia, Italy, 224 pp., 1990.

Rahmstorf, S.: Influence of mediterranean outflow on climate, EOS T. Am. Geophys. Un., 79, 281–281, https://doi.org/10.1029/98eo00208, 1998.

Reiniger, R. and Ross, C.: A method of interpolation with application to oceanographic data, Deep-Sea Res., 15, 185–193, https://doi.org/10.1016/0011-7471(68)90040-5, 1968.

Rixen, M., Beckers, J.-M., Brankart, J.-M., and Brasseur, P.: A numerically efficient data analysis method with error map generation, Ocean Modell., 2, 45–60, https://doi.org/10.1016/S1463-5003(00)00009-3, 2000.

Rixen, M., Beckers, J.-M., and Allen, J.: Diagnosis of vertical velocities with the QG Omega equation: a relocation method to obtain pseudo-synoptic data sets, Deep-Sea Res. Pt. I, 48, 1347–1373, https://doi.org/10.1016/S0967-0637(00)00085-6, 2001.

Rixen, M., Beckers, J.-M., Levitus, S., Antonov, J., Boyer, T., Maillard, C., Fichaut, M., Balopoulos, E., Iona, S., Dooley, H., Garcia, M.-J., Manca, B., Giorgetti, A., Manzella, G., Mikhailov, N., Pinardi, N., Zavatarelli, M., and the Medar Consortium: The Western Mediterranean Deep Water: a proxy for global climate change, Geophys. Res. Lett., 32, L12608, https://doi.org/10.1029/2005GL022702, 2005.

Robinson, A., Leslie, W., Theocharis, A., and Lascaratos, A.: Mediterranean Sea Circulation, Encyclopedia of Ocean Sciences, 1689–1705, https://doi.org/10.1006/rwos.2001.0376, 2001.

Roether, W., Manca, B. B., Klein, B., Bregant, D., Georgopoulos, D., Beitzel, V., Kovacevic, V., and Luchetta, A.: Recent Changes in Eastern Mediterranean Deep Waters, Science, 271, 333–335, https://doi.org/10.1126/science.271.5247.333, 1996.

Schlitzer, R.: Interactive analysis and visualization of geoscience data with Ocean Data View, Comput. Geosci., 28, 1211–1218, https://doi.org/10.1016/S0098-3004(02)00040-7, 2002.

Schröder, K., Gasparini, G. P., Tangherlini, M., and Astraldi, M.: Deep and intermediate water in the western Mediterranean under the influence of the Eastern Mediterranean Transient, Geophys. Res. Lett., 33, https://doi.org/10.1029/2006gl027121, 2006.

Schroeder, K., Chiggiato, J., Bryden, H. L., Borghini, M., and Ben Ismail, S.: Abrupt climate shift in the Western Mediterranean Sea, Scientific Reports, 6, 23009, https://doi.org/10.1038/srep23009, 2016.

SeaDataNet Group: Data Quality Control Procedures, Tech. Rep. Version 2.0, SeaDataNet consortium, availabel at: https://www.seadatanet.org/content/download/596/3118/file/SeaDataNet_QC_procedures_V2_(May_2010).pdf?version=1, (last access: 9 July 2018), 2010.

Simoncelli, S., Tonani, M., Grandi, A., Coatanoan, C., Myroshnychenko, V., Sagen, H., Back, O., Scory, S., Schlitzer, R., and Fichaut, M.: First Release of the SeaDataNet Aggregated Data Sets Products, WP10 Second Year Report – DELIVERABLE D10.2, https://doi.org/10.13155/49827, 2014.

Simoncelli, S., Coatanoan, C., Myroshnychenko, V., Sagen, H., Back, O., Scory, S., Grandi, A., Barth, A., and Fichaut, M.: SeaDataNet, First Release of Regional Climatologies, WP10 Third Year Report – DELIVERABLE D10.3, available at: https://doi.org/10.13155/50381, 2015a.

Simoncelli, S., Coatanoan, C., Myroshnychenko, V., Sagen, H., Back, O., Scory, S., Grandi, A., Schlitzer, R., and Fichaut, M.: Second release of the SeaDataNet aggregated data sets products, WP10 Fourth Year Report – DELIVERABLE D10.4, https://doi.org/10.13155/50382, 2015b.

Simoncelli, S., Grandi, A., and Iona, S.: New Mediterranean Sea climatologies, Bollettino di Geofisica, Vol. 57 – Supple-ment, 2016, 252 pp., available at: https://imdis.seadatanet.org/Previous-editions/IMDIS-2016/Proceedings (last access: 9 July 2018), 2016.

Troupin, C., Machín, F., Ouberdous, M., Sirjacobs, D., Barth, A., and Beckers, J.-M.: High-resolution Climatology of the North-East Atlantic using Data-Interpolating Variational Analysis (Diva), J. Geophys. Res., 115, C08005, https://doi.org/10.1029/2009JC005512, 2010.

Troupin, C., Sirjacobs, D., Rixen, M., Brasseur, P., Brankart, J.-M., Barth, A., Alvera-Azcárate, A., Capet, A., Ouberdous, M., Lenartz, F., Toussaint, M.-E., and Beckers, J.-M.: Generation of analysis and consistent error fields using the Data Interpolating Variational Analysis (Diva), Ocean Modell., 52-53, 90–101, https://doi.org/10.1016/j.ocemod.2012.05.002, 2012.

Troupin, C., Watelet, S., Barth, A., and Beckers, J.-M.: Diva-User-Guide: v1.0 (Version v1.0). Zenodo., Tech. rep., GeoHydrodynamics and Environment Research, University of Liège, Liège, Belgium, https://doi.org/10.5281/zenodo.836723, 2017.

Watelet, S., Ouberdous, M., Troupin, C., Barth, A., and Beckers, J.-M.: DIVA Version 4.6.11, Tech. rep., GeoHydrodynamics and Environment Research, University of Liège, Liège, Belgium, https://doi.org/10.5281/zenodo.400970, 2015a.

Watelet, S., Ouberdous, M., Troupin, C., Barth, A., and Beckers, J.-M.: DIVA Version 4.6.9, Tech. rep., GeoHydrodynamics and Environment Research, University of Liège, Liège, Belgium, https://doi.org/10.5281/zenodo.400968, 2015b.

Watelet, S., Troupin, C., Beckers, J.-M., Barth, A., and Ouberdous, M.: DIVA Version 4.7.1, Tech. rep., GeoHydrodynamics and Environment Research, University of Liège, Liège, Belgium, https://doi.org/10.5281/zenodo.836727, 2017.

Wijffels, S. E., Willis, J., Domingues, C. M., Barker, P., White, N. J., Gronell, A., Ridgway, K., and Church, J. A.: Changing Expendable Bathythermograph Fall Rates and Their Impact on Estimates of Thermosteric Sea Level Rise, J. Climate, 21, 5657–5672, https://doi.org/10.1175/2008jcli2290.1, 2008.

Zweng, M., Reagan, J., Antonov, J., Locarnini, R., Mishonov, A., Boyer, T., Garcia, H., Baranova, O., Johnson, D., Seidov, D., and Biddle, M.: World Ocean Atlas 2013, Volume 2: Salinity, in: NOAA Atlas NESDIS, edited by: Levitus, S. and Mishonov, A., 74, 39 pp., NOAA, available at: http://data.nodc.noaa.gov/woa/WOA13/DOC/woa13_vol2.pdf (last access: 9 July 2018), 2013.

9

A new phase in the production of quality-controlled sea level data

Graham D. Quartly[1], Jean-François Legeais[2], Michaël Ablain[2], Lionel Zawadzki[2], M. Joana Fernandes[3,4], Sergei Rudenko[5,6], Loren Carrère[2], Pablo Nilo García[7], Paolo Cipollini[8], Ole B. Andersen[9], Jean-Christophe Poisson[2], Sabrina Mbajon Njiche[10], Anny Cazenave[11,12], and Jérôme Benveniste[13]

[1]Plymouth Marine Laboratory, Plymouth, PL1 3DH, UK
[2]CLS, 31520 Ramonville-Saint-Agne, France
[3]Faculdade de Ciências, Universidade do Porto, 4169-007, Porto, Portugal
[4]Centro Interdisciplinar de Investigação Marinha e Ambiental (CIIMAR), 4450-208 Matosinhos, Portugal
[5]Deutsches Geodätisches Forschungsinstitut, Technische Universität München, 80333 Munich, Germany
[6]Helmholtz Centre Potsdam GFZ German Research Centre for Geosciences, Telegrafenberg 14473 Potsdam, Germany
[7]isardSAT, 08042 Barcelona, Spain
[8]National Oceanography Centre, Southampton, SO14 3ZH, UK
[9]DTU Space, 2800 Kongens Lyngby, Denmark
[10]CGI, Leatherhead, KT22 7LP, UK
[11]LEGOS, 31400 Toulouse, France
[12]ISSI, 3912 Bern, Switzerland
[13]ESA/ESRIN, 00044 Frascati, Italy

Correspondence to: Graham D. Quartly (gqu@pml.ac.uk)

Abstract. Sea level is an essential climate variable (ECV) that has a direct effect on many people through inundations of coastal areas, and it is also a clear indicator of climate changes due to external forcing factors and internal climate variability. Regional patterns of sea level change inform us on ocean circulation variations in response to natural climate modes such as El Niño and the Pacific Decadal Oscillation, and anthropogenic forcing. Comparing numerical climate models to a consistent set of observations enables us to assess the performance of these models and help us to understand and predict these phenomena, and thereby alleviate some of the environmental conditions associated with them. All such studies rely on the existence of long-term consistent high-accuracy datasets of sea level. The Climate Change Initiative (CCI) of the European Space Agency was established in 2010 to provide improved time series of some ECVs, including sea level, with the purpose of providing such data openly to all to enable the widest possible utilisation of such data. Now in its second phase, the Sea Level CCI project (SL_cci) merges data from nine different altimeter missions in a clear, consistent and well-documented manner, selecting the most appropriate satellite orbits and geophysical corrections in order to further reduce the error budget. This paper summarises the corrections required, the provenance of corrections and the evaluation of options that have been adopted for the recently released v2.0 dataset (https://doi.org/10.5270/esa-sea_level_cci-1993_2015-v_2.0-201612). This information enables scientists and other users to clearly understand which corrections have been applied and their effects on the sea level dataset. The overall result of these changes is that the rate of rise of global mean sea level (GMSL) still equates to $\sim 3.2\,\mathrm{mm\,yr^{-1}}$ during 1992–2015, but there is now greater confidence in this result as the errors associated with several of the corrections have been reduced. Compared with v1.1 of the SL_cci dataset, the new rate of change is $0.2\,\mathrm{mm\,yr^{-1}}$ less during 1993 to 2001 and $0.2\,\mathrm{mm\,yr^{-1}}$ higher during 2002 to 2014. Application of new correction models brought a reduction of altimeter crossover variances for most corrections.

1 Introduction

Sea level is widely recognised as an essential climate variable (ECV) that has a significant impact on mankind. An accelerated rise in global mean sea level (GMSL) shows the integrated effect of increased ocean heat content and the enhanced melting of glaciers and ice sheets. Many major conurbations are sited on the coast and vulnerable to long-term sea level rise. This is also critical for low-lying islands (such as the Maldives) and highly populated river deltas (such as the Brahmaputra in Bangladesh) where continued sea level rise threatens the lives of many.

The issue of sea level rise thus has aspects that are global, regional and local. Global mean sea level rise is related to increased forcing within the global climate through increased ocean warming and land ice loss. Satellite altimetry also reveals significant regional variability, with some regions experiencing greater rates of sea level rise. The sea level in a region also responds to ocean circulation changes associated with various modes of climatic variability, which may temporally ameliorate or exacerbate the effects of global change; such regional climatic oscillations need to be better measured, modelled and understood. An accurate robust record of regional changes can help to provide the "fingerprint" to distinguish between different models of the Earth's response to enhanced climate forcing (Hasselmann, 1997). At the coast what is important to the population is the combined effects of large-scale climate variations, local changes in waves and currents, and vertical land motion. In many regions the ground is subsiding in response to increased sediment load in deltas or ground water depletion near megacities. Also, the land masses are still undergoing a delayed response to the removal of their burden from the last ice age (a phenomenon known as "glacial isostatic adjustment"). Together these effects and sea level rise amplify the vulnerability of coastal regions, producing major societal impacts. Finally, sea level variations need to be precisely monitored at the mesoscale (50–200 km) as the variability associated with eddies and current fluctuations provides many of the mechanisms for transporting and mixing water masses, with attendant effects on primary productivity.

The European Space Agency (ESA) set up the Climate Change Initiative (CCI) in 2010 to develop consistent long-term datasets of many of the recognised essential climate variables (ECVs), with one using satellite altimetry to provide sea level data over most of the open ocean, with the aim of addressing part of the aforementioned wide range of scientific and societal needs. The initial (v1.0) dataset spanned 1993–2010 (Ablain et al., 2015, 2016); the second phase of

the CCI (2014–2016) has not only extended the data duration (up to end of 2015) but also revisited many aspects of the data processing and corrections to improve the quality of the dataset for global, regional and mesoscale applications. This paper details the processing options selected for the production of the v2.0 dataset.

The whole dataset is based on the concept of altimetry, i.e. that a satellite flying in a near-polar orbit measures the ocean surface topography by recording the time taken for radar pulses emitted by the satellite to reflect off the surface and be recorded on the satellite. There are many technical details to the measurement of this distance to within a few centimetres from a satellite ~ 720–$1350\,\mathrm{km}$ above the Earth's surface, which are described in Chelton et al. (1989), Fu and Cazenave (2001) and Escudier et al. (2017). Range is then computed by multiplying half the time delay by the speed of light in vacuo, and then applying corrections for the components of the return path where speed is slightly less – these are the dry tropospheric correction (DTC), wet tropospheric correction (WTC), and the ionospheric correction (Iono). Subtracting this altimetric range from a well-modelled orbit height then gives a value for the sea surface height relative to some reference surface. To give a measure that is useful for oceanographic applications, the value needs to be adjusted for the effect of changes in atmospheric conditions (dynamic atmosphere correction, DAC) and tides. Finally there is an empirical correction, sea state bias (SSB), accounting for various effects related to the wind and wave conditions. Thus the required oceanographic parameter, the sea level anomaly (SLA), is defined as

$$
\begin{aligned}
\mathrm{SLA} = {}& \mathrm{Orbit} - (\mathrm{Range} + \mathrm{DTC} + \mathrm{WTC} + \mathrm{Iono}) - \mathrm{DAC} \\
& - \mathrm{Tides} - \mathrm{SSB} - \mathrm{MSS},
\end{aligned} \tag{1}
$$

where the mean sea surface (MSS) is the sum of the geoid (the geopotential surface indicating the level that would be recorded for a motionless ocean) and the mean dynamic topography (MDT), which corresponds to the topographic variations associated with the mean circulation of the ocean. Values for these corrections are supplied in the geophysical data records (GDRs) provided by the space and meteorological agencies; however, there is a need to review whether new ones are more accurate, and also to establish a consistent selection across all missions used.

Gridded altimeter products combine information from two sets of altimeters – the "reference missions" (TOPEX/Poseidon, Jason-1, Jason-2 etc.) in a high-altitude ($\sim 1336\,\mathrm{km}$) orbit, with a 9.92-day repeat cycle, and the "complementary missions", which are in a lower orbit,

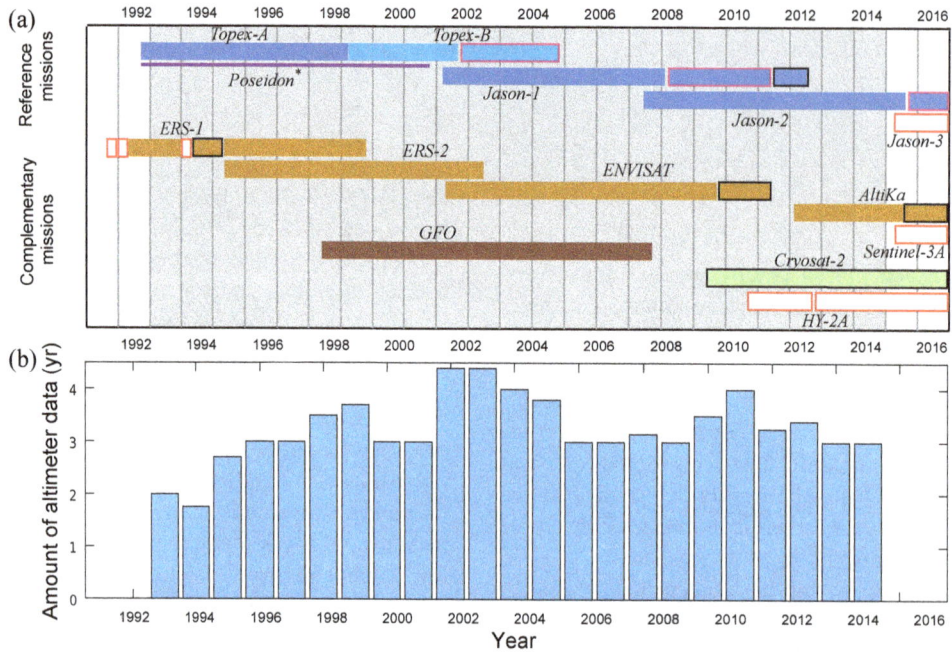

Figure 1. (a) Gantt chart of the available altimetry missions. (Full names of satellites and other commonly used abbreviations are given in Appendix A.) * The spacecraft TOPEX/Poseidon had two separate altimeters, with the experimental Poseidon instrument on for ∼ 10 % of the time, during which TOPEX did not operate. The "reference missions" all commenced in the same orbit, with 66° orbit inclination and a repeat period of 9.92 days; subsequent phases of those missions were then in a 9.92-day interleaved orbit (pink outline) or a long-repeat (geodetic) orbit (black outline). The missions highlighted in orange were principally in another common orbit (98.5° inclination and 35-day repeat), except for geodetic phases (black outline) and short periods in a 3-day repeat (ERS-1). The other complementary missions are GFO (72° inclination, 17.05-day) and CryoSat-2 (88° inclination, geodetic orbit). The periods indicated by white bars with red outlines are not used in the production of CCI v2.0 product. **(b)** Annual amount of independent altimeter data used in the production of the v2.0 dataset. (Note, for example, that during the 6-month "tandem" phases between successive "reference missions" the contribution of one of the pair to the sea level record is redundant.)

several of which (ERS-1, ERS-2, Envisat) have been in a 35-day repeat orbit. In progressing from the Sea Level CCI (SL_cci) v1.1 product to the v2.0 product, the length of the dataset has been extended and two new sources of altimeter data have been included (SARAL/AltiKa and CryoSat-2; see Fig. 1), and all the corrections have been reappraised to ensure that they are the most appropriate for establishing a consistent and stable long-term record for use at global, regional and mesoscale. Note that the SL_cci Algorithms Theoretical Basis Document (Ablain et al., 2016) provides the details on all algorithms used to compute the 1 Hz along-track measurements. This paper deals with each of these correction terms, documenting the selections made and their justification; subsequent papers will exploit the SL_cci v2.0 data to improve our understanding of present-day sea level variations at global and regional scales, and their causes.

The assessment of new corrections has been carried out by a formal validation protocol using a common set of diagnoses defined to fulfil the sea level accuracy and precision requirements, as defined by the Global Climate Observing System (GCOS, 2011). This protocol consists of comparing new al-

timeter corrections with previous ones through their impact on the sea level calculation. The validation diagnoses are distributed into three distinct families allowing the assessment of altimetry data with complementary objectives.

1. "Global internal analyses", which check the internal consistency of a specific mission-related altimetry system by analysing the computed sea level, its instrumental parameters (from altimeter and radiometer) and associated geophysical corrections,

2. "Global multi-mission comparisons", which evaluate the coherence between two different altimetry systems through comparison of SLA data,

3. "Altimetry comparison with in situ data", which computes differences between altimeter SLA data and those from in situ sea level measurements, e.g. tide gauges or Argo-based steric sea level data (Legeais et al., 2016a); this third approach allows for the detection of potential drifts or jumps in the long-term sea level time series.

2 Orbits and range

Orbital height and altimeter range are the two large terms that are differenced in the calculation of SLA. The former term refers to the height of the satellite above the reference ellipsoid, whilst the range is the measurement from the radar altimeter to the ocean surface. The orbit is not measured everywhere but rather calculated from a sophisticated numerical theory of satellite motion using a well-defined reference frame and taking into account various forces acting on a satellite, such as gravitational fields of the Earth, Moon, Sun and major planets of the Solar System; drag in the Earth's atmosphere; and radiation from the Sun and the Earth. The orbit computation for the various altimetry satellites uses a variety of data – precise satellite laser ranging from ground stations, GNSS locations from navigation satellites that are in a much higher orbit, and radio-positioning information from DORIS and PRARE – although not all sources are available for every satellite. The calculation of altimeter range includes waveform retracking (i.e. fitting a model to the shape of the radar echo) and compensation for an altimeter bias specific to the instrument (Ablain et al., 2017; Escudier et al., 2017).

2.1 Modelled orbits

As the orbital height of the satellites needs to be known to centimetric accuracy (i.e. one part in 10^8), the Earth's gravity field requires a detailed representation usually expressed in spherical harmonic coefficients, typically to degree and order 90–120 for satellites at altitudes between 700 and 1400 km. Terrestrial gravimeters and geodetic satellites, such as LAGEOS, and more recently the space gravimetry mission GRACE (Tapley et al., 2004) revealed that the Earth's gravity field changes with time. Detailed analysis of the observations of satellites in low Earth orbit, in particular, from the missions designed to observe the Earth's gravity field, such as CHAMP (2000–2010), GRACE (2002–present) and GOCE (2009–2013) has significantly improved knowledge about the Earth's static and time-variable gravity. Time variations in the gravity field include the mass redistribution within and between the Earth's atmosphere, hydrosphere, ocean and cryosphere, on a variety of timescales, from subseasonal to multidecadal. Ollivier et al. (2012) and Rudenko et al. (2014) showed that ignoring a time-variable (secular) part of the geopotential causes up to 3 mm yr^{-1} east–west errors in the regional sea level trends. Additionally, ignoring non-tidal high-frequency atmospheric and oceanic mass variations can lead to errors of up to 7 mm in sea level and up to 0.25 mm yr^{-1} in the regional trend (Rudenko et al., 2016). Achieving precise orbits also requires an accurate model of the spacecraft itself in order to understand the drag terms from a very tenuous atmosphere, the effects of solar radiation pressure and relativistic effects.

New VER11 orbit solutions of ERS-1, ERS-2, Envisat, TOPEX/Poseidon, Jason-1 and Jason-2 have been gener-

ated at GFZ (Rudenko et al., 2017). Additionally, a new orbit version (POE-E) has been computed at CNES for Jason-1, Jason-2, AltiKa and CryoSat-2, and finally, a new orbit version (GSFC std1504) has been derived at GSFC for TOPEX/Poseidon, Jason-1 and Jason-2 (Lemoine et al., 2017). All these orbit solutions have been derived in the extended ITRF2008 reference frame (Altamimi et al., 2011) by using SLRF2008 (Pavlis, 2009), DPOD2008 (Willis et al., 2016) and IGS08 (Rebischung et al., 2012) station solutions and are based on the GDR-E orbit standards (Dumont et al., 2017) or similar standards. The main differences of these standards with respect to the previous GDR-D (Dumont et al., 2017) orbit standards consist of (i) using a more refined Earth time-variable gravity field model EIGEN-GRGS.RL03-v2.MEAN-FIELD including time-variable geopotential terms up to degree and order 80 (instead of 50 in the previous standards), (ii) increased expansion of the atmospheric gravity model (from degree and order 20 to 70), (iii) modelling of tidal and nontidal geocentre variations, (iv) improved modelling of nongravitational forces for some satellites, (v) improvements in the troposphere correction model for DORIS observations, and (vi) using Earth orientation parameters consistent with the ITRF2008 reference frame.

A validation of these new orbit solutions has been performed with respect to those selected for the SL_cci v1.0 product (Table 1 of Ablain et al., 2015). The main criteria for the selection are a reduction of the SLA crossover variance differences and minimum absolute difference of the mean sea level computed using ascending and descending passes. As a result of this validation, the following orbit solutions have been selected: GFZ VER11 orbits for ERS-1, ERS-2 and Envisat; CNES POE-E orbits for Jason-1, Jason-2, AltiKa and CryoSat-2; and GSFC std1504 orbit for TOPEX/Poseidon. Consequently, using the GSFC std1504 orbit for TOPEX/Poseidon instead of the GSFC std1204 orbit (used for the SL_cci v1.0 product) reduces the mean of sea surface height (SSH) crossovers from 0.34 to 0.24 cm. The standard deviation of these crossovers shows an improvement from 4.99 to 4.96 cm for Jason-1, from 4.91 to 4.87 cm for Jason-2, and, from 5.55 to 5.51 cm for Cryosat-2, when using the CNES POE-E orbit instead of the CNES POE-D orbit. Since no new orbit solution has become available for GFO, the same (GSFC std08; Lemoine et al., 2006) orbit was used for the generation of the SL_cci v2.0 product, as for its predecessor. Couhert et al. (2015) showed that using Jason-1/2 orbits derived with SLR and DORIS measurements may cause up to 0.3 mm yr^{-1} decadal and 1 mm yr^{-1} interannual regional errors when employing ITRF2005 reference frame instead of ITRF2008 one for orbit computations. Since no DORIS data were used to derive GFO GSFC std08 orbit, the impact of using this orbit on the regional sea level may be larger, when using just one mission. However, since regional sea level is derived in the SL_cci v2.0 product using data from nine altimetry missions over the time span 1993–2015,

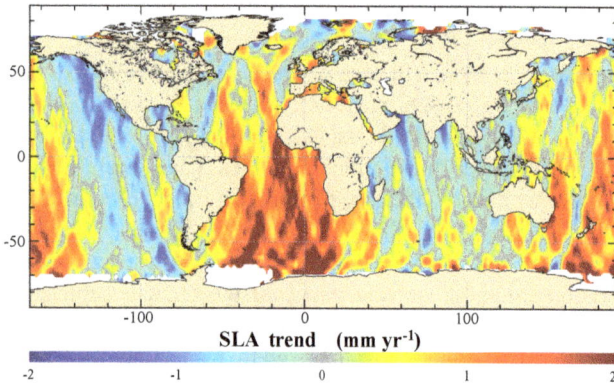

Figure 2. Difference in sea level trends for ERS-1 data (October 1992 to June 1996) computed using GFZ VER11 orbit and the REAPER combined orbit (which was used in an earlier CCI sea level product; Ablain et al., 2015).

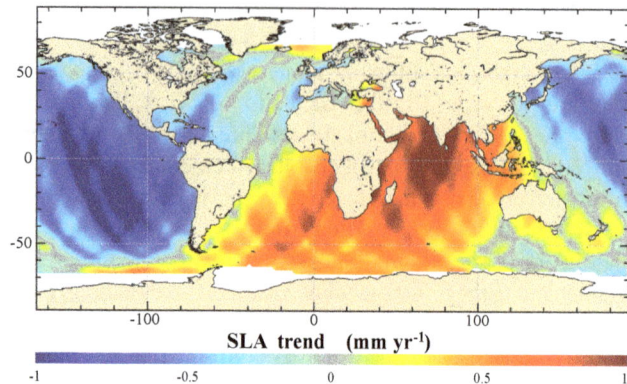

Figure 3. Change in SLA trend for Jason-1 sea level upon a switch from CNES orbit POE-D to POE-E.

the impact of using GFO orbit derived in ITRF2005, while the orbits of the other eight missions are in the ITRF2008, is rather small. There is no impact of the GFO orbit on the global mean sea level (GMSL), since GFO is not included in the reference missions used to derive that in the SL_cci v2.0 product.

The SL_cci v1.1 product used the REAPER combined orbit for ERS-1 and ERS-2 (Rudenko et al., 2012), whilst GFZ VER11 orbit was used for the new (v2.0) product detailed here. The differences in the regional sea level trends computed using these two different orbits reach ± 2.0 mm yr^{-1} (Fig. 2). A switch from CNES POE-D orbit to POE-E orbit for Jason-1 caused changes in the SLA trend of up to ± 1.5 mm yr^{-1} (Fig. 3). The broad dipole pattern corresponds to errors in the modelling of geocentre motion, whilst individual tracks are prominent where changes to the gravity field have a more local effect.

2.2 Precise determination of the altimeter range

A waveform, i.e. the full radar echo recorded on board the altimeter, corresponds to the radar return from a disc a few kilometres across on the sea surface. Provided the surface is homogeneous, the shape of the waveform will conform to the Brown model (Brown, 1977; Hayne, 1980). In such circumstances, the position of the waveform (and thus the range) may be very accurately extracted; these values are stored in the GDR provided by the space agencies. In general, the sea level CCI project has not attempted to perform its own retracking of all the different missions but has assessed the quality of those available. In particular, the v2.0 product makes use of the latest ERS-1 and ERS-2 reprocessings from the REAPER project (Brockley et al., 2017), and incorporates the new GDR (version E) for Jason-1, which includes improved estimates of internal errors. The TOPEX waveform data show a sawtooth effect plus various data spikes associated with specific waveform bins (Hayne et al., 1994) and some of the waveform bins are averaged in pairs or groups of four, making the variability statistics complicated (Quartly et al., 2001). There has also been a degradation of the point target response of the "side A" instrument, heading to significant changes in wave height (Queffeulou, 2004), signal amplitude (Quartly, 2000) and derived range (Chambers et al., 2003). No new product for that mission was available in time for the reprocessed SL_cci v2.0 product, although Dieng et al. (2017) have recently suggested that a new correction for that period would yield a slightly smaller rate of sea level rise (see also Watson et al., 2015; Chen et al., 2017).

As part of the Level 1b processing, corrections are applied to the range for changes in the point target response (PTR) in response to ageing of the instrument, and also any drift in the ultra-stable oscillator (USO) that controls the on-board timing of pulses. Within the early years of the SL_cci project it had been found that Envisat's PTR waveform needed to be reversed in the Level 1b processing at Ku band (García and Roca, 2010); this change caused a notable impact on range, leading to better agreement of the long-term trends between Envisat and the reference missions. During the second phase, the S-band signal (used to compute the ionospheric correction) was assessed, but no change was made because there was no discernible benefit.

During the first phase of the SL_cci project, the coastal zone and the Arctic had been recognised as two areas requiring special effort because the waveforms were not "Brown-like" due to inhomogeneities within the full instrument footprint. Waveforms in coastal regions may contain early contributions from land or "bright target" responses from glassy seas in sheltered regions (Gómez-Enrí et al., 2010; Cipollini et al., 2017). The SL_cci project has been assessing two methodologies to overcome such anomalous waveforms: including a Gaussian peak within the shape model (Halimi et al., 2013) or focussing the shape-fitting mainly on the leading edge (Passaro et al., 2014). In the Arctic, the inhomogeneities are due to a mix of ice floes and thin leads (gaps within the ice exposing very calm waters). Poisson et al. (2017) have developed a processing scheme for classifying the data according to reflecting surface and retracking the waveforms from leads using an extended Brown model. So far, only data from the Envisat and SARAL/AltiKa missions have been processed, which has led to the production of a promising Arctic sea level product now available for the users. However, both the coastal and Arctic work are part of ongoing research, and additional efforts are required so that these retracked data could be included in a future SL_cci product.

3 Corrections to atmospheric propagation

The main atmospheric retardation of the radar signal, the dry tropospheric correction (DTC), is simply due to the mass of neutral dry air that it propagates through, and that can be retrieved from atmospheric pressure at sea level. As that cannot be measured from space, what is required is a good atmospheric model that incorporates measurements, i.e. a reanalysis product. The wet tropospheric component (WTC), representing the extra delay from atmospheric water vapour and liquid water, can also be extracted from an assimilating model, but the scales of temporal and spatial variations of the water vapour are usually not adequately resolved by global reanalyses, so some direct measurements of water vapour and liquid water are beneficial. Most altimetric satellites carry a nadir-viewing microwave radiometer (MWR) to record relevant emissions for WTC retrieval; however, CryoSat-2 has no such package, as its focus is on polar latitudes, where the WTC may largely be neglected. However, microwave radiometers are not reliable in the coastal zone due to their large footprint (typically 20–40 km) and global atmospheric models lack the resolution to incorporate coastal processes. An alternative data source is provided by shore-based GNSS stations, as the WTC derived from their L-band measurements is also valid at Ku and Ka band, since the troposphere is a non-dispersive medium at these frequencies.

The ionospheric delay is a retardation of the passage of radio waves by free electrons, which get accelerated. Such an effect predominantly occurs on the Sun-facing side of the Earth, and is strongest in two bands near the tropics. It is proportional to the columnar total electron content (TEC) divided by the square of the radar frequency. The TOPEX, Jason and Envisat spacecraft were designed with dual-frequency altimeters specifically to allow an estimation of the pertinent ionospheric correction from the difference in range delay recorded at the two frequencies. This was because the early ionospheric models were not deemed to be accurate enough to support the high precision required from the reference missions, and indeed the measuring and modelling of the ionospheric correction is a topic that still needs further development. However, there have been marked improvements in the ionospheric models in the past decade. Since AltiKa operates at Ka band, the size of this correction

is only one-seventh of that for the other instruments (which operate at Ku) and so operation at multiple frequencies was not justified.

3.1 Dry tropospheric correction

Dry tropospheric corrections (DTC) were calculated (Ablain et al., 2016) according to three different numerical models: ECMWF operational, ERA-Interim and JRA-55. Analysis of the sea level variance at crossovers and investigation of trends were performed for sea level data computed with each correction (ASM, 2015b). Although the operational version of the ECMWF model has the highest spatial resolution for recent years, giving it a superior performance to the others, it is not consistent for the whole 20+ year period; thus, the atmospheric model reanalyses are better suited for the present climate purpose. The ERA-Interim correction led to a smaller variance of crossover differences than when using the JRA-55 model, especially at southern latitudes, where the pressure variability is higher, which indicates a better performance for the ERA-Interim model. Thus, considering the long-period reanalyses for climate purposes, the ERA-Interim corrections were the ones adopted for SL_cci v2.0 for all altimeter instruments.

3.2 Wet tropospheric correction

The University of Porto has developed a robust method for determining the WTC by data combination through space–time objective analysis of various data types: valid measurements from the on-board MWR (whenever available) and third-party observations from GNSS and scanning imaging MWR. The latest version of these corrections, designated GNSS-derived Path Delay Plus (GPD+; see Fernandes and Lázaro, 2016), includes improved calibration of all radiometers on altimetric satellites by comparing them with the known stable performance of the SSMI and SSMIS. In addition to the calibration with respect to SSMI and SSMIS, the original GPD solution (Fernandes et al., 2015) has been augmented by adding new datasets (from scanning imaging radiometers) and improved selection criteria for selecting valid MWR observations. The GPD+ correction is implemented in SL_cci v2.0 for all missions except GFO, although similar corrections have subsequently become available for this satellite (Fernandes and Lázaro, 2016). In SL_cci v2.0, the WTC for GFO is calculated from its MWR for observations located > 50 km from the coast, and from the ECMWF operational model for data between 10 and 50 km from coast. There were problems with the radiometer during GFO cycles 135–137, 166, 181, 189 and after 201; in such cases ECMWF values were used for all observations.

The GPD+ correction allows the recovery of a significant number of altimeter measurements, ensuring the continuity and consistency of the correction in the transition region between the open ocean and coastal zone, and also at high lat-

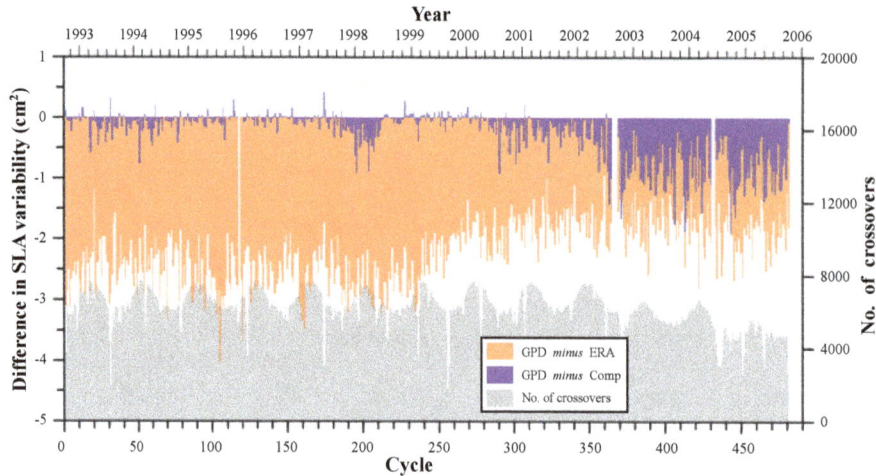

Figure 4. Difference in variance at TOPEX/Poseidon crossovers for SLA calculated with different WTC. Orange compares GPD+ with ERA-Interim and purple with the composite WTC. Negative values indicate an improvement (i.e. reduction) in crossovers for GPD+.

itudes. Figure 4 illustrates the improved performance of the GPD+ correction over that from ERA-Interim and the composite correction present in the AVISO products.

3.3 Ionospheric correction

Within the SLOOP project (Faugere et al., 2010), there has been considerable effort to develop an improved ionospheric correction using an iterative filtering scheme applied to the dual-frequency altimeter missions (TOPEX, Jason-1, Jason-2 and Envisat). This has been independently evaluated by a round-robin comparison with previous ionospheric corrections, and it was found that the SLOOP set of corrections led to an improvement in the recovery of mesoscale signals and increased data gain (due to less flagging of suspect data).

For the missions that do not have a second frequency (including Envisat after the loss of S-band data), a model is required. The one used in SL_cci v2.0 is GIM (Iijima et al., 1999), which is based on measurements from GPS satellites. However, prior to 1998 there were relatively few GPS data, so for ERS-1 and ERS-2 we use an interpretation based on the NIC09 climatology (Scharroo and Smith, 2010) modified by contemporaneous TOPEX records of global mean TEC. The corrections for Poseidon are based on the measurements from the DORIS system on board its satellite.

4   Corrections for sea state bias

Sea state bias (SSB) is a correction term encompassing three different effects: electro-magnetic (EM) bias, skewness and tracker bias. A wave field is not usually uniformly covered with identical reflecting facets – the surface tends to be smoother in the troughs of waves than at the crests, so there will be a proportionately stronger response from the lower-lying facets. This effect, the EM bias, will depend upon the

radar frequency. Most altimetric retrackers are designed to locate the mid-power point of the leading edge of the waveform; this equates to the median height of reflecting surfaces, rather than the mean. Thus a second effect, the skewness, relates to the difference between the heights of mean and median surfaces, which is a property of the ocean, independent of the radar frequency used for the sensing. The third effect relates to the algorithms used to find the range – this effect will vary with each retracker implemented, but should be the same for identical instruments. However, there are always slight differences between sister instruments, e.g. ERS-1 and ERS-2, so the overall sea state bias model is usually determined independently for each altimeter plus retracker. In practice, all three of these effects scale roughly with wave height, so the overall sea state bias is expressed as a multiplier of wave height that is a weakly varying function of sea state conditions.

Although the first two components of SSB should be the same for all Ku-band observing systems, a separate total SSB solution has to be derived for each individual altimeter. For each dataset, minimisation procedures are used to express SSB in terms of wave height and wind speed, leading to the least variance at crossovers. Many of these solutions remain as defined at the end of their respective missions, i.e. once all available data have been analysed. However, as these are optimisations based on observational data, improvements to the orbits or a change in the modelled PTR or the retracker applied could necessitate a revision to the SSB model.

Early solutions for SSB expressed the SSB coefficient in terms of two key parameters: wave height and wind speed. Those parametric forms are still used for ERS-1 (Gaspar and Ogor, 1994) and Poseidon (Gaspar et al., 1996). A non-parametric form, offering a better fit to the data, can be achieved for later missions for which there are greater volumes of more precise data. The non-parametric models

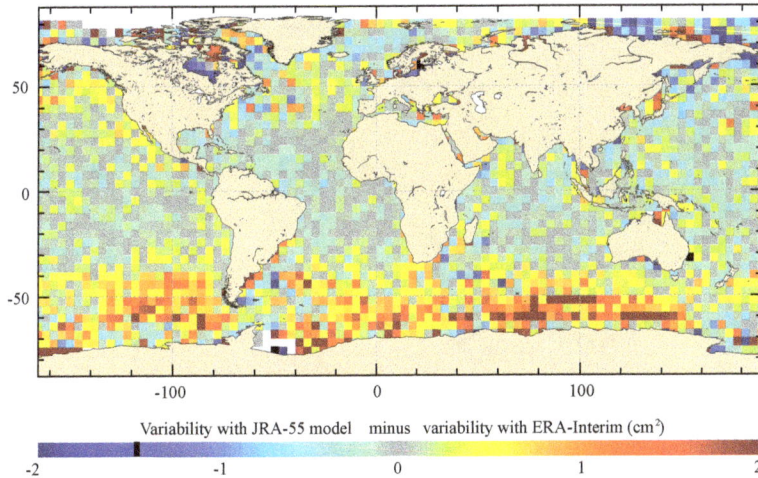

Figure 5. Change in crossover differences between processing TOPEX/Poseidon mission with IB calculated using JRA-55 or ERA-Interim. Positive values indicate greater variance with corrections from JRA-55.

adopted within SL_cci v2 are for ERS-2 (Mertz et al., 2005), TOPEX (Tran et al., 2010), Jason-1 & 2 (Tran et al., 2012), Envisat (Tran, 2015), GFO (N. Tran and S. Labroue, personal communication, 2009) and AltiKa (from the PEACHI project). The Cryosat-2 data used in this product are solely those in low-resolution mode; at the time that algorithm selection was completed, the most appropriate choice was that derived from Jason-1 GDR-C products, although ones based on CryoSat-2 data have subsequently become available. The changes from the previous product, slcci_v1.1, are the use of the Tran et al. (2012) for the Jason instruments and Tran (2015) for Envisat to replace the versions on their GDRs.

5   Corrections for short-term atmospheric and oceanographic phenomena

Our concern within the sea level CCI project is to provide the best dataset for observing climate scale variations in sea level and changes associated with geostrophic currents. The temporal sampling by altimeters is insufficient to resolve all timescales, so high-frequency ocean variability is aliased to longer timescales, thus polluting climate estimations if not adequately corrected. Thus, short-term effects have to be removed using accurate physical ocean models, which are expected to be independent of satellite missions.

5.1   Atmospheric pressure correction

Early altimeter processing included an "inverse barometer effect" (IB; see Fu and Pihos, 1994) whereby the sea surface was deemed to be depressed by 1 cm for each increase in atmospheric pressure by 1 mbar, with this computed effect being removed from the data to give the sea level expected in the absence of atmospheric effects. Instead a dy-

namic atmospheric correction (DAC) was introduced, based on a barotropic global ocean model forced by instantaneous atmospheric pressure and winds fields, and taking into account the ocean dynamic response to atmospheric forcing at high frequencies (Carrère and Lyard, 2003) and keeping the IB for low frequencies.

Several atmospheric models have been used to compute the IB and the DAC corrections (ECMWF, ERA-Interim, NCEP, JRA-55) in order to find the atmospheric reanalysis most suitable for the present climate analysis. The comparison of input weather models is another exercise of finding which correction (here, IB and DAC), when applied to altimeter measurements, leads to the greatest consistency between ascending and descending passes and thus reduces the altimeter crossover variance. Figure 5 shows that, for TOPEX/Poseidon data, the sea level anomalies calculated using the JRA-55 model produces greater crossover differences than the SLA using ERA-Interim, with much greater variance in the high southern latitudes, where the variability in the atmospheric forcing is strong. Moreover, using a DAC forced by ERA-Interim significantly reduced the crossover variance compared with the operational DAC forced by ECMWF analysis (Carrère et al., 2016). Based on such crossover variance analysis, ERA-Interim is the preferred model to force the DAC for all missions (ASM, 2015b).

5.2   Tides

There are five separate phenomena linked under the label "tides": ocean tide, ocean loading tide, solid Earth tide, pole tide and internal tides. The ocean tide is usually by far the largest, but all aspects need to be included in order to discern correctly regional variations and long-term trends. An ocean tide model will include many harmonics (not just M2 and S2) and may be an empirical fit to altimetric sea level data or pro-

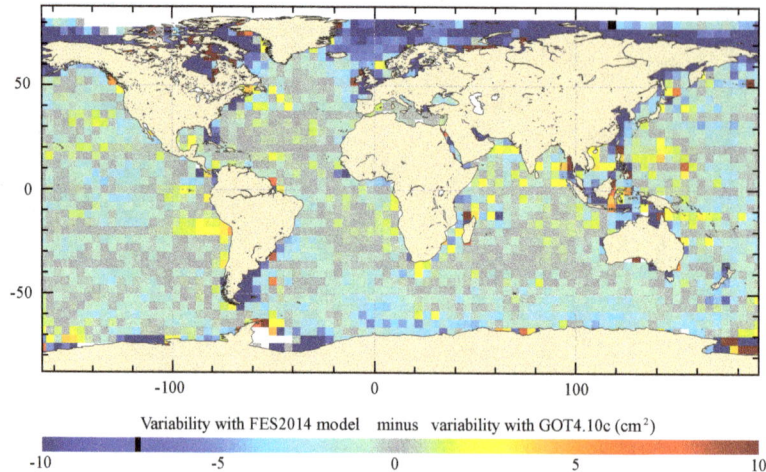

Figure 6. Change in crossover differences between processing Envisat mission with FES2014 tides or GOT4.10. Negative values indicate reduced variance with FES2014.

duced by a high-resolution fluid flow model or a combination of both. Early in the altimetry era there could be as many as 12 independent models to be assessed (Andersen et al., 1995), with, more recently, Stammer et al. (2014) evaluating seven data-constrained models. However, there are presently two main families of solutions to be compared, termed GOT (Ray, 2013) and FES (Carrère et al., 2012; Lyard et al., 2017). The GOT4.10 solution is mostly based on Jason data, excluding those from TOPEX/Poseidon because of poorly understood effects occurring at the S2 alias period (59 days). Figure 6 shows that the variance for Envisat data is reduced with the FES2014 model, especially in the Arctic. This model is also very effective in reducing the 59-day signal noted with some of the reference missions (Zawadzki et al., 2017).

The second aspect is the loading tide, which corresponds to the flexing of the Earth in response to the weight of water lying on it. For this we adopt the solution of Ray (2013), which, at the time of algorithm selection, was the only one consistent with the FES2014 ocean tide. The third aspect is the Earth tide, i.e. the changes in the Earth's topography due to the changing gravitational attraction of the moon and sun – here the long established solutions by Cartwright and Tayler (1971), modified by Cartwright and Edden (1973), continue to be applied.

Next, there is the "pole tide", a term describing the small long-period oscillations associated with the movement of the Earth's rotational axis. The recent advance by Desai et al. (2015) takes into account self-gravitation, loading, conservation of mass, and geocentre motion. Moreover, this new model includes a bias and a drift, which means that the new computed pole tide does not include the effects of the Earth's displacement response to that mean pole drift. Removing the long-term mean pole drift has a significant impact on the regional MSL trend estimation; this impact has been validated by comparisons with an Argo database over the time span

of the Envisat mission (ASM, 2015a; Legeais et al., 2017). Thus the recent model of Desai et al. (2015) is the one implemented in SL_cci v2.0. At present, there is no satisfactory model of the internal tides, so there is no correction for the effect of this phenomenon.

6 Reference surfaces

For some applications, it is useful to estimate sea level anomalies with respect to the mean sea surface (MSS), which is the sum of the geoid and mean dynamic topography (see Eq. 1). Frequent updates of the MSS are provided as new data become available, in particular from CryoSat-2 at high latitudes, and from the "end of life" geodetic phases of recent missions.

The SL_cci v2.0 is referenced to the DTU15 MSS (ASM, 2015e), and corresponds to a mean over the period 1993–2012. The DTU datasets provide a complete global coverage (including the high-latitude Arctic). This version is an improvement on earlier versions (DTU10 & DTU13, see Andersen et al., 2015) in that it makes use of 4 years of CryoSat-2 data but gives less weighting to data from IceSat (whose large errors gave an unrealistic stripiness to derived MSS fields). Thus the major improvements within DTU15 are the increased data coverage in the high latitudes (both Arctic and Antarctic) and the Mediterranean, and the finer scales resolved due to the use of shorter correlation scales in the interpolation. The inter-annual content of the reprocessed v2.0 product will change compared with the previous version due to the evolution of the reference period (1993–2008 for DTU10 in the SL_cci v1.1 product). This will affect assimilating models since these systems are sensitive to the reference period used.

CryoSat-2 has contributed significantly in the band 82–88° N not sampled by the other radar altimeters and, due to

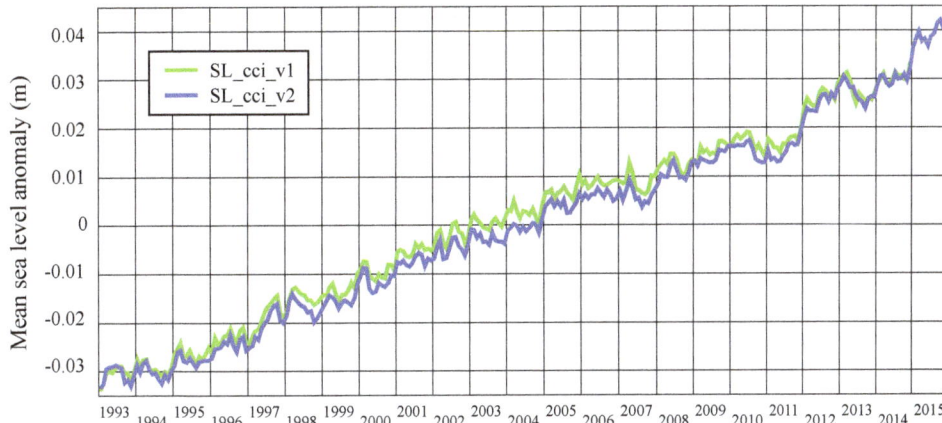

Figure 7. Comparison of time series of global mean sea level (seasonal signal removed). The v1.1 dataset had been updated until the end of 2014 and has a mean trend of 3.18 mm yr^{-1}; v2.0, described in this paper, now extends to end of 2015 and has a trend of 3.21 mm yr^{-1} over the same period as v1.1.

its long repeat orbit, provides finer longitudinal resolution than the ERS-1, ERS-2 and Envisat instruments for latitudes south of 82° N. (The DTU15 MSS no longer utilises data from the geodetic phases of ERS-1 and Geosat, as those measurements were noisy; consequently less spatial filtering is required leading to a higher resolution product.) The delay-Doppler mode of CryoSat-2 makes its measurements more resilient to stray reflections from nearby land; thus CryoSat-2 data have led to marked improvements in the MSS in many coastal areas, particularly those around the Mediterranean and the Bay of Fundy.

7   Editing and gridding

The production of the SL_cci v2.0 product uses the same procedures as for the previous version v1 (Ablain et al., 2015). An overview of the different processing steps to produce the Sea Level CCI products can be found in Ablain and Legeais (2014). In brief, these are to acquire and pre-process data, perform input checks and quality control (data are discarded if flagged for rain, land or ice), inter-calibrate and unify the multi-satellite measurements, and generate along-track and gridded merged products.

In addition to the reduction of the global and regional biases between two successive altimeter missions (thanks to the calibration phase during which both satellites observe the same ocean), the unification also involves a further orbit error reduction. This is first carried out for the "reference missions" (TOPEX/Poseidon, Jason-1 and Jason-2) by minimising the crossover differences between ascending and descending tracks. These missions have all been in the same 9.92-day orbital cycle and have a high altitude (1336 km), making their trajectories less sensitive to higher-order terms of the Earth's gravity field and to the drag effects. Then the "complementary missions" are adjusted to

minimise crossovers with data from the reference missions (Le Traon and Ogor, 1998). Thus, the reference missions are used to ensure the stability of the ECV. The global MSL estimation and large-scale changes rely on these reference missions. The complementary missions (adjusted on the reference missions) contribute to increase the spatial resolution of the grids and to increase their accuracy. This adjustment towards the orbits of the reference missions also overcomes a spurious SLA drift during Envisat's first year of operation.

Finally, output checks and quality control are performed and the multi-satellite along-track data are mapped to generate gridded sea level products. The sensitivity of the gridded products to the mapping algorithms is described in detail in Pujol (2012). Different mapping methods were tested in order to assess their ability to accurately reproduce climate signals. This evaluation has been carried out separating the different temporal and spatial scales related to climate applications. A monthly optimal interpolation is applied (including additional weighted information from part of the previous and following months) to produce maps of sea level on a 0.25° grid for the middle of each month. Note that this approach differs from the one used in the production of the DUACS dataset (Pujol et al., 2016) (daily optimal interpolation with different parameters) as the SL_cci approach has been designed to better answer the needs of climate users.

8   Data availability

The gridded monthly files of sea level anomaly at 0.25° resolution (https://doi.org/10.5270/esa-sea_level_cci-MSLA-1993_2015-v_2.0-201612; Legeais et al., 2016b) are freely available (upon email application to info-sealevel@esa-sealevel-cci.org). The Sea Level CCI website (http://www.esa-sealevel-cci.org/products) also contains derived products suitable for some climate studies:

Table 1. Summary of the data sources and the corrections applied to each altimeter instrument.

	TOPEX	Poseidon	Jason-1	Jason-2	ERS-1	ERS-2	Envisat	AltiKa	GFO	CryoSat-2
Orbit	GSFC std1504	GSFC std1504	CNES POE-E	CNES POE-E	GFZv11 (Rudenko et al., 2017)	GFZv11 (Rudenko et al., 2017)	GFZv11 (Rudenko et al., 2017)	CNES POE-E	GSFC std08	CNES POE-E
Data source (Retracker)	RGDR (least squares)	MLE-3	GDR-E (MLE-4)	GDR-E (MLE-4)	REAPER	REAPER	GDR (Ocean-1)	GDR (Ocean-3)	on-board α-β	GDR (SAMOSA 2.5.0)
Dry trop.	ERA-Interim	ERA-Interim	ERA-Interim	ERA-Interim	ERA-Interim	ERA-Interim	ERA-Interim	ERA-Interim	ERA-Interim	ERA-Interim
Wet trop.	GPD+	GPD+	GPD+	GPD+	GPD+	GPD+	GPD+	GPD+	MWR/ECMWF	GPD+
Iono	SLOOP	DORIS	SLOOP	SLOOP	NIC09	NIC09/GIM	SLOOP/GIM	GIM	GIM	GIM
SSB	Tran et al. (2010)	Gaspar et al. (1996)	Tran et al. (2012)	Tran et al. (2012)	Gaspar and Ogor (1994)	Mertz et al. (2005)	Tran (2015)	PEACHI	N. Tran and S. Labroue (personal communication, 2009)	Tran et al. (2012)
DAC	ERA-Interim	ERA-Interim	ERA-Interim	ERA-Interim	ERA-Interim	ERA-Interim	ERA-Interim	ERA-Interim	ERA-Interim	ERA-Interim
Ocean tide	FES2014	FES2014	FES2014	FES2014	FES2014	FES2014	FES2014	FES2014	FES2014	FES2014
Loading tide	GOT4v8AC	GOT4v8AC	GOT4v8AC	GOT4v8AC	GOT4v8AC	GOT4v8AC	GOT4v8AC	GOT4v8AC	GOT4v8AC	GOT4v8AC
Earth tide	Cartwright–Tayler–Edden									
Pole tide	Desai et al. (2015)									
MSS	DTU MSS 2015									

GDR is the geophysical data record, which is the standard product providing altimeter data, with some recommended corrections.

- Global Mean Sea Level temporal evolution (https://doi.org/10.5270/esa-sea_level_cci-IND_MSL_MERGED-1993_2015-v_2.0-201612).

- Regional Mean Sea Level trend (https://doi.org/10.5270/esa-sea_level_cci-IND_MSLTR_MERGED-1993_2015-v_2.0-201612).

- Amplitude and Phase of annual cycle (https://doi.org/10.5270/esa-sea_level_cci-IND_MSLAMPH_MERGED-1993_2015-v_2.0-201612).

9 Conclusions

During phase 2 of the ESA Sea Level CCI project, the consortium has reappraised all the corrections to be used in the production of the v2.0 dataset. In some cases, e.g. Earth tide, there has been no change in the recommended correction; in others, such as the pole tide, a new model has become available that is readily endorsed since it significantly improves the accuracy. For many other terms, there was a choice of two or three corrections: the project evaluated these through a variety of techniques including minimisation of mono-mission crossovers, comparison between different altimeter missions, and validation with in situ data. This paper has documented the choices made (Table 1).

The v2.0 dataset was released in December 2016, with details provided at http://www.esa-sealevel-cci.org/products. This will provide a consistent unbiased estimate of sea level spanning 1993–2015, which should greatly enhance the potential for climatic studies of sea level. The SL_cci ECV v2.0 products and their validation results are described in Legeais et al. (2017). In terms of the GMSL, the change from v1.1 to v2.0 products has led to changes of the order of 0.1 mm that persist for many months to years, but has not led to a significantly different long-term trend (~ 3.2 mm yr^{-1}; see Fig. 7). The changes that have had the most impact on derived trends are those for orbits and for wet tropospheric correction. Improvements to the Earth's time-variable gravity field model have led to major changes in the regional mean sea level trends (> 0.5 mm yr^{-1}; ASM, 2015c). Through its revision of the calibration of the MWR on altimetric satellites, the GPD+ solution has a significant impact on the trend of GMSL during the first and second decades of continuous altimetry: -0.2 mm yr^{-1} during 1993–2001 and $+0.2$ mm yr^{-1} during 2002–2014 (ASM, 2015d).

Appendix A: Abbreviations used

This appendix provides details of the abbreviations not expanded in the main text, because doing so would adversely affect the readability.

CHAMP	CHAllenging Minisatellite Payload
CNES	Centre National d'Etudes Spatiales
DORIS	Doppler Orbitography and Radiopositioning Integrated by Satellite
DPOD	DORIS terrestrial reference frame for precise orbit determination
DTU	Danish Technical University
ECMWF	European Centre for Medium-Range Weather Forecasts
Envisat	Environmental Satellite
ERA	ECMWF Reanalysis
ERS	European Remote-sensing Satellite
GFO	GEOSAT Follow-On (satellite)
GNSS	Global Navigation Satellite System
GOCE	Gravity field and steady-state Ocean Circulation Explorer
GPS	Global positioning by satellite
GRACE	Gravity Recovery and Climate Experiment
GSFC	Goddard Space Flight Center
IGS	International GNSS Service
ITRF	International Terrestrial Reference Frame
JRA	Japanese Meteorological Agency Reanalysis
LAGEOS	Laser Geodynamics Satellite
NCEP	National Centers for Environmental Prediction
POE	precise orbit ephemeris
PRARE	Precise Range And Range-Rate Equipment
REAPER	REprocessing of Altimeter Products for ERS
SARAL	Satellite with ARgos and ALtiKa
SLOOP	a Step forward aLtimetry Open Ocean Products
SLR	satellite laser ranging
SLRF	satellite laser ranging frame
SSMI	Special Sensor Microwave Imager
SSMIS	Special Sensor Microwave Imager/Sounder
TOPEX	Ocean Topography Experiment
VER11	version 11

Author contributions. Phase 2 of the Sea Level CCI project was managed by JFL, who oversaw the evaluation and selection of corrections. The initial draft of the paper was written by GQ. All other authors contributed through their derivation and evaluation of the suite of possible corrections, provision of figures and revision of the text.

Competing interests. The authors declare that they have no conflict of interest.

Acknowledgements. The authors acknowledge the support of ESA in the frame of the Sea Level CCI project, launched and co-ordinated by technical officer Jérôme Benveniste.

Edited by: Robert Key

References

Ablain, M. and Legeais, J.-F.: Detailed Processing Model for Sea-Level CCI system, Ref. CLS-DOS-NT-13-248, nomenclature SLCCI-DPM-33, Issue 1.1, available at: http://www.esa-sealevel-cci.org/webfm_send/239 (last access: 10 August 2017), 2014.

Ablain, M., Cazenave, A., Larnicol, G., Balmaseda, M., Cipollini, P., Faugère, Y., Fernandes, M. J., Henry, O., Johannessen, J. A., Knudsen, P., Andersen, O., Legeais, J., Meyssignac, B., Picot, N., Roca, M., Rudenko, S., Scharffenberg, M. G., Stammer, D., Timms, G., and Benveniste, J.: Improved sea level record over the satellite altimetry era (1993–2010) from the Climate Change Initiative project, Ocean Sci., 11, 67–82, https://doi.org/10.5194/os-11-67-2015, 2015.

Ablain, M., Zawadzki, L., and Legeais, J.-F.: ATBD v3.3: SL_cci Algorithm Theoretical Basis Document, Ref. CLS-SLCCI-16-0008, Nomenclature SLCCI-ATBDv1-016, Issue 3.3, available at: http://www.esa-sealevel-cci.org/webfm_send/538 (last access: 10 August 2017), 2016.

Ablain, M., Legeais, J.-F., Prandi, P., Marcos, M., Fenoglio-Marc, L., Dieng, H.-B., Benveniste, J., and Cazenave, A.: Satellite altimetry-based sea level at global and regional scales, Surv Geophys., 38, 7–31, https://doi.org/10.1007/s10712-016-9389-8, 2017.

Altamimi, Z., Collilieux, X., and Metivier, L.: ITRF2008: An improved solution of the International Terrestrial Reference Frame, J. Geodesy, 85, 457–473, https://doi.org/10.1007/s00190-011-0444-4, 2011.

Andersen, O. B., Woodworth, P. L., and Flather, R. A.: Intercomparison of recent ocean tide models, J. Geophys. Res., 100, 25261–25282, https://doi.org/10.1029/95JC02642, 1995.

Andersen, O. B., Knudsen, P., and Stenseng, L.: The DTU13 MSS (mean sea surface) and MDT (mean dynamic topography) from 20 years of satellite altimetry, International Association of Geodesy Symposia, Springer, 111–121, https://doi.org/10.1007/1345_2015_182, 2015.

ASM: Sl_cci phase 2 Algorithm Selection Meeting. Tides Selection, Toulouse, 26 November 2015, available at: http://www.esa-sealevel-cci.org/webfm_send/389 (last access: 10 August 2017), 2015a.

ASM: Sl_cci phase 2 Algorithm Selection Meeting. Atmospheric Corrections Selection, Toulouse, 26 November 2015, available at: http://www.esa-sealevel-cci.org/webfm_send/384 (last access: 10 August 2017), 2015b.

ASM: Sl_cci phase 2 Algorithm Selection Meeting. Selection of new Orbit Solutions, Toulouse, 26 November 2015, available at: http://www.esa-sealevel-cci.org/webfm_send/382 (last access: 10 August 2017), 2015c.

ASM: Sl_cci phase 2 Algorithm Selection Meeting. Wet Troposphere correction Selection, Toulouse, 26 November 2015, available at: http://www.esa-sealevel-cci.org/webfm_send/385 (last access: 10 August 2017), 2015d.

ASM: Sl_cci phase 2 Algorithm Selection Meeting. MSS Model Selection, Toulouse, 26 November 2015, available at: http://www.esa-sealevel-cci.org/webfm_send/390 (last access: 10 August 2017), 2015e.

Brockley, D., Baker, S., Féménias, P., Martinez, B., Massmann, F.-H., Otten, M., Paul, F., Picard, B., Prandi, P., Roca, M., Rudenko, S., Scharroo, R., and Visser, P.: REAPER: Reprocessing 12 years of ERS-1 and ERS-2 altimeter and microwave radiometer data, IEEE T. Geosci. Remote. Sens., https://doi.org/10.1109/TGRS.2017.2709343, in press, 2017.

Brown, G.: The average impulse response of a rough surface and its applications, IEEE T. Antennas Propag., 25, 67–74, https://doi.org/10.1109/JOE.1977.1145328, 1977.

Carrère, L. and Lyard, F.: Modeling the barotropic response of the global ocean to atmospheric wind and pressure forcing – comparisons with observations, Geophys. Res. Lett., 30, 1275, https://doi.org/10.1029/2002GL016473, 2003.

Carrère, L., Lyard, F., Cancet, M., Guillot, A., and Roblou, L.: FES2012: A new global tidal model taking advantage of nearly twenty years of altimetry, Proceedings of the 20 Years of Progress in Radar Altimetry Symposium (Venice, Italy), 1–20, 2012.

Carrère, L., Faugère, Y., and Ablain, M.: Major improvement of altimetry sea level estimations using pressure-derived corrections based on ERA-Interim atmospheric reanalysis, Ocean Sci., 12, 825–842, https://doi.org/10.5194/os-12-825-2016, 2016.

Cartwright, D. E. and Edden, A. C.: Corrected tables of tidal harmonics, Geophys. J. Internat., 33, 253–264, https://doi.org/10.1111/j.1365-246X.1973.tb03420.x, 1973.

Cartwright, D. E. and Tayler, R. J.: New computations of the tide-generating potential, Geophys. J. Internat., 23, 45–73, https://doi.org/10.1111/j.1365-246X.1971.tb01803.x, 1971.

Chambers, D. P., Hayes, S. A., Ries, J. C., and Urban, T. J.: New TOPEX sea state bias models and their effect on global mean sea level, J. Geophys. Res., 108, 3305, https://doi.org/10.1029/2003JC001839, 2003.

Chelton, D. B., Walsh, E. J., and MacArthur, J. L.: Pulse compression and sea level tracking in satellite altimetry, J. Atmos. Ocean. Tech., 6, 407–438, https://doi.org/10.1175/1520-0426(1989)006<0407:PCASLT>2.0.CO;2, 1989.

Chen, X., Zhang, X., Church, J. A., Watson, C. S., King, M. A., Monselesan, D., Legresy, B., and Harig, C.: The increasing rate of global mean sea-level rise during 1993–2014, Nat. Clim.

Change, 7, 492–495, https://doi.org/10.1038/nclimate3325, 2017.

Cipollini, P., Calafat, F. M., Jevrejeva, S., Melet, A., and Prandi, P.: Monitoring sea level in the coastal zone with satellite altimetry and tide gauges, Surv. Geophys., 38, 35–59, https://doi.org/10.1007/s10712-016-9392-0, 2017.

Couhert, A., Cerri, L., Legeais, J. F., Ablain, M., Zelensky, N. P., Haines, B. J., Lemoine, F. G., Bertiger, W. I., Desai, S. D., and Otten, M.: Towards the 1 mm/y stability of the radial orbit error at regional scales, Adv. Space Res., 55, 2–23, https://doi.org/10.1016/j.asr.2014.06.041, 2015.

Desai, S., Wahr, J., and Beckley, B.: Revisiting the pole tide for and from satellite altimetry, J. Geodesy, 89, 1233–1243, https://doi.org/10.1007/s00190-015-0848-7, 2015.

Dieng, H. B., Cazenave, A., Meyssignac, B., and Ablain, M.: New estimate of the current rate of sea level rise from a sea level budget approach, Geophys. Res. Lett., 44, 3744–3751, https://doi.org/10.1002/2017GL073308, 2017.

Dumont, J. P., Rosmorduc, V., Carrère, L., Picot, N., Bronner, E., Couhert, A., Desai, S., Bonekamp, H., Scharroo, R., and Leuliette, E.: OSTM/Jason-2 Products Handbook, rev. 11, available at: https://www.aviso.altimetry.fr/fileadmin/documents/data/tools/hdbk_j2.pdf (last access: 10 August 2017), 2017.

Escudier, P., Ablain, M., Amarouche, L., Carrère, L., Couhert, A., Dibarboure, G., Dorandeu, J., Dubois, P., Mallet, A., Mercier, F., Picard, B., Richard, J., Steunou, N., Thibaut, P., Rio, M.-H., and Tran, N.: Satellite radar altimetry: principle, accuracy & precision, in: Satellite Altimetry Over Oceans and Land Surfaces, edited by: Stammer, D. and Cazenave, A., CRC Press, in press, 2017.

Faugere, Y., Rio, M.-H., Labroue, S., Rosmorduc, V,. Thibaut, P., Amarouche, L., Obligis, E., Ablain, M., Legeais, J.-F., Carrere, L., Schaeffer, P., Tran, N., Pujol, I., Dufau, C., Dibarboure, G., Lux, M., Bronner, E., and Picot, N.: The SLOOP project: Preparing the next generation of altimetry products for open ocean, OSTST meeting 2010, Lisbon, Portugal, available at: http://www.aviso.altimetry.fr/fileadmin/documents/OSTST/2010/Faugere_SLOOP.pdf (last access: 10 August 2017), 2010.

Fernandes, M. J., Lázaro, C., Ablain, M., and Pires, N.: Improved wet path delays for all ESA and reference altimetric missions, Remote Sens. Environ., 169, 50–74, https://doi.org/10.1016/j.rse.2015.07.023, 2015.

Fernandes, M. J. and Lázaro, C.: GPD+ wet tropospheric corrections for CryoSat-2 and GFO altimetry missions, Remote Sens., 8, 851, https://doi.org/10.3390/rs8100851, 2016.

Fu, L.-L. and Cazenave, A. (Eds.): Satellite altimetry and Earth sciences: A handbook of techniques and applications, Academic Press, San Diego, 463 pp., 2001.

Fu, L.-L. and Pihos, G.: Determining the response of sea level to atmospheric pressure using TOPEX/POSEIDON data, J. Geophys. Res., 99, 24633–24642, https://doi.org/10.1029/94JC01647, 1994.

García, P. and Roca, M.: On-board PTR processing analysis: MSL drift differences, Technical Note, ISARD_ESA_L1B_ESL_CCN_PRO_064, isardSAT under ESA Contract (RA-2 L1b ESL), 2010.

Gaspar, P., Ogor, F., and Escoubes, C.: Nouvelles calibration et analyse du biais d'etat de mer des altimètres TOPEX et POSEIDON, Technical note 96/018 of CNES Contract 95/1523, 1996.

Gaspar, P. and Ogor, F.: Estimation and analysis of the sea state bias of the ERS-1 altimeter, Rapport technique, Report of task B1-B2 of IFREMER Contract no. 94/2.426 016/C.84, 1994.

GCOS: Systematic observation requirements for satellite-based data products for climate (2011 update) – supplemental details to the satellite-based component of the "Implementation plan for the global observing system for climate in support of the UNFCCC (2010 update)", GCOS-154 (WMO), available at: https://library.wmo.int/opac/doc_num.php?explnum_id=3710 (last access: 10 August 2017), 2011.

Gómez-Enrí, J., Vignudelli, S., Quartly, G. D., Gommenginger, C. P., Cipollini, P., Challenor, P. G., and Benveniste, J.: Modeling Envisat RA-2 waveforms in the coastal zone: Case-study of calm water contamination, IEEE Geosci. Remote Sens. Lett., 7, 474–478, https://doi.org/10.1109/LGRS.2009.2039193, 2010.

Halimi, A., Mailhes, C., Tourneret, J.-Y., Thibaut, P., and Boy, F.: Parameter estimation for peaky altimetric waveforms, IEEE T. Geosci. Remote Sens., 51, 1568–1577, https://doi.org/10.1109/TGRS.2012.2205697, 2013.

Hasselmann, K.: Multi-pattern fingerprint method for detection and attribution of climate change, Clim Dynam., 13, 601–611, https://doi.org/10.1007/s003820050185, 1997.

Hayne, G. S.: Radar altimeter mean return waveforms from near-normal-incidence ocean surface scattering, IEEE T. Antennas Propag., 28, 687–692, https://doi.org/10.1109/TAP.1980.1142398, 1980.

Hayne, G. S., Hancock, D. W., Purdy, C. L., and Callahan, P. S.: The corrections for significant wave height and attitude effects in the TOPEX radar altimeter, J. Geophys. Res., 99, 24941–24955, https://doi.org/10.1029/94JC01777, 1994.

Iijima, B. A., Harris, I. L., Ho, C. M., Lindqwister, U. J., Mannucci, A. J., Pi, X., Reyes, M. J., Sparks, L. C., and Wilson, B. D.: Automated daily process for global ionospheric total electron content maps and satellite ocean altimeter ionospheric calibration based on Global Positioning System data, J. Atmos. Sol.-Terr. Phys., 61, 1205–1218, https://doi.org/10.1016/S1364-6826(99)00067-X, 1999.

Legeais, J.-F., Prandi, P., and Guinehut, S.: Analyses of altimetry errors using Argo and GRACE data, Ocean Sci., 12, 647–662, https://doi.org/10.5194/os-12-647-2016, 2016a.

Legeais, J.-F., et al.: Sea Level CCI Phase 2. Time series of gridded Sea Level Anomalies (SLA), https://doi.org/10.5270/esa-sea_level_cci-1993_2015-v_2.0-201612, 2016b.

Legeais, J.-F., Cazenave, A., Ablain, M., Zawadzki, L., Zuo, H., Johannessen, J. A., Scharffenberg, M. G., Fenoglio-Marc, L., Fernandes, M. J., Andersen, O., Rudenko, S., Cipollini, P., Quartly, G., Passaro, M., and Benveniste, J.: An accurate and homogeneous altimeter sea level record: The reprocessed ESA Essential Climate Variable, Nature Scientific Data, in preparation, 2017.

Lemoine, F. G., Zelensky, N. P., Chinn, D. S., Beckley, B. D., and Lillibridge, J. L.: Towards the GEOSAT Follow-On precise orbit determination goals of high accuracy and near-real-time processing, AIAA/AAS Astrodynamics Specialist Conference and Exhibit, 21–24 August 2006, Keystone, Colorado, available at: https://arc.aiaa.org/doi/pdfplus/10.2514/6.2006-6402 (last access: 10 August 2017), 2006.

Lemoine, F. G., Zelensky, N. P., Chinn, D. S., Beckley, B. D., Rowlands, D. D., and Pavlis, D. E.: A new time series of orbits (std1504) for TOPEX/Poseidon. Jason-1, and Jason-2 (OSTM), Ocean Surface Topography Science Team Meeting, Reston, available at: http://meetings.aviso.altimetry.fr/fileadmin/user_upload/tx_ausyclsseminar/files/OSTST2015/POD-02-Lemoine.pdf_01.pdf, last access: 18 May 2017.

Le Traon, P.-Y. and Ogor, F.: ERS-1/2 orbit improvement using TOPEX/POSEIDON: The 2 cm challenge, J. Geophys. Res., 103, 8045–8057, https://doi.org/10.1029/97JC01917, 1998.

Lyard, F., Carrère, L., Cancet, M., Guillot, A., and Picot, N.: FES2014, a new finite elements tidal model for global ocean, Ocean Dynam., in preparation, 2017.

Mertz, F., Mercier, F., Labroue, S., Tran, N., and Dorandeu, J.: ERS-2 OPR data quality assessment; Long-term monitoring – particular investigation, CLS.DOS.NT-06.001, available at: http://www.aviso.altimetry.fr/fileadmin/documents/calval/validation_report/E2/annual_report_e2_2005.pdf (last access: 10 August 2017), 2005.

Ollivier, A., Faugere, Y., Picot, N., Ablain, M., Fémenias, P., and Benveniste, J.: Envisat ocean altimeter becoming relevant for mean sea level trend studies, Mar. Geodesy, 35, 118–136, https://doi.org/10.1080/01490419.2012.721632, 2012.

Pavlis, E. C.: SLRF2008: The ILRS reference frame for SLR POD contributed to ITRF2008, Ocean Surface Topography Science Team 2009, Seattle, Washington, available at: http://www.aviso.oceanobs.com/fileadmin/documents/OSTST/2009/poster/Pavlis_2.pdf (last access: 10 August 2017), 2009.

Passaro, M., Cipollini, P., Vignudelli, S., Quartly, G. D., and Snaith, H. M.: ALES: A multi-mission adaptive subwaveform retracker for coastal and open ocean altimetry, Remote Sens. Environ., 145, 173–189, https://doi.org/10.1016/j.rse.2014.02.008, 2014.

Poisson, J.-C., Quartly, G. D., Kurekin, A., Thibaut, P., Hoang, D., and Nencioli, F.: Development of an ENVISAT altimetry processor ensuring sea level continuity between open ocean and Arctic leads, IEEE T. Geosci. Remote Sens., submitted, 2017.

Pujol I.: WP2500 mapping methods. CLS-DOS-NT-11-229, SALP-NT-MA-EA-22007-CLS, Issue 1.1, SL_cci validation report, available at: www.esa-sealevel-cci.org/webfm_send/184 (last access: 10 August 2017), 2012.

Pujol, M.-I., Faugère, Y., Taburet, G., Dupuy, S., Pelloquin, C., Ablain, M., and Picot, N.: DUACS DT2014: the new multimission altimeter data set reprocessed over 20 years, Ocean Sci., 12, 1067–1090, https://doi.org/10.5194/os-12-1067-2016, 2016.

Quartly, G. D.: Monitoring and cross-calibration of altimeter σ^0 through dual-frequency backscatter measurements, J. Atmos. Ocean. Tech., 17, 1252–1258, https://doi.org/10.1175/1520-0426(2000)017<1252:MACCOA>2.0.CO;2, 2000.

Quartly, G. D., Srokosz, M. A., and McMillan, A. C.: Analyzing altimeter artifacts: Statistical properties of ocean waveforms, J. Atmos. Ocean. Tech., 18, 2074–2091, https://doi.org/10.1175/1520-0426(2001)018<2074:AAASPO>2.0.CO;2, 2001.

Queffeulou, P.: Long-term variation of wave height measurements from altimeters, Mar. Geodesy, 27, 495–510, https://doi.org/10.1080/01490410490883478, 2004.

Ray, R. D.: Precise comparisons of bottom-pressure and altimetric ocean tides, J. Geophys. Res., 118, 4570–4584, https://doi.org/10.1002/jgrc.20336, 2013.

Rebischung, P., Griffiths, J., Ray, J., Schmid, R. Collilieux, X., and Garayt, B.: IGS08: the IGS realization of ITRF2008, GPS Solut., 16, 483, https://doi.org/10.1007/s10291-011-0248-2, 2012.

Rudenko, S., Otten, M., Visser, P., Scharroo, R., Schöne, T., and Esselborn, S.: New improved orbit solutions for the ERS-1 and ERS-2 satellites, Adv. Space Res., 49, 1229–1244, https://doi.org/10.1016/j.asr.2012.01.021, 2012.

Rudenko, S., Dettmering, D., Esselborn, S., Schöne, T., Förste, Ch., Lemoine, J.-M., Ablain, M., Alexandre, D., and Neumayer, K.-H.: Influence of time variable geopotential models on precise orbits of altimetry satellites, global and regional mean sea level trends, Adv. Space Res., 54, 92–118, https://doi.org/10.1016/j.asr.2014.03.010, 2014.

Rudenko, S., Dettmering, D., Esselborn, S., Fagiolini, E., and Schöne, T.: Impact of atmospheric and oceanic de-aliasing Level-1B (AOD1B) products on precise orbits of altimetry satellites and altimetry results, Geophys. J. Int., 204, 1695–1702, https://doi.org/10.1093/gji/ggv545, 2016.

Rudenko, S., Neumayer, K.-H., Dettmering, D., Esselborn, S., Schöne, T., and Raimondo, J.-C.: Improvements in precise orbits of altimetry satellites and their impact on mean sea level monitoring, IEEE T. Geosci. Remote Sens., 55, 3382–3395, https://doi.org/10.1109/TGRS.2017.2670061, 2017.

Scharroo, R. and Smith, W. H. F.: A global positioning system–based climatology for the total electron content in the ionosphere, J. Geophys. Res., 115, A10318, https://doi.org/10.1029/2009JA014719, 2010.

Stammer, D., Ray, R. D., Andersen, O. B., Arbic, B. K., Bosch, W., Carrère, L., Cheng, Y., Chinn, D. S., Dushaw, B. D., Egbert, G. D., Erofeeva, S. Y., Fok, H. S., Green, J. A. M., Griffiths, S., King, M. A., Lapin, V., Lemoine, F. G., Luthcke, S. B., Lyard, F., Morison, J., Muller, M., Padman, L., Richman, J. G., Shriver, J. F., Shum, C. K., Taguchi, E., and Yi, Y.: Accuracy assessment of global barotropic ocean tide models, Rev. Geophys., 52, 243–282, https://doi.org/10.1002/2014RG000450, 2014.

Tapley, B., Bettadpur, S., Watkins, M., and Reigber, C.: The gravity recovery and climate experiment: Mission overview and early results, Geophys. Res. Lett., 31, L09607, https://doi.org/10.1029/2004GL019920, 2004.

Tran, N.: Envisat Phase-F: Sea State Bias, Technical Report CLS-DOS-NT-15-031, ESA Contract, ENVISAT RA-2 and MWR ESL and prototypes maintenance support (level 1b and level 2), 2015.

Tran, N., Labroue, S., Philipps, S., Bronner, E., and Picot, N.: Overview and update of the sea state bias corrections for the Jason-2, Jason-1 and TOPEX Missions, Mar. Geodesy, 33, 348–362, https://doi.org/10.1080/01490419.2010.487788, 2010.

Tran, N., Philipps, S., Poisson, J.-C., Urien, S., Bronner, E., and Picot, N.: Impact of GDR-D standards on SSB corrections, Aviso, OSTST, available at: http://www.aviso.altimetry.fr/fileadmin/documents/OSTST/2012/oral/02_friday_28/01_instr_processing_I/01_IP1_Tran.pdf (last access: 10 August 2017), 2012.

Watson, C. S., White, N. J., Church, J. A., King, M. A., Burgette, R. J., and Legresy, B.: Unabated global mean sea-level rise over the satellite altimeter era, Nat. Clim. Change, 5, 565–568, https://doi.org/10.1038/NCLIMATE2635, 2015.

Willis, P., Zelensky, N. P., Ries, J., Soudarin, L., Cerri, L., Moreaux, G., Lemoine, F. G., Otten, M., Argus, D. F., and Heflin, M. B.: DPOD2008, a DORIS-oriented terrestrial reference frame for precise orbit determination, IAG Symposia series, 143, 175–181, https://doi.org/10.1007/1345_2015_125, 2016.

Zawadzki, L., Ablain, M., Carrère, L., Ray, R. D., Zelensky, N. P., Lyard, F., Guillot, A., and Picot, N.: Reduction of the 59-day error signal in the Mean Sea Level derived from TOPEX/Poseidon, Jason-1 and Jason-2 data with the latest FES and GOT ocean tide models, IEEE T. Geosci. Remote Sens., submitted, 2017.

10

A compilation of global bio-optical in situ data for ocean-colour satellite applications

André Valente[1], Shubha Sathyendranath[2], Vanda Brotas[1], Steve Groom[2], Michael Grant[2], Malcolm Taberner[3], David Antoine[4,5], Robert Arnone[6], William M. Balch[7], Kathryn Barker[8], Ray Barlow[9], Simon Bélanger[10], Jean-François Berthon[11], Şükrü Beşiktepe[12], Vittorio Brando[13,14], Elisabetta Canuti[11], Francisco Chavez[15], Hervé Claustre[16], Richard Crout[17], Robert Frouin[18], Carlos García-Soto[19,20], Stuart W. Gibb[21], Richard Gould[17], Stanford Hooker[22], Mati Kahru[18], Holger Klein[23], Susanne Kratzer[24], Hubert Loisel[25], David McKee[26], Brian G. Mitchell[18], Tiffany Moisan[27], Frank Muller-Karger[28], Leonie O'Dowd[29], Michael Ondrusek[30], Alex J. Poulton[31], Michel Repecaud[32], Timothy Smyth[2], Heidi M. Sosik[33], Michael Twardowski[34], Kenneth Voss[35], Jeremy Werdell[22], Marcel Wernand[36], and Giuseppe Zibordi[11]

[1]Marine and Environmental Sciences Centre (MARE), University of Lisbon, Lisbon, Portugal
[2]Plymouth Marine Laboratory, Plymouth, PL1 3DH, UK
[3]EUMETSAT, Eumetsat-Allee 1, 64295 Darmstadt, Germany
[4]Sorbonne Universités, UPMC Univ. Paris 06, CNRS, Laboratoire d'Océanographie de Villefranche, Villefranche-sur-mer, 06238, France
[5]Remote Sensing and Satellite Research Group, Department of Physics, Astronomy and Medical Radiation Sciences, Curtin University, Perth, WA 6845, Australia
[6]University of Southern Mississippi, Stennis Space Center, Kiln, MS, USA
[7]Bigelow Laboratory for Ocean Sciences, East Boothbay, ME, USA
[8]ARGANS Ltd, Plymouth, UK
[9]Bayworld Centre for Research and Education, Cape Town, South Africa
[10]Département de biologie, chimie et géographie, Université du Québec à Rimouski, Rimouski (Québec), Canada
[11]European Commission, Joint Research Centre, Ispra, Italy
[12]Institute of Marine Science and Technology, Dokuz Eylul University, Izmir, Turkey
[13]CSIRO Oceans and Atmosphere, Canberra, Australia
[14]CNR IREA, Milan, Italy
[15]Monterey Bay Aquarium Research Institute, Moss Landing, CA, USA
[16]Laboratoire d'Océanographie de Villefranche (LOV), Sorbonne Universités, UPMC Univ Paris 06, INSU-CNRS, 181 Chemin du Lazaret, 06230 Villefranche-sur-Mer, France
[17]Naval Research Laboratory, Stennis Space Center, Kiln, MS, USA
[18]Scripps Institution of Oceanography, University of California, San Diego, CA, USA
[19]Spanish Institute of Oceanography (IEO), Corazón de María 8, 28002 Madrid, Spain
[20]Plentziako Itsas Estazioa/Euskal Herriko Unibetsitatea (PIE/EHU), Areatza z/g, 48620 Plentzia, Spain
[21]Environmental Research Institute, North Highland College, University of the Highlands and Islands, Thurso, Scotland, UK
[22]NASA Goddard Space Flight Center, Greenbelt, Maryland, USA
[23]Operational Oceanography Group, Federal Maritime and Hydrographic Agency, Hamburg, Germany
[24]Department of Ecology, Environment and Plant Sciences, Frescati Backe, Stockholm University, 106 91 Stockholm, Sweden
[25]Laboratoire d'Océanologie et de Géosciences, Université du Littoral – Côte d'Opale, Maison de la Recherche en Environnement Naturel, Wimereux, France
[26]Physics Department, University of Strathclyde, Glasgow G4 0NG, Scotland, UK

[27]NASA Goddard Space Flight Center, Wallops Flight Facility, Wallops Island, VA, USA

[28]Institute for Marine Remote Sensing/ImaRS, College of Marine Science, University of South Florida, St. Petersburg, FL, USA

[29]Fisheries and Ecosystem Advisory Services, Marine Institute, Rinville, Oranmore, Galway, Ireland

[30]NOAA/NESDIS/STAR/SOCD, College Park, MD, USA

[31]Ocean Biogeochemistry and Ecosystems, National Oceanography Centre, Waterfront Campus, Southampton, UK

[32]IFREMER Centre de Brest, Plouzane, France

[33]Biology Department, Woods Hole Oceanographic Institution, Woods Hole, MA, USA

[34]Harbor Branch Oceanographic Institute, Fort Pierce, FL, USA

[35]Physics Department, University of Miami, Coral Gables, FL, USA

[36]Physical Oceanography, Marine Optics & Remote Sensing, Royal Netherlands Institute for Sea Research, Texel, Netherlands

Correspondence to: André Valente (adovalente@fc.ul.pt)

Abstract. A compiled set of in situ data is important to evaluate the quality of ocean-colour satellite-data records. Here we describe the data compiled for the validation of the ocean-colour products from the ESA Ocean Colour Climate Change Initiative (OC-CCI). The data were acquired from several sources (MOBY, BOUSSOLE, AERONET-OC, SeaBASS, NOMAD, MERMAID, AMT, ICES, HOT, GeP&CO), span between 1997 and 2012, and have a global distribution. Observations of the following variables were compiled: spectral remote-sensing reflectances, concentrations of chlorophyll a, spectral inherent optical properties and spectral diffuse attenuation coefficients. The data were from multi-project archives acquired via the open internet services or from individual projects, acquired directly from data providers. Methodologies were implemented for homogenisation, quality control and merging of all data. No changes were made to the original data, other than averaging of observations that were close in time and space, elimination of some points after quality control and conversion to a standard format. The final result is a merged table designed for validation of satellite-derived ocean-colour products and available in text format. Metadata of each in situ measurement (original source, cruise or experiment, principal investigator) were preserved throughout the work and made available in the final table. Using all the data in a validation exercise increases the number of matchups and enhances the representativeness of different marine regimes. By making available the metadata, it is also possible to analyse each set of data separately.

1 Introduction

Currently, there are several bio-optical in situ datasets worldwide suitable for validation of ocean-colour satellite data. While some are managed by the data producers, others are in international repositories with contributions from multiple scientists. Many have rigid quality controls and are built specifically for ocean-colour validation. The use of only one of these datasets would limit the number of data in validation exercises. It would therefore be useful to acquire and merge all these datasets into a single unified dataset to maximise the number of matchups available for validation and their distribution in time and space and consequently reduce the uncertainties in the validation exercise. However, merging several datasets together can be a complicated task. First, it is necessary to acquire and harmonise all datasets into a single standard format. Second, during the merging, the du-

plicates between datasets have to be identified and removed. Third, the metadata should be propagated throughout the process and made available in the final merged product. Ideally, the compiled dataset would be made available as a simple text table, to facilitate ease of access and manipulation. In this work such unification of multiple datasets is presented. This was done for the validation of the ocean-colour products from the ESA Ocean Colour Climate Change Initiative (OC-CCI), but with the intent to serve the broad user community as well.

A merged dataset is not without drawbacks: it is likely to be large and so not always easy to manipulate; because the merging is done on pre-existing, processed databases, one does not have full command of the whole processing chain; and the dataset would be a compilation of observations collected by several investigators using different instruments, sampling methods and protocols, which might eventually have been modified by the processing routines used by the repositories or archives. Nevertheless, to minimise these potential drawbacks, we have, for the most part, incorporated only datasets that have emerged from the long-term efforts of the ocean-colour and biological oceanographic communities to provide scientists with high-quality in situ data and implemented additional quality checks on the data to enhance confidence in the quality of the merged product.

In Sect. 2 the methodologies used to harmonise and integrate all data, as well as a description of individual datasets acquired, are provided. In Sect. 3 the geographic distribution and other characteristics of the final merged dataset are shown. Section 4 provides an overview of the data.

2   Data and methods

2.1   Preprocessing and merging

The compiled global set of bio-optical in situ data described in this work has an emphasis, though not exclusively, on open-ocean data from all geographic regions. It is comprised of the following variables: remote-sensing reflectance (rrs), chlorophyll a concentration (chla), algal pigment absorption coefficient (aph), detrital and coloured dissolved organic matter absorption coefficient (adg), particle backscattering coefficient (bbp) and diffuse attenuation coefficient for downward irradiance (kd). A similar effort of compiling bio-optical in situ data from different sources has been recently published by Nechad et al. (2015). Given their focus on selected coastal regions, most of the data presented here are not part of their compilation. The variables rrs, aph, adg, bbp and kd are spectrally dependent, and this dependence is hereafter implied. The data were compiled from 10 sources of in situ data (MOBY, BOUSSOLE, AERONET-OC, SeaBASS, NOMAD, MERMAID, AMT, ICES, HOT, GeP&CO), each described in Sect. 2.2. The compiled in situ observations have a global distribution and cover the recent period of satellite ocean-colour data between 1997 and 2012. The listed variables were chosen as they are the operational satellite ocean-colour products of the ESA OC-CCI project, which currently focuses on the use of three ocean-colour satellite platforms to create a time series of satellite data: the Medium Resolution Imaging Spectrometer (MERIS) of ESA, the Moderate Resolution Imaging Spectroradiometer (MODIS) of NASA, and the Sea-viewing Wide Field-of-view Sensor (SeaWiFS) of NASA, .

Rrs is a primary ocean-colour product routinely produced by several space agencies. It is defined as rrs $=$ Lw / Es, where Lw is the upward water-leaving radiance and Es is the total downward irradiance at sea level. Remote-sensing reflectance is related to irradiance reflectance (Rw) approximately through rrs $=$ Rw / Q, where Q ranges from 3 to 5 in natural waters and is equal to π for an isotropic (Lambertian) light field. Another quantity that is often required is the "normalised" water-leaving radiance (nLw) (Gordon

and Clark, 1981), which is related to remote-sensing reflectance via rrs $=$ nLw / Fo, where Fo is the top-of-the-atmosphere solar irradiance. If not directly available, remote-sensing reflectance was calculated through the equations described above, depending on the format of the original data. The original data were acquired in an advanced form (e.g. time-averaged, extrapolated to surface) from six data sources particularly designed for ocean-colour validation (MOBY, BOUSSOLE, AERONET-OC, SeaBASS, NOMAD, MERMAID), therefore only requiring the conversion to a common format. In the processing made by the space agencies, the quantity rrs is normalised to a single Sun-viewing geometry (Sun at zenith and nadir viewing) taking in account the bidirectional effects as described in Morel and Gentili (1996) and Morel et al. (2002). Thus, for consistency with the satellite rrs product, only in situ rrs that included the latter normalisation was included in the compilation.

Chlorophyll a concentration is the traditional measure for phytoplankton biomass and one of the most widely used satellite ocean-colour products (IOCCG, 2008). To validate satellite-derived chlorophyll a concentration, two different variables were compiled: one of these represents chlorophyll a measurements made through fluorometric or spectrophotometric methods, referred to hereafter as chla_fluor and the other is the chlorophyll concentration derived from HPLC (high-performance liquid chromatography) measurements, referred to hereafter as chla_hplc. The chlorophyll data were compiled from eight data sources: BOUSSOLE, SeaBASS, NOMAD, MERMAID, AMT, ICES, HOT and GeP&CO. One requirement for chla_fluor measurements was that they were made using in vitro methods (i.e. based on extractions of chlorophyll a). Although this severely decreased the number of observations, since in situ fluorometry (e.g. fluorometers mounted on CTDs) is widely available in oceanographic databases, it was decided to exclude such data because of potential problems with the calibration of in situ fluorometers. The variable chla_hplc was calculated by summing all reported chlorophyll a derivatives, including divinyl chlorophyll a, epimers, allomers and chlorophyllide a. The two chlorophyll variables are retained separately in the database to facilitate their use. HPLC measurements are considered of higher quality, but fluorometric measurements are more abundant. Thus one option for users is to use chla_fluor only when there are no chla_hplc measurements available. To be consistent with satellite-derived chlorophyll values, which are derived from the light emerging from the upper layer of the ocean, all chlorophyll observations found in the top 10 m (replicates at the same depth or measurements at multiple depths) were averaged if the coefficient of variation among observations was less than 50 %; otherwise they were discarded. The averages were then assigned to the surface. The depth of 10 m was chosen as a compromise between clear oligotrophic and turbid eutrophic waters. Other methods, such as chlorophyll depth averages using local attenuation conditions (Morel and Maritorena, 2001), require observations at multiple depths, which, given our decision to use only in vitro measurements, would have reduced considerably the final number of observations.

With regard to the inherent optical properties (aph, adg, bbp), if not already calculated and provided in the contributed datasets, they were computed from related variables that were available: particle absorption (ap), detrital absorption (ad), coloured dissolved organic matter (CDOM) absorption (ag) and total backscattering (bb). The following equations were used $adg = ad + ag$, $ap = aph + ad$ and $bb = bbp + bbw$. For the latter equation, the variable bbw was computed using $bbw = bw / 2$, where bw is the scattering coefficient of seawater derived from Zhang et al. (2009). The diffuse attenuation coefficient for downward irradiance (kd) did not require any conversion and was compiled as originally acquired. Observations of inherent optical properties (surface values) and diffuse attenuation coefficient for downward irradiance, were acquired from three data sources particularly designed for ocean-colour validation (SeaBASS, NOMAD, MERMAID) and were thus already subject to the processing routines of these datasets.

The merged dataset was compiled from 10 sets of in situ data, which were obtained individually either from archives that incorporate data from multiple contributors (SeaBASS, NOMAD, MERMAID and ICES) or from particular measurement programs or projects (MOBY, BOUS-SOLE, AERONET-OC, HOT, GeP&CO, AMT) and were subsequently homogenised and merged. Data contributors are listed in Table 2. There were methodological differences between datasets. Therefore, after acquisition, and prior to any merging, each set of data was preprocessed for quality control and conversion to a common format. During this process, data were discarded if they had (1) unrealistic or missing date, time and geographic coordinate fields; (2) poor quality (e.g. original flags) or a method of observation that did not meet the criteria for the dataset (e.g. in situ fluorescence for chlorophyll concentration); and (3) spuriously high or low data. For the latter, the following limits were imposed: for chla_fluor and chla_hplc $[0.001–100]\,\mathrm{mg\,m^{-3}}$; for rrs $[0–0.15]\,\mathrm{sr^{-1}}$; for aph, adg and bbp $[0.0001–10]\,\mathrm{m^{-1}}$; for kd $[aw(\lambda)–10]\,\mathrm{m^{-1}}$, where aw is the pure water absorption coefficients derived from Pope and Fry (1997). Also during this stage, three metadata strings were attributed to each observation: dataset, subdataset and pi. The dataset contains the name of the original set of data, and can only be one of the following: "aoc", "boussole", "mermaid", "moby", "nomad", "seabass", "hot", "ices", "amt" or "gepco". The subdataset starts with the dataset identifier and is followed by additional information about the data, in the format <dataset>_<cruise/station/site>) (e.g. seabass_car71). The pi contains the name of the principal investigator(s). An effort was made to homogenise the names of principal investigators from the different sets of data. These three metadata are the link to trace each observation to its origin and were propagated throughout the processing. Finally, this processing

stage ended with each set of data being scanned for replicate variable data and replicate station data, which when found, were averaged if the coefficient of variation was less than 50 %; otherwise they were discarded. Replicates were defined as multiple observations of the same variable, with the same date, time, latitude, longitude and depth. Replicate station data were defined as multiple measurements of the same variable, with the same date, time, latitude and longitude. For the latter case, a search window of 5 min in time and 200 m in distance was given, to account for station drift. A small number of observations that were identified as replicates had different subdataset identifiers (i.e. a different cruise name). These observations were considered suspicious if the values were different and were discarded. If the values were the same, one of the observations was retained. This possibly originated from the same group of data being contributed to an archive by two different principal investigators.

Once each set of data was homogenised, all data were integrated into a unique table. This final merging focused on the removal of duplicates between the sets of data. Although some duplicates are known (e.g. MOBY, BOUS-SOLE, AERONET-OC and NOMAD data are found in SeaBASS and MERMAID sets of data), others are unknown (e.g. how much of GeP&CO, ICES, AMT, HOT is within NOMAD, SeaBASS and MERMAID). Therefore, duplicates were identified using the metadata (dataset and subdataset) when possible and temporal–spatial matches as an additional precaution. For temporal–spatial matches, several thresholds were used, but typically 5 min and 200 m were taken to be enough to identify most duplicated data, which reflected small differences in time, latitude and longitude, between the different sets of data. Larger thresholds were used in some cases as a cautionary procedure. This was the case when searching for NOMAD data in other datasets because NOMAD includes a few cases where merging of radiometric and pigment data was done with large spatial-temporal thresholds (Werdell and Bailey, 2005). With regard to all data, if duplicates were found, data from the NOMAD dataset were selected first, followed by data from individual projects (MOBY, BOUSSOLE, AERONET-OC, AMT, HOT and GeP&CO) and finally for the remaining datasets (SeaBASS, MERMAID and ICES). This procedure was chosen to preserve the NOMAD dataset as a whole, since it is widely used in ocean-colour validation. After all data were free of duplicates, they were merged consecutively by variable in the final table. During this process, we also searched for rows (stations) that were separated from each other by time differences less than 5 min and horizontal spatial differences of less than 200 m. When such rows were found, the observations in those rows were merged into a single row. The compiled merged data were compared with the original sets to certify that no errors occurred during the merging. As a final step, a water-column (station) depth was recorded for each observation, which was the closest water-column depth from the ETOPO1 global relief model (National Geophysical Data Center ETOPO1; Amante and Eakins, 2009). For observations where the closest water depth was above sea level (e.g. data collected very near the coast), it was given the value of zero.

Data processing thus included two major steps: preprocessing and merging. The first step was related to each set of contributing datasets in particular and aimed to identify problems and convert the data of interest to a standard format. The second step dealt with the integration of all data into one unique file and included the elimination of duplicated data between the individual sets of data acquired. In the next subsections a brief overview of each original set of data is provided.

2.2 Preprocessing of each set of data

2.2.1 Marine Optical Buoy (MOBY)

The Marine Optical Buoy (MOBY) is a fixed mooring system operated by the National Oceanic and Atmospheric Administration (NOAA) that provides a continuous time series of water-leaving radiance and surface irradiance in the visible region of the spectra from 1997 onwards. The site is located a few kilometres west of the Hawaiian island of Lāna'i where the water depth is about 1200 m. Since its deployment, MOBY measurements have been the primary basis for the on-orbit vicarious calibrations of the SeaWiFS and MODIS ocean-colour sensors. A full description of the MOBY system and processing is provided in Clark et al. (2003). Data are freely available for scientific use at the MOBY Gold directory. The products of interest are the Scientific Time Series files, which refer to MOBY data averaged over sensor-specific wavelengths and particular hours of the day (around 20:00–23:00 UTC). For this work, the satellite band-average products for SeaWiFS, MODIS AQUA and MERIS were compiled from the January 2005 reprocessing for the early data and from the latest reprocessing for data after 2011. The "inband" average subproduct was used, and to maintain the highest quality, only data determined from the upper two arms (Lw1) and flagged "good" quality were acquired. Data from the MOBY203 deployment were discarded due to the absence of surface irradiance data. The compiled variable was the remote-sensing reflectance, rrs, which was computed from the original water-leaving radiance (Lw) and surface irradiance (Es). The water-leaving radiances were corrected for the bidirectional nature of the light field (Morel and Gentili, 1996; Morel et al., 2002) using the same look-up table and method as that used in the SeaWiFS Data Analysis System (SeaDAS) processing code. As mentioned before, the MOBY data compiled in this work are sensor-specific. Therefore, attention is necessary to use the correct MOBY data when validating a particular sensor. The way MOBY data are stored in the final merged table is consistent with the original wavelengths; however, these wavelengths can differ from what is sometimes expected to be the central

wavelength of a given band and sensor. Irrespective of the wavelength where MOBY data are stored in the final table, for validation of bands 1–6 of SeaWiFS, MOBY data stored in the final merged table at 412, 443, 490, 510, 555 and 670 nm, respectively, should be used. For validation of bands 1–6 of MODIS AQUA, MOBY data stored in the final merged table at 416, 442, 489, 530, 547 and 665 nm, respectively, should be used. Finally, for validation of bands 1–7 of MERIS, MOBY data stored in the final merged table at 410.5, 440.4, 487.8, 507.7, 557.6, 617.5 and 662.4 nm, respectively, are the appropriate data.

2.2.2 BOUée pour l'acquiSition de Séries Optiques à Long termE (BOUSSOLE)

The BOUée pour l'acquiSition de Séries Optiques à Long termE (BOUSSOLE) project started in 2001 with the objective of establishing a time series of bio-optical properties in oceanic waters to support the calibration and validation of ocean-colour satellite sensors (Antoine et al., 2006). The project is composed of a monthly cruise program and a permanent optics mooring (Antoine et al., 2008). The mooring collects radiometry and inherent optical properties (IOPs) in continuous mode every 15 min at two depths (4 and 9 m nominally). The monthly cruises are devoted to the mooring servicing, to the collection of vertical profiles of radiometry and IOPs, and to water sampling at 11 depths from the surface down to 200 m, for subsequent analyses including phytoplankton pigments, particulate absorption, CDOM absorption and suspended particulate matter load. The BOUSSOLE mooring is in the western Mediterranean Sea at a water depth of 2400 m. All pigment (2001–2012) and radiometric (2003–2012) data were provided by the principal investigator. The compiled variables were rrs and chla_hplc. Observations of the diffuse attenuation coefficient (kd) were not included in the present compilation, as they were under internal quality revision at the time of data acquisition. Remote-sensing reflectance was computed from the original "fully normalised" water-leaving radiance (nLw_ex), which is the "normalised" water-leaving radiance (nLw previously described), with a correction for the bidirectional nature of the light field (Morel and Gentili, 1996; Morel et al., 2002). The solar irradiance (Fo) was computed from two available variables in the original set of data – nLw and rrs – using the equation $Fo = nLw / rrs$. Only radiometric observations that meet the following criteria were used: (1) tilt of the buoy was less than 10°; (2) the buoy was not lowered by more than 2 m as compared to its nominal water line (to ensure the Es reference sensor is above water and exempt from sea spray); and (3) the solar irradiance was within 10 % of its theoretical clear-sky value (determined from Gregg and Carder, 1990). The latter criterion was used to select clear skies only. An additional quality control was to remove observations that were 50 % higher or lower than the daily average. This removed a small number of spikes in the time series. The final quality control step was to remove days where the standard deviation was more than half of the daily average. This was meant to identify days with high variability. Very few days ($N = 2$) were removed with this test. These quality control criteria were applied per wavelength, which resulted in some observations with an incomplete spectrum.

2.2.3 AErosol RObotic NETwork-Ocean Color (AERONET-OC)

The AErosol RObotic NETwork-Ocean Color (AERONET-OC) is a component of AERONET, including sites where sun photometers operate with a modified measurement protocol leading to the determination of the fully normalised water-leaving radiance (Zibordi et al., 2006, 2009). The result of collaboration between the Joint Research Centre (JRC) and NASA, this component has been specifically developed for the validation of ocean-colour radiometric products. The strength of AERONET-OC is "the production of standardised measurements that are performed at different sites with identical measuring systems and protocols, calibrated using a single reference source and method, and processed with the same codes" (Zibordi et al., 2006, 2009). All high-quality data (level 2) were acquired from the project website for 11 sites: Abu_Al_Bukhoosh ($\sim 25°$ N, $\sim 53°$ E), COVE_SEAPRISM ($\sim 36°$ N, $\sim 75°$ W), Gloria ($\sim 44°$ N, $\sim 29°$ E), Gustav_Dalen_Tower ($\sim 58°$ N, $\sim 17°$ E), Helsinki Lighthouse ($\sim 59°$ N, $\sim 24°$ E), LISCO ($\sim 40°$ N, $\sim 73°$ W), Lucinda ($\sim 18°$ S, $\sim 146°$ E), MVCO ($\sim 41°$ N, $\sim 70°$ W), Palgrunden ($\sim 58°$ N, $\sim 13°$ E), Venice ($\sim 45°$ N, $\sim 12°$ E) and WaveCIS_Site_CSI_6 ($\sim 28°$ N, $\sim 90°$ W). The compiled variable was rrs. Remote-sensing reflectance was computed from the original fully normalised water-leaving radiance (see Sect. 2.2.2 for definition). The solar irradiance (Fo), which is not part of the AERONET-OC data, was computed from the Thuillier (2003) solar spectrum irradiance by averaging Fo over a wavelength-centred 10 nm window. Data were compiled for the exact wavelengths of each record, which can change over time for a given site depending on the specific instrument deployed.

2.2.4 SeaWiFS Bio-optical Archive and Storage System (SeaBASS)

The SeaWiFS Bio-optical Archive and Storage System (SeaBASS) is one of the largest archives of in situ marine bio-optical data (Werdell et al., 2003). It is maintained by NASA's Ocean Biology Processing Group (OBPG) and includes measurements of optical properties, phytoplankton pigment concentrations, and other related oceanographic and atmospheric data. The SeaBASS database consists of in situ data from multiple contributors, collected using a variety of measurement instruments with consistent, community-vetted protocols, from several marine platforms such as fixed buoys, hand-held radiometers and profiling instruments. Quality

control of the received data includes a rigorous series of protocols that range from file format verification to the inspection of the geophysical data values (Werdell et al., 2003). Radiometric data were acquired through the Validation search tool, which provided in situ data with matchups for particular ocean-colour sensors (Bailey and Werdell, 2006). The criteria in the search query were defined to have the minimal flag conditions in the satellite data in order to retrieve a greater number of matchups and therefore in situ data. Regarding phytoplankton pigment data, they were acquired through the Pigment search tool, which provides pigment data directly from the archives. As stated in the SeaBASS website (see Pigment tab at http://seabass.gsfc.nasa.gov/seabasscgi/search.cgi), the Pigment search tool was originally designed to return only in vitro fluorometric measurements, which is consistent with our approach, but over time chlorophyll a measurements made using other methods (e.g. in situ fluorometry) were included in the retrieved pigment data. In the pigment data used in this work, a large number of in situ fluorometric measurements from continuous underway instruments were identified and discarded. These data were firstly identified from cruises with more than 50 observations per day and then re-checked on the SeaBASS website to confirm whether indeed they were continuous underway measurements. A total of 148 015 such measurements were identified and discarded. Given the large volume of this group of data, it is possible that some chlorophyll a observations from in situ methods may have escaped the scrutiny and made it into the final merged dataset. In the future, the SeaBASS plans to add ancillary information to the extractions, which will enable users to distinguish the different types of chlorophyll measurements. The compiled variables from SeaBASS data were: rrs, chla_hplc, chla_fluor, aph, adg, bbp and kd. No conversion was necessary since all variables were acquired in the desired format.

2.2.5 NASA bio-Optical Marine Algorithm Data set (NOMAD)

The NASA bio-Optical Marine Algorithm Data set (NOMAD) is a publicly available dataset compiled by the NASA OBPG at the Goddard Space Flight Center. It is a high-quality global dataset of coincident radiometric and phytoplankton pigment observations for use in ocean-colour algorithm development and satellite-data product-validation activities (Werdell and Bailey, 2005). The source bio-optical data are the SeaBASS archive; therefore, many dependencies exist between these two datasets, which were addressed during the merging. The current version (Version 2.0 ALPHA, 2008) includes data from 1991 to 2007 and an additional set of observations of inherent optical properties. The current version was used in this work, but with an additional set of columns of remote-sensing reflectance corrected for the bidirectional effects (Morel and Gentili, 1996; Morel et al., 2002). This additional set of columns was provided directly by the NOMAD creators. The compiled variables were rrs, chla_hplc, chla_fluor, aph, adg, bbp and kd. Conversion was only necessary for aph, adg and bbp and followed the procedures described in Sect. 2.1. For the calculation of bbp the variable bb was used with a smooth fitting to remove noise. A portion of the NOMAD data were optically weighted (for methods, see Werdell and Bailey, 2005). These data are not consistent with the protocols chosen in this work, but these observations were retained since NOMAD is a widely used dataset in ocean-colour validation.

2.2.6 MERIS Match-up In situ Database (MERMAID)

The MERIS Match-up In situ Database (MERMAID) provides in situ bio-optical data matched with concurrent and comparable MERIS Level 2 satellite ocean-colour products (Barker, 2013a, b). The MERMAID in situ database consists of data from multiple contributors, measured using a variety of instruments and protocols, from several marine platforms such as fixed buoys, hand-held radiometers and profiling instruments. Comprehensive quality control and protocols are used by MERMAID to integrate all the data into a common and comparable format (Barker, 2013a, b). Access to MERMAID data is limited to the MERIS Validation Team, the MERIS Quality Working Group and to the in situ data contributors. For this work, access has been granted to the MERMAID database through a signed service-level agreement. The MERMAID data include subsets of several datasets used in this compilation (MOBY, AERONET-OC, BOUSSOLE, NOMAD). These observations were removed from the MERMAID dataset to avoid duplication (as discussed in Sect. 2.1). The compiled variables were rrs, chla_hplc, chla_fluor, aph, adg, bbp and kd. Remote-sensing reflectance was calculated by dividing by π the original irradiance reflectance provided. Conversion was also necessary for aph, adg and bbp and followed the procedures described in Sect. 2.1.

2.2.7 Hawaii Ocean Time-series (HOT)

The Hawaii Ocean Time-series (HOT) programme provides repeated comprehensive observations of the hydrography, chemistry and biology of the water column at a station located 100 km north of O'ahu, Hawai'i, from October 1988 onwards (Karl and Michaels, 1996). This site is representative of the North Pacific subtropical gyre. Cruises are made approximately once a month to the deep-water station ALOHA (A Long-Term Oligotrophic Habitat Assessment; 22°45′ N, 158°00′ W). Pigment data (chla_hplc and chla_fluor) were extracted directly from the project website. Radiometric measurements from the HOT project are also available, but observations of rrs and kd from the HOT project were acquired in this work as part of the SeaBASS dataset.

2.2.8 Geochemistry, Phytoplankton, and Color of the Ocean (GeP&CO)

The Geochemistry, Phytoplankton, and Color of the Ocean (GeP&CO) is part of the French PROcessus Océaniques et Flux (PROOF) programme and aims to describe and understand the variability of phytoplankton populations and to assess its consequences for the geochemistry of the oceans (Dandonneau and Niang, 2007). It is based on the quarterly travels of the merchant ship *Contship London* from France to New Caledonia. A scientific observer embarked on each travel and operated the sampling for surface water, filtration and various measurements at several hours of each day. The experiment started in October 1999 and finished in July 2002. Pigment data were extracted from the project website. The compiled variable was chla_hplc.

2.2.9 Atlantic Meridional Transect (AMT)

The Atlantic Meridional Transect (AMT) is a multidisciplinary programme, which undertakes biological, chemical and physical oceanographic research during an annual voyage between the UK and destinations in the South Atlantic (Robinson et al., 2006). The programme was established in 1995 and since then has completed 23 research cruises. Pigment data between 1997 (AMT5) and 2005 (AMT17) were provided by the British Oceanographic Data Centre (BODC) following a specific request. For any interest in the original data, the BODC is the point of contact, which ensures that if there are any updates, the most recent data are supplied. The compiled variables are chla_hplc and chla_fluor.

2.2.10 International Council for the Exploration of the Sea (ICES)

The International Council for the Exploration of the Sea (ICES) is a network of more than 4000 scientists from almost 300 institutes, with 1600 scientists participating in activities annually. The ICES Data Centre manages a number of large dataset collections related to the marine environment covering the North East Atlantic, Baltic Sea, Greenland Sea and Norwegian Sea. The majority of data originate from national institutes that are part of the ICES network of member countries. Data were provided (on 28 April 2014) from the ICES database on the marine environment (Copenhagen, Denmark) following a specific request. The ICES data were made available under the ICES data policy and if there is any conflict between this and the policy adopted by the users, then the ICES policy applies. The compiled variables were chla_hplc and chla_fluor.

3   Results

In this work several sets of bio-optical in situ data were acquired, homogenised and merged into a single table. The table is comprised of in situ observations between 1997 and 2012, with a global distribution, and include the following variables: remote-sensing reflectance (rrs), chlorophyll a concentration (chla), algal pigment absorption coefficient (aph), detrital and coloured dissolved organic matter absorption (adg), particle backscattering coefficient (bbp), and diffuse attenuation coefficient for downward irradiance (kd). All observations in the table were processed in such a way that they can be compared directly with satellite-derived ocean-colour data. The table consists of 80 524 rows and 267 columns. Each row represents a unique station in space and time, separated from each other by at least 5 min and 200 m. For each observation in a given station, there are three metadata strings: dataset, subdataset and pi. The columns of the table take the form described in Table 1. The contributors of data in the table are shown in Table 2. Regarding spectral variables, all original wavelengths were preserved, which requires a large number of unique wavelengths to be maintained in the database. No band shifting was performed (though some archived data in SeaBASS and MERMAID may have been merged with nearby wavelengths) and no minimum number of wavelengths per observation was imposed. This allows further manipulation of the table for different purposes. In the following paragraphs, the table is analysed and the final group of observations is described for each contributing dataset; however, the numbers reported here do not reflect the original numbers in each dataset, since duplicates across contributing datasets were removed (e.g. removed NOMAD and others from MERMAID).

Observations of remote-sensing reflectance are available at 134 unique wavelengths (i.e. columns), between 405 and 1022.1 nm (Fig. 1). In total there are 44 191 observations (i.e. rows) with remote-sensing reflectance in the table. The total number of observations are partitioned per contributing datasets as follows: AERONET-OC (17 405), BOUSSOLE (17 364), MOBY (4513), NOMAD (3326), MERMAID (885) and SEABASS (698). The data from AERONET-OC, BOUSSOLE and MOBY correspond to continuous time series and hence the higher number of observations. The data distribution at $44X$ nm and $55X$ nm is provided in Fig. 2a and b, respectively. Data were first searched at 445 and 555 nm and then with a search window of up to 8 nm to also include data at 547 nm. Median values at $44X$ nm range from $0.003\,\mathrm{m}^{-1}$ (AERONET-OC) and $0.009\,\mathrm{m}^{-1}$ (MOBY), whereas at $55X$ nm the median values lie between $0.001\,\mathrm{m}^{-1}$ (MOBY) and $0.004\,\mathrm{m}^{-1}$ (AERONET-OC). The observations are evenly distributed on a monthly basis in the Northern Hemisphere (Fig. 3). In the Southern Hemisphere, where the number of stations is smaller, there is a decrease in the number of observations during the austral winter months (Fig. 3). For additional analysis, rrs band ratios were plotted against each other (490 : 555 vs. 412 : 443, Fig. 4). Most points are within the boundaries of the NOMAD dataset, but some scattered points were found. These points were retained in the table to allow further manipulation with different quality con-

Table 1. The standard variables, nomenclatures and units in the final table.

Variable/column	Description and units
time	GMT, <YYYY-MM-DD>T<HH:MM:SS>Z
lat	Decimal degree, $-90 : 90$, south negative
lon	Decimal degree, $-180 : 180$, west negative
depth_water	Sampling depth (m) – all assigned to zero
chla_hplc	Total chlorophyll a concentration determined from HPLC method ($\mathrm{mg\,m}^{-3}$)
chla_fluor	Chlorophyll a concentration determined from fluorometric or spectrophotometric methods ($\mathrm{mg\,m}^{-3}$)
rrs_<band>	Remote-sensing reflectance (sr^{-1})
aph_<band>	Algal pigment absorption coefficient (m^{-1})
adg_<band>	Detrital plus CDOM absorption coefficient (m^{-1})
bbp_<band>	Particle backscattering coefficient (m^{-1})
kd_<band>	Diffuse attenuation coefficient for downward irradiance (m^{-1})
etopo1	Water depth from ETOPO1 (m)
chla_hplc_dataset	Metadata string for chla_hplc
chla_hplc_subdataset	Metadata string for chla_hplc
chla_hplc_pi	Metadata string for chla_hplc
chla_fluor_dataset	Metadata string for chla_fluor
chla_fluor_subdataset	Metadata string for chla_fluor
chla_fluor_pi	Metadata string for chla_fluor
rrs_dataset	Metadata string for rrs
rrs_subdataset	Metadata string for rrs
rrs_pi	Metadata string for rrs
aph_dataset	Metadata string for aph
aph_subdataset	Metadata string for aph
aph_pi	Metadata string for aph
adg_dataset	Metadata string for adg
adg_subdataset	Metadata string for adg
adg_pi	Metadata string for adg
bbp_dataset	Metadata string for bbp
bbp_subdataset	Metadata string for bbp
bbp_pi	Metadata string for bbp
kd_dataset	Metadata string for kd
kd_subdataset	Metadata string for kd
kd_pi	Metadata string for kd

trol criteria. Complementary analysis of remote-sensing reflectance data is made when other variables are concurrently available and discussed further on in the text (see Figs. 11 and 16). The geographic distribution of remote-sensing reflectance observations (Fig. 5) shows a higher number of observations in some coastal regions, such as those of North America and Northern Europe. The central regions of the ocean show a lower number of observations, with the Atlantic Ocean having the highest density in relation to the other oceans. Best geographic coverage is provided by the NOMAD database. Data from SeaBASS are smaller in number but are still important. Data from MERMAID are mainly located along the coasts of Europe, North America and the central region of the North Atlantic Ocean.

For chlorophyll a concentration, two types of observations were compiled, one measured by fluorometric or spectrophotometric methods (chla_fluor), and the other measured

Table 2. Original sets of data and data contributors in the final table.

Data source	Description	Data contributors
Marine Optical Buoy (MOBY)	Daily observations of remote-sensing reflectance, measured by a fixed mooring system, located west of the Hawaiian island of Lāna'i. Data compiled between 1997–2012. Data were obtained from the MOBY website. Compiled standard variable: rrs.	Paul DiGiacomo, Kenneth Voss
BOUée pour l'acquiSition de Séries Optiques à Long termE (BOUSSOLE)	High frequency (15 min) observations of remote-sensing reflectance, from a fixed mooring system, located in the western Mediterranean Sea. Measurements of chlorophyll a concentration are also available at the mooring locations. Remote-sensing reflectance and chlorophyll a data were compiled between 2003–2012 and 2001–2012, respectively. Data were provided by David Antoine. Compiled standard variables: rrs, chla_hplc.	David Antoine
AErosol RObotic NETwork-Ocean Color (AERONET-OC)	Daily observations of remote-sensing reflectance, measured by modified sun photometers. Data compiled between 2002–2012. Sites included: Abu_Al_Bukhoosh ($\sim 25°$ N, $\sim 53°$ E), COVE_SEAPRISM ($\sim 36°$ N, $\sim 75°$ W), Gloria ($\sim 44°$ N, $\sim 29°$ E), Gustav_Dalen_Tower ($\sim 58°$ N, $\sim 17°$ E), Helsinki Lighthouse ($\sim 59°$ N, $\sim 24°$ E), LISCO ($\sim 40°$ N, $\sim 73°$ W), Lucinda ($\sim 18°$ S, $\sim 146°$ E), MVCO ($\sim 41°$ N, $\sim 70°$ W), Palgrunden ($\sim 58°$ N, $\sim 13°$ E), Venice ($\sim 45°$ N, $\sim 12°$ E) and WaveCIS_Site_CSI_6 ($\sim 28°$ N, $\sim 90°$ W). Data were obtained from the AERONET-OC website. Compiled standard variable: rrs.	Robert Arnone (WaveCIS), Sam Ahmed (LISCO), Vittorio Brando (Lucinda), Dick Crout (WaveCIS), Hui Feng (MVCO), Alex Gilerson (LISCO), Rick Gould (WaveCIS), Brent Holben (COVE-SEAPRISM), Susanne Kratzer (Palgruden), Heidi M. Sosik (MVCO), Giuseppe Zibordi (Abu Al Bukhoosh & Gloria & Gustav Dalen Tower & Helsinki Lighthouse & Venice)
SeaWiFS Bio-optical Archive and Storage System (SeaBASS)	Global archive of in situ marine data from multiple contributors. Bio-optical global data between 1997–2012 were extracted from the SeaBASS website. Pigment data were extracted using the Data Search tool, which provides data directly from the archives. Radiometric data were extracted using the Validation tool, which only provides in situ data with matchups for ocean-colour sensors. Compiled standard variables: rrs, chla_hplc, chl_fluor, aph, adg, bbp and kd.	Robert Arnone, Kevin Arrigo, William Balch, Ray Barlow, Mike Behrenfeld, Şükrü Beşiktepe, Emmanuel Boss, Chris Brown, Douglas Capone, Ken Carder, Francisco Chavez, Alex Chekalyuk, Jay-Chung Chen, Dennis Clark, Hervé Claustre, Jorge Corredor, Glenn Cota, Yves Dandonneau, Heidi Dierssen, David Eslinger, Piotr Flatau, Robert Frouin, Carlos García, Joaquim Goes, Gwo-Ching Gong, Rick Gould, Larry Harding, Jon Hare, Stan Hooker, Chuanmin Hu, Sung-Ho Kang, Gary Kirkpatrick, Oleg Kopelevich, Sam Laney, Zhongping Lee, Ricardo Letelier, Marlon Lewis, Antonio Mannino, John Marra, Chuck McClain, Christophe Menkes, Mark Miller, Greg Mitchell, Ru Morrison, James Mueller, Frank Muller-Karger, James Nelson, Norman Nelson, Mary Jane Perry, David Phinney, John Porter, Collin Roesler, David Siegel, Mike Sieracki, Jeffrey Smart, Raymond Smith, Heidi M. Sosik, James Spinhirne, Dariusz Stramski, Rick Stumpf, Ajit Subramaniam, Chuck Trees, Michael Twardowski, Kenneth Voss, Marcel Wernand, Ronald Zaneveld, Eric Zettler, Giuseppe Zibordi, Richard Zimmerman
NASA bio-Optical Marine Algorithm Data set (NOMAD)	High-quality global dataset of coincident bio-optical in situ data. The dataset was built upon SeaBASS archive. The current version (version 2.0 ALPHA, 2008) was used with an additional set of columns of remote-sensing reflectance corrected for the bidirectional nature of the light field, provided by NOMAD creators. Data compiled between 1997–2007. Compiled standard variables: rrs, chla_hplc, chl_fluor, aph, adg, bbp and kd.	Robert Arnone, Kevin Arrigo, William Balch, Ray Barlow, Mike Behrenfeld, Chris Brown, Douglas Capone, Ken Carder, Francisco Chavez, Dennis Clark, Hervé Claustre, Jorge Corredor, Glenn Cota, David Eslinger, Piotr Flatau, Robert Frouin, Rick Gould, Larry Harding, Stan Hooker, Oleg Kopelevich, Marlon Lewis, Antonio Mannino, John Marra, Mark Miller, Greg Mitchell, Tiffany Moisan, Ru Morrison, Frank Muller-Karger, James Nelson, Norman Nelson, David Siegel, Raymond Smith, Timothy Smyth, James Spinhirne, Dariusz Stramski, Rick Stumpf, Ajit Subramaniam, Kenneth Voss
MERIS Match-up In situ Database (MERMAID)	Global database of in situ bio-optical data matched with concurrent MERIS Level 2 satellite ocean-colour products. The Extract matchup tool was used to acquire data. Data were compiled between 2002 and 2012. Access was granted through a signed Service Level Agreement. Compiled standard variables: rrs, chla_hplc, chl_fluor, aph, adg, bbp and kd.	Simon Bélanger, Jean-François Berthon, Vanda Brotas, Elisabetta Canuti, Pierre Yves Deschamps, Annelies Hommersom, Mati Kahru, Holger Klein, Hubert Loisel, David McKee, Greg Mitchell, Michael Ondrusek, Michel Repecaud, David Siegel, Giuseppe Zibordi
Atlantic Meridional Transect (AMT)	Multidisciplinary programme that makes biological, chemical and physical oceanographic measurements during an annual voyage between the United Kingdom and destinations in the South Atlantic. It has compiled observations of chlorophyll a concentration between 1997 (AMT5) and 2005 (AMT17). Data were provided by the British Oceanographic Data Centre (BODC). Compiled standard variables: chla_hplc, chl_fluor.	Ray Barlow, Stuart Gibb, Victoria Hill, Patrick Holligan, Gerald Moore, Leonie O'Dowd, Alex Poulton, Emilio Suarez

Table 2. Continued.

International Council for the Exploration of the Sea (ICES)	Database of several collections of data related to the marine environment. It has compiled observations of chlorophyll *a* concentration in the northern European seas, between 1997 and 2012. Data were provided by the ICES database on the marine environment (2014, Copenhagen, Denmark). Compiled standard variables: chla_hplc, chl_fluor.	Not available
Hawaii Ocean Time-series (HOT)	Multidisciplinary programme that makes repeated biological, chemical and physical oceanographic observations near O'ahu, Hawai'i. Measurements of chlorophyll *a* concentration between 1997–2012 were extracted from the project website. Compiled standard variables: chla_hplc, chl_fluor.	Bob Bidigare, Matthew Church, Ricardo Letelier, Jasmine Nahorniak
Geochemistry, Phytoplankton, and Color of the Ocean (GeP&CO)	Program of in situ data collection aboard merchant ship from France to New Caledonia, between 1999 and 2002. Measurements of chlorophyll *a* concentration were obtained from the project website. Compiled standard variables: chla_hplc.	Yves Dandonneau

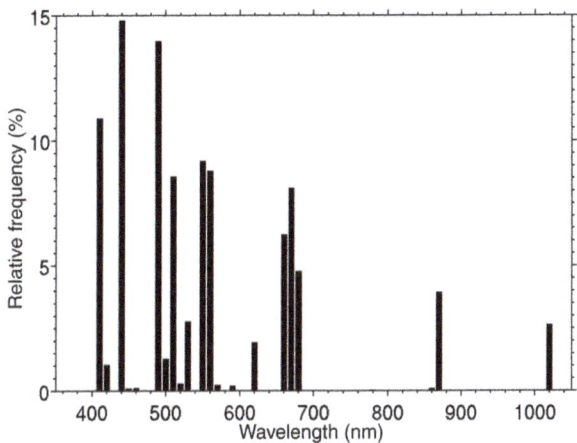

Figure 1. Relative spectral frequency of remote-sensing reflectance in the final table, using 10 nm wide class intervals, defined as the ratio of the number of observations at a particular waveband to the total number of observations at all wavebands, multiplied by 100 to report results in percentage. Data at a total of 134 unique wavelengths, between 405 and 1022.1 nm, were compiled.

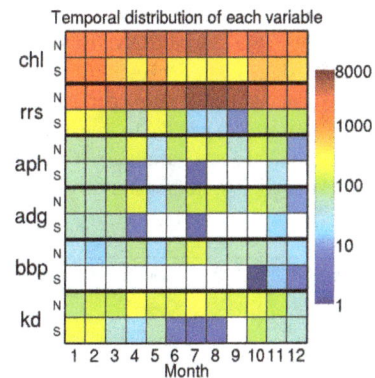

Figure 3. Temporal distribution of chlorophyll *a* concentration (chl), remote-sensing reflectance (rrs), algal pigment absorption coefficient (aph), detrital plus CDOM absorption coefficient (adg), particle backscattering coefficient (bbp) and the diffuse attenuation coefficient for downward irradiance (kd) in the final table. All chlorophyll data were considered, but for a given station HPLC data were selected if available. Colours indicate the number of stations available for each variable, as a function of months and the hemispheres of data acquisition (N – Northern Hemisphere; S – Southern Hemisphere).

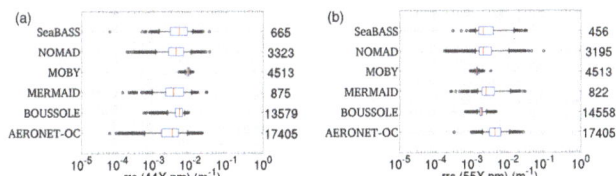

Figure 2. The distribution of (**a**) rrs at $44X$ nm and (**b**) rrs at $55X$ nm. Data were first searched at 445 and 555 nm and then with a search window of up to 8 nm to include data at 547 nm. The black boxes delimit the percentiles 0.25 and 0.75 of the data and the black horizontal lines show the extension of up to the 5th and 95th percentiles. The red line represents the median value and the black circles the values below (and above) the percentile 0.05 (0.95). The number of measurements of each dataset is reported on the right axis of the graph.

by HPLC methods (chla_hplc). A comparison of both measurements (Fig. 6), when available at the same station shows good agreement (Trees et al., 1985). As stated before, the analysis was done on the final merged table; thus, no data were filtered and the good relation can in part be explained by the quality control implemented by the data providers and curators of repositories such as NOMAD and SeaBASS (Werdell and Bailey, 2005). The total number of rows with concurrent chla_fluor and chla_hplc is 2002, with contributions from NOMAD (32 %), SeaBASS (47 %), MERMAID (11 %), HOT (7 %), AMT (2 %) and GeP&CO (1 %). The chla_fluor observations are available at 27 933 stations (rows), with values ranging from 0.0011 to 100 mg m^{-3} (Fig. 7). They are from NOMAD (2350), SeaBASS (15 728),

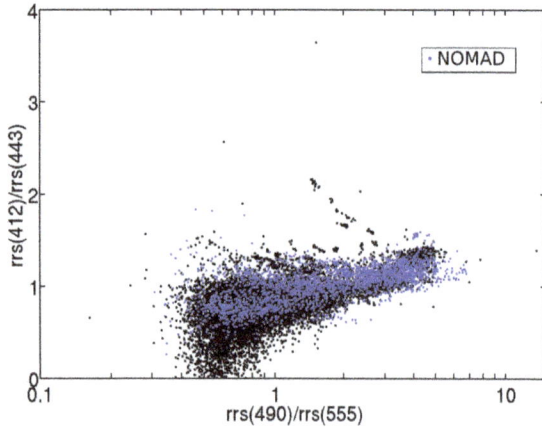

Figure 4. Ranges of remote-sensing reflectance band ratios (412 : 443 and 490 : 555) for all data. The points from the NOMAD dataset are shown in blue for reference. The total number of points is divided between MOBY (4513), AERONET-OC (17 293), BOUS-SOLE (3533), NOMAD (3120), SeaBASS (432) and MERMAID (677). To maximise the number of ratios per dataset, a search window of up to 12 nm was used when the four wavelengths (412, 443, 490, 555) were not simultaneously available. The effect of different search windows was negligible in the ratio distribution.

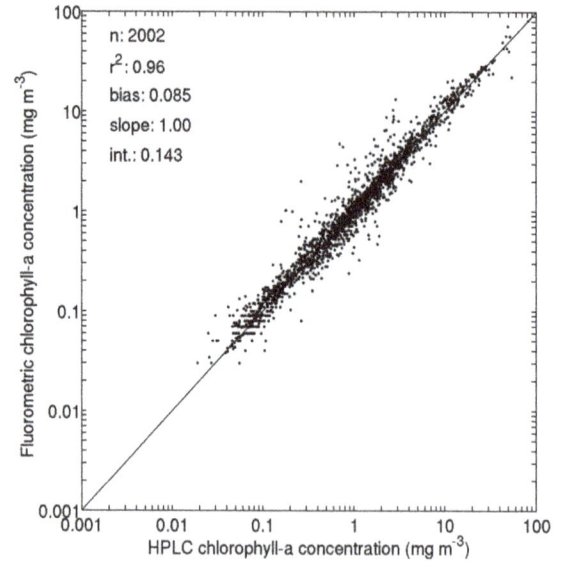

Figure 6. Comparison of coincident observations of chlorophyll *a* concentration derived with different methods (chla_fluor and chla_hplc). The data were transformed prior to regression analysis to account for their log-normal distribution.

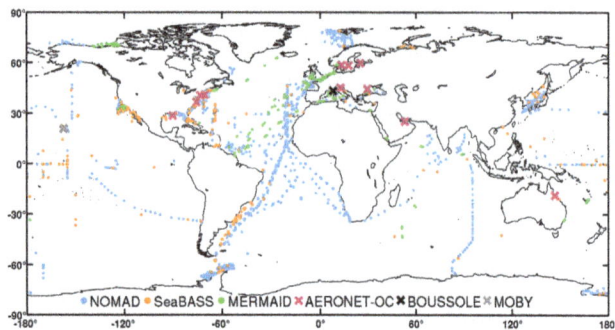

Figure 5. Global distribution of remote-sensing reflectance per dataset in the final table. The data sources are identified with different colours. Points show locations where at least one observation is available. Crosses show sites from where time series data of remote-sensing reflectance are available.

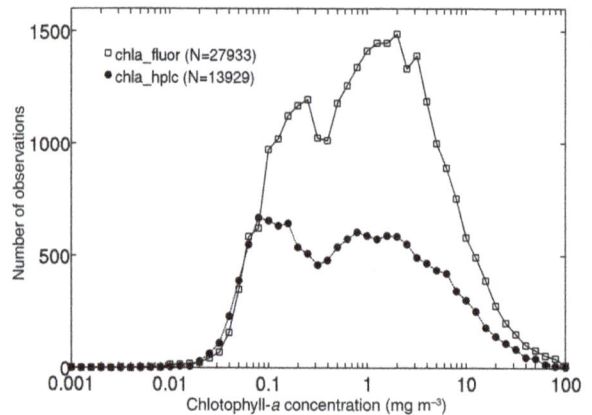

Figure 7. Number of observations per chlorophyll *a* concentration acquired with different methods (chla_fluor and chla_hplc).

MERMAID (3711), ICES (5421), HOT (559) and AMT (164). The total number of chla_hplc observations is 13 918, ranging from 0.006 to 99.8 mg m^{-3} (Fig. 7), with contributions from NOMAD (1309), SeaBASS (5920), MERMAID (707), ICES (2994), HOT (153), GeP&CO (1536), BOUS-SOLE (397) and AMT (902). The combined chlorophyll dataset (all chlorophyll data considered, but for a given station HPLC data were selected if available) has a total of 39 849 observations, with 11, 41 and 48 % from oligotrophic (< 0.1 mg m^{-3}), mesotrophic (0.1–1 mg m^{-3}) and eutrophic (> 1 mg m^{-3}) waters, respectively. When compared with the proportions of the world ocean in these trophic classes, i.e. 56 % oligotrophic, 42 % mesotrophic and 2 % eutrophic (An-

toine et al., 1996), oligotrophic waters are under-represented and eutrophic waters are over-represented in the compilation. The combined chlorophyll dataset is evenly distributed between each month of the year in the Northern Hemisphere, but in the Southern Hemisphere there are relatively few data points during the winter months compared with the rest of the year (Fig. 3). The spatial distribution of the chlorophyll values for the combined dataset (Fig. 8) shows a good agreement with known biogeographical features, such as low chlorophyll values in the subtropical gyres and high values in temperate, coastal and upwelling regions. Many regions show a good spatial coverage (e.g. Atlantic and Pacific oceans), while others are poorly sampled (e.g. Southern and

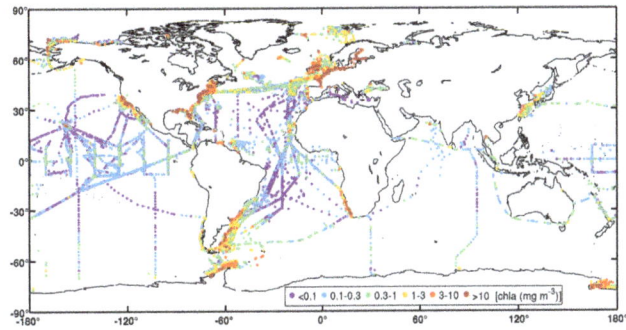

Figure 8. Global distribution of chlorophyll a concentration per intervals of the observed value. All chlorophyll data were considered, but for a given station HPLC data were selected if available.

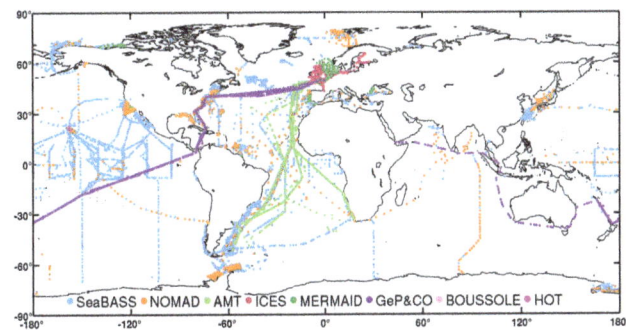

Figure 9. Global distribution of chlorophyll a concentration per dataset in the final table. All chlorophyll data were considered, but for a given station HPLC data were selected if available.

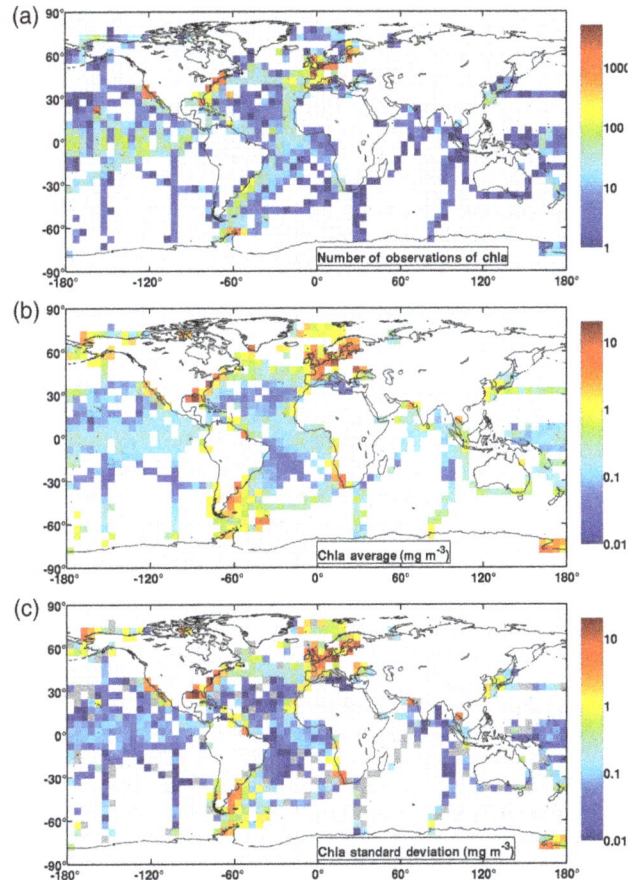

Figure 10. The chlorophyll a ($mg\,m^{-3}$) data partitioned into $5° \times 5°$ boxes showing the **(a)** number of observations, **(b)** average value and **(c)** standard deviation in each box. All chlorophyll data were considered, but for a given station HPLC data were selected if available. In the standard deviation plot, grey represents boxes with zero standard deviation (i.e. one observation).

Indian oceans). Of the contributing datasets, NOMAD and SeaBASS provide a good spatial coverage in many regions (Fig. 9). The ICES and MERMAID data are mainly located along the coastal regions of Europe. The AMT data cover the central part of the Atlantic Ocean. For additional analysis and as an example of the applications of the compiled dataset, the combined chlorophyll data (chla_fluor and chla_hplc) were partitioned into $5° \times 5°$ boxes and for each box the number of observations, average value and standard deviation were computed (Fig. 10 a, b and c, respectively). The number of observations can be very high (> 1000) in some boxes along the European and North American coastlines and relatively low (< 20) in oceanic regions. Again there is an appearance in the average value map (Fig. 10b) of well-known biogeographical features, such has the lower chlorophyll in the subtropical gyres and high values in coastal and upwelling areas. There is a close correspondence between the spatial patterns of the averaged and standard deviation maps (Fig. 10 b and c), which may be an indicator of the data quality.

Coincident observations of chlorophyll a concentration and remote-sensing reflectance are available at 3562 stations. These observations are mostly from NOMAD (85 %), MERMAID (10 %) and SeaBASS (5 %). The maximum of three band ratios of remote-sensing reflectance is plotted against chlorophyll a concentration (Fig. 11). The chla values used are the combined HPLC and fluorometric chlorophyll a, and for rrs the closest spectral observation within 2 nm was used. The maximum band ratios were calculated using the maximum value from [rrs(443) / rrs(555), rrs(490) / rrs(555), rrs(510) / rrs(555)] or [rrs(443) / rrs(560), rrs(490) / rrs(560), rrs(510) / rrs(560)] if rrs(555) was not available. The relationship between maximum band ratio and chlorophyll is close to the NASA OC4 and OC4E v6 standard algorithm (http://oceancolor.gsfc.nasa.gov/cms/atbd/chlor_a), equally based on maximum band ratios, providing confidence in the quality of the compiled data.

The inherent optical properties (aph, adg and bbp) are available at 27 unique wavelengths between 405 and 683 nm. There are a total of 1276, 1123 and 638 observations for aph, adg and bbp, respectively. For aph the total number of observations is distributed among NOMAD (1190),

Table 3. Summary of median values for aph, adg and bbp at $44X$ and $55X$ nm for each dataset (as shown in Fig. 12a–f). Data were first searched at 445 and 555 nm and then with a search window of up to 8 nm to include data at 547 nm.

	Median aph		Median adg		Median bbp	
	$44X$ nm	$55X$ nm	$44X$ nm	$55X$ nm	$44X$ nm	$55X$ nm
SeaBASS	0.0549	0.0074	0.0711	0.0222	0.0035	0.0025
MERMAID	0.0353	0.0046	0.0515	0.0112	0.0030	0.0022
NOMAD	0.0282	0.0052	0.1149	0.0286	0.0080	0.0052

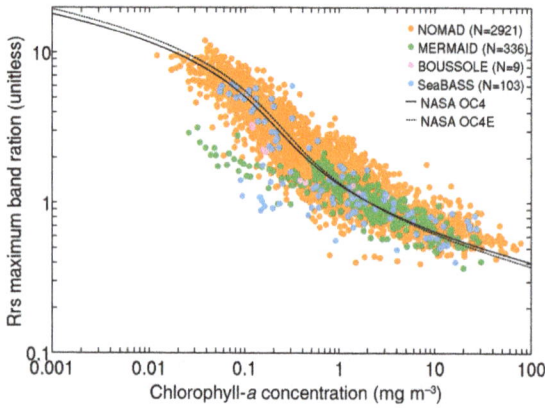

Figure 11. A remote-sensing reflectance maximum band ratio (as defined in text) ([443,490,510] / 555 or [443,490,510] / 560 if 555 not available) as a function of chlorophyll a concentration. All chlorophyll data were considered, but for a given station HPLC data were selected if available. Data within 2 nm of the wavelengths were used. For reference the solid and dotted line show the NASA OC4 and OC4E v6 standard algorithms, respectively (http://oceancolor.gsfc.nasa.gov/cms/atbd/chlor_a). The total number of points was 3369, of which 86% were from NOMAD.

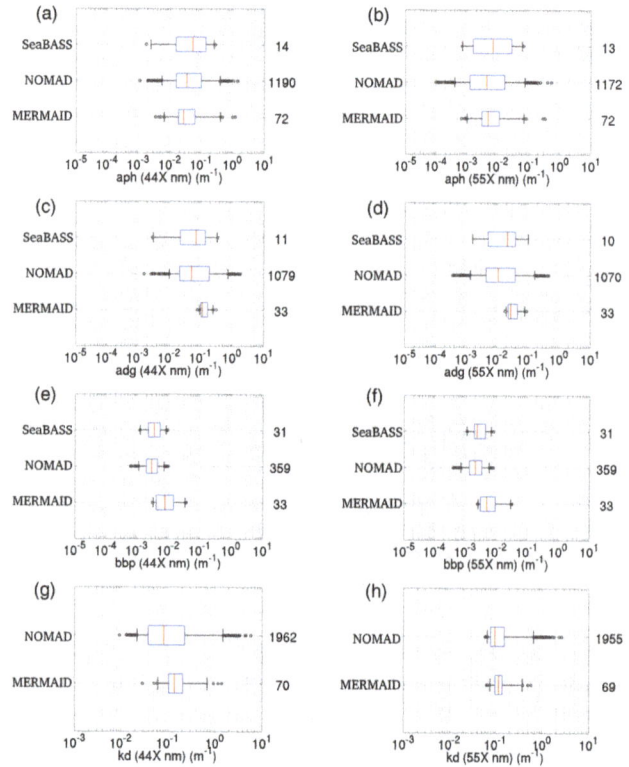

Figure 12. The distribution of **(a)** aph at $44X$ nm, **(b)** aph at $55X$ nm, **(c)** adg at $44X$ nm, **(d)** adg at $55X$ nm, **(e)** bbp at $44X$ nm, **(f)** bbp at $55X$ nm, **(g)** kd at $44X$ nm and **(h)** kd at $55X$ nm. Data were first searched at 445 and 555 nm and then with a search window of up to 8 nm to include data at 547 nm. The graphical convention is identical to Fig. 2.

SeaBASS (14) and MERMAID (72). For adg the contributions are as follows: NOMAD (1079), SeaBASS (11) and MERMAID (33). The bbp observations come from NOMAD (371), SeaBASS (32) and MERMAID (235). Data distribution of aph, adg and bbp at $44X$ and $55X$ nm for each dataset is provided in Fig. 12a–f. Median values of aph, adg and bbp at $44X$ and $55X$ nm for each dataset are summarised in Table 3. For additional analysis, the following band ratios for the absorption coefficients were calculated: aph(490) / aph(443), aph(412) / aph(443), adg(443) / adg(490) and adg(412) / adg(443). Data within 2 nm of the wavelengths were used to maximise the number of points. The distribution of the ratios is shown in Fig. 13. Several observations were found to be above the thresholds used in the IOCCG report 5 (IOCCG, 2006) for quality control (see dotted vertical black lines in Fig. 13). These points are highlighted here for information but retained in the database, as these were mostly from NOMAD and there was an interest to preserve this dataset as a whole. Also, not discarding this data allows further manipulation with different

quality control criteria. On the annual scale, the observations of the inherent optical properties are strongly underrepresented in the Southern Hemisphere where there is a complete absence of data in several months of the year (Fig. 3). Overall, the geographic coverage for observations of aph, adg and bbp (Fig. 14) is poor, with most open-ocean regions not being sampled, with the exception of the Atlantic Ocean. Small clusters of data are located in particular coastal regions.

Finally, for the diffuse attenuation coefficient for downward irradiance (kd), there are 25 unique wavelengths between 405 and 709 nm. There are a total of 2454 obser-

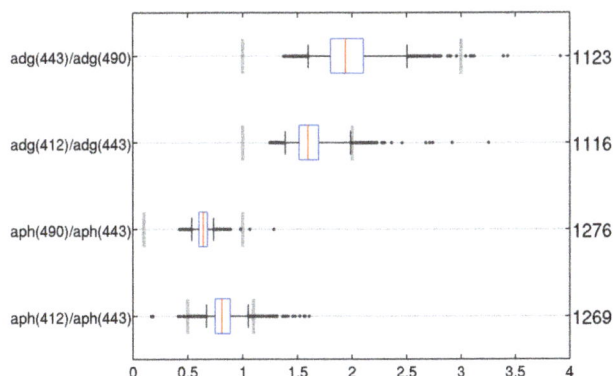

Figure 13. The distribution of absorption coefficient band ratios: adg(443) / adg(490), adg(412) / adg(443), aph(490) / aph(443) and aph(412) / aph(443). Data within 2 nm of the wavelengths were used. The graphical convention is identical to Fig. 2. The vertical dashed lines show the lower and upper thresholds used for quality control in the IOCCG report 5 (IOCCG, 2006). The total number of points are divided between NOMAD (93–96 %), MERMAID (3–6 %) and SeaBASS (1 %).

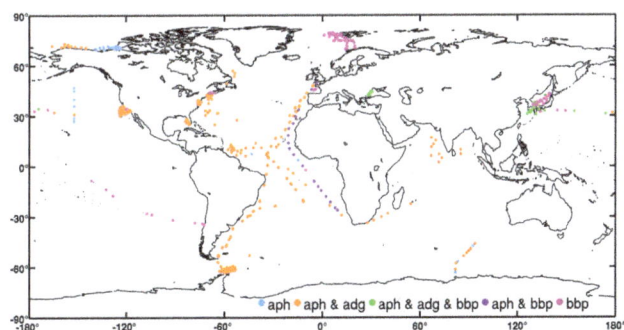

Figure 14. Global distribution of observations of inherent optical properties (algal pigment absorption coefficient aph, detrital plus CDOM absorption coefficient adg and particle backscattering coefficient bbp) in the final table.

vations, from NOMAD (2266), SeaBASS (118) and MERMAID (70). Data distribution of kd at $44X$ and $55X$ nm for each dataset is shown in Fig. 12g and h. No kd data at these wavelengths were available for the SeaBASS dataset (only at 490 nm). Median values of kd at $44X$ nm span between $0.08\,\mathrm{m}^{-1}$ (NOMAD) to $0.1\,\mathrm{m}^{-1}$ (MERMAID), whereas at $55X$ nm the kd values are approximately $0.1\,\mathrm{m}^{-1}$ (NOMAD and MERMAID). NOMAD provides the best geographic coverage (Fig. 15), with a higher coverage in the Atlantic, compared with other oceans. With the exception of the coastal regions of North America and the Japan Sea, most coastal regions are not sampled. In the Northern Hemisphere, kd is distributed roughly evenly across all months of the year, but in the Southern Hemisphere there are few data points during the austral winter and none at all in September (Fig. 3).

Although most of the stations with concurrent variables are mainly from the NOMAD dataset, for complete-

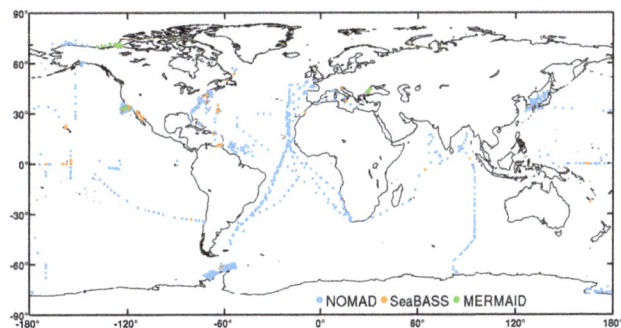

Figure 15. Global distribution of diffuse attenuation coefficient for downward irradiance (kd) per dataset in the final table.

ness, an examination of bio-optical relationships is provided (Fig. 16). The relation between aph at 443 nm and chlorophyll a (Fig. 16a) agrees with the relation proposed by Bricaud et al. (2004). A total of 1070 points exist with these two variables available (93 % from NOMAD). The relation between the sum of aph and adg at 443 nm and rrs at 443 nm (Fig. 16b) shows a similar dispersion, with the exception of some scattered points, to an equivalent analysis on the IOCCG report 5 (see their Fig. 2.3). Again, the scattered data were retained in the final table to preserve the NOMAD dataset. A total of 1112 points exist for which these three variables are available (97 % from NOMAD). The relation between the ratio rrs(490) / rrs(555) and kd(490) (Fig. 16c) shows a good agreement with the NASA KD2S standard algorithm (http://oceancolor.gsfc.nasa.gov/cms/atbd/kd_490). A total of 2280 points exist for which these three variables are available (93 % from NOMAD). The relation between the ratio rrs(490) / rrs(555) and bbp at 555 nm (Fig. 16c) shows a good agreement with the relation suggested by Tiwari and Shanmugam (2013). A total of 357 points exist for which these three variables are available (91 % from NOMAD).

4 Conclusions

A compilation of bio-optical in situ data is presented in this work. The compiled data have global coverage and span from 1997 to 2012, covering the recent period of ocean-colour satellite observations. They resulted from the acquisition, homogenisation and integration of several sets of data obtained from different sources. Minimal changes were made to the original data, except for those necessary for conversion to standard format and quality control. In situ measurements of the following variables were compiled: remote-sensing reflectance, chlorophyll a concentration, algal pigment absorption coefficient, detrital and coloured dissolved organic matter absorption coefficient, particle backscattering coefficient and diffuse attenuation coefficient for downward irradiance.

The final set of data consists of a substantial number of in situ observations, available in a simple text table, and processed in a way that could be used directly for the evaluation

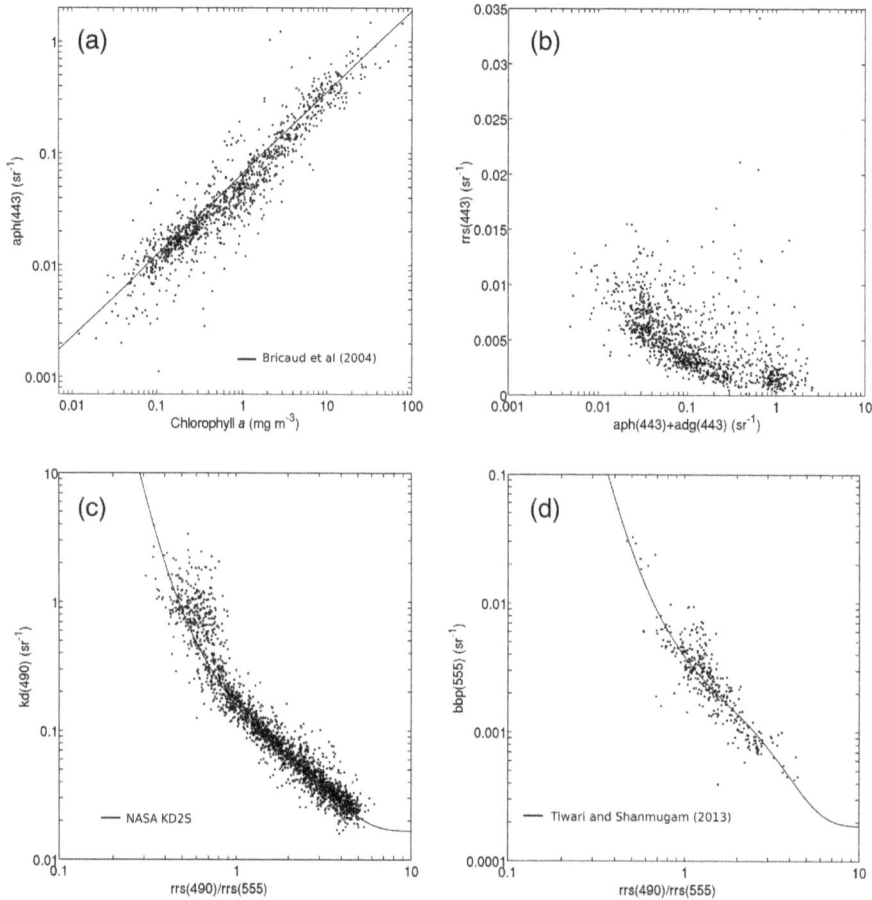

Figure 16. Examples of bio-optical relationships in the final merged table. Panel (**a**): aph(443) vs. chlorophyll a. Total number of points (1070) is divided between MERMAID (70), NOMAD (991) and SeaBASS (9). For reference the solid line show the regression from Bricaud et al. (2004). Panel (**b**): [aph(443) + adg(443)] vs. rrs(443). Total number of points (1112) is divided between MERMAID (33) and NOMAD (1079). Panel (**c**): [rrs(490) / rrs(555)] vs. kd(490). The total number of points (2280) is divided between MERMAID (62), NOMAD (2117) and SeaBASS (101). For reference the solid line show the NASA KD2S standard algorithm (http://oceancolor.gsfc.nasa.gov/cms/atbd/kd_490). Panel (**d**): [rrs(490) / rrs(555)] vs. bbp(555). The total number of points (357) is divided between MERMAID (33) and NOMAD (324). For reference the solid line show the relation proposed by Tiwari and Shanmugam (2013). A search window of 2 nm was used for panels (**a**) and (**b**), and a search window of 5 nm was used for panels (**c**) and (**d**) to include data at 560 nm when not available at 555 nm.

of satellite-derived ocean-colour data. The major advantages of this compilation are that it merges six commonly used data sources in ocean-colour validation (MOBY, BOUSSOLE, AERONET-OC, SeaBASS, NOMAD, MERMAID) and four additional sets of chlorophyll a concentration data (AMT, ICES, HOT and GeP&CO) into a simple text table free of duplicated observations. This compilation was initially created with the intention of evaluating the quality of the satellite ocean-colour products from the ESA OC-CCI project. The objective of publishing the compilation is to make it easy for the broader community to use it.

Appendix A: Notation

ad	Detrital absorption coefficient (m^{-1})
adg	Detrital plus CDOM absorption coefficient (m^{-1})
AERONET-OC	AErosol RObotic NETwork-Ocean Color
ag	CDOM absorption coefficient (m^{-1})
AMT	Atlantic Meridional Transect
ap	Particle absorption coefficient (m^{-1})
aph	Algal pigment absorption coefficient (m^{-1})
aw	Pure water absorption coefficient (m^{-1})
bb	Total backscattering coefficient (m^{-1})
bbp	Particle backscattering coefficient (m^{-1})
bbw	Backscattering coefficient of seawater (m^{-1})
BOUSSOLE	BOUée pour l'acquiSition de Séries Optiques à Long termE
CDOM	Coloured dissolved organic matter
chla	Chlorophyll a concentration ($mg\,m^{-3}$)
chla_fluor	Chlorophyll a concentration determined from fluorometric or spectrophotometric methods ($mg\,m^{-3}$)
chla_hplc	Total chlorophyll a concentration determined from HPLC method ($mg\,m^{-3}$)
Es	Surface irradiance (or above-water downwelling irradiance) ($mW\,cm^{-2}\,\mu m^{-1}$)
ESA	European Space Agency
Fo	Top-of-the-atmosphere solar irradiance ($mW\,cm^{-2}\,\mu m^{-1}$)
GeP&CO	Geochemistry, Phytoplankton, and Color of the Ocean
HOT	Hawaii Ocean Time-series
HPLC	High-performance liquid chromatography
ICES	International Council for the Exploration of the Sea
kd	Diffuse attenuation coefficient for downward irradiance (m^{-1})
Lw	Water-leaving radiance (or above-water upwelling radiance) ($mW\,cm^{-2}\,\mu m^{-1}\,sr^{-1}$)
MERIS	Medium Resolution Imaging Spectrometer
MERMAID	MERIS Match-up In situ Database
MOBY	Marine Optical Buoy
MODIS	Moderate Resolution Imaging Spectroradiometer
NASA	National Aeronautics and Space Administration
nLw	Normalised water-leaving radiance ($mW\,cm^{-2}\,\mu m^{-1}\,sr^{-1}$)
nLw_ex	nLw with a correction for bidirectional effects ($mW\,cm^{-2}\,\mu m^{-1}\,sr^{-1}$)
NOMAD	NASA bio-Optical Marine Algorithm Data set
OC-CCI	Ocean Colour Climate Change Initiative
rrs	Remote-sensing reflectance (sr^{-1})
SeaBASS	SeaWiFS Bio-optical Archive and Storage System
SeaWiFS	Sea-viewing Wide Field-of-view Sensor
Rw	Irradiance reflectance (dimensionless)

Appendix B: Data table

The compiled data are available at doi:10.1594/PANGAEA.854832. The merged text table described in this work, considered here as the main table, is accompanied by four extra files. One extra file is a csv table with detailed information about the number of observations per variable, dataset, subdataset and pi. Two other extra files are text tables generated from the main table, and are provided to help with the analysis of spectral data. These two files contain the spectral data aggregated within ±2 and ±6 nm, respectively, of SeaWiFS, MODIS AQUA and MERIS sensor bands. The files are generated by assigning, in each row of the main table, the closest spectral observation within 2 nm (or 6 nm) of a sensor band. The centre wavelength of each band and sensor used in the generation of the files is the following: SeaWiFS bands 1–8 were centred at [412, 443, 490, 510, 555, 670, 765, 865] nm, respectively; MODIS-AQUA bands 1–9 were centred at [412, 443, 488, 531, 547, 667, 678, 748, 869] nm, respectively; and MERIS bands 1–13 were centred at [412, 442, 490, 510, 560, 620, 665, 681, 709, 753, 779, 865, 885] nm, respectively. An exception to this procedure was made to confirm that the correct MOBY data are stored in the files (see Sect. 2.2.1. for discussion on how MOBY wavelengths are stored in the main file). Finally, a readme file is provided to help the user.

Author contributions. The first six authors belong to the ESA OC-CCI team and contributed to the design of the compilation. The remaining authors are listed alphabetically and are data contributors (see their respective dataset on Table 2) or individuals responsible for the development of a particular dataset (Jeremy Werdell for NO-MAD and Katherine Barker for MERMAID). All data contributors (listed on Table 2) were contacted for authorisation of data publishing and offered co-authorship. In the case of the ICES dataset the permission for publishing was given by the ICES team. All the authors have critically reviewed the manuscript.

Acknowledgements. This paper is a contribution to the ESA OC-CCI project. This work is also a contribution to project PEst-OE/MAR/UI0199/2014. We are grateful for the efforts of the teams responsible for the collection of the data in the field and of the teams responsible for processing and storing the data in archives, without which this work would not be possible. We thank Tamoghna Acharyya and Robert Brewin at Plymouth Marine Laboratory for their initial contribution to this work. We thank the NOAA (US) for making available the MOBY data and Yong Sung Kim for the help with questions about MOBY data. BOUSSOLE is supported and funded by the European Space Agency (ESA), the Centre National d'Etudes Spatiales (CNES), the Centre National de la Recherche Scientifique (CNRS), the Institut National des Sciences de l'Univers (INSU), the Université Pierre et Marie Curie (UPMC) and the Observatoire Océanologique de Villefranche-sur-mer (OOV). We thank ACRI-ST, ARGANS and ESA for access to the MERMAID Database (http://hermes.acri.fr/mermaid). We thank Annelies Hommersom, Pierre Yves Deschamps and David Siegel for allowing the use of MERMAID data for which they are principal investigators. We thank the British Oceanographic Data Centre (BODC) for access to AMT data and in particular Polly Hadziabdic and Rob Thomas for their help with questions about the AMT dataset. We thank Victoria Hill, Patrick Holligan, Gerald Moore and Emilio Suarez for the use of AMT data for which they are principal investigators. We thank Sam Ahmed, Hui Feng, Alex Gilerson and Brent Holben for allowing the use of the AERONET-OC data for which they are principal investigators. We thank also the AERONET staff and site support people. We thank Bob Bidigare, Matthew Church, Ricardo Letelier and Jasmine Nahorniak for making the HOT data available, and the National Science Foundation for support of the HOT research (grant OCE 09-26766). We thank Yves Dandonneau for allowing the use of GeP&CO data. We thank the ICES database on the marine environment (Copenhagen, Denmark, 2014) for allowing the use of their archived data, and Marilynn Sørensen for the help with questions about the ICES dataset. We thank all ICES contributors for their data. We thank Eric Zettler and the SEA Education Association. We thank NASA, SeaBASS and the Ocean Biology Processing Group (OBPG) for access to SeaBASS and NOMAD data. We thank NASA for project funding for data collection. We thank Chris Proctor from SeaBASS for his valuable and prompt help with a variety of questions. Finally, we are deeply thankful to the data contributors of NOMAD and SeaBASS: Kevin Arrigo, Mike Behrenfeld, Emmanuel Boss, Chris Brown, Douglas Capone, Ken Carder, Alex Chekalyuk, Jay-Chung Chen, Dennis Clark, Jorge Corredor, Glenn Cota, Yves Dandonneau, Heidi Dierssen, David Eslinger, Piotr Flatau, Joaquim Goes, Gwo-Ching Gong, Larry Harding, Jon Hare, Chuanmin Hu, Sung-Ho Kang, Gary Kirkpatrick, Oleg Kopelevich, Sam Laney, Zhongping Lee, Ricardo Letelier, Marlon Lewis, Antonio Mannino, John Marra, Chuck McClain, Christophe Menkes, Mark Miller, Ru Morrison, James Mueller, James Nelson, Norman Nelson, Mary Jane Perry, David Phinney, John Porter, Collin Roesler, David Siegel, Mike Sieracki, Jeffrey Smart, Raymond Smith, James Spinhirne, Dariusz Stramski, Rick Stumpf, Ajit Subramaniam, Chuck Trees, Ronald Zaneveld, Eric Zettler and Richard Zimmerman.

Edited by: D. Carlson

References

Amante, C. and Eakins, B. W.: ETOPO1, 1 Arc-Minute Global Relief Model: Procedures, Data Sources and Analysis, NOAA Technical Memorandum NESDIS NGDC-24. National Geophysical Data Center, NOAA, available at: https://catalog.data.gov/dataset/etopo1-1-arc-minute-global-relief-model/resource/9e53be26-60cb-4139-b70c-51a2c4561bbb (last access: 3 June 2016), 2009.

Antoine, D., André, J. M., and Morel, A.: Oceanic primary production: 2. Estimation at global scale from satellite (CZCS) chlorophyll, Global Biogeochem. Cy., 10, 57–70, 1996.

Antoine, D., Chami, M., Claustre, H., D'Ortenzio, F., Morel, A., Bécu, G., Gentili, B., Louis, F., Ras, J., Roussier, E., Scott, A. J., Tailliez, D., Hooker, S. B.,Guevel, P., Desté, J.-F., Dempsey, C., and Adams, D.: BOUSSOLE : a joint CNRS-INSU, ESA, CNES and NASA Ocean Color Calibration And Validation Activity. NASA Technical memorandum No. 2006-214147, 61 pp., available at: http://ntrs.nasa.gov/archive/nasa/casi.ntrs.nasa.gov/20070028812.pdf (last access: 3 June 2016), 2006.

Antoine, D., Guevel, P., Desté, J.-F., Bécu, G., Louis, F., Scott, A., and Bardey, P.: The "BOUSSOLE" Buoy – A New Transparent-to-Swell Taut Mooring Dedicated to Marine Optics: Design, Tests, and Performance at Sea. J. Atmos. Oceanic Technol., 25, 968–989, 2008.

Bailey, S. W. and Werdell, P. J.: A multi-sensor approach for the on-orbit validation of ocean color satellite data products, Remote Sens. Environ., 102, 12–23, 2006.

Barker, K.: In-situ Measurement Protocols. Part A: Apparent Optical Properties, Issue 2.0, Doc. no: CO-SCI-ARG-TN-0008, ARGANS Ltd., p. 126, available at: http://mermaid.acri.fr/dataproto/CO-SCI-ARG-TN-0008_In-situ_Measurement_Protocols-AOPs_Issue2_Mar2013.pdf (last access: 3 June 2016), 2013a.

Barker, K.: In-situ Measurement Protocols. Part B: Inherent Optical Properties and in-water constituents, Issue 1.0, Doc. no: CO-SCI-ARG-TN-0008, ARGANS Ltd., p. 39, available at: http://mermaid.acri.fr/dataproto/CO-SCI-ARG-TN-0008_In-situ_Measurement_Protocols-IOPs-Constituents_Issue1_Mar2013.pdf (last access: 3 June 2016), 2013b.

Bricaud, A., Claustre, H., Ras, J., and Oubelkheir, K.: Natural variability of phytoplanktonic absorption in oceanic waters: Influence of the size sctructure of algal populations, J. Geophys. Res., 109, C11010, doi:10.1029/2004JC002419, 2004.

Clark, D. K., Yarborough, M. A., Feinholz, M. E., Flora, S., Broenkow, W., Kim, Y. S., Johnson, B. C., Brown, S. W., Yuen,

M., and Mueller, J. L.: MOBY, A Radiometric Buoy for Performance Monitoring and Vicarious Calibration of Satellite Ocean Colour Sensors: Measurements and Data Analysis Protocols, in: Ocean Optics Protocols for Satellite Ocean Colour Sensor Validation, edited by: Muller, J. L., Fargion, G., and McClain, C., NASA Technical Memo. 2003-211621/Rev4, Vol. VI, 3–34, NASA/GSFC, Greenbelt, MD, USA, 2003.

Dandonneau, Y. and Niang, A.: Assemblages of phytoplankton pigments along a shipping line through the North Atlantic and Tropical Pacific, Prog. Oceanogr., 73, 127–144, 2007.

Gordon, H. R. and Clark, D. K.: Clear water radiances for atmospheric correction of coastal zone color scanner imagery, Appl. Optics, 20, 4175–4180, 1981.

Gregg, W. W. and Carder, K. L.: A simple spectral solar irradiance model for cloudless maritime atmospheres, Limnol. Oceanogr., 35, 1657–1675, 1990.

IOCCG: Report 5: Remote Sensing of Inherent Optical Properties: Fundamentals, Tests of Algorithms, and Applications, in: Reports of the International Ocean-Colour Coordinating Group, edited by: Lee, Z.-P., No. 5. vol. 5, IOCCG, 2006, Dartmouth, Canada, p. 126, 2006.

IOCCG: Why Ocean Colour? The Societal Benefits of Ocean-Colour Technology, in: Reports of the International Ocean-Colour Coordinating Group, edited by: Platt, T., Hoepffner, N., Stuart, V., and Brown, C., No. 7, IOCCG, Dartmouth, Canada, 2008.

Karl, D. M. and Michaels, A. F.: The Hawaiian Ocean Time-series (HOT) and Bermuda Atlantic Time-series Study (BATS) – Preface, Deep-Sea Res. II, 43, 127–128, 1996.

Morel, A. and Gentilli, B.: Diffuse Reflectance of Oceanic Waters. 3. Implications of Bidirectionality for the Remote-Sensing Problem, Appl. Optics, 35, 4850–4862, 1996.

Morel, A. and Maritorena, S.: Bio-optical properties of oceanic waters: A reappraisal, J. Geophys. Res., 106, 7163–7180, 2001.

Morel, A., Antoine, D., and Gentilli, B.: Bidirectional reflectance of oceanic waters: accounting for Raman emission and varying particle scattering phase function, Appl. Optics, 41, 6289–6306, 2002.

Nechad, B., Ruddick, K., Schroeder, T., Oubelkheir, K., Blondeau-Patissier, D., Cherukuru, N., Brando, V., Dekker, A., Clementson, L., Banks, A. C., Maritorena, S., Werdell, P. J., Sá, C., Brotas, V., Caballero de Frutos, I., Ahn, Y.-H., Salama, S., Tilstone, G., Martinez-Vicente, V., Foley, D., McKibben, M., Nahorniak, J., Peterson, T., Siliò-Calzada, A., Röttgers, R., Lee, Z., Peters, M., and Brockmann, C.: CoastColour Round Robin data sets: a database to evaluate the performance of algorithms for the retrieval of water quality parameters in coastal waters, Earth Syst. Sci. Data, 7, 319–348, doi:10.5194/essd-7-319-2015, 2015.

Pope, R. and Fry, E.: Absorption spectrum (380–700 nm) of pure

waters: II. Integrating cavity measurements, Appl. Optics, 36, 8710–8723, 1997.

Robinson, C., Poulton, A. J., Holligan, P. M., Baker, A. R., Forster, G., Gist, N., Jickells, T. D., Malin G., Upstill-Goddard, R., Williams, R. G., Woodward, E. M. S., and Zubkov, M. V.: The Atlantic Meridional Transect (AMT) Programme: a contextual view 1995–2005, Deep-Sea Res. II, 53, 1485–1515, doi:10.1016/j.dsr2.2006.05.015, 2006.

Thuillier, G., Hersé, M., Labs, D., Foujols, T., Peetermans, W., Gillotay, D., Simon, P. C., and Mandel, H.: The solar spectral irradiance from 200 nnm to 2400 nm as measured by the SOLSPEC spectrometer from the ATLAS 1-2-3 and EURECA missions, Sol. Phys., 214, 1–22, 2003.

Tiwari, S. P. and Shanmugam, P.: An optical model for deriving the spectral particulate backscattering coefficients in oceanic waters, Ocean Sci., 9, 987–1001, 2013.

Trees, C. C., Kennicutt II, M. C., and Brooks, J. M.: Errors associated with the standard fluorimetric determination of chlorophylls and phaeopigments, Mar. Chem., 17, 1–12, 1985.

Valente, A., Sathyendranath, S., Brotas, V., Groom, S., Grant, M., Taberner, M., Antoine, D., Arnone, R., Balch, W. M., Barker, K., Barlow, R., Bélanger, S., Berthon, J.-F., Besiktepe, S., Brando, V., Canuti, E., Chavez, F. P., Claustre, H., Crout, R., Frouin, R., García-Soto, C., Gibb, S., Gould, R., Hooker, S., Kahru, M., Klein, H., Kratzer, S., Loisel, H., McKee, D,. Mitchell, G., Moisan, T., Muller-Karger, F. E., O'Dowd, L., Ondrusek, M., Poulton, A. J., Repecaud, M., Smyth, T. J., Sosik, H., Twardowski, M. S., Voss, K., Werdell, P. J., Wernand, M. R., and Zibordi, G.: A compilation of global bio-optical in situ data for ocean-colour satellite applications, doi:10.1594/PANGAEA.854832, 2015.

Werdell, P. J. and Bailey, S. W.: An improved bio-optical data set for ocean color algorithm development and satellite data product validation, Remote Sens. Environ., 98, 122–140, 2005.

Werdell, P. J., Bailey, S., Fargion, G., Pietras, C., Knobelspiesse, K., Feldman, G., and McClain, C.: Unique data repository facilitates ocean color satellite validation, EOS Transactions AGU, 84, 377–387, 2003.

Zhang, X., Hu, L., and He, M.-X.: Scattering by pure seawater: Effect of Salinity, Opt. Express, 17, 5698–5710, 2009.

Zibordi, G., Holben, B. N., Hooker, S. B., Mélin, F., Berthon, J.-F., Slutsker, I., Giles, D., Vandemark, D., Feng, H., Rutledge, K., Schuster, G., and Al Mandoos, A.: A network for standardized ocean color validation measurements, EOS Trans. Am. Geophys. Union, 87, 293–297, 2006.

Zibordi, G., Holben, B. N., Slutsker, I., Giles, D., D'Alimonte, D., Mélin, F., Berthon, J.-F., Vandemark, D., Feng, H., Schuster, G., Fabbri, B. E., Kaitala, S., and Seppälä, J.: AERONET-OC: A network for the validation of ocean color primary radiometric products, J. Atmos. Ocean. Tech., 26, 1634–1651, 2009.

11

Mediterranean Sea climatic indices: monitoring long-term variability and climate changes

Athanasia Iona[1,2], Athanasios Theodorou[2], Sarantis Sofianos[3], Sylvain Watelet[4], Charles Troupin[4], and Jean-Marie Beckers[4]

[1] Hellenic Centre for Marine Research, Institute of Oceanography, Hellenic National Oceanographic Data Centre, 46,7 km Athens Sounio, Mavro Lithari P.O. Box 712 19013 Anavissos, Attica, Greece

[2] University of Thessaly, Department of Ichthyology & Aquatic Environment, Laboratory of Oceanography, Fytoko Street, 38445, Nea Ionia Magnesia, Greece

[3] Ocean Physics and Modelling Group, Division of Environmental Physics and Meteorology, University of Athens, University Campus, Phys–5, 15784 Athens, Greece

[4] University of Liège, GeoHydrodynamics and Environment Research, Quartier Agora, Allée du 6-Août, 17, Sart Tilman, 4000 Liège 1, Belgium

Correspondence: Athanasia Iona (sissy@hnodc.hcmr.gr)

Abstract. We present a new product composed of a set of thermohaline climatic indices from 1950 to 2015 for the Mediterranean Sea such as decadal temperature and salinity anomalies, their mean values over selected depths, decadal ocean heat and salt content anomalies at selected depth layers as well as their long time series. It is produced from a new high-resolution climatology of temperature and salinity on a 1/8° regular grid based on historical high-quality in situ observations. Ocean heat and salt content differences between 1980–2015 and 1950–1979 are compared for evaluation of the climate shift in the Mediterranean Sea. The two successive periods are chosen according to the standard WMO climate normals. The spatial patterns of heat and salt content shifts demonstrate that the climate changes differently in the several regions of the basin. Long time series of heat and salt content for the period 1950 to 2015 are also provided which indicate that in the Mediterranean Sea there is a net mean volume warming and salinification since 1950 that has accelerated during the last two decades. The time series also show that the ocean heat content seems to fluctuate on a cycle of about 40 years and seems to follow the Atlantic Multidecadal Oscillation climate cycle, indicating that the natural large-scale atmospheric variability could be superimposed onto the warming trend. This product is an observation-based estimation of the Mediterranean climatic indices. It relies solely on spatially interpolated data produced from in situ observations averaged over decades in order to smooth the decadal variability and reveal the long-term trends. It can provide a valuable contribution to the modellers' community, next to the satellite-based products, and serve as a baseline for the evaluation of climate-change model simulations, thus contributing to a better understanding of the complex response of the Mediterranean Sea to the ongoing global climate change.

1 Introduction

During the twentieth century the Mediterranean Sea has undergone profound and rapid changes. Temperature and salinity have increased with accelerating trends in recent decades (Rohling and Bryden, 1992; Vargas-Yáñez et al., 2010a; Schroeder et al., 2017), reflecting apparently the global warming tendency (Levitus et al., 2012). Because of its geographical position, its small size (reduced volume to area size ratio) and its being enclosed between continents, the Mediterranean Sea is very sensitive and responds faster and more strongly to climate changes than the open ocean, e.g. changes to atmospheric forcings and/or anthropogenic influences (Béthoux et al., 1999; Schroeder et al., 2017). Moreover, the Mediterranean region has been identified as one of the hotspots for future climate change in the world (Giorgi, 2006) where changes are expected to be largest. According to the IPCC (2014) 5th assessment report, the observed global mean sea level (GMSL) has changed since the mid-nineteenth century, with a larger rate than the mean rate during the previous two millennia (high confidence). It is very likely that the mean rate of global averaged sea level rise was 1.7 [1.5 to 1.9] mm yr^{-1} between 1901 and 2010, 2.0 [1.7 to 2.3] mm yr^{-1} between 1971 and 2010, and 3.2 [2.8 to 3.6] mm yr^{-1} between 1993 and 2010. The most important contributions to global and regional mean sea level rise are a) increase in the ocean volume as a result of increase in the mass of the water (due to melting of ice sheets and shrinking of glaciers), and b) increase in the ocean volume as a result of decrease in ocean water density (the ocean expands as it warms). However, ocean observations indicate that the ocean is getting saltier and an increase in density should compensate for the thermal expansion. Recent studies suggest that the water cycle has been amplified because of the global warming, contributing to a saltier ocean (Skliris et al., 2016; Durack and Wijffels, 2010; Durack et al., 2012; Zika et al., 2018). The projected future changes show that the GMSL will continue to rise during the twenty-first century with a rate that will very likely exceed that observed during 1971 to 2010 due to increased ocean warming and increased loss of mass from glaciers and ice sheets. Sea level rise will not be uniform. In the Mediterranean region, climate model projections show an acceleration of warming, salinification as well as sea level rise during the twenty-first century (Somot et al., 2008; Mariotti et al., 2008; Giorgi, 2006; Giorgi and Lionello, 2008; Adloff et al., 2015; IPCC, 2014) with a potential strong impact on the marine environment, its effective management and thus human welfare (IPCC, 2014; Füssel et al., 2017).

In turn, the Mediterranean Sea plays an essential role in influencing the water formation processes and thermohaline circulation in the North Atlantic (Lozier et al., 1995; Béthoux et al., 1998; Rahmstorf, 1998). As a concentration basin (where evaporation exceeds precipitation) it exports at intermediate depths salty water through the Strait of Gibraltar to the Atlantic, a major site of dense water formation for the global thermohaline circulation. In this context, monitoring the changes of the ocean heat content (OHC) and ocean salt content (OSC) of the Mediterranean Sea is of fundamental importance.

The ocean is the dominant component of the Earth's heat balance, and most of the total warming caused by climate change is manifested in increased OHC. Good estimates of past changes in OHC are essential for understanding the role of the oceans in past climate change and for assessing future climate change (IPCC, 2014). However, accurate assessments of the OHC are still a challenge, mainly because of insufficient and irregular data coverage.

The Mediterranean Sea (Fig. 1) has a very high spatial and temporal variability at all scales, from small turbulence to basin-scale processes (Fusco et al., 2003). Three main water masses are found, the surface, intermediate and deep waters, which form a special flow regime characterized by an active thermohaline (overturning) circulation: (a) one shallow cell that extends over the two basins and communicates directly with the Atlantic Ocean and consists of the inflowing Atlantic Water and the return flow of saltier Mediterranean Water, and (b) two separate deep overturning cells, in the western and eastern basins with several sites of deep water formation, e.g. in the Gulf of Lions in the western and southern Adriatic, and the Aegean Sea in the eastern basin (Tsimplis et al., 2006, and references therein). Complexity arises from multiple driving forces, strong topographic and coastal influences and internal dynamical processes that interact on several temporal and spatial scales (basin, sub-basin and mesoscale) to form an extremely complex and variable circulation. The seasonal, interannual and decadal variabilities are associated with the internal variability of the climatic system. The variability of the atmospheric circulation patterns induces variations in the water masses either by changing temperature and salinity properties through freshwater and heat fluxes or indirectly by changing the main circulation pathways which in turn can produce changes in the preconditioning phases previous to intermediate and deep water production or redistributing salt and heat content in the water column (Schroeder et al., 2012, and references therein).

A major abrupt change has been recorded in the Mediterranean in the last decades which induced important changes to the heat and salt contents. Between the late 1980s and middle 1990s an interannual variation, the Eastern Mediterranean Transient (EMT), strongly influenced the intermediate and deep water masses' pathways and characteristics (Malanotte-Rizzoli et al., 1999; Lascaratos et al., 1999; Klein et al., 1999; Roether et al., 1996). During that event, the circulation of the eastern Mediterranean experienced a dramatic change from the surface layers to the bottom. Dense water of Aegean origin replaced the resident Eastern Mediterranean Deep Water (EMDW) of Adriatic origin. Inducing the uplifting of the Ionian deep waters, the EMT significantly modified the characteristics of the water masses flowing through the Sicily Strait, while the remarkable presence of salty Cretan Intermediate Water (CIW) (Klein et al., 1999) in the Ionian Sea enhanced the salt export from the eastern to western Mediterranean at the end of the 1990s. The EMT affected not only the nearby Tyrrhenian Sea, but also the Western Mediterranean Deep water production (Schröder et al., 2006; Schroeder et al., 2017). After the late 1990s, the dense waters of Aegean origin were no longer dense enough to reach the bottom layer and the Adriatic Sea regained its role as the primary source of dense water (Theocharis et al., 2002; Manca, 2003).

Heat and salt contents are calculated from temperature and salinity differences in relation to mean climatological reference values integrated over a particular reference depth and study area (see the next section for more details). To detect their long-term tendency, long time series extending to more than a few decades are needed in order to identify the natural climate long-term oscillations and quantify any remaining trends related to global warming. In small areas where the data coverage is sufficient, OHC/OSC changes are calculated directly from the in situ measurements. But at the large basin scale, where the coverage is not good enough, we need to interpolate the data to fill the gaps. In such cases, the noise from the interpolation schemes is an additional source of uncertainty. Levitus (2000), in using the World Ocean Database (https://www.nodc.noaa.gov/OC5/WOD/pr_wod.html, last access: 9 October 2018), was the first who spoke about the warming of the global oceans and quantified the interannual-to-decadal variability of the heat content. Since then, periodical updates are released based on additional data, updated estimations of corrections for the time-varying systematic bias in expendable bathythermograph data and corrections of some ARGO float data. The first publication on the time-dependent warm bias of the bathythermograph data was by Gouretski and Koltermann (2007). The proposed corrections were included in the World Ocean Database and in Levitus et al. (2009). Levitus showed that the proposed corrections of bathythermographs reduce the interdecadal variability but that the long-term trends remain similar (Levitus et al., 2009). An analogous study in the Mediterranean showed that including or not the bathythermographs in the OHC estimates of the western Mediterranean does not significantly change the results (Vargas-Yáñez et al., 2010b). Levitus et al. (2012) reported that for the period 1955–2010, the heat content of the world ocean for the 0–2000 m layer increased by $24.0 \pm 1.9 \times 10^{22}$ J, corresponding to a rate of 0.39 W m^{-2} (per unit area of the world ocean) and a volume mean warming of 0.09 °C. This warming corresponds to a rate of 0.27 W m^{-2} per unit area of the Earth's surface. The heat content of the world ocean for the 0–700 m layer increased by $16.7 \pm 1.6 \times 10^{22}$ J, corresponding to a rate of 0.27 W m^{-2} (per unit area of the world ocean) and a volume mean warming of 0.18 °C. They also reported that the 0–700 m ocean layer accounted for approximately one-third of the warming of the 0–2000 m layer of the world ocean

(Levitus et al., 2012). It is worth mentioning that the ARGO array of profiling floats (their deployment started in 2000) improved significantly the in situ observations' spatial coverage and the subsequent assessments for 0–2000 m, but there are still many regional seas uncovered (observations in these seas come mainly from hydrographic cruises).

In the Mediterranean, many works since the late 1980s have been carried out trying to quantify the trends of temperature and salinity and determine which causes underlie these (such as global warming or anthropogenic climate change due to main rivers damming). Table 1 in Vargas-Yáñez et al. (2008) and Manca et al. (2004), and Table 1 in Skliris et al. (2018), summarize the main findings. An analysis of these results shows that there are differences between them arising from (a) the input data (in situ or interpolated data or model or satellite), (b) their spatial and temporal variability, (c) the choice of the climatological reference, (d) the quality control procedures, (e) the instruments' accuracy, and (f) the mapping techniques, e.g. the gridding and infilling methodologies such as optimal interpolation or variational inverse methods used to fill the data gaps and obtain a gridded 3-D continuous field and time series thereafter as well as which assumptions are made in areas of missing data (Jordà et al., 2017).

Some of the above findings are outlined below. The increasing trend is more evident in the salinity than the temperature. The temperature and salinity of the deep waters of the western Mediterranean are increasing. In the eastern Mediterranean, for the intermediate layer there is no general consensus. Rixen et al. (2005), using the MEDATLAS climatology (MEDAR Group, 2002), found an increase in OHC and OSC of about [1.3–1.5] 10^{21} J and [1.4–1.6] 10^{14} PSU m^3, respectively, over the whole Mediterranean for the period 1950–2000, corresponding to volume mean T and S anomalies of about [0.09–0.10] °C and [0.035–0.04], respectively. During the last decades, the western Mediterranean OHC and OSC have been increasing with an accelerating tendency of the western deep waters towards higher temperatures and salinities since the 1950s, with the process accelerating after the second half of the 1980s. The variation of the intermediate layers is attributed to decadal variability. Skliris et al. (2018) identified a strong basin-scale multi-decadal salinification, particularly in the intermediate and deep layers of order 0.015 practical salinity scale(pss) decade^{-1}, by analysing the inter-annual objectively analysed gridded fields from EN4 from the Met Office Hadley Centre (subversion En4.1.1., http://www.metoffice.gov.uk/hadobs/en4, last access: 9 October 2018) and MEDAR/MEDATLAS climatology (MEDAR Group, 2002), for two reference periods, 1950–2002 and 1950–2015. Schroeder et al. (2016), analysing in situ data, found that over the period 1950–2010, the deep Western Mediterranean Deep Water heat and salt contents increased almost steadily, with an acceleration after the mid-1980s. Below 1000 m, the

Mediterranean underwent the strongest salinity gain in the world ocean (Skliris et al., 2014).

The objective of this work is to provide estimates of T/S and OHC/OSC variations using the latest SeaDataNet historical data sets combined with a modern, numerically efficient interpolation technique that takes into account constraints such as physical boundaries. The new product is expected to give a more detailed insight into the spatial pattern of the changes for the whole Mediterranean and the decadal variability of OHC/OSC. The originality of this product compared to the existing ones is that we (a) use a higher spatial and temporal resolution for gridded fields of T/S anomalies, (b) provide a large, basin-scale spatial pattern for the trends of the decadal T/S and OHC/OSC anomalies, (c) provide long-term time series of the decadal anomalies, and (d) provide 30-year averages for evaluating the climate shift in the Mediterranean. The finer spatial resolution of the input data climatology filters out the noise induced by the mesoscale features, but at the same time is such that it smoothes less the large-scale features. The temporal resolution is such that it smoothes the strong seasonal, interannual and decadal variability so that the final product is able to resolve in more detail the climatic variability and identify possible warming trends. Three layers were considered in this work as representative of the main water masses found in the Mediterranean: 0–150, 150–600, and 600–4000 m, respectively, for the surface, intermediate and deep waters, as in Robinson et al. (2001), and one additional 0–4000 m for the whole water column and volume assessments.

2 Data and methods

2.1 Data sources

Gridded horizontal fields from a new high-resolution climatology of temperature and salinity for the Mediterranean (Iona et al., 2018) were used as input data. These fields were produced using the SeaDataNet temperature/salinity historical data collection V2 (Simoncelli et al., 2015, available at http://sextant.ifremer.fr/record/8c3bd19b-9687-429c-a232-48b10478581c/, last access: 9 October 2018). The SeaDataNet collection comprises 213 542 temperature and 138 691 salinity profiles from in situ measurements for the 1911 to 2015 period. The gridded fields cover the geographical region 6.25° W–36.5° E, 30–46° N on 31 standard depth levels from 0 to 4000 m: [0, 5, 10, 20, 30, 50, 75, 100, 125, 150, 200, 250, 300, 400, 500, 600, 700, 800, 900, 1000, 1100, 1200, 1300, 1400, 1500, 1750, 2000, 2500, 3000, 3500, 4000]. The spatial resolution is $1/8° \times 1/8°$. The seasonal scale is winter (January–March), spring (April–June), summer (July–September), and autumn (October–December). The gridding of the in situ observations was done with the Data Interpolating Variational Analysis (DIVA) software tool that allows the spatial interpolation of data in an optimal way, comparable to optimal interpolation (OI) using a finite-element method (Beckers et al., 2014; Troupin et al., 2012).

The time filtering applied to the in situ observations (decadal averaging) results in spatial correlations found in the data of the order 300–350 km, much larger than the Rossby radius of the deformation scale (10–15 km) associated with mesoscale motions. The mesoscale features and other smaller patches are therefore filtered out and the climatologies used for the indices' calculations focus on the variability of the large-scale features. The influence of the uneven distribution in space where a large number of data points are concentrated in a very small area and within a very short period is controlled by applying different weights (and lower than 1) to each of these data points because such points cannot be considered independent in a climatological analysis. The characteristic length of weighting was set to be equal to 0.08° (in the same units as the data locations) and the characteristic time of weighting was set to be equal to 90 days (3 months). Detrending was applied in the observations used for the reference climatologies in order to remove the uneven spatial distributions in time.

The input gridded data are listed below. They are stored in netCDF files and are accessible from the Zenodo platform, a research platform where papers, data, software codes or any other object contributing to the reproducibility of scientific results can be uploaded and then cited using a digital object identifier (DOI).

1. *Annual climatology (reference)*, obtained by analysing all data (regardless of month or season) for the whole period from 1950 to 2015. This climatology is used as a mean reference that is subtracted from the annual decadal climatology to obtain the T/S anomalies. It is available here: https://doi.org/10.5281/zenodo.1146976 (Iona, 2018a).

2. *Annual decadal climatology*, obtained by analysing all data regardless of month or season for each of the 57 running decades from 1950–1959 to 2006–2015. It is available here: https://doi.org/10.5281/zenodo.1146957 (Iona, 2018b).

3. *Seasonal climatology (reference)*, obtained by analysing all data of the whole period from 1950 to 2015 falling within each season. This climatology is used as a mean reference that is subtracted from the seasonal decadal climatologies to obtain the T/S anomalies. It is available here: https://doi.org/10.5281/zenodo.1146953 (Iona, 2018c).

4. *Seasonal decadal climatology*, obtained by analysing all data falling within each season for each of the 57 running decades from 1950–1959 to 2006–2015. It is available here: https://doi.org/10.5281/zenodo.1146938 (Iona, 2018d).

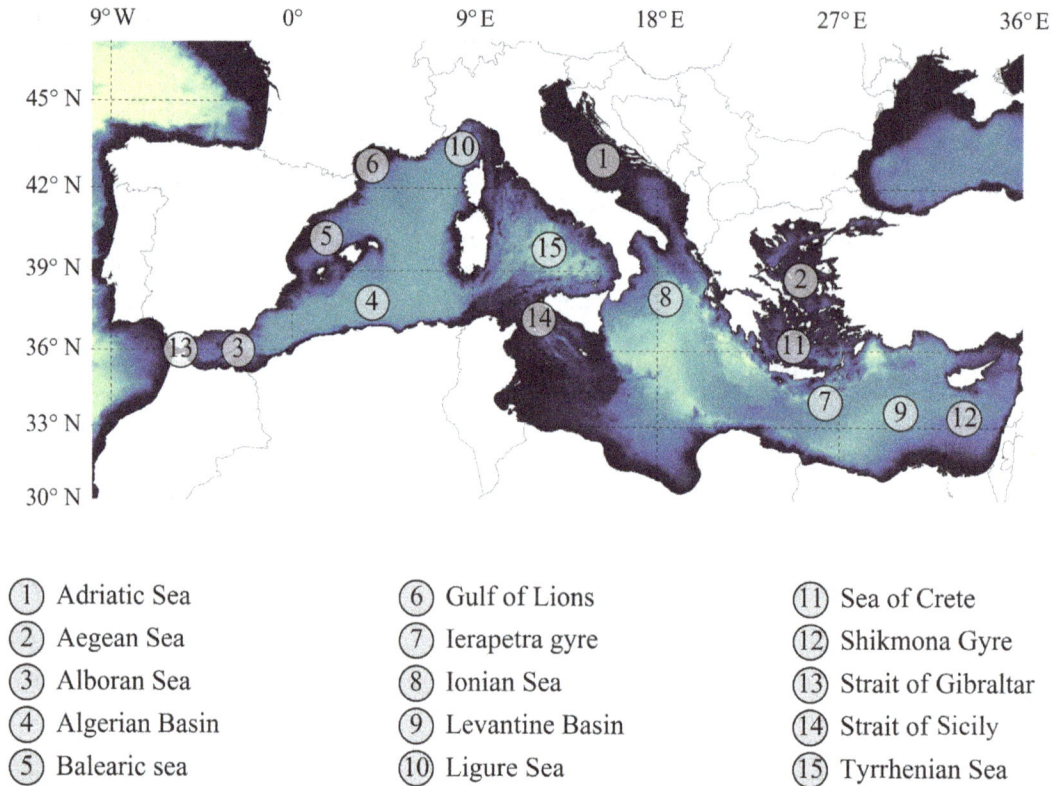

Figure 1. The Mediterranean Sea and its main regions.

The input data used for the current work have already been evaluated with existing comparable products in the region such as SeaDataNet 2015 and WOA13 monthly climatologies, along with MEDAR/MEDATLAS 2002 monthly and decadal climatologies, and are of higher spatial and temporal resolution (Iona et al., 2018). It is important to note that in the used input climatology, each gridded field is accompanied by an error field that allows one to assess the reliability of the input data. This helps to objectively identify areas with poor data coverage, mask them and exclude them from further processing.

2.2 Definitions

Anomalies In all products, temperature and salinity anomalies have been used. Anomaly is defined as the difference between the value of a grid point and a mean climatological reference.

Mean climatological references – Annual climatology used as a reference for the annual decadals.

– Seasonal climatology used as a reference for the seasonal decadals.

Climates The World Meteorological Organization (WMO) recommendation of using 30-year averages (climate normals) to describe climate conditions was used in this study (WMO, 2011). Climate shift is defined as the difference between two successive 30-year averages.

Linear trends They were computed by linear regression with a constant term.

It is noted that in the climate shifts presented in this work, the period 1950 to 1979 contains 3 decades and the period 1980 to 2015 contains 6 years more because the period from 2000 to 2015 is treated as a decade. This was done for two reasons: (a) not to exclude the recent data from the representations of the regional patterns of the climate shifts (or the oldest ones if the study period was shifted later than 1950), and (b) the averaging of the additional recent years actually does not change the qualitative results of the comparison of the two successive periods. Concerning the quantitative differences, Tables 1 and 2 below show, for the two different averaging periods, the mean values of the climate shifts for OHC and OSC areal density over the whole Mediterranean. It can be seen that the inclusion of the additional recent years (about 15 000 additional T/S on the about 150 000 T and 100 000 S profiles of the period 1980–2009) actually reduces the T changes of the first 600 m. The user of course can choose between any period and average the decades according to their needs of each study since the available product includes all 57 running decades from where the climates are computed.

Table 1. Mean values for the whole Mediterranean for areal density of ocean heat content in 10^9 J m^{-2}.

Layer (m)	[1980–2015]–[1950–1979]	[1980–2009]–[1950–1979]
5–150	−0.063933	−0.191992
150–600	−0.087625	−0.168138
600–4000	0.575278	0.567194
5–4000	0.371213	0.349208

Table 2. Mean values for the whole Mediterranean for areal density of ocean salt content in 10^2 ppt m.

Layer (m)	[1980–2015]–[1950–1979]	[1980–2009]–[1950–1979]
5–150	1.803160	1.727270
150–600	0.871111	0.814890
600–4000	0.353313	0.342651
5–4000	0.437761	0.420287

2.3 Process outline

First, seasonal and annual decadal fields of temperature and salinity anomalies (T/S) at each standard depth were generated. Next, T/S vertical averages were calculated for the four layers, 0–150 m (and 5–150 m in the case of the annual fields), 150–600, 600–4000, and 0–4000 m. The thickness of the layers was used as weights for the vertical averaging calculated as half of the distance between adjacent depths. The following weights were used for the 31 standard depth levels: 2.5, 5, 7.5, 10, 15, 22.5, 25, 25, 25, 37.5, 50, 50, 75, 100, 100, 100, 100, 100, 100, 100, 100, 100, 100, 100, 175, 250, 375, 500, 500, 500, 500. In the case of the annual fields, 30 weights are used starting from the second value (5). The used weights are available with the netCDF files used for the computation of the indices. For the estimation of the OHC anomalies the following methodology was used. Each T/S anomaly at each standard depth is associated with a volume which consists of the area of the $1/8° \times 1/8°$ longitude–latitude grid multiplied by the thickness of each layer, e.g. the vertical weights. By multiplying the volume by the T anomalies, by the density of seawater, and by the specific heat, we obtain the OHC anomaly of a specific grid point at each standard depth. By integrating over a depth layer and over all of the analysis area, we obtain the OHC anomaly (in Joules) for the whole Mediterranean Sea according to the following equation:

$$\text{OHC} = \rho\, C_p \sum_{i=1}^{n} \mathrm{d}x\,\mathrm{d}y \int_{z_1}^{z_2} \Delta T \mathrm{d}z. \tag{1}$$

The areal density of OHC (in J m^{-2}) is obtained by integrating the vertically averaged T anomaly over a depth layer according to the equation

$$\text{areal density OHC} = \rho\, C_p \int_{z_1}^{z_2} \Delta T \mathrm{d}z, \tag{2}$$

where $\rho = 1028$ kg m^{-3} is the density of reference seawater, $C_p = 3985$ J kg^{-1} °C the specific heat of seawater, n the number of grid cells and almost the same as the number of grid points as $nx = \text{length(lon)}$, $ny = \text{length(lat)}$ and $n = (nx - 1) \times (ny - 1)$, $\mathrm{d}x = 10951.1$ m, $\mathrm{d}y = 13897.2$ m, ΔT the temperature anomaly, and z_1 and z_2 the upper and lower depths. In the current climatologies density and specific heat of seawater are not calculated separately, but it would be possible to derive them from T and S gridded fields. Such calculations will be available in future releases of the indices. The $\mathrm{d}x$, $\mathrm{d}y$ are the longitude (1/8°), latitude (1/8°) steps of the output grid transformed from degrees to metres. A mean basin volume is estimated at 3.86×10^{15} m^3 and corresponds to the mean wet volume of the analysis grid of the interpolation. For the OSC, the same methodology is used except that we do not multiply by ($\rho\, C_p$), the term that converts temperature to thermal energy (heat). The OSC (in ppt m^3) is given from the equation

$$\text{OSC} = \sum_{i=1}^{n} \mathrm{d}x\,\mathrm{d}y \int_{z_1}^{z_2} \Delta S \mathrm{d}z, \tag{3}$$

and the areal density of OSC (in ppt m) from the following equation:

$$\text{areal density OSC} = \int_{z_1}^{z_2} \Delta S \mathrm{d}z. \tag{4}$$

2.4 Climatic index content

The produced climatic indices for the whole Mediterranean Sea (6.25° W–36.5° E, 30–46° N) consist of the following.

– Annual and seasonal T/S anomalies at 31 standard depths, for 57 running decades from 1950–1959 to 2006–2015.

– Annual and seasonal T/S vertical averaged anomalies at four layers (surface, intermediate, deep and whole column), for 57 running decades from 1950–1959 to 2006–2015.

– Annual and seasonal areal density of OHC/OSC anomalies in four layers (surface, intermediate, deep and whole column), for 57 running decades from 1950–1959 to 2006–2015.

– Annual and seasonal linear trends of T/S, OHC/OSC anomalies at four layers (surface, intermediate, deep and whole column) for all 57 decades.

– Annual and seasonal time series of T/S, OHC/OSC anomalies at four layers (surface, intermediate, deep and whole column) over the whole Mediterranean Sea.

– Differences of two 30-year averages of annual and seasonal T/S anomalies at 31 standard depths for the period 1950 to 2015.

– Differences of two 30-year averages of annual and seasonal T/S, OHC/OHC anomalies for the period 1950 to 2015, at four layers (surface, intermediate, deep and whole column).

All data are stored in netCDF files and are accessible using the following DOIs.

– Annual and seasonal T/S anomalies: https://doi.org/10.5281/zenodo.1408832 (Iona, 2018e).

– Annual and seasonal T/S vertical averaged anomalies: https://doi.org/10.5281/zenodo.1408929 (Iona, 2018f).

– Annual and seasonal areal density of OHC/OSC anomalies: https://doi.org/10.5281/zenodo.1408877 (Iona, 2018g).

– Annual and seasonal linear trends of T/S, OHC/OSC anomalies: https://doi.org/10.5281/zenodo.1408917 (Iona, 2018h).

– Annual and seasonal time series of T/S, OHC/OSC anomalies: https://doi.org/10.5281/zenodo.1411398 (Iona, 2018i).

– Differences of two 30-year averages of annual and seasonal T/S, OHC/OSC anomalies: https://doi.org/10.5281/zenodo.1408903 (Iona, 2018j).

3 Results

We outline below some of the capabilities of the new product. The explanation of the long-term variability patterns that are revealed and attribution of possible causes is out of the scope of this work. A short overview of these was given in the introduction to facilitate the viewing of the products for those readers who are not familiar with the Mediterranean complex dynamics. The comparison of two successive 30-year averages of heat and salt content anomalies for the period 1950 to 2015 can be used for the evaluation of the Mediterranean Sea climate changes. The 30-year periods are averages of three successive decades: the first one refers to the decades 1950–1959 to 1970–1979 and the second to 1980–1989 to 2000–2015. The two 30-year successive periods were selected for consistency with the World Meteorological Organization's recommendation of using as climate normals 30-year periods. Figure 2 illustrates the geographical distribution over the whole Mediterranean of the 30-year climate shift as the

OHC differences between the period 1980–2015 and 1950–1979 in the upper 5–150 m (Fig. 2a), 150–600 m (Fig. 2b), 600–4000 m (Fig. 2c), and 5–4000 m (Fig. 2d).

From the surface layer down to 150 m the climate shift is not uniform. The western Mediterranean surface layer (Fig. 2a) has experienced warming almost everywhere expect the Gulf of Lions and the northern Tyrrhenian–Ligurian eastern basins. The surface layer of the eastern Mediterranean (Fig. 2a) is cooling with a noticeable warming spot at the Ierapetra gyre. We observe the same with surface patterns at the intermediate layers of 150–600 m but with about half the strength of the surface (Fig. 2b). The deep waters are warming almost everywhere except the southern Adriatic, the southern Levantine and the south-western Ionian basin (Fig. 2c).

Regarding salinity, it is important to notice that in the whole western Mediterranean there is a clear OSC increase throughout the whole water column (Fig. 3a–d), while in the eastern basin we see that the spatial pattern is not uniform and a notable salt content increase is observed in the areas of deep water formation, e.g. the southern Adriatic and Aegean Sea. According to the bibliography, this is the Eastern Mediterranean Transit (EMT) signature on the intermediate and deep waters, not only in the eastern Mediterranean, but also in the whole basin (Klein et al., 1999; Theocharis et al., 1999; Rixen et al., 2005; Schroeder et al., 2016). Notable salt increases are found at the Shikmona gyre and the south-western Ionian Sea, following the patterns of the heat content. To illustrate the temporal variability of the thermohaline content, the annual OHC and OSC anomalies for six discrete periods, for the three layers (5–150, 150–600, 600–4000 m) and the whole water column (5–4000 m) are shown in Fig. 4a–b. We observe that apart from the strong spatial variability shown in Figs. 2 and 3, there is a similar irregular pattern from one decade to another. One remarkable feature in these distributions is the acceleration and the substantial heat and salt gain of the deep layer (600–4000 m) starting from 1990. It is also found that the correlations (significant at the 95 % confidence level) between the decadal Atlantic Multidecadal Oscillation (AMO) index and the decadal OHC averages are 0.69 for 0–150 m, 0.64 for 150–600 m, 0.63 for 600–4000 m and 0.76 for the whole column: 5–4000 m. The corresponding correlations with the North Atlantic Oscillation (NAO) index are 0.23, −0.22, 0.50 and 0.38. These findings seem to be in agreement with the bibliographical references, where observed acceleration from satellite data of the Mediterranean waters' warming during the 1990s could be attributed to the positive phase of AMO (Macias et al., 2013).

To get a more detailed insight into the long-term fluctuations we show in Fig. 4c, time series of the decadal OHC and OSC anomalies were integrated over the whole column depth and area of the Mediterranean Sea. There are changes, slowdowns and accelerations throughout the study period and we can distinguish three main periods: (a) from 1960 to the late 1970s with an increasing trend in salt but a decreasing heat

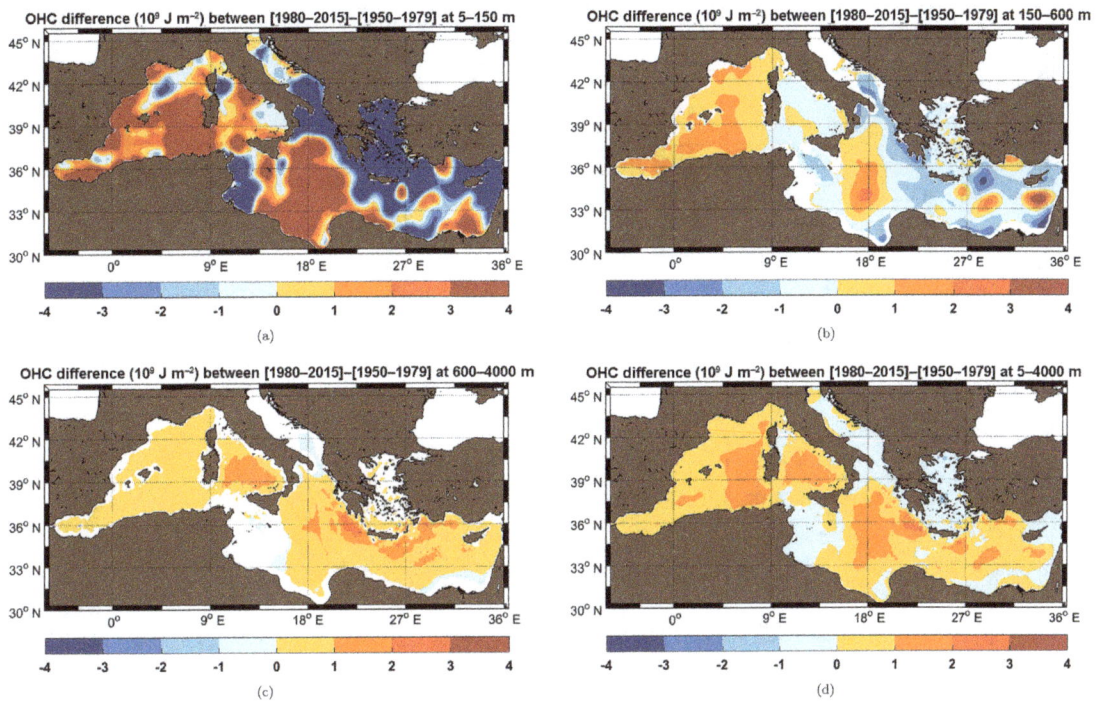

Figure 2. Climate shift of areal density of ocean heat content in $10^9\,\mathrm{J\,m^{-2}}$ between two 30-year periods 1980–2015 and 1950–1979 for **(a)** 5–150 m, **(b)** 150–600 m, **(c)** 600–4000 m, and **(d)** 5–4000 m.

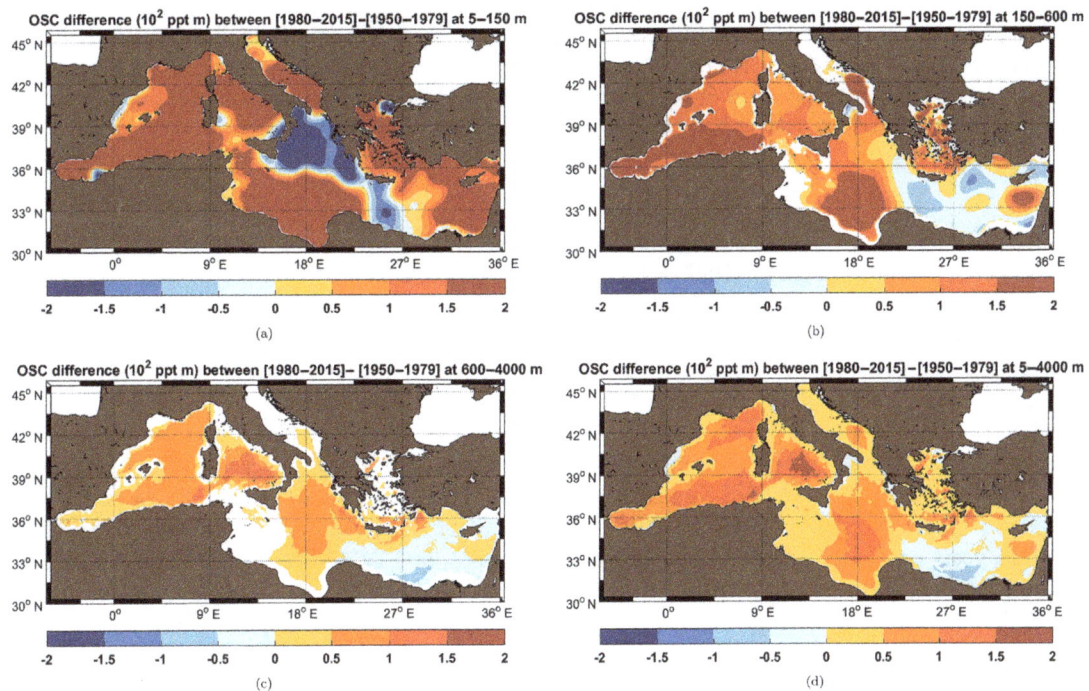

Figure 3. As in Fig. 2 for the climate shift of areal density of ocean salt content in $10^2\,\mathrm{ppt\,m}$ between two 30-year periods 1980–2015 and 1950–1979 for **(a)** 5–150 m, **(b)** 150–600 m, **(c)** 600–4000 m, and **(d)** 5–4000 m.

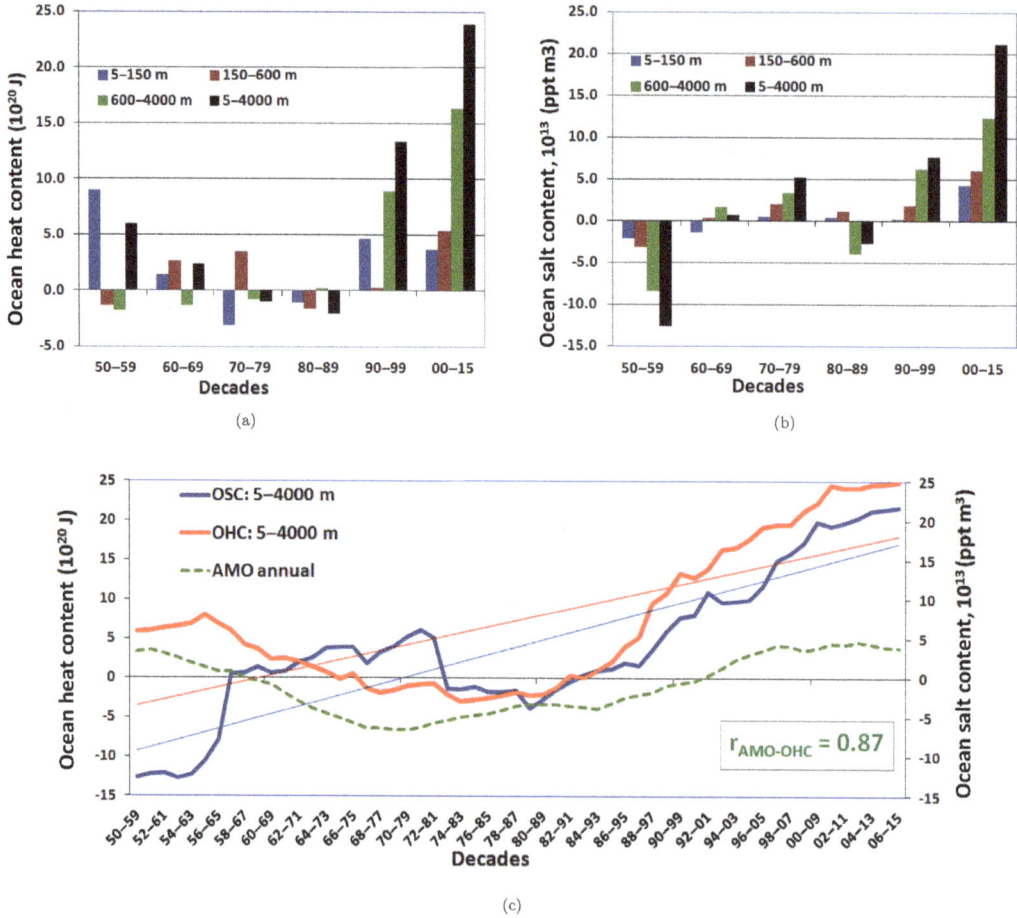

Figure 4. Decadal anomalies of OHC **(a)** and OSC **(b)** in 5–150, 150–600, 600–4000, and 5–4000 m. OHC anomalies are in 10^{20} J and OSC in 10^{13} ppt m^3. **(c)** Volume integrals of OHC (10^{20} J) and OSC (10^{13} ppt m^3) anomalies at 5–4000 m over the whole Mediterranean Sea. Trend values (per decade) are given for OHC (in red) and OSC (in blue). AMO annual values (multiplied by 25 to resemble the OHC shape) are shown with green dots. The correlation between annual AMO (normal and non-multiplied values) and decadal OHC significant at the 95 % confidence level is shown in green.

trend, (b) from 1980 to 1990 with no significant changes, and (c) from 1990 to 2015 with strong OHC/OSC increasing trends (Skliris et al., 2018).

For the study period there is an overall change in heat and salt content of about +18.9 (10^{20} J) and +34.2 (10^{13} ppt m^3) between the last decade 2006–2015 and the first decade 1950–1959.

Finally we computed a 57-year trend for the period 1950–1959 to 2000–2015 based on the decadal T/S anomalies averaged over the four depth layers. As reference, the annual climatology of all years was used. Figure 5 illustrates the statistically significant spatial pattern of the linear trends for temperature (°C decade^{-1}, Fig. 5, left-hand side) and salinity (ppt decade^{-1}, Fig. 5), right-hand side). The trend for the whole water column (Fig. 5g and h) reveals that for the salinity (Fig. 5h) there is a positive trend everywhere. The temperature pattern reveals two main areas of long-term decreasing trends (Fig. 5d), the Aegean and southern Adriatic Sea.

The results show a noisy and patchy spatial patterns of the temperature anomaly trend at the first 600 m (Fig. 5a and c) which are more noisy than the corresponding ones for the salinity (Fig. 5b and d). At the surface (Fig. 5b) the salinity trend is positive almost everywhere in the Mediterranean Sea, while in the intermediate depths (Fig. 5d) we distinguish the strong positive trends at the areas of deep water formation at the eastern Mediterranean, southern Adriatic and Aegean Sea. This strong signal can also be traced out at the Alboran Sea and the southern Algerian basin (Fig. 5d), the outflow path of the LIW towards the Atlantic Ocean, a result that is in agreement with the bibliography (Millot et al., 2006).

Comparing with the spatial salinity linear trends at three layers 0–150, 150–600, and 600–400, presented in Skliris et al. (2018), we observe the following. There are similarities to the patterns of the MEDATLAS 2002 climatology of 1/4° horizontal resolution expected from the refreshing areas at the surface and intermediate layers of the northern Aegean,

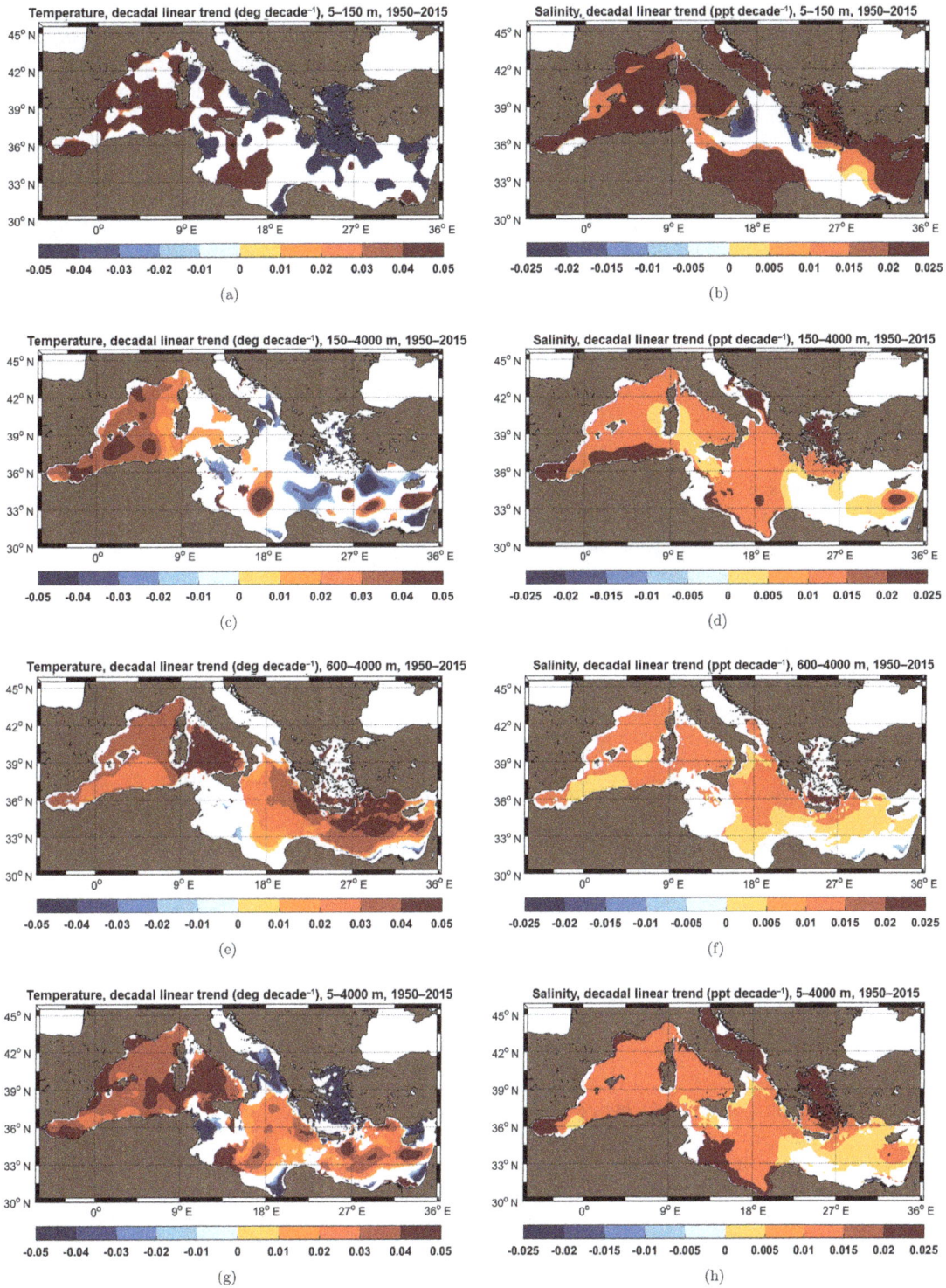

Figure 5. 57-year liner trend of temperature **(a, c, e, g)** and salinity anomalies **(b, d, f, h)** averaged over 5–150, 150–600, 600–4000, and 5–4000 m, in °C decade^{-1} and ppt decade^{-1}, respectively. Regions where the linear trend is not significant at the 95 % confidence level are not plotted.

northern Adriatic, and Gulf of Lions. Also, there are more spatial maskings in the current indices because of the statistical significance of the linear trend. SeaDataNet V2 data collection on which the current product is based has almost double more salinity profiles for the period 1950–2015 than the MEDATLAS collection of the period 1950–2002.

Compared with the patterns of the EN4 Met Office climatology of 1° horizontal resolution, in the current product there are areas with decreasing trends at the central Mediterranean only and more spatial variability at the intermediate layers. EN4 climatology is based to a great extent on the World Ocean Database and the latter for the common period 1950–2015 with the current product has about 14 % more salinity profiles for the common period 1950–2015 with the current product.

4   Code and data availability

The netCDF Operators (NCO) command-line programs and the mathematical and statistical algorithms of the GSL (the GNU Scientific Library) were used for the manipulation and analysis of the netCDF gridded fields of temperature and salinity of the Mediterranean Atlas (functions gsl_fit_linear, gsl_stats_covariance, gsl_stats_sd) NCO toolkit is available here: http://nco.sourceforge.net/ (last access: 9 October 2018). The GSL is available here: http://www.gnu.org/software/gsl (last access: 9 October 2018). The climatic indices are distributed through Zenodo at the following links: https://doi.org/10.5281/zenodo.1408832, https://doi.org/10.5281/zenodo.1408929, https://doi.org/10.5281/zenodo.1408877, https://doi.org/10.5281/zenodo.1408917, https://doi.org/10.5281/zenodo.1411398, https://doi.org/10.5281/zenodo.1408903 (Iona, 2018e, f, g, h, i, j). The DIVA interpolation software tool is distributed through Zenodo (https://zenodo.org/record/836727, Watelet et al., 2017) and GitHub (https://github.com/gher-ulg/DIVA, last access: 9 October 2018).

5   Conclusions

We presented a new product of climatic indices for the Mediterranean Sea oriented to the description and study of the long-term variability and climate change of the area. The assessment of the T/S and OHC/OSC changes is a key priority for monitoring the climate changes in a focal region such as the Mediterranean. So far, the insufficient spatial and temporal coverage of historical in situ data has induced large uncertainties and differences among the used approaches, especially in large basin-scale estimations. Thanks to data repositories such as SeaDataNet, which are improving continuously in terms of abundance, quality and state-of-the-art mapping techniques (implemented by the DIVA software tool), we were able to interpolate in an optimal way and pro-duce high-resolution products. These products can fill data gaps and can be used in a more efficient way by many applications, for the study of the past, present and future climate changes. There is a total increase of 7 % in the number of profiles in the latest SeaDataNet version V2 (2015) used in this study compared to the previous version V1.1 (8 % increase in T and 15 % increase in S profiles). SeaDataNet infrastructure includes data of more than 100 data providers which are quality controlled, archived in data centres and distributed into the infrastructure by the SeaDataNet participants. The data aggregation and validation are performed by regional experts and in close collaboration with the data originators for ensuring the highest quality of the delivered data sets. To avoid duplicates, data not belonging to the SeaDataNet consortium are not included in the repository. There is therefore a significant amount of data such as bathythermographs (more than 100 000 T profiles, mainly navy data) which were used in the World Ocean Atlas 2013 but not in this study. In the next version of this product, these additional sources will be combined with the SeaDataNet data. Future improvements include the use of density climatological fields instead of a constant value at the OHC estimations. Such a density gridded field is not currently available as input since we interpolate T/S separately, but it would be possible to derive it from the T/S gridded fields. Another improvement concerns the correction of the historical bathythermograph data, although previous studies indicated that it does not alter the final results.

Author contributions. AI created the climatic index product, wrote the first version of the manuscript and prepared the figures. JMB, SW, CT, AT and SS reviewed the manuscript. AI, CT and SW formatted the document in LaTeX. CT prepared Fig. 1 and the netCDF files for the time series.

Competing interests. The authors declare that they have no conflict of interest.

Acknowledgements. Data were provided through the SeaDataNet Pan-European infrastructure for ocean and marine data management (http://www.seadatanet.org, last access: 9 October 2018). The DIVA development received funding from the European Union Sixth Framework Programme (FP6/2002–2006) under grant agreement no. 026212, SeaDataNet, the Seventh Framework Programme (FP7/2007–2013) under grant agreement no. 283607, SeaDataNet II, SeaDataCloud and EMODnet (MARE/2008/03 – Lot 3 Chemistry – SI2.531432) from the Directorate-General for Maritime Affairs and Fisheries.

Edited by: Giuseppe M. R. Manzella

References

Adloff, F., Somot, S., Sevault, F., Jordà, G., Aznar, R., Déqué, M., Herrmann, M., Marcos, M., Dubois, C., Padorno, E., Alvarez-Fanjul, E., and Damia, G.: Mediterranean Sea response to climate change in an ensemble of twenty first century scenarios, Clim. Dynam., 45, 2775–2802, https://doi.org/10.1007/s00382-015-2507-3, 2015.

Beckers, J.-M., Barth, A., Troupin, C., and Alvera-Azcárate, A.: Some approximate and efficient methods to assess error fields in spatial gridding with DIVA (Data Interpolating Variational Analysis), J. Atmos. Ocean. Tech., 31, 515–530, https://doi.org/10.1175/JTECH-D-13-00130.1, 2014.

Béthoux, J.-P., Gentili, B., and Tailliez, D.: Warming and freshwater budget change in the Mediterranean since the 1940s, their possible relation to the greenhouse effect, Geophys. Res. Lett., 25, 1023–1026, https://doi.org/10.1029/98gl00724, 1998.

Béethoux, J. P., Gentili, B., Morin, P., Nicolas, E., Pierre, C., and Ruiz-Pino, D.: The Mediterranean Sea: a miniature ocean for climatic and environmental studies and a key for the climatic functioning of the North Atlantic, Prog. Oceanogr., 44, 131–146, https://doi.org/10.1016/s0079-6611(99)00023-3, 1999.

Durack, P. J. and Wijffels, S. E.: Fifty-Year Trends in Global Ocean Salinities and Their Relationship to Broad-Scale Warming, J. Climate, 23, 4342–4362, https://doi.org/10.1175/2010jcli3377.1, 2010.

Durack, P. J., Wijffels, S. E., and Matear, R. J.: Ocean Salinities Reveal Strong Global Water Cycle Intensification During 1950 to 2000, Science, 336, 455–458, https://doi.org/10.1126/science.1212222, 2012.

Fusco, G., Manzella, G. M. R., Cruzado, A., Gacic, M., Gasparini, G. P., Kovacevic, V., Millot, C., Tziavos, C., Velasquez, Z. R., Walne, A., Zervakis, V., and Zodiatis, G.: Variability of mesoscale features in the Mediterranean Sea from XBT data analysis, Ann. Geophys., 21, 21–32, https://doi.org/10.5194/angeo-21-21-2003, 2003.

Füssel, H.-M., Jol, A., Marx, A., and Hildén, M.: Climate change, impacts and vulnerability in Europe 2016, An indicator-based report, Tech. rep., European Environment Agency, https://doi.org/10.2800/534806, 2017.

Giorgi, F.: Climate change hot-spots, Geophys. Res. Let., 33, L08707, https://doi.org/10.1029/2006gl025734, 2006.

Giorgi, F. and Lionello, P.: Climate change projections for the Mediterranean region, Global Planet. Change, 63, 90–104, https://doi.org/10.1016/j.gloplacha.2007.09.005, 2008.

Gouretski, V. and Koltermann, K. P.: How much is the ocean really warming?, Geophys. Res. Lett., 34, L01610, https://doi.org/10.1029/2006gl027834, 2007.

Iona, A.: Mediterannean Sea-Temperature and Salinity Annual Climatology [Data set], Zenodo, https://doi.org/10.5281/zenodo.1146976, 2018a.

Iona, A.: Mediterannean Sea-Temperature and Salinity Annual Climatology for 57 running decades from 1950–1959 to 2006–2015 [Data set], Zenodo, https://doi.org/10.5281/zenodo.1146957, 2018b.

Iona, A.: Mediterannean Sea-Temperature and Salinity Seasonal Climatology [Data set], Zenodo, https://doi.org/10.5281/zenodo.1146953, 2018c.

Iona, A.: Mediterannean Sea-Temperature and Salinity Seasonal Climatology for 57 running decades from 1950–1959 to 2006–2015 [Data set], Zenodo, https://doi.org/10.5281/zenodo.1146938, 2018d.

Iona, A.: Mediterranean Sea Climatic Indices – T/S Anomalies [Data set], Zenodo, https://doi.org/10.5281/zenodo.1408832, 2018e.

Iona, A.: Mediterranean Sea Climatic Indices – T/S Vertical Averages [Data set], Zenodo, https://doi.org/10.5281/zenodo.1408929, 2018f.

Iona, A.: Mediterranean Sea Climatic Indices – Areal density of OHC/OSC [Data set], Zenodo, https://doi.org/10.5281/zenodo.1408877, 2018g.

Iona, A.: Mediterranean Sea Climatic Indices – Linear trends of T/S, OHC/OSC Anomalies [Data set], Zenodo, https://doi.org/10.5281/zenodo.1408917, 2018h.

Iona, A.: Mediterranean Sea Climatic Indices – Time Series of T/S, OHC/OSC Anomalies [Data set], Zenodo, https://doi.org/10.5281/zenodo.1411398, 2018i.

Iona, A.: Mediterranean Sea Climatic Indices – Differences of two 30-years averages of T/S, OHC/OSC Anomalies [Data set], Zenodo, https://doi.org/10.5281/zenodo.1408903, 2018j.

Iona, A., Theodorou, A., Watelet, S., Troupin, C., Beckers, J.-M., and Simoncelli, S.: Mediterranean Sea Hydrographic Atlas: towards optimal data analysis by including time-dependent statistical parameters, Earth Syst. Sci. Data, 10, 1281–1300, https://doi.org/10.5194/essd-10-1281-2018, 2018.

IPCC: Climate Change 2013 – The Physical Science Basis: Working Group I Contribution to the Fifth Assessment Report of the Intergovernmental Panel on Climate Change, chap. Technical Summary, Cambridge University Press, https://doi.org/10.1017/CBO9781107415324, 2014.

Jordà, G., Von Schuckmann, K., Josey, S., Caniaux, G., García-Lafuente, J., Sammartino, S., Özsoy, E., Polcher, J., Notarstefano, G., Poulain, P.-M., and Adloff, F.: The Mediterranean Sea heat and mass budgets: Estimates, uncertainties and perspectives, Prog. Oceanogr., 156, 174–208, https://doi.org/10.1016/j.pocean.2017.07.001, 2017.

Klein, B., Roether, W., Manca, B. B., Bregant, D., Beitzel, V., Kovacevic, V., and Luchetta, A.: The large deep water transient in the Eastern Mediterranean, Deep-Sea Res. Pt. I, 46, 371–414, https://doi.org/10.1016/s0967-0637(98)00075-2, 1999.

Lascaratos, A., Roether, W., Nittis, K., and Klein, B.: Recent changes in deep water formation and spreading in the eastern Mediterranean Sea: a review, Prog. Oceanogr., 44, 5–36, https://doi.org/10.1016/s0079-6611(99)00019-1, 1999.

Levitus, S.: Warming of the World Ocean, Science, 287, 2225–2229, https://doi.org/10.1126/science.287.5461.2225, 2000.

Levitus, S., Antonov, J. I., Boyer, T. P., Locarnini, R. A., Garcia, H. E., and Mishonov, A. V.: Global ocean heat content 1955–2008 in light of recently revealed instrumentation problems, Geophys. Res. Lett., 36, L07608, https://doi.org/10.1029/2008gl037155, 2009.

Levitus, S., Antonov, J. I., Boyer, T. P., Baranova, O. K., Garcia, H. E., Locarnini, R. A., Mishonov, A. V., Reagan, J. R., Seidov, D., Yarosh, E. S., and Zweng, M. M.:

World ocean heat content and thermosteric sea level change (0–2000 m), 1955–2010, Geophys. Res. Lett., 39, L10603, https://doi.org/10.1029/2012gl051106, 2012.

Lozier, M. S., Owens, W. B., and Curry, R. G.: The climatology of the North Atlantic, Prog. Oceanogr., 36, 1–44, https://doi.org/10.1016/0079-6611(95)00013-5, 1995.

Macias, D., Garcia-Gorriz, E., and Stips, A.: Understanding the Causes of Recent Warming of Mediterranean Waters. How Much Could Be Attributed to Climate Change?, PLoS ONE, 8, e81591, https://doi.org/10.1371/journal.pone.0081591, 2013.

Malanotte-Rizzoli, P., Manca, B. B., d'Alcala, M. R., Theocharis, A., Brenner, S., Budillon, G., and Ozsoy, E.: The Eastern Mediterranean in the 80s and in the 90s: the big transition in the intermediate and deep circulations, Dynam. Atmos. Oceans, 29, 365–395, https://doi.org/10.1016/s0377-0265(99)00011-1, 1999.

Manca, B., Burca, M., Giorgetti, A., Coatanoan, C., Garcia, M.-J., and Iona, A.: Physical and biochemical averaged vertical profiles in the Mediterranean regions: an important tool to trace the climatology of water masses and to validate incoming data from operational oceanography, J. Marine Syst., 48, 83–116, https://doi.org/10.1016/j.jmarsys.2003.11.025, 2004.

Manca, B. B.: Evolution of dynamics in the eastern Mediterranean affecting water mass structures and properties in the Ionian and Adriatic Seas, J. Geophys. Res., 108, 8102, https://doi.org/10.1029/2002jc001664, 2003.

Mariotti, A., Zeng, N., Yoon, J.-H., Artale, V., Navarra, A., Alpert, P., and Li, L. Z. X.: Mediterranean water cycle changes: transition to drier 21st century conditions in observations and CMIP3 simulations, Environ. Res. Lett., 3, 044001, https://doi.org/10.1088/1748-9326/3/4/044001, 2008.

MEDAR Group: MEDATLAS/2002 database, Mediterranean and Black Sea database of temperature salinity and bio-chemical parameters, Climatological Atlas, Tech. rep., Ifremer, 4 Cdroms, 2002.

Millot, C., Candela, J., Fuda, J.-L., and Tber, Y.: Large warming and salinification of the Mediterranean outflow due to changes in its composition, Deep-Sea Res. Pt. I, 53, 656–666, https://doi.org/10.1016/j.dsr.2005.12.017, 2006.

Rahmstorf, S.: Influence of mediterranean outflow on climate, Eos, Transactions American Geophysical Union, 79, 281–281, https://doi.org/10.1029/98eo00208, 1998.

Rixen, M., Beckers, J.-M., Levitus, S., Antonov, J., Boyer, T., Maillard, C., Fichaut, M., Balopoulos, E., Iona, S., Dooley, H., Garcia, M.-J., Manca, B., Giorgetti, A., Manzella, G., Mikhailov, N., Pinardi, N., Zavatarelli, M., and the Medar Consortium: The Western Mediterranean Deep Water: a proxy for global climate change, Geophys. Res. Lett., 32, L12608, https://doi.org/10.1029/2005GL022702, 2005.

Robinson, A., Leslie, W., Theocharis, A., and Lascaratos, A.: Mediterranean Sea Circulation, Encyclopedia of Ocean Sciences, 1689–1705, https://doi.org/10.1006/rwos.2001.0376, 2001.

Roether, W., Manca, B. B., Klein, B., Bregant, D., Georgopoulos, D., Beitzel, V., Kovacevic, V., and Luchetta, A.: Recent Changes in Eastern Mediterranean Deep Waters, Science, 271, 333–335, https://doi.org/10.1126/science.271.5247.333, 1996.

Rohling, E. J. and Bryden, H. L.: Man-Induced Salinity and Temperature Increases in Western Mediter-

ranean Deep Water, J. Geophys. Res., 97, 11191–11198, https://doi.org/10.1029/92jc00767, 1992.

Schröder, K., Gasparini, G. P., Tangherlini, M., and Astraldi, M.: Deep and intermediate water in the western Mediterranean under the influence of the Eastern Mediterranean Transient, Geophys. Res. Lett., 33, L21607, https://doi.org/10.1029/2006gl027121, 2006.

Schroeder, K., García-Lafuente, J., Josey, S. A., Artale, V., Nardelli, B. B., Carrillo, A., Gačić, M., Gasparini, G. P., Herrmann, M., Lionello, P., and Ludwig, W.: Circulation of the Mediterranean Sea and its Variability, The Climate of the Mediterranean Region, 187–256, https://doi.org/10.1016/b978-0-12-416042-2.00003-3, 2012.

Schroeder, K., Chiggiato, J., Bryden, H. L., Borghini, M., and Ben Ismail, S.: Abrupt climate shift in the Western Mediterranean Sea, Sci. Rep.-UK, 6, 23009, https://doi.org/10.1038/srep23009, 2016.

Schroeder, K., Chiggiato, J., Josey, S. A., Borghini, M., Aracri, S., and Sparnocchia, S.: Rapid response to climate change in a marginal sea, Sci. Rep.-UK, 7, 4065, https://doi.org/10.1038/s41598-017-04455-5, 2017.

Simoncelli, S., Coatanoan, C., Myroshnychenko, V., Sagen, H., Bäck, O., Scory, S., Grandi, A., Barth, A., and Fichaut, M.: SeaDataNet, First Release of Regional Climatologies. WP10 Third Year Report – DELIVERABLE D10.3, Tech. rep., SeaDataNet, https://doi.org/10.13155/50381, 2015.

Skliris, N., Marsh, R., Josey, S. A., Good, S. A., Liu, C., and Allan, R. P.: Salinity changes in the World Ocean since 1950 in relation to changing surface freshwater fluxes, Clim. Dynam., 43, 709–736, https://doi.org/10.1007/s00382-014-2131-7, 2014.

Skliris, N., Zika, J. D., Nurser, G., Josey, S. A., and Marsh, R.: Global water cycle amplifying at less than the Clausius-Clapeyron rate, Sci. Rep.-UK, 6, 38752, https://doi.org/10.1038/srep38752, 2016.

Skliris, N., Zika, J. D., Herold, L., Josey, S. A., and Marsh, R.: Mediterranean sea water budget long-term trend inferred from salinity observations, Clim. Dynam., 51, 2857–2876, https://doi.org/10.1007/s00382-017-4053-7, 2018.

Somot, S., Sevault, F., Déqué, M., and Crépon, M.: 21st century climate change scenario for the Mediterranean using a coupled atmosphere–ocean regional climate model, Global Planet. Change, 63, 112–126, https://doi.org/10.1016/j.gloplacha.2007.10.003, 2008.

Theocharis, A., Nittis, K., Kontoyiannis, H., Papageorgiou, E., and Balopoulos, E.: Climatic changes in the Aegean Sea influence the eastern Mediterranean thermohaline circulation (1986–1997), Geophys. Res. Lett., 26, 1617–1620, https://doi.org/10.1029/1999gl900320, 1999.

Theocharis, A., Klein, B., Nittis, K., and Roether, W.: Evolution and status of the Eastern Mediterranean Transient (1997–1999), J. Marine Syst., 33-34, 91–116, https://doi.org/10.1016/s0924-7963(02)00054-4, 2002.

Troupin, C., Sirjacobs, D., Rixen, M., Brasseur, P., Brankart, J.-M., Barth, A., Alvera-Azcárate, A., Capet, A., Ouberdous, M., Lenartz, F., Toussaint, M.-E., and Beckers, J.-M.: Generation of analysis and consistent error fields using the Data Interpolating Variational Analysis (Diva), Ocean Model., 52–53, 90–101, https://doi.org/10.1016/j.ocemod.2012.05.002, 2012.

Tsimplis, M. N., Zervakis, V., Josey, S. A., Peneva, E. L., Struglia, M. V., Stanev, E. V., Theocharis, A., Lionello, P., Malanotte-Rizzoli, P., Artale, V., Tragou, E., and Oguz, T.: Chapter 4 Changes in the oceanography of the Mediterranean Sea and their link to climate variability, in: Developments in Earth and Environmental Sciences, edited by: Lionello, P., Malanotte-Rizzoli, P., and Boscolo, R., Elsevier, 227–282, https://doi.org/10.1016/s1571-9197(06)80007-8, 2006.

Vargas-Yáñez, M., Moya, F., Tel, E., García-Martínez, M. C., Guerber, E., and Bourgeon, M.: Warming and salting in the western Mediterranean during the second half of the 20th century: inconsistencies, unknowns and the effect of data processing, Sci. Mar., 73, 7–28, https://doi.org/10.3989/scimar.2009.73n1007, 2008.

Vargas-Yáñez, M., Moya, F., García-Martínez, M., Tel, E., Zunino, P., Plaza, F., Salat, J., Pascual, J., López-Jurado, J., and Serra, M.: Climate change in the Western Mediterranean Sea 1900–2008, J. Marine Syst., 82, 171–176, https://doi.org/10.1016/j.jmarsys.2010.04.013, 2010a.

Vargas-Yáñez, M., Zunino, P., Benali, A., Delpy, M., Pastre, F., Moya, F., García-Martínez, M. d. C., and Tel, E.: How much is the western Mediterranean really warming and salting?, J. Geophys. Res., 115, C04001, https://doi.org/10.1029/2009jc005816, 2010b.

Watelet, S., Troupin, C., Beckers, J.-M., Barth, A., and Ouberdous, M.: gher-ulg/DIVA: v4.7.1 (Version v4.7.1), Zenodo, https://doi.org/10.5281/zenodo.836727, 2017.

WMO: Guide to Climatological Practices, Tech. rep., World Meteorological Organization, Geneva, Switzerland, available at: https://library.wmo.int/pmb_ged/wmo_100_en.pdf (last access: 9 October 2018), ISBN 978-92-63-10100-6, 2011.

Zika, J. D., Skliris, N., Blaker, A. T., Marsh, R., Nurser, A. J. G., and Josey, S. A.: Improved estimates of water cycle change from ocean salinity: the key role of ocean warming, Environ. Res. Lett., 13, 074036, https://doi.org/10.1088/1748-9326/aace42, 2018.

12

Meteorological buoy measurements in the Iceland Sea

Guðrún Nína Petersen

Icelandic Meteorological Office, Bústaðavegi 9, 108 Reykjavík, Iceland

Correspondence to: Guðrún Nína Petersen (gnp@vedur.is)

Abstract. The Icelandic Meteorological Office (IMO) conducted meteorological buoy measurements in the central Iceland Sea in the time period 2007–2009, specifically in the northern Dreki area on the southern segment of the Jan Mayen Ridge. Due to difficulties in deployment and operations, in situ measurements in this region are sparse. Here the buoy, deployment and measurements are described with the aim of giving a future user of the data set information that is as comprehensive as possible.

1 Introduction

The Icelandic Meteorological Office (IMO) conducted measurements in the central Iceland Sea in the time period 2007–2009, specifically in the northern Dreki area on the southern segment of the Jan Mayen Ridge. The deployment was a part of a governmental preparation project for the potential exploration for oil and gas on the Icelandic continental shelf and the purpose of the measurements was to obtain information on the local weather. The measurements were conducted with a meteorological buoy making basic in situ meteorological measurements as well as some oceanographic measurements. After the end of the deployment the government phased out the project and the data remained untouched and unused for several years.

Lately the data have become of interest to the scientific community. One reason is the discovery of the North Icelandic Jet (Jonsson and Valdimarsson, 2004), a deep-reaching ocean current hypothesized to originate in the Iceland Sea (Våge et al., 2013), although the exact source and related water mass transformation processes are not known. The Iceland Sea is a local heat flux minimum and thus the oceanic deep convection is not driven by the average large-scale atmospheric circulation (Moore, 2012). However,

Harden et al. (2015) used the buoy data presented here to examine the surface meteorological condition in the central Iceland Sea and concluded that although on average the heat flux was low, on shorter timescales the Iceland Sea frequently experienced high heat flux events in the wintertime.

The buoy measurements are unique due to the sparsity of in situ observations over the ocean, especially in this region, and Dukhovskoy et al. (2017) used it in an evaluation of ocean surface winds from reanalysis data sets and scatterometer-derived gridded products.

Although measurements from the data set have been used in both Harden et al. (2015) and Dukhovskoy et al. (2017) they have not been made publicly available earlier. However, now the data set has been quality-checked, suspect data removed and made publicly available from the PANGAEA Data Publisher.

The purpose of this article is to give a full description of the buoy, deployment and measurements for any future users of the data set.

In the following section there is a description of the buoy, parameters measured and the deployment. Section 3 contains information on the measurements, the availability of data and, where possible, reasons for periods of missing data. Lastly, final remarks are in Sect. 5.

Table 1. The instrumentation of the buoy: parameters measured, instrument, range and accuracy according to manufacturers.

Parameter	Instrument	Range	Accuracy
Air pressure	Vaisala PTB 220A	500–1100 hPa	± 0.15 hPa
Air temperature	OCEANOR/Omega 300006	-40–$+75\,^\circ$C	$\pm 0.1\,^\circ$C
Relative humidity	Vaisala HMP 45A	0.8–100 %	± 2–3 %
Wind speed	Gill WindSonic	0–60 m s^{-1}	± 2 %
Wind direction	Gill WindSonic	0–359°	± 3 %
Wave height	Fugro OCEANOR Wavesense	0–20 m	± 0.05 m
Wave period	Fugro OCEANOR Wavesense	1–30 s	± 0.15 s
Wave direction	Fugro OCEANOR Wavesense	0–360°	$\pm 1^\circ$
Current speed	Aanderaa DCS 4100R	0–300 cm s^{-1}	± 1 %
Current direction	Aanderaa DCS 4100R	0–360° magnetic	± 5–7.5 %
Water temperature	Aanderaa DCS 4100R	-10–$+43\,^\circ$C	$\pm 0.16\,^\circ$C
Location	OCEANOR Jupiter 12 GPS receiver	n/a	± 5 m
Orientation	Precision Navigation TCM 2.5 electronic compass	n/a	$\pm 1^\circ$

Figure 1. The meteorological buoy on board the Icelandic Marine Research Institute's vessel *Bjarni Sæmundsson* on 22 November 2007, the day before deployment. Photo: Sighvatur K. Pálsson.

2 The buoy and the deployment

The buoy was a SEAWATCH Wavescan wave buoy from Fugro OCEANOR that measured wave, current and meteorological parameters (Fugro OCEANOR, 2017). The hull was discus-shaped with a keel mounted at the bottom to prevent capsizing of the buoy. The meteorological sensors and the antennae were mounted on a mast (see Fig. 1). The hull had a diameter of 2.8 m and total height (mast to keel) of 6.75 m. It had solar panels and sealed lead acid backup batteries. Due to the low sun radiation condition during winter at the measurement site the buoy was also supplied with lithium batteries. Table 1 lists the instrumentation of the buoy and, for each parameter, range and accuracy according to the manufacturers (Fugro OCEANOR AS, 2007).

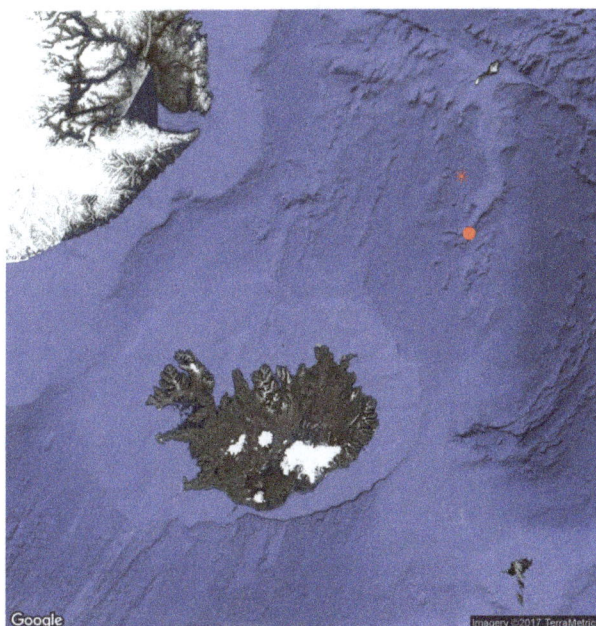

Figure 2. The location of the buoy, November 2007–April 2009. The retrieval location on 21 August 2009 is marked with a star.

The buoy was deployed in the northern Dreki area[1] on the southern segment of the Jan Mayen Ridge, anchored at 68.47° N, 9.27° W from 23 November 2007 to 21 August 2009 (see Fig. 2), drifting inside a circle with a diameter of approximately 2 km. It was serviced once during the deployment, on 7 June 2008. The internal compass and the GPS sensor failed on 17 April 2009 between 08:00 and 09:00 UTC. Thus after that time the exact location of the buoy is not known and there are no measurements of current speed and

---

[1] Dreki is Icelandic for dragon and thus the buoy is known by the Icelandic meteorological and oceanographic community as the dragon buoy.

Table 2. Buoy measurements, units, resolution and short names used in the hourly data set.

Name	Parameter	Units	Resolution	Short name in data set
Location				
Latitude	lat	degrees	0.01	Latitude
Longitude	lon	degrees	0.01	Longitude
Meteorological				
Sea level pressure	P	hPa	0.1	PPPP
Air temperature at 3.5 m height	T	°C	0.001	TTT
Relative humidity at 3.5 m height	RH	%	0.01	RH
Wind direction at 4 m height	Wdir	degrees	0.01	dd
Wind speed at 4 m height	WSP	$\mathrm{m\,s^{-1}}$	0.001	ff
Wind gust at 4 m height	WGS	$\mathrm{m\,s^{-1}}$	0.001	ff gust
Oceanographic				
Water temperature at 1.5 m depth	T_w	°C	0.01	Temp
Current direction at 1.5 m depth	Cdir	degrees	0.1	DIR
Current speed at 1.5 m depth	CSP	$\mathrm{cm\,s^{-1}}$	0.001	V
Height of highest wave	H_s	m	0.001	Wave h max
Period of the highest wave	T_{H_s}	s	0.001	PwPw
Heave parameters computed from spectral analysis				
Significant wave height (Hs), estimate	H_s	m	0.001	Wave h
Significant wave height (Hs), estimate lower frequency band	H_sa	m	0.001	Wave h
Significant wave height (Hs), estimate mid-frequency band	H_sb	m	0.001	Wave h
Mean wave period (Tz), estimate 1	T_{z1}	s	0.001	PwPw
Mean wave period (Tz), estimate 2	T_{z2}	s	0.001	PwPw
Mean wave period (Tz), estimate 2, lower frequency band	T_{z2a}	s	0.01	PwPw
Mean wave period (Tz), estimate 2, mid-frequency band	T_{z2b}	s	0.001	PwPw
Period of spectral peak	T_p	s	0.001	Time
Directional wave parameters computed from spectral analysis				
Mean spectra wave direction	Mdir	degrees	0.01	Wave dir spr
Mean spectra wave direction, lower frequency band	$\mathrm{Mdir_a}$	degrees	0.01	Wave dir spr
Mean spectra wave direction, mid-frequency band	$\mathrm{Mdir_b}$	degrees	0.01	Wave dir spr
Directional wave parameters				
High-frequency mean wave period	T_hhf	s	0.01	Time
Wave spreading at spectral peak period	$\mathrm{SPS}_{T_\mathrm{p}}$	degrees	0.001	Wave dir spr

Table 3. List of periods where measurements are missing or clearly erroneous, parameters affected and reasons.

Time	Parameter	Reason
31 Dec 2007 00:00 UTC–6 Jun 2008 18:00 UTC	Location	Errors in location records
6 Jun 2008 19:00 UTC–7 Jun 2008 12:00 UTC	All data	The buoy was serviced
28 Oct 2008 01:00 UTC–1 Feb 20:09 UTC	P	The pressure measurements were too high*
27 Nov 2008 12:00 UTC–27 Nov 2008 16:00 UTC	Relative humidity	Relative humidity > 100 %
3 Dec 2008 09:00 UTC–11 Dec 2008 10:00 UTC	Relative humidity	Relative humidity > 100 %
12 Dec 2008 18:00 UTC	Location	Errors in location records
2 Feb 2009 08:00 UTC–1 Mar 2009 02:00 UTC	Relative humidity	Relative humidity > 100 %
20 Feb 2009 12:00 UTC–24 Feb 2009 09:00 UTC	All data	Measurements were missing
17 Apr 2009 09:00 UTC–end of deployment	Location and current parameters	GPS sensor failed
17 Apr 2009 09:00 UTC–end of deployment	Water temperature	Temperature sensor failed

* Pressure measurements were compared to ECMWF operational forecasts.

direction. After the failure the buoy broke free. Using the locations of the satellites retrieving information from the buoy it can be seen that this happened in May and the buoy then started drifting northward. It was rescued by the Marine Research Institute on 21 August 2009 at 69.40° N, 9.62° W, having drifted about 104 km to the north from the mooring location; see location in Fig. 2.

Table 4. Monthly average values, as well as maximum and minimum values where appropriate, for most of the meteorological parameters. The dew point, T_d, and the vapour pressure, P_v, are calculated using an equation from WMO (2012). Insufficient data is represented with –.

	$\overline{T}$ °C	$\overline{T_{max}}$ °C	T_{max} °C	$\overline{T_{min}}$ °C	T_{min} °C	$\overline{T_d}$ °C	$\overline{P_v}$ hPa	$\overline{P}$ hPa	P_{max} hPa	P_{min} hPa	$\overline{WSP}$ m s^{-1}	WSP_{max} m s^{-1}	WGS_{max} m s^{-1}
Dec 2007	1.3	2.7	5.1	−0.3	−6.4	−1.0	5.8	999.7	1024.4	966.2	8.7	16.3	25.8
Jan 2008	−0.2	1.1	3.9	−1.3	−7.7	−2.7	5.2	997.3	1019.3	971.1	8.5	16.1	22.9
Feb 2008	−0.6	0.7	3.6	−2.1	−9.5	−3.1	5.1	998.9	1047.4	949.5	9.6	18.3	27.0
Mar 2008	−2.1	−0.8	2.5	−3.4	−7.3	−5.2	4.4	1007.5	1020.4	977.8	8.7	16.4	25.6
Apr 2008	−0.4	0.4	2.4	−1.4	−4.6	−2.4	5.3	1016.7	1035.2	999.8	7.0	14.5	22.1
May 2008	1.6	2.4	4.6	0.8	−3.4	0.3	6.4	1024.2	1039.2	1007.8	5.1	14.0	17.7
Jun 2008	4.3	5.0	6.9	3.5	1.3	2.7	7.5	1016.0	1024.5	1007.4	6.1	11.5	15.5
Jul 2008	6.9	7.6	10.7	6.1	2.7	6.2	9.6	1012.9	1028.5	987.9	5.7	11.2	16.4
Aug 2008	8.7	9.2	10.9	8.0	5.7	6.8	10.0	1009.2	1028.9	984.0	5.9	14.6	20.8
Sep 2008	7.6	8.4	10.7	6.7	2.0	5.5	9.3	1007.6	1021.6	975.3	7.6	16.3	27.7
Oct 2008	2.9	4.3	7.1	1.4	−4.8	0.3	6.6	992.3	1014.1	940.6	9.0	17.8	29.8
Nov 2008	−0.3	1.4	6.1	−2.1	−8.0	−3.2	–	–	–	–	8.8	16.9	25.8
Dec 2008	0.7	1.8	4.0	−0.6	−6.1	−1.7	–	–	–	–	8.2	16.1	25.1
Jan 2009	0.0	1.3	3.6	−1.6	−8.0	−1.9	–	–	–	–	7.6	17.2	24.4
Feb 2009	−2.0	−0.6	2.6	−3.3	−8.5	−2.3	5.3	1026.6	1033.0	994.7	7.9	17.0	23.4
Mar 2009	−0.5	0.8	2.5	−1.8	−5.7	−2.1	5.3	1004.5	1019.9	980.5	8.5	19.3	25.3
Apr 2009	0.7	1.7	3.4	−0.4	−5.2	−0.9	5.9	1010.2	1027.9	993.3	7.4	14.7	21.9
May 2009	2.9	3.6	4.8	1.9	−1.3	2.0	7.1	1010.1	1034.1	983.4	6.6	14.1	20.5
Jun 2009	3.7	4.5	8.4	2.9	0.0	1.6	7.1	1019.7	1030.6	994.9	4.3	14.8	18.7
Jul 2009	7.1	7.9	9.6	6.2	3.6	5.8	9.3	1015.3	1025.2	996.2	5.3	11.6	16.0
Aug 2009*	8.6	9.2	10.0	7.8	6.1	7.3	10.4	1010.4	1021.3	978.0	5.7	14.1	25.6

* Only 21 days as the buoy was retrieved on 21 August at 20:00 UTC.

Table 5. Monthly average values for a few of the oceanographic parameters: water temperature, current speed, mean wave period, height of highest wave, significant wave height and period of spectral peak. The parameter $\overline{\Delta(T_w - T)}$ is the monthly averaged difference between the water temperature and the air temperature. Insufficient data is represented with −.

	T_{Water} °C	$\overline{\Delta(T_w - T)}$ °C	$\overline{CSP}$ cm s^{-1}	$\overline{T_Z}$ s	$\overline{H_{max}}$ m	$\overline{H_s}$ m	$\overline{T_p}$ s
Dec 2007	3.0	1.7	14.7	8.3	5.7	3.9	11.0
Jan 2008	2.2	2.3	14.5	8.0	5.2	3.5	10.8
Feb 2008	1.4	2.0	17.2	8.1	5.9	4.0	10.7
Mar 2008	1.2	3.3	14.7	7.4	4.9	3.3	9.6
Apr 2008	0.8	1.2	11.7	6.7	3.3	2.2	9.0
May 2008	2.2	0.5	8.8	6.1	2.2	1.5	8.2
Jun 2008	5.1	0.8	11.2	5.9	2.3	1.6	7.8
Jul 2008	7.4	0.5	12.0	5.6	2.1	1.5	7.3
Aug 2008	9.5	0.8	13.2	5.8	2.3	1.6	7.6
Sep 2008	8.5	0.9	14.6	6.7	3.8	2.6	8.8
Oct 2008	5.2	2.3	15.8	7.9	5.6	3.9	10.3
Nov 2008	3.0	3.3	15.1	7.6	5.2	3.5	10.3
Dec 2008	2.1	1.4	13.1	7.8	5.0	3.4	10.5
Jan 2009	1.6	1.6	11.2	8.2	5.1	3.5	10.8
Feb 2009	1.3	3.3	10.7	7.8	4.6	3.1	10.5
Mar 2009	1.1	1.6	11.4	7.3	4.5	3.1	9.7
Apr 2009	1.7	1.3	11.1	6.8	3.7	2.5	8.9
May 2009	–	–	–	6.4	3.0	2.0	8.4
Jun 2009	–	–	–	5.6	1.7	1.2	7.1
Jul 2009	–	–	–	5.7	2.0	1.3	7.6
Aug 2009*	–	–	–	6.2	2.5	1.7	8.2

* Only 21 days as the buoy was retrieved on 21 August at 20:00 UTC.

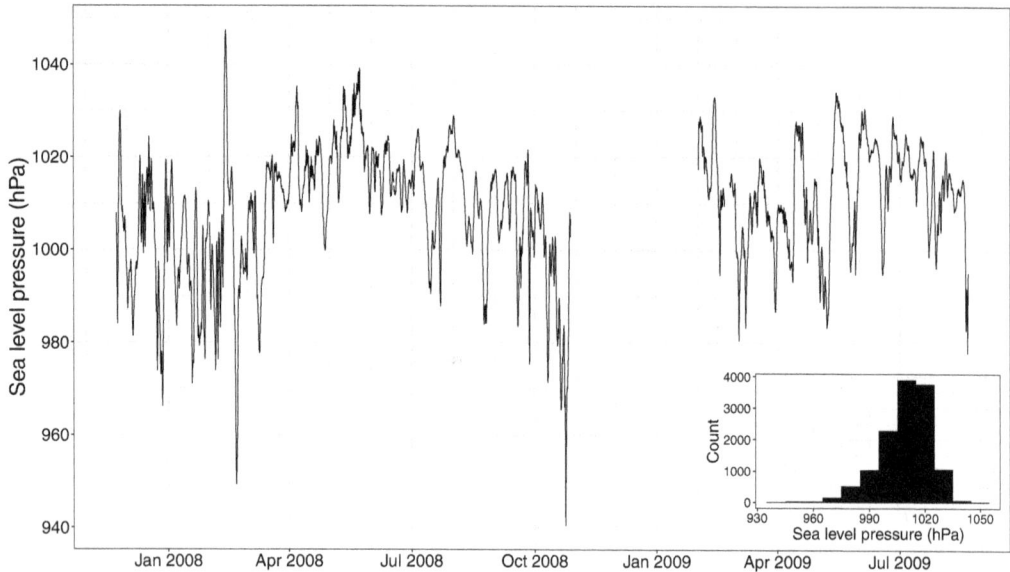

Figure 3. Measurements of mean sea level pressure (hPa). The gap from 28 October 2008 to 1 February 2009 is due to measurement errors. A histogram of measurements is inset.

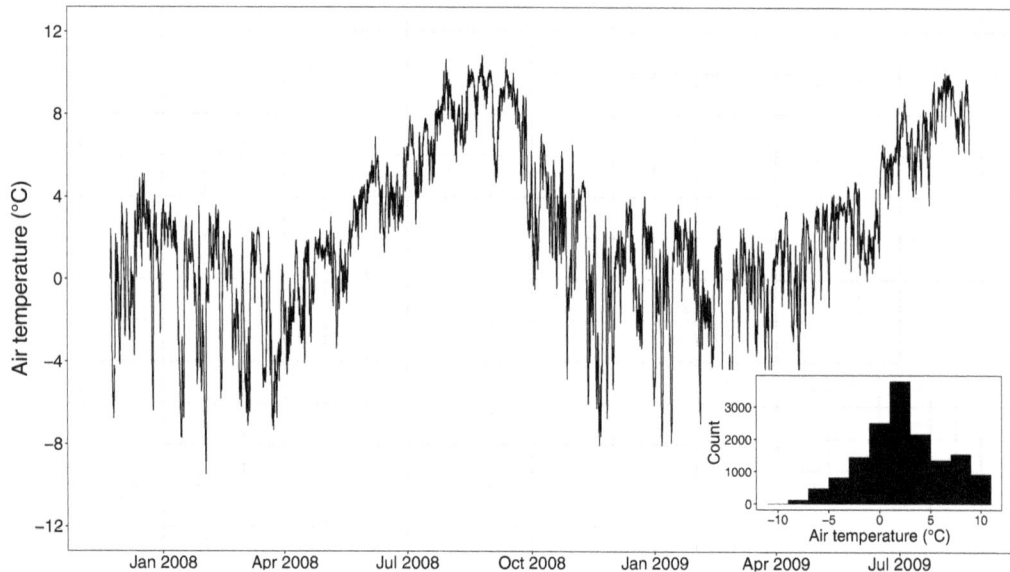

Figure 4. Measurements of air temperature (°C) at 3.5 m height. A histogram of measurements is inset.

3   The measurements

A list of measurements conducted, units, resolution and short names in the data set can be found in Table 2. The data were recorded every hour. The measurement time period was 23 November 2007 03:00 UTC–21 August 2009 20:00 UTC and Table 3 contains a list of times when some or all of the measurements were missing and the reason when known. In addition other obvious errors were removed, such as spurious longitude = 0°.

Table 4 contains averages for most of the meteorological parameters measured, and maxima and minima where appropriate, while Table 5 contains averages for the main oceanic parameters. The daily values and average, maximum and minimum values are calculated and then the monthly averages of the daily values. In addition to measured values the dew point, T_d, the vapour pressure, P_v, and the difference between the water temperature and the air temperature are calculated and the mean values included in the table.

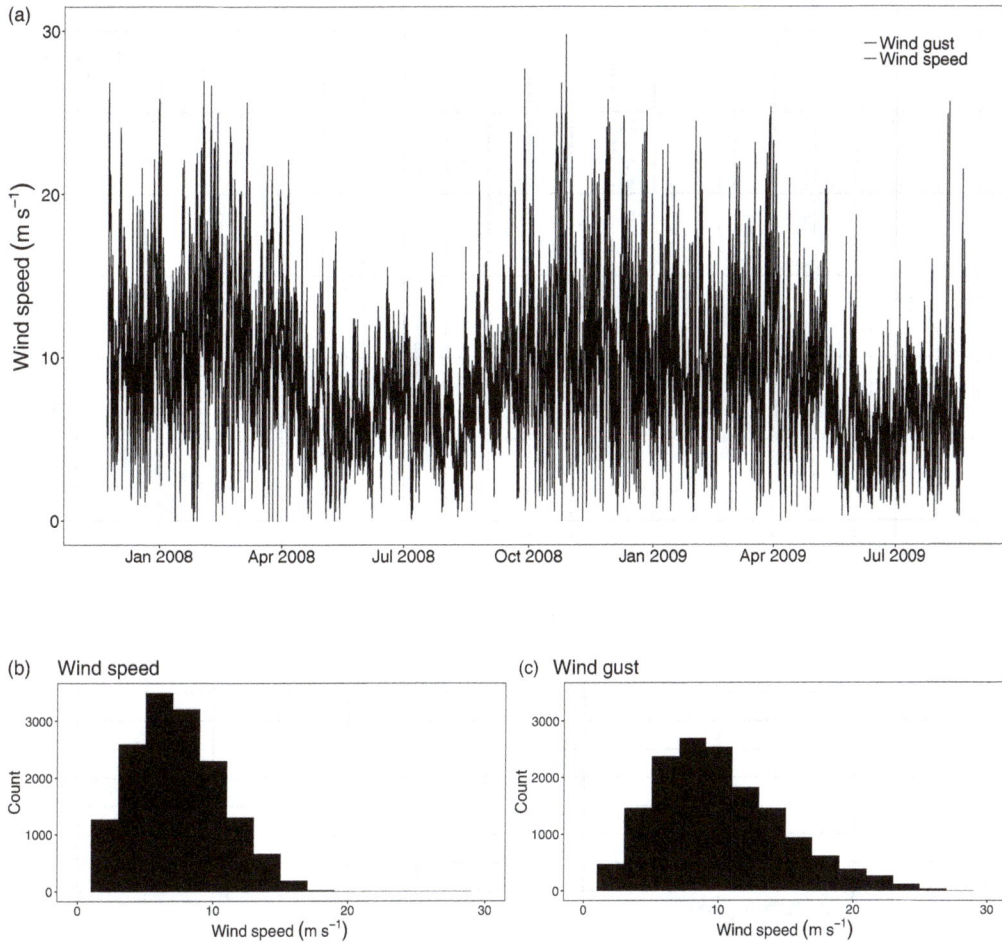

Figure 5. Measurements of wind speed and wind gust $(\mathrm{m\,s^{-1}})$ at 4 m height: time series (**a**) and histograms of wind speed measurements (**b**) and of wind gust measurements (**c**).

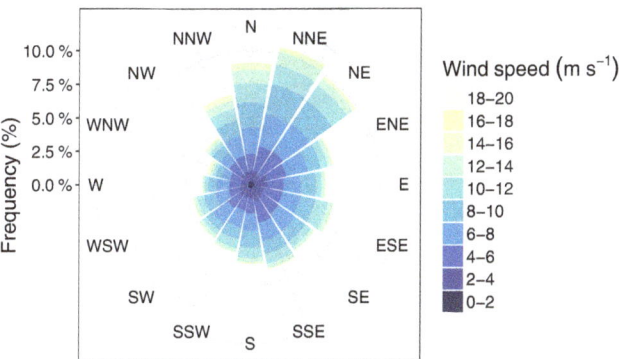

Figure 6. A wind rose, showing the frequency (%) and wind speed $(\mathrm{m\,s^{-1}})$ for different wind directions at 4 m height (22.5° bins).

3.1 Air pressure

Air pressure was measured with a Vaisala PTB220A digital barometer designed to operate over a wide pressure and temperature range (Fugro OCEANOR AS, 2007). A comparison of the pressure data to ECMWF operational analysis confirmed the suspicion that the measurements were off for months after the passing of the deepest low, on 24 October 2008, possibly due to water in the sensor inlet. The error was not just a bias error; in addition, during the first half the pressure was too high, while during the second half the magnitudes of extreme were too small and the extrema were not all accounted for. From 1 February 2009 the measurements were in agreement with ECMWF analysis (not shown). To avoid any confusion the pressure data from 24 October 2008 to 1 February 2009 are removed from the data set. The buoy measurements of mean sea level pressure are shown in Fig. 3, with the exception of the erroneous data. During the deployment, the average pressure was 1009 hPa, varying from a minimum of 941 hPa to a maximum of 1047 hPa. Note that the maximum was measured on 13 February 2008 and a local minimum of 950 hPa a week later, on 21 February 2008,

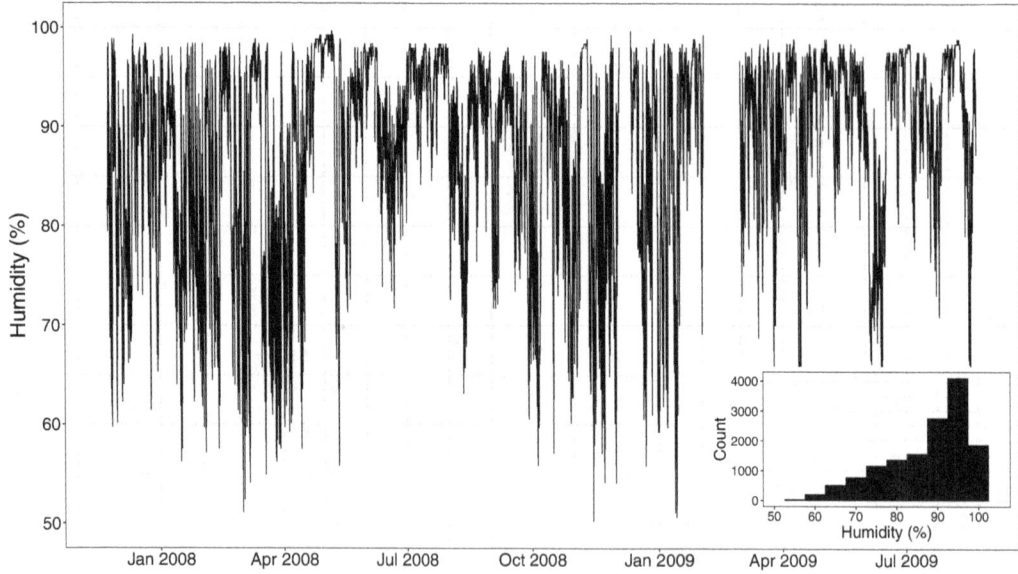

Figure 7. Measurements of relative humidity (%) at 3.5 m height. A histogram of measurements is inset.

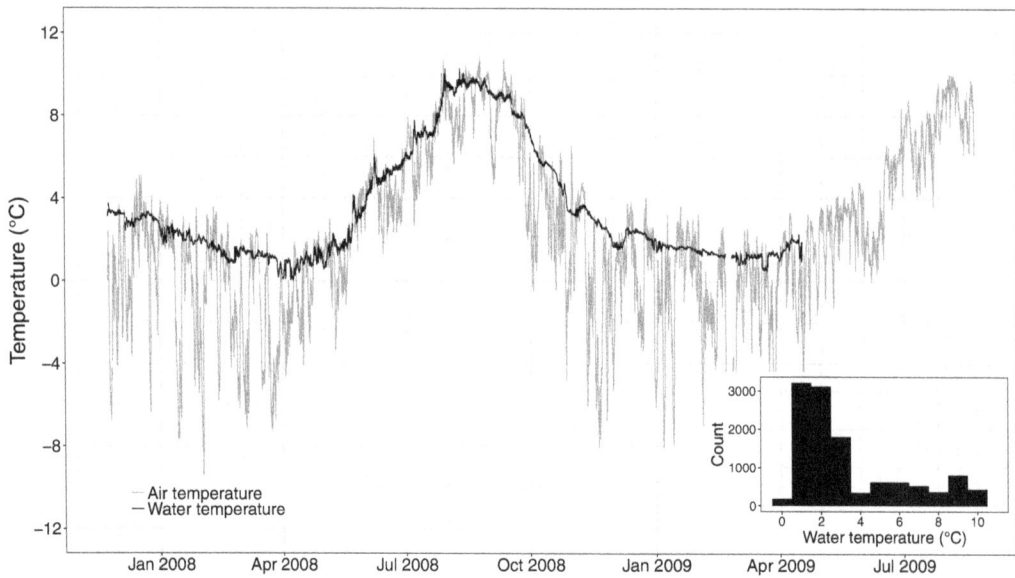

Figure 8. Measurements of water temperature (°C) at 1.5 m depth. Air temperature is shown in the background. A histogram of water temperature measurements is inset.

emphasizing the variability of the synoptic situation in the region.

3.2 Air temperature

Air temperature was measured at 3.5 m height with a OCEANOR/Omega temperature sensor, a robust and compact sensor. The thermistor element was protected by a sun radiation screen (Fugro OCEANOR AS, 2007). The lowest temperature measurement was −9.5 °C and the highest 10.9 °C, a span of about 20 °C (see Fig. 4). During late autumn, winter and spring the temperature variations were much greater than during the summer and early autumn. The temperature variations during the cool seasons are related to the variation in weather regimes, from northerly cold air outbreaks to warm air advection by synoptic cyclones moving into the area from the south (Harden et al., 2015). The highest temperatures were measured in late summer, mainly in August and September 2008, while temperatures below −5 °C were measured most frequently in February, followed closely by January and March.

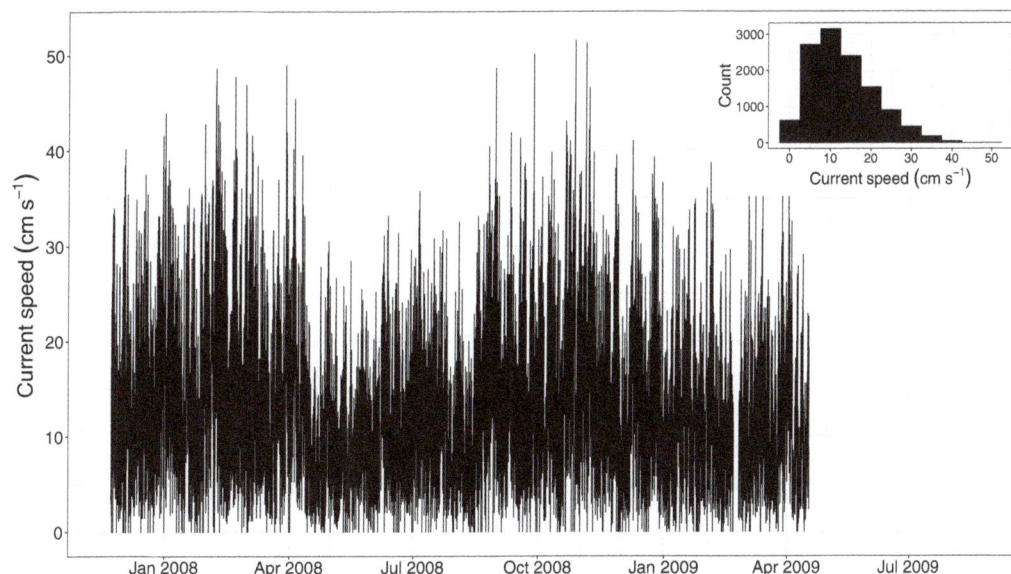

Figure 9. Measurements of current speed (cm s^{-1}) at 1.5 m depth. A histogram of measurements is inset.

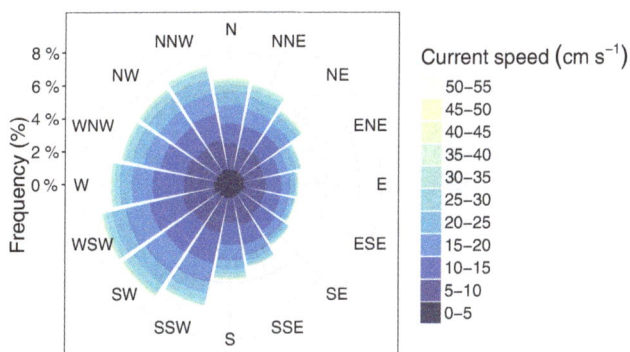

Figure 10. A frequency rose, showing the frequency (%) and current speed (cm s^{-1}) for different current directions (22.5° bins) at 4 m height. Note that current direction is defined as the direction the current is streaming toward, which is opposite to the convention of wind direction.

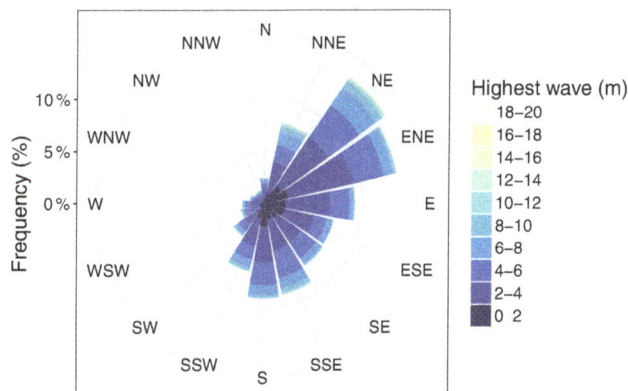

Figure 11. A frequency rose, showing the frequency (%) and height of highest wave (m) for different wave directions (22.5° bins), with wave directions defined in the same way as wind directions.

3.3 Wind speed and wind direction

The wind speed, wind gust and wind direction were measured at 4 m height with a Gill WindSonic wind sensor (Fugro OCEANOR AS, 2007). It has been shown that during rough seas, due to sheltering effects and elevation changes, wind measurement by buoys can be negatively biased (e.g. Large et al., 1995; Zeng and Brown, 1998). Here, no attempt is made to compensate for a potential bias in the data set; that is left to the user.

The maximum measured wind was 19.3 m s^{-1} and the maximum gust 29.8 m s^{-1}. The gust factor was in general below 1.5. There was a significant lower wind speed during the summer months than the winter months (see Fig. 5). The

monthly mean wind speed was the highest in February 2008 and lowest in June 2009 (see Table 4).

The wind rose in Fig. 6 shows that the most common wind directions were northerly to northeasterly, approximately 30 % of the time, and the least common wind directions were westerly to northwesterly. This is in accordance with wind directional frequency in Iceland (not shown).

3.4 Relative humidity

The relative humidity was measured at 3.5 m height with a Vaisala HMP 45A relative humidity sensor based on the capacitive thin film polymer HUMICAP 180 (Fugro OCEANOR AS, 2007). As the measurements were made over the sea the relative humidity was in general high with

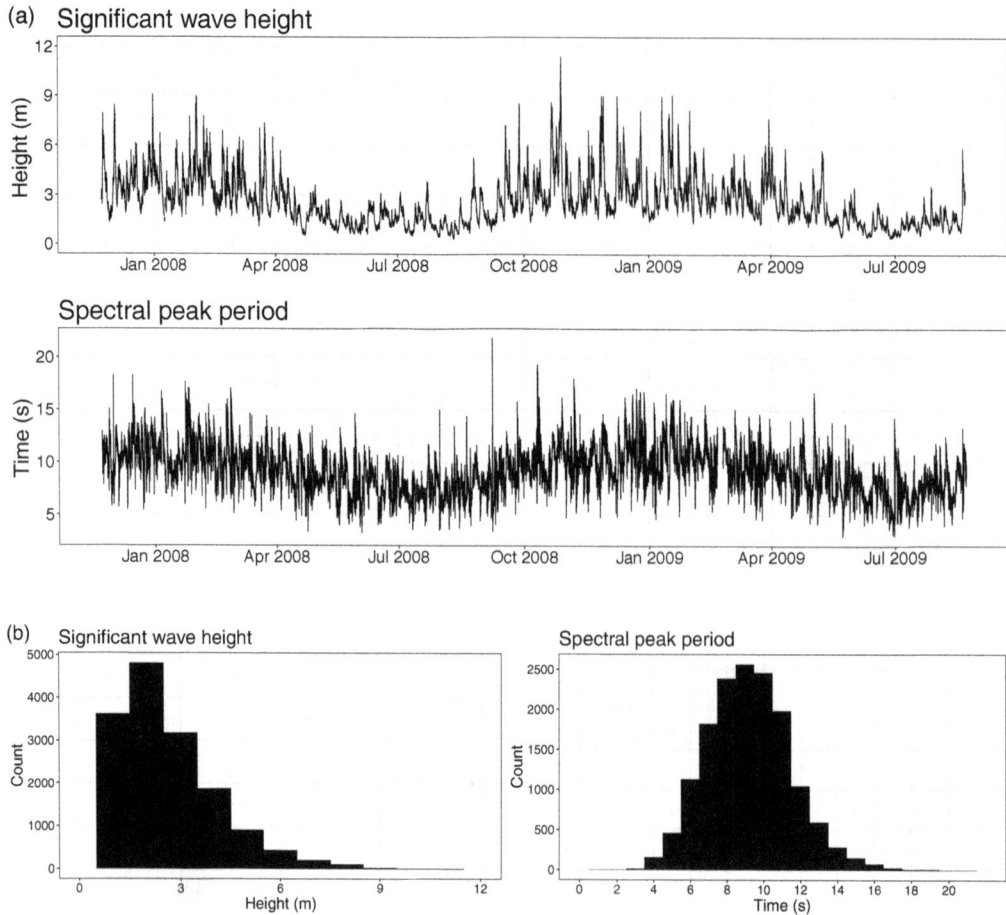

Figure 12. Measurements of significant wave height (m) and the spectral peak period (s). **(a)** Time series and **(b)** histograms.

the lowest measurement at 50 % (see Fig. 7). In all cases of relative humidity below 55 % the air temperature was below freezing and in most cases winds were light. The mean relative humidity was 87 %.

3.5   Water temperature

Water temperature was measured at 1.5 m depth with the Aanderaa DCS 4100R current speed and water temperature sensor. The temperature was measured using a temperature-dependent crystal-oscillator circuit (Fugro OCEANOR AS, 2007). After 17 April 2009 all measurements from the current sensor were invalid. The water temperature measurements were thus made over 16 consecutive months. The lowest temperature was measured in spring and the highest from the end of July until mid-September. The rise in temperature during late spring and early summer was slower than the fall in the autumn (see Fig. 8). The lowest and highest monthly means were in April and August 2008. The air temperature is also shown in the figure. The amplitude of the water temperature variations is much less than that of the air temperature. The differences between the water temperature and the air

temperature, $T_w - T$, were calculated, and on a monthly basis the difference varied between 0.5 and 3.3 °C, with the least difference during summer months but large variation during the winter months, related to different air mass impacting the area.

3.6   Current speed and direction

Current speed and direction were measured at 1.5 m depth with the Aanderaa DCS 4100R current speed and water temperature sensor. The sensor used the Doppler shift principle and together with orientation from the internal compass determined current speed and direction (Fugro OCEANOR AS, 2007). As mentioned earlier, after the compass and the GPS sensor broke down on 17 April 2009 all current measurements are missing. As stated in Sect. 2 the buoy was anchored at 68.47° N, 9.27° W and drifted inside a circle with a diameter of approximately 2 km. Here, no adjustments are made to the measured ocean currents, but a user of the ocean data may want to consider doing corrections. The mean current speed was 13 cm s^{-1} and the maximum 52 cm s^{-1}. The monthly mean current speed varied from 8.8 cm s^{-1} in May 2008 to

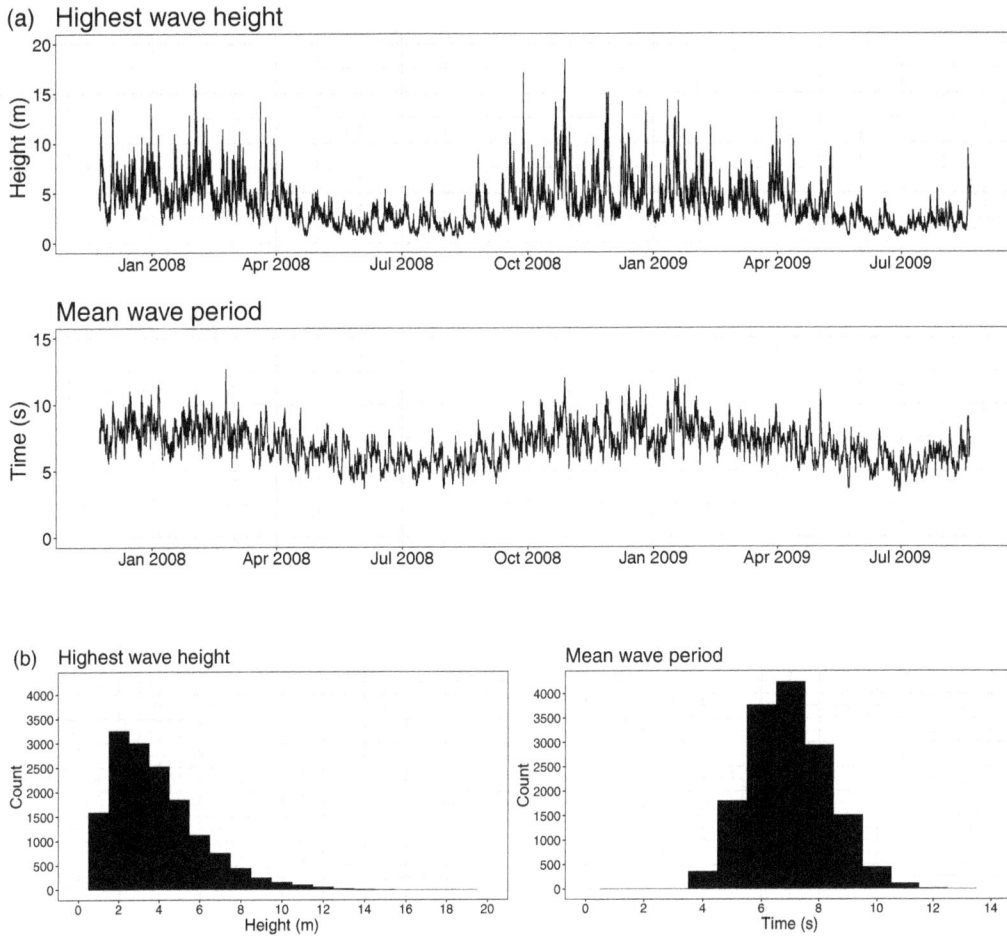

Figure 13. Measurements of the highest wave height (m) and the mean wave period (s). **(a)** Time series and **(b)** histograms.

17.2 cm s^{-1} in February 2008. By convention, current directions are defined opposite to wind directions; northerly current moves toward north while northerly wind is coming from north. The most common current directions were southwesterly (toward southwest) and easterly directions (toward east) least common (see Fig. 10).

3.7   Wave parameters

The wave parameters are measured with Fugro OCEANOR Wavesense, a robust integrated wave sensor and data logger. Accelerometers, rate gyros and magnetometers were mounted orthogonally to provide the basic data. These data were then used as input to algorithms, which calculated heave, roll, pitch, surge, sway and compass time series. Significant wave height, wave direction, wave period and a number of other statistical parameters were then found from these time series (Fugro OCEANOR AS, 2007). Figure 11 shows the frequency of wave directions. By convention, wave directions are defined in the same way as wind directions. The most common directions were northeasterly to east-

northeasterly, a slight clockwise rotation from the most common wind directions (see Fig. 6). The least common wave direction was northwesterly. Figure 12 shows the significant wave height (the mean height of the highest one-third of the waves) and the spectral peak period (the wave period with the highest energy) and Fig. 13 the highest wave height and the mean wave period (the mean of all wave periods). Both wave height and wave periods have an annual variation with minimum during summer and maximum during winter, as well as more variability during winter. The monthly mean significant wave height varied from 1.2 to 3.9 m and the monthly mean maximum wave height from 1.7 to 5.9 m.

4   Data availability

The processed data set and a selection of daily and monthly mean parameters are available from the PANGAEA Data Publisher (https://doi.org/10.1594/PANGAEA.876206).

5 Conclusions

The Icelandic Meteorological Office deployed a meteorological buoy in the northern Dreki area on the southern segment of the Jan Mayen Ridge for 21 months, from 23 November 2007 to 21 August 2009. This is a region of the North Atlantic with few in situ measurements, and thus the data set is unique. The data set has been quality-checked and is now publicly available. This short paper presents the data set, which parameters were measured and at which height, as well as data gaps. The figures in this paper are meant to give a potential user of the data set a quick view of the data. It is the hope of the author that the measurements can be of use for scientists studying the meteorology and oceanography of the northern North Atlantic as well.

Competing interests. The author declares that they have no conflict of interest.

Acknowledgements. The author would like to thank Sigvaldi Árnason for information on the deployment of the buoy. The guidance of Gísli Viggóson and Ingunn Erna Jónsdóttir regarding the oceanographic parameters is greatly appreciated.

Edited by: David Carlson

References

Dukhovskoy, D. S., Bourassa, M. A., Petersen, G. N., and Steffen, J.: Comparison of the ocean surface vector winds from atmospheric reanalysis and scatterometer-based wind products over the Nordic Seas and the northern North Atlantic and their application for ocean modeling, J. Geophys. Res.-Oceans, 122, 1943–1973, https://doi.org/10.1002/2016JC012453, 2017.

Fugro OCEANOR AS: User Manual Wavescan Buoy, Fugro OCEANOR AS, Trondheim, Norway, 2007.

Fugro OCEANOR: SEAWATCH Wavescan Buoy, available at: https://www.fugro.com/about-fugro/our-expertise/technology/seawatch-metocean-buoys-and-sensors, last access: 28 August 2017.

Harden, B. E., Renfrew, I. A., and Petersen, G. N.: Meteorological buoy observations from the central Iceland Sea, J. Geophys. Res.-Atmos., 120, 3199–3208, 2014JD022584, https://doi.org/10.1002/2014JD022584, 2015.

Jonsson, S. and Valdimarsson, H.: A new path for the Denmark Strait overflow water from the Iceland Sea to Denmark Strait, Geophys. Res. Lett., 31, L03305, https://doi.org/10.1029/2003GL019214, 2004.

Large, W., Morzel, J., and Crawford, G. B.: Comparison of three weather prediction models with buoy and aircraft measurements under cyclone conditions in Fram Strait, J. Phys. Oceanogr., 25, 2959–2971, https://doi.org/10.1175/1520-0485(1995)025<2959:AFSWDO>2.0.CO;2, 1995.

Moore, G. W. K.: A new look at Greenland flow distortion and its impact on barrier flow, tip jets and coastal oceanography, Geophys. Res. Lett., 39, L22806, https://doi.org/10.1029/2012GL054017, 2012.

Våge, K., Pickart, R. S., Spall, M. A., Moore, G. W. K., Valdimarsson, H., Torres, D. J., Erofeeva, S. Y., and Nilsen, J. E. Ø.: Revised circulation scheme north of the Denmark Strait, Deep-Sea Res. Pt. I, 79, 20–39, https://doi.org/10.1016/j.dsr.2013.05.007, 2013.

WMO: Guide to Meteorological Instruments and Methods of Observation, World Meteorological Organization, WMO-No 8, annex 4.B, 2012.

Zeng, L. and Brown, R. A.: Scatterometer observations at high wind speeds, J. Appl. Meteorol., 37, 1412–1420, https://doi.org/10.1175/1520-0450(1998)037<1412:SOAHWS>2.0.CO;2, 1998.

13

From pole to pole: 33 years of physical oceanography onboard R/V *Polarstern*

Amelie Driemel[1], Eberhard Fahrbach[1,†], Gerd Rohardt[1], Agnieszka Beszczynska-Möller[2],
Antje Boetius[1], Gereon Budéus[1], Boris Cisewski[4], Ralph Engbrodt[3], Steffen Gauger[3], Walter Geibert[1],
Patrizia Geprägs[5], Dieter Gerdes[1], Rainer Gersonde[1], Arnold L. Gordon[6], Hannes Grobe[1],
Hartmut H. Hellmer[1], Enrique Isla[7], Stanley S. Jacobs[6], Markus Janout[1], Wilfried Jokat[1],
Michael Klages[8], Gerhard Kuhn[1], Jens Meincke[9,*], Sven Ober[10], Svein Østerhus[11], Ray G. Peterson[12,†],
Benjamin Rabe[1], Bert Rudels[13], Ursula Schauer[1], Michael Schröder[1], Stefanie Schumacher[1],
Rainer Sieger[1], Jüri Sildam[14], Thomas Soltwedel[1], Elena Stangeew[3], Manfred Stein[4,*], Volker H Strass[1],
Jörn Thiede[1,*], Sandra Tippenhauer[1], Cornelis Veth[10,*], Wilken-Jon von Appen[1], Marie-France Weirig[3],
Andreas Wisotzki[1], Dieter A. Wolf-Gladrow[1], and Torsten Kanzow[1]

[1] Alfred-Wegener-Institut Helmholtz-Zentrum für Polar- und Meeresforschung, Bremerhaven, Germany
[2] Institute of Oceanography Polish Academy of Science, Sopot, Poland
[3] independent researcher
[4] Thünen-Institut: Seefischerei, Hamburg, Germany
[5] MARUM – Zentrum für Marine Umweltwissenschaften, Bremen, Germany
[6] Lamont-Doherty Earth Observatory, Columbia University, New York, NY, USA
[7] Institute of Marine Sciences-CSIC, Barcelona, Spain
[8] University of Gothenburg, Department of Marine Sciences, Gothenburg, Sweden
[9] Institut für Meereskunde, Hamburg, Germany
[10] Royal Netherlands Institute for Sea Research, 't Horntje, the Netherlands
[11] Uni Research Climate, Bergen, Norway
[12] Scripps Institution of Oceanography, UC San Diego, USA
[13] University of Helsinki, Helsinki, Finland
[14] Centre for Maritime Research and Experimentation, La Spezia, Italy
[*] retired
[†] deceased

Correspondence to: Gerd Rohardt (Gerd.Rohardt@awi.de)

Abstract. Measuring temperature and salinity profiles in the world's oceans is crucial to understanding ocean dynamics and its influence on the heat budget, the water cycle, the marine environment and on our climate. Since 1983 the German research vessel and icebreaker *Polarstern* has been the platform of numerous CTD (conductivity, temperature, depth instrument) deployments in the Arctic and the Antarctic. We report on a unique data collection spanning 33 years of polar CTD data. In total 131 data sets (1 data set per cruise leg) containing data from 10 063 CTD casts are now freely available at doi:10.1594/PANGAEA.860066. During this long period five CTD types with different characteristics and accuracies have been used. Therefore the instruments and processing procedures (sensor calibration, data validation, etc.) are described in detail. This compilation is special not only with regard to the quantity but also the quality of the data – the latter indicated for each data set using defined quality codes. The complete data collection includes a number of repeated sections for which the quality code can be used to investigate and evaluate long-term changes. Beginning with 2010, the salinity measurements presented here are of the highest quality possible in this field owing to the introduction of the OPTIMARE Precision Salinometer.

1   Introduction

Our oceans are always in motion – huge water masses are circulated not only by winds but also by global seawater density gradients. These gradients result from differences in water temperatures and salinities and the water movement transports heat, oxygen, CO_2, and nutrients among latitudes (Stewart, 2009). Measuring the ocean's temperature and salinity is therefore essential not only to understand the ecology of the world oceans but also the influence of the oceans on our climate.

According to Stewart (2009) the first water samples from depths down to around 1600 m were taken in the tropical Atlantic aboard the *Earl of Hallifax* in 1750/1751 with a special bucket and a thermometer (Hales, 1751). Even then, the results (a stable cold water layer beneath the warm surface) hinted at an inflow of deep water from the polar regions (Stewart, 2009). Until the 1970s, measurements of ocean temperatures and salinities were conducted primarily using reversing mercury thermometers and Nansen water bottles (Warren, 2008). Due to the usually limited number of Nansen bottles and thermometers on board, the number of depth levels which could be sampled was also limited, which resulted in a rather coarse vertical resolution of temperature and salinity. With the development of submersible electrical instruments for temperature and salinity (conductivity) measurements in the 1950s, high-resolution measurements of temperature and salinity profiles became possible (Brown, 1991; Stewart, 2009). During the 1970s and 80s, the use of CTDs (conductivity, temperature, depth instruments) replaced the formerly used method almost completely. Numerous manufacturers produced a variety of sensors and instruments. For example, in 1974 Neil Brown formed Neil Brown Instrument Systems, Inc. and manufactured the Mark III CTD[1] (Brown, 1991).

R/V *Polarstern* is a research icebreaker operated by the Alfred Wegener Institute (AWI) in Bremerhaven (Germany), which has operated since 1982 in Antarctica (austral summer) and the Arctic (northern summer)[2]. The first CTD used on *Polarstern* was the aforementioned Neil Brown, Mark IIIB CTD. It was deployed for the first time on cruise leg ANT-II/3, during the ship's second trip to Antarctica in November/December 1983. This giant step for AWI oceanographers, which was supervised by Gerd Rohardt (Rohardt, 2010a), ended abruptly when that same probe accidentally "flew" overboard a month later during ANT-II/4. Despite many efforts to regain it, the probe was lost, which is why the respective data set of this leg (Rohardt, 2010b) contains CTD as well as Nansen-bottle-derived data. The latter

---

[1] Later it was manufactured by EG&G Ocean Instruments and after that by General Oceanics.

[2] See also the description in Driemel et al. (2016) about *Polarstern* history and cruise characteristics.

was only possible due to the fact that guest researcher Manfred Stein (Institut für Seefischerei, Hamburg) had brought Nansen bottles and reversing thermometers as a backup for his ME-OTS-CTD on board during leg ANT-II/2 (no data). This anecdote clearly demonstrates that 1983 was still a transition period for hydrographic observations to electronic devices.

Despite this rather unfortunate start, a Neil Brown, Mark IIIB CTD was successfully used on *Polarstern* until 1996. Starting in 1992, a Sea-Bird SBE 911plus CTD was in use by Kees Veth (Royal Netherlands Institute for Sea Research, data included here). A year later, Gereon Budéus was the first AWI researcher to use the SBE 911plus on *Polarstern*, testing the behavior of the probe in cold conditions. The instrument has been used routinely on *Polarstern* since then, in parallel with Neil Brown equipment. On four cruise legs (1995–1999) *Polarstern* was equipped with the direct successor of the Mark IIIB, called the ICTD and manufactured by Falmouth Scientific. Additionally, during two legs (1986–1987), guest researchers deployed a ME-OTS-CTD. The SEA-BIRD SBE 911plus is probably the most widely used CTD type currently, and has been the only type used on *Polarstern* since 1999 (see also Fig. 1).

In the following, we describe a data compilation of 33 years (1983–2016) of CTD measurements from R/V *Polarstern*. In Sect. 2 we provide details on the CTD types used, the parameters measured, and on data processing. A focus is set on the improvement of the salinity measurements over time and the reasons thereof. In Sect. 3 we describe the data sets in respect to composition, extent, access, and quality.

2   Methods

A CTD directly measures conductivity, temperature, and pressure of water during its down- and up-cast, resulting in a profile from the water surface to the bottom and back. Derived variables are salinity, density, and water depth. The CTDs onboard *Polarstern* were typically deployed in combination with a water sampler construction, holding 12, 24, or 36 bottles (named rosette or carousel, depending on the manufacturer; see Fig. 2). The CTD is mounted inside the frame of the water sampler in a way that the sensors measure the undisturbed water during the down-cast. The down-cast CTD profile is displayed on board in realtime to allow the CTD operator to choose the water layers from which water samples for subsequent chemical and biological analyses are to be taken during the up-cast.

Due to the mounting technique, the measurements taken during the up-cast are not from undisturbed water but are influenced by water parcels from deeper layers which are dragged upwards by the CTD/rosette. Therefore, mostly only

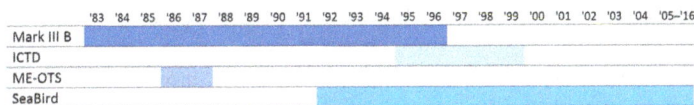

Figure 1. Overview of the period of deployment of different CTD types onboard *Polarstern*, with first line denoting the years. Sea-Bird CTD sondes were here combined into a single bar.

Figure 2. Picture of a typical CTD/rosette system used by the Alfred Wegener Institute (picture by Gerd Rohardt).

the down-cast CTD profile is used and archived (for details, see Sect. 2.5).

2.1 Instruments and specifications

Five different CTD types have been used onboard *Polarstern* from 1983 until the present. As the instruments have changed, so have the range, accuracy, stability, resolution, and response of the sensors. Table 1 shows in detail the manufacturers' specifications of the instruments, and the periods of use are shown in Fig. 1 (Sea-Bird probes combined). The table also indicates the accuracy limits officially adopted for the World Ocean Circulation Experiment (WOCE). Using the OPTIMARE Precision Salinometer (OPS) has provided accuracies even better than those required by WOCE (see Sect. 2.6). However, we would like to stress here that regular servicing and calibration is required to keep the instrument at least within the accuracy given by the manufacturer.

2.2 Laboratory calibration of instruments

In order to obtain precise hydrographic data, frequent calibrations of the sensors and careful inspection and preparation of the instruments (CTD, water sampler and bottles) is necessary. From 1983 until 1986, Neil Brown Mark IIIB CTDs were calibrated by the manufacturer. Each sensor had its own electronic board with the calibration stored on it. Changing a sensor thus required installing the corresponding electronic board as well. When Ray Weiss from Scripps Institution of Oceanography (SIO) participated on *Polarstern* cruise ANT-V/3 in 1986, he suggested including the AWI-CTDs into the SIO calibration process. Since that time, the AWI-CTDs have been calibrated by SIO before and after each campaign. The first calibration revealed that the AWI Mark IIIB showed the same behavior as the SIO Mark IIIB: (a) the pressure sensor showed strong hysteresis depending on the maximum pressure and (b) the temperature readout showed a step-like discontinuity near $0\,°C$ which further depended on the direction of the temperature change, i.e., whether the temperature increased or decreased (R. Williams, SIO, personal communication, 1986). Because a temperature correction of such a behavior is fairly complicated, a few years later SIO modified the electronic boards, shifting the discontinuity from about 0 to $+3\,°C$.

The Falmouth Triton ICTDs from AWI were also shipped to SIO for calibration. This continued support made the change from Mark IIIB to ICTD much easier, and underlined the advantages of the new instrument: the SIO calibration confirmed that the pressure showed negligible levels of hysteresis and that the temperature correction was only small, with no stepwise behavior from -2 to $30\,°C$.

The long lasting collaboration between AWI and the calibration laboratory of SIO ended after completely switching over to Sea-Bird SBE911plus because Sea-Bird Electronics themselves performed high-level calibration of their instruments. In general, ever since the SBE 911plus was introduced, the CTD operators' job on board became much easier. The SBE 911plus featured dual sensors (two for both temperature and conductivity) and software, which displayed the sensor differences. This allowed identifying and changing sensors which became faulty. Replacing faulty sensors early prevents losing valuable data. With the introduction of dual sensors and the use of special software, in situ calibrations were still executed (see Sect. 2.4), but the number of samples could be reduced.

Table 1. Sensor types and the manufacturers' specifications of CTDs used on board *Polarstern*.

Instrument and manufacturer	Period of use	Specifications	Pressure	Temperature	Conductivity
		WOCE accuracy limits	± 3 dbar	$\pm 0.001\,°\mathrm{C}$	$\pm 0.003\,\mathrm{mS\,cm^{-1}}$
Multisonde* ME-OTS-CTD Meerestechnik Elektronik, Trappenkamp	1986 to 1987	Sensor: Range: Accuracy: Stability: Resolution: Response:	Strain gauge bridge 0 to 6000 dbar 0.35 % f.s. – 0.2 dbar –	Platinum resistance -2 to $35\,°\mathrm{C}$ $\pm 0.005\,°\mathrm{C}$ $\pm 0.001\,°\mathrm{C\,month^{-1}}$ $0.001\,°\mathrm{C}$ $60\,\mathrm{ms}$	Symmetric electrode cell 5 to $55\,\mathrm{mS\,cm^{-1}}$ $\pm 0.005\,\mathrm{mS\,cm^{-1}}$ $0.002\,\mathrm{mS\,cm^{-1}\,month^{-1}}$ $0.001\,\mathrm{mS\,cm^{-1}}$ –
Mark IIIB Neil Brown Instruments later: EG&G Marine Instruments/ General Oceanics	1983 to 1996	Sensor: Range: Accuracy: Stability: Resolution: Response:	Strain gauge bridge 0 to 6500 dbar ± 6.5 dbar 0.1 % month^{-1} 0.1 dbar –	Platinum Thermistor -3 to $32\,°\mathrm{C}$ $\pm 0.005\,°\mathrm{C}$ $0.001\,°\mathrm{C\,month^{-1}}$ $0.0005\,°\mathrm{C}$ –	Four-electrode cell 1 to $65\,\mathrm{mS\,cm^{-1}}$ $\pm 0.005\,\mathrm{mS\,cm^{-1}}$ $0.003\,\mathrm{mS\,cm^{-1}\,month^{-1}}$ $0.001\,\mathrm{mS\,cm^{-1}}$ –
Triton ICTD Falmouth Scientific Product line continued by Teledyne RD Instruments	1995 to 1999	Sensor: Range: Accuracy: Stability: Resolution: Response:	Precision-machined Si 0 to 7000 dbar ± 0.01 % f.s. ± 0.002 % f.s. month^{-1} 0.0004 % f.s. 25 ms	Platinum Thermistor -2 to $35\,°\mathrm{C}$ $\pm 0.002\,°\mathrm{C}$ $\pm 0.0002\,°\mathrm{C\,month^{-1}}$ $0.00005\,°\mathrm{C}$ $150\,\mathrm{ms}$	Inductive cell 1 to $70\,\mathrm{mS\,cm^{-1}}$ $\pm 0.002\,\mathrm{mS\,cm^{-1}}$ $\pm 0.0005\,\mathrm{mS\,cm^{-1}\,month^{-1}}$ $0.0001\,\mathrm{mS\,cm^{-1}}$ $5\,\mathrm{cm}$ at $1\,\mathrm{m\,s^{-1}}$
SBE911plus Sea-Bird Electronics	1992 to present	Sensor: Range: Accuracy: Stability: Resolution: Response:	Paroscientific Digiquartz 0 to 6800 dbar ± 0.015 % f.s. ± 0.0015 % f.s. month^{-1} 0.001 % f.s. 15 ms	Thermistor -5 to $35\,°\mathrm{C}$ $\pm 0.001\,°\mathrm{C}$ $\pm 0.0002\,°\mathrm{C\,month^{-1}}$ $0.0002\,°\mathrm{C}$ $65\,\mathrm{ms}$	Three-electrode cell 1 to $70\,\mathrm{mS\,cm^{-1}}$ $\pm 0.003\,\mathrm{mS\,cm^{-1}}$ $\pm 0.003\,\mathrm{mS\,cm^{-1}\,month^{-1}}$ $0.00001\,\mathrm{mS\,cm^{-1}}$ $65\,\mathrm{ms}$
SBE19 self-recording Sea-Bird Electronics	1997 to 2003	Sensor: Range: Accuracy: Stability: Resolution: Response:	Strain gauge 0 to 10 000 psi 0.15 % f.s. – 0.015 % f.s. –	Thermistor -5 to $35\,°\mathrm{C}$ $\pm 0.01\,°\mathrm{C}$ – $0.001\,°\mathrm{C}$ –	Three-electrode cell 0 to $70\,\mathrm{mS\,cm^{-1}}$ $\pm 0.01\,\mathrm{mS\,cm^{-1}}$ – $0.001\,\mathrm{mS\,cm^{-1}}$ –

* operated by guest institutes; f.s.: full scale; –: no data.

2.3 Water samplers

With the exception of the self-contained probe SBE19, all CTDs were used in combination with a water sampler. The Neil Brown Mark IIIB was combined with a General Oceanics (GO) rosette. The GO rosette required taking numerous samples for checking conductivity measurements and also using reversing thermometers to verify that bottles were closed at the desired depth. The reason is that GO used a non robust mechanical release to close the water samplers. Often the mechanics failed, which resulted in the closure of two or more samplers during one release command. This problem was solved with the introduction of the ICTD because Falmouth Scientific (FSI) supplied a new release module which confirmed successful or non-successful release commands.

Later the complete GO hardware was replaced by a release unit from FSI, which used a release system similar to the one used in the SBE32 carousel water sampler, confirming the release command and thus making water sampling more reliable. This positive development (1992 onwards) affected the in situ calibration, rendering the usage of reversing thermometers obsolete.

2.4 In situ calibration

Laboratory calibration of instruments (see Sect. 2.2) is crucial to maintain the sensors and obtain comparable results. It is not sufficient, however to anticipate how a sensor behaves at sea under tough environmental conditions, especially during deep casts. Also sensor drift is not necessarily a continuous process. For this purpose in situ calibrations are essential.

Temperature: the Mark IIIB CTD was equipped with one temperature (and one conductivity) sensor only. Therefore reversing thermometers attached to the bottles of the water sampler had to be used to verify the quality of the temperature data. The Triton ICTD was equipped with a redundant temperature sensor which allowed for much better control of temperature data than the reversing thermometers. Lastly, the SBE911plus features double sensors, both for temperature and conductivity measurements, allowing the plotting of the difference between both sensors versus depth, which eases identification of individual sensor problems and pressure effects. Additionally, a SBE35 Deep Ocean Standards Thermometer was attached to the water sampler, recording the temperature every time a water sample was taken. However, the comparison of the CTD and SBE3plus temperature values to the SBE35 temperature values is only possible if the water temperature is relatively stable, i.e., if the values do not vary much.

Conductivity: for the in situ calibration of the conductivity sensor (Mark IIIB and Triton ICTD) or the conductivity double sensors (SBE911plus), water samples were taken and measured on board with the laboratory salinometer Guildline Autosal 8400a/b and, from 2010, with the OPS (see Sect. 2.6). The samples were taken from deep (> 3000 m) and shallow depths (ca. 500–1000 m) regularly during the CTD deployments in order to reveal pressure effects of the conductivity sensor and its temporal shift.

2.5 Data processing

The data processing procedures were substantially dependent on the development of the CTD and the computer generation. In 1983, CTD data were recorded on nine-track magnetic tape. The station data (location, water depth, date, and time) were noted on a sheet of paper. An HP 9825B computer was used to visualize the temperature and salinity profile on a connected plotter. The data processing was performed at the institute. Due to the fact that, for safety reasons, the magnetic tapes always came back to Bremerhaven with *Polarstern*, the data processing often only started several months after the end of the cruise leg. Later (around 1986), EG&G – who took over the production of the Mark IIIB in 1984 – transferred the FORTRAN code of the data acquisition and processing routines of the Woods Hole Oceanographic Institution (WHOI) (Millard and Yang, 1993) for use on PCs. A similar software package was also provided for the ICTD from Falmouth Scientific. This made the data acquisition and visualization as well as the transfer of raw data to AWI much easier. The substructure of the software for applying the SIO calibration came from R. Williams and F. Delahoyde (personal communication, 1990).

Sea-Bird Electronics provided the data acquisition software SEASAVE and developed a package especially for their pumped CTD SBE911plus, SBE DataProcessing. This software became the primary tool for CTD data processing at the AWI. Also, the raw data were routinely stored on the onboard computer and transferred to the AWI in an automatic workflow.

The data processing workflow can be divided into four parts, as explained in the following subsections:

2.5.1 Data cropping and handling

Data recording started before the actual profile began (starting point at the lowering of the CTD/rosette to the water surface). Thus, one of the first tasks was the truncation of the unused beginning (the depth of the first "used" data point depends on the wave height). Converting the raw file into readable engineering units was the next step as well as the separation between the down- and up-cast, if both had been saved in one file. Afterwards, the station information was added to the data file. In the past this information was manually edited from handwritten station protocols. With the inauguration of the DSHIP electronic station book (http://www.werum.de/en/platforms/DSHIP.jsp) station details were directly merged with the CTD data.

2.5.2 Correction of measurement errors

Physical properties of the sensors and environmental influences on them, as well as disturbances of the data transmission between sensors and recording units on deck, can create measurement errors. These were reduced using suitable software in the following ways:

- Spikes: spikes in the pressure measurements resulting, for example, from winch cable or slip ring problems, were removed. The procedure is called "par" in Sea-Bird's SBEDataProcessing software package.

- Response time/time lag correction: salinity was computed from conductivity, temperature, and pressure. The response time of a temperature sensor, however, is higher than the response time of the pressure and conductivity sensors. If left uncorrected, this would result in salinity spikes in layers with strong gradients. A precise correction for this time lag would require a constant lowering speed of the CTD, which is not possible on a moving ship. Sea-Bird solved this problem by pumping water with a constant speed through the temperature and conductivity sensors. For Mark IIIB and ICTD the time lag was adjusted/corrected by minimizing the salinity spikes and evaluated visually based on profile plots.

- Pressure hysteresis: Mark IIIB strain gauge pressure sensors did not respond linearly to increasing pressure and additionally exhibited a lagged response during decreasing pressure. This behavior also depended on the maximum pressure. A laboratory calibration (see

Sect. 2.2) revealed this behavior and provided the coefficients for the software to apply the correction. However, the software was rather tricky because it only used hysteresis correction for the maximum pressure (6500 dbar) to calculate the correction for all profile depths (R. Williams and F. Delahoyde, personal communication, 1990). A second calibration up to 1500 dbar was recorded to verify the algorithm of the software. ICTDs did not show this behavior and only a minor offset had to be applied. A Digiquatz® pressure sensor from Paroscientific was used in the SBE911plus. This sensor was stable, operating without hysteresis, so no frequent calibration was necessary.

- Compression and thermal effect: the ICTD with its inductive conductivity sensor had a known pressure dependency (compression of the cell ceramics), which was corrected by SBE software. In addition a thermal mass correction[3] was applied for the ICTD and the SBE911plus conductivity cell.

2.5.3 Creation of a uniform profile

- Monotonic increasing pressure: as a ship is always pitching and rolling, the constant lowering speed of the winch is superimposed by the ships motion. Rejecting all records with pressure reversals is thus one of the standard procedures in CTD data processing, and was also applied on *Polarstern* data.

- Averaging: the SBE911plus CTD sampled with a frequency of 24 Hz. A typical lowering speed of $0.8\,\mathrm{m\,s^{-1}}$ resulted in a vertical resolution of around 3 cm. This sample rate was needed to apply the time lag correction reliably and also to guarantee that, although lots of records were rejected, a monotonic increasing pressure record could be created. In the end, the profile was smoothed by averaging on 1 dbar levels (i.e., P, T and C were averaged between ≥ 1.5 and < 2.5; between ≥ 2.5 and < 3.5 dbar, and so on). As this will not necessarily result in an averaged pressure record for $2.0, 3.0, \ldots$ dbar (more probable in $1.97, 3.05, \ldots$ dbar), a linear interpolation was applied for temperature and conductivity, so that the values could be centered on exactly $2.0, 3.0, \ldots$ dbar. Only after this procedure was the salinity calculated.

2.5.4 Final correction and validation

- Drift, stability, and pressure dependency: the physical characteristics of sensors change continuously through

time. This behavior becomes visible as a slight change of their sensitivity. The order of this change is given by the manufacturer ("stability"; see Table 1). But the stability depends on the environmental conditions as well. For example, by conducting many deep casts, an additional sensor drift could be induced due to an a priori unknown pressure dependency. Also marine growth inside the conductivity cell will change the drift. Additionally, *Polarstern* CTDs were deployed even in rough weather conditions meaning that the instruments could bump against the ship's hull or experienced hard impacts on deck. These events could result in a visible step-like change. The station log sheets (which essentially contain descriptions of special occurrences), the pre-, post-, and in situ calibration helped to reconstruct the history of a sensor during a cruise and to identify which $T\text{--}C$ sensor pair should be used. General plotting software can be used to visualize the in situ calibrations versus pressure or versus time to investigate the dependency (drift), and to then apply and verify the corrections.

- Validation: all profiles were imported into Ocean Data View (Schlitzer, 2015) which provides various plots (profiles, scatter, and sections) for a visual inspection. When a suspicious profile was found, the processing steps mentioned above were repeated from the necessary level onwards. Additionally, these profiles were compared to profiles from previous cruises. The working database included a number of regularly repeated transects, which allowed consistency checks and quality confirmation.

2.6 OPTIMARE precision salinometer

Since 1985, laboratory measurements of salinity have been conducted on water samples taken with a rosette/carousel multi-bottle sampler to cross-validate the in situ CTD measurements. These laboratory salinity measurements were taken with salinometers. Salinometer measurements have several advantages compared to in situ measurements. For one, the salinometer measurements are controlled directly with the primary standard International Association for the Physical Sciences of the Oceans (IAPSO) Standard Seawater, which means that the salinometer is closer to the primary standard. Furthermore, the SBE911plus salinity sensor (SBE4) is calibrated using a bath of nearly constant salinity and varying temperatures, leading to different conductivities. The salinometer, however, is calibrated by using different salt concentrations, which makes the salinometer measurements more accurate for salinities varying around the typical open-ocean value of 35 PSU.

A Guildline Autosal 8400a/b salinometer was in use until 2010. Since then, it has been replaced by a new laboratory salinometer, the OPS developed by AWI scientists and en-

---

[3] A cell which is lowered from a warm into a cold layer needs some time to reach the same temperature as the water. That means that heat from the cell is transferred into the water and the water becomes slightly warmer resulting in higher conductivity.

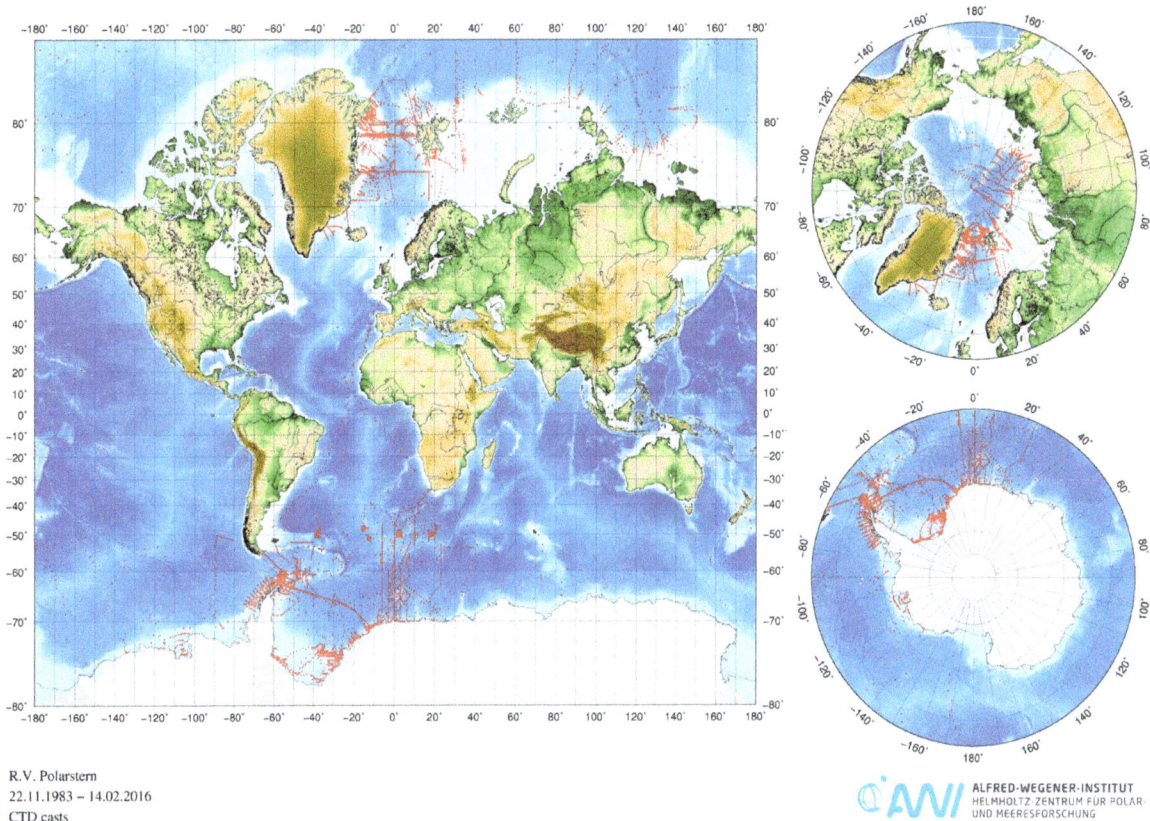

Figure 3. Map showing all sites where CTD data were collected with *Polarstern* from 1983 to 2016.

gineers and manufactured by OPTIMARE[4] (Budéus, 2011, 2015). The highly accurate OPS lab measurements have been in use since June 2010 to cross-calibrate in situ salinity data measured by the CTD. As a result, beginning with campaign ANT-XXV/1 (2010/06) the accuracy of the salinity measurements improved tremendously and the resulting data sets are of the highest quality possible for these kinds of measurements.

3 Resulting data sets

In total 131 data sets (1 data set per cruise leg) containing data from 10 063 CTD casts have been produced on *Polarstern* in the course of 33 years (22 November 1983 to 14 February 2016, Fig. 3) and are archived in the PANGAEA (Data Publisher for Earth and Environmental Science, www.pangaea.de) database. The data sets can be accessed at http://doi.pangaea.de/10.1594/PANGAEA.860066 (Rohardt et al., 2016). This link leads to the central page which contains all meta-information of the respective cruise legs (name of leg, start/end, area, link to cruise report), the number of CTD casts, the CTD type used, the overall quality

of the data, the link to a map displaying all CTD stations, and the link to the data set of the specific leg.

When clicking on the link to a data set of one cruise leg (see, e.g., Rohardt, 2010c, doi:10.1594/PANGAEA.733664) the data set page contains metadata, a Google map of all sample sites, and on the bottom the actual data. On the top of the page the citation of the data set is given, followed by the citation of the respective cruise report (if available). The CTD type used is indicated in the "Method" column of the Parameter(s) overview table of the page. The data table opens by clicking on "View dataset as html". Here, the position, date/time (at maximum depth) of sampling, and the water depth precede the actual data. The "Elevation" is the bathymetric depth relative to sea level and is therefore negative. It can be used, for example, to extract information on how close to the seafloor the CTD measurements ended (comparing water depth of the last measurement with the elevation). The "Number of observations" is the number of measurements included in one averaging step (see Sect. 2.5.3). With programs like Ocean Data View (Schlitzer, 2015) and Pan2Applic (Sieger and Grobe, 2005) the data can be visualized easily (for more information, see https://wiki.pangaea.de/wiki/ODV). With respect to CTD type, the 131 data sets are composed of 27 legs with Mark IIIB CTD data, 4 legs with ICTD data, 2 legs with ME-OTS data, 5 legs with data

---

[4]Optimare Sensorsysteme GmbH & Co. KG

Table 2. Quality code details for *Polarstern* CTD data sets in PANGAEA.

Quality code	Description	Comment	Possible use (example)
A	Highest accuracy and quality possible	SBE911plus with double sensors; pre- and post-calibration applied, salinity samples measured during the cruise	Investigate long-term changes of temperature and salinity
B	Within WOCE accuracy and quality limits	SBE911plus without double sensors, Mark IIIB or ICTD; pre- and post-calibration applied, salinity samples measured during the cruise	Investigate long-term changes of temperature
C	Accuracy and quality of the data is rather low or unknown	Without pre- and post-calibration, no salinity samples, or no detailed documentation of data processing	Hydrography for the specific cruise only

from a Sea-Bird self-recording CTD, and 93 legs with the SBE911plus. Most of the data sets (1992 onwards) contain additional measurements of oxygen concentration, light transmission/attenuation, and/or chlorophyll fluorescence.

3.1 Several remarks on the best use of *Polarstern* CTD data

– If available, the respective cruise report is linked to the data set. It contains valuable information on the cruise itinerary, the scientific purpose, and on the quality of the CTD data or the calibration applied.

– We defined a column on the overall quality of the data of each leg in Rohardt et al. (2016) called "Quality code". Here we use flags "A", "B", and "C" to classify the data with A being high quality data (see Table 2 for details).

– In general, the number of decimals in the data sets is at least $n + 1$, with n being the last significant decimal. This was done deliberately, as we experienced that for calculations (in models), the actual (unrounded) number of the last significant decimal can be essential.

– You can search for specific parameters, regions, etc., in the data sets described here using the www.pangaea. de search engine and adding "PSctd". You can then define a geographic bounding box in the map (right side of search page) to search for specific regions (e.g., Arctic data) and press "apply". Or you can try "PSctd + parameter:oxygen" to get all data sets with oxygen measurements. We also added an overview in .xls format of these additional measurements at http://doi.pangaea.de/10.1594/PANGAEA.860066 (under "Further details") which contains information (where available) on whether or not these measurements were calibrated, and during which campaign which additional measurements were taken.

– To download several or all data sets at once, you can either use the Data Warehouse integrated into PANGAEA, or you can use a program especially designed for this purpose called PanGet. Data Warehouse: log in to PANGAEA (or create an account) at www. pangaea.de, then search for "PSctd" (or, for example, "PSctd + oxygen"). On the top right corner you can click on Data Warehouse (above the Google map). Here you can choose which parameters to download, followed by clicking on "Start Data Warehouse Query". Please be aware that downloading all files requires over 1.5 GB which might take some time to download. How to download files with PanGet is described at https: //wiki.pangaea.de/wiki/PanGet.

– A CTD file downloaded from PANGAEA can easily be imported into Ocean Data View (Schlitzer, 2015) using Pan2Applic (Sieger and Grobe, 2005). Open the downloaded file in Pan2Applic, click on "Convert ⇒ Ocean Data View" and click "OK", then choose "Select data (2:)" and "Select geocode (3:)" and press "OK". The data are now loaded into ODV and you can, for example, visualize it with different modes at "View ⇒ Layout templates".

– For a detailed geographical search of the available data we created a Google .kmz file containing all CTD casts. When clicking on a single cast, a small window opens up displaying metadata details and a link to the data in PANGAEA. The .kmz file can be found under "Further details" at http://doi.pangaea.de/10.1594/PANGAEA.860066 or at http://hdl.handle.net/10013/epic.50376.d001.

4 Data availability

All data are accessible via http://doi.pangaea.de/10.1594/PANGAEA.860066. The data sets are freely available and can be directly downloaded. A moratorium is still in place for the latest campaign (PS96), but the data are available upon request.

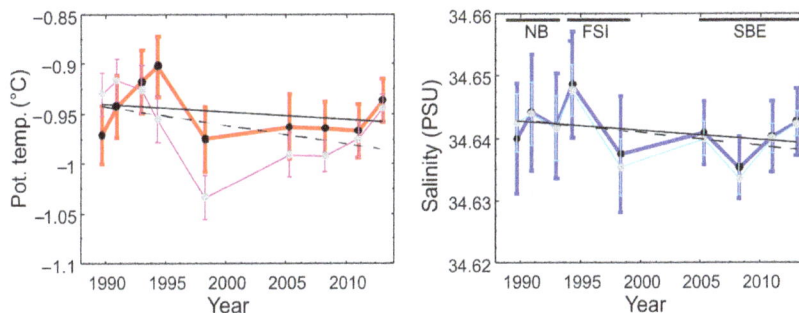

Figure 4. Mean potential temperature (left) and mean salinity (right) of the Weddell Sea Bottom Water calculated from nine repeated CTD sections at the tip of the Antarctic Peninsula (T. Kanzow, personal communication, 2016). The thin curves – magenta and cyan – include the seasonal effect. In the thick red and blue curves, the seasonal influence is eliminated. Linear regression lines are shown in black: dashed lines with seasonality included and solid lines with seasonality removed. The CTD type used is shown as follows: NB, Neil Brown; FSI, Triton ICTD; and SBE, SBE911plus.

5 Conclusions

Even small changes in sea-water density might affect vertical layering of water masses in the ocean (Olbers et al., 2012). Especially at low temperatures, small salinity changes affect the density much more than temperature changes of the same order (Schott et al., 1993). Therefore precise salinity measurements are needed, especially in polar regions. Based on repeated measurements, long-term changes of water mass properties can be studied (see, e.g., Fahrbach et al., 2011). Figure 4 shows the mean potential temperature and mean salinity of the Weddell Sea Bottom Water from nine repeated CTD sections at the tip of the Antarctic Peninsula. While the temperature shows similar errors of the mean, the errors of the mean salinity have become much smaller since 2005, which coincides with the use of the SBE911plus CTD on *Polarstern*. This illustrates clearly that when analyzing long-term trends from CTD data, the CTD type has to be taken into account. Additionally, in situ onboard calibration, regular servicing and laboratory calibration, data processing procedures, and experienced operators are required for precise data. CTD data therefore are of the highest value only if they come with proper documentation. One ambitious project analyzing and describing the complete set of available Arctic CTD data in respect to quality is currently taking place and will hopefully be published soon (Behrendt et al., 2017).

Competing interests. The authors declare that they have no conflict of interest.

Acknowledgements. We would like to thank Wolfgang Cohrs for creating Fig. 3. Many thanks to the numerous students who joined the CTD watches. Without their help it would have been impossible to run CTDs in 24 h shifts. Sometimes it was necessary to use external CTD operators and we thus appreciate the help of the Bremerhaven firms OPTIMARE and FIELAX, who both did an excellent job. Well-prepared instruments are the foundation for precise measurements. We therefore extend our gratitude to all CTD technicians, first and foremost to Ekkehard "Ekki" Schütt, who was not only dedicated to his job but also had the uncanny talent of identifying impending failures before they happened. Many thanks also to our current technicians: Matthias Monsees, Rainer Graupner and Carina Engicht. Last but not least, we are indebted to the crew of R/V *Polarstern*. During these 33 years crew members changed, of course, but it was always an exceptional team determined to make every campaign a success. Additionally, we would like to also thank the German Bundesministerium für Bildung und Forschung (BMBF) for placing *Polarstern* at the disposal of science. This ship has enabled polar research in conditions with up to force 8 winds. It even allowed for CTD transects during extremely heavy sea ice conditions with the ships powerful thrusters keeping the water surface ice-free while getting the CTD/rosette into the sea and back on deck safely.

Edited by: G. M. R. Manzella

References

Behrendt, A., Sumata, H., Rabe, B., and Schauer, U.: A comprehensive, quality-controlled and up-to-date data set of Arctic Ocean temperature and salinity, in preparation, 2017.

Brown, N.: The history of salinometers and CTD Sensor:systems, Oceanus, 34, 61–66, 1991.

Budéus, G. T.: Bringing laboratory salinometry to modern standards, Sea Technology Magazine, 52, 45–48, http://www.sea-technology.com/features/2011/1211/salinometry.php, last access: 10 January 2017, 2011.

Budéus, G. T.: FAQ to Optimare Precision Salinometer (OPS), Alfred Wegener Institute, Helmholtz Center for Polar and Marine Research, Bremerhaven, hdl:10013/epic.49362.d001, 2015.

Driemel, A., Loose, B., Grobe, H., Sieger, R., and König-Langlo, G.: 30 years of upper air soundings on board of R/V *POLARSTERN*, Earth Syst. Sci. Data, 8, 213–220, doi:10.5194/essd-8-213-2016, 2016.

Fahrbach, E., Hoppema, M., Rohardt, G., Boebel, O., Klatt, O., and Wisotzki, A.: Warming of deep Water masses along the Greenwich meridian on decadal time scales: The Weddell gyre as a heat buffer, Deep-Sea Res. Pt. II, 58, 2509–2523, doi:10.1016/j.dsr2.2011.06.007, 2011.

Hales, S.: A Letter to the Rev. Dr. Hales, F. R. S. from Captain Henry Ellis, F. R. S. Dated Jan. 7, 1750–51, at Cape Monte Africa, Ship Earl of Hallifax, Philosophical Transactions, 47, 211–216, 1751.

Millard, R. C. and Yang, K.: CTD Calibration and Processing Methods used at Woods Hole Oceanographic Institution, Woods Hole Oceanographic Institution, Technical Report, WHOI-93-44, doi:10.1575/1912/638, 1993.

Olbers, D., Willenbrand, J., and Eden, C.: Ocean Dynamics, Springer Berlin Heidelberg, 703 pp., doi:10.1007/978-3-642-23450-7, 2012.

Rohardt, G.: Physical oceanography during POLARSTERN cruise ANT-II/3, Alfred Wegener Institute, Helmholtz Center for Polar and Marine Research, Bremerhaven, doi:10.1594/PANGAEA.734969, 2010a.

Rohardt, G.: Physical oceanography during POLARSTERN cruise ANT-II/4, Alfred Wegener Institute, Helmholtz Center for Polar and Marine Research, Bremerhaven, doi:10.1594/PANGAEA.734972, 2010b.

Rohardt, G.: Physical oceanography during POLARSTERN cruise ANT-XXII/3, Alfred Wegener Institute, Helmholtz Center for Polar and Marine Research, Bremerhaven, doi:10.1594/PANGAEA.733664, 2010c.

Rohardt, G., Fahrbach, E., Beszczynska-Möller, A., Boetius, A., Brunßen, J., Budéus, G., Cisewski, B., Engbrodt R., Gauger, S., Geibert, W., Geprägs, P., Gerdes, D., Gersonde, R., Gordon, A. L., Hellmer, H. H., Isla, E., Jacobs, S. S., Janout, M., Jokat, W., Klages, M., Kuhn, G., Meincke, J., Ober, S., Østerhus, S., Peterson, R. G., Rabe, B., Rudels, B., Schauer, U., Schröder, M., Sildam, J., Soltwedel, T., Stangeew, E., Stein, M., Strass, V. H., Thiede, J., Tippenhauer, S., Veth, C., von Appen, W.-J., Weirig, M.-F., Wisotzki, A., Wolf-Gladrow, D. A., and Kanzow, T.: Physical oceanography on board of POLARSTERN (1983-11-22 to 2016-02-14), Alfred Wegener Institute, Helmholtz Center for Polar and Marine Research, Bremerhaven, doi:10.1594/PANGAEA.860066, 2016.

Schott, F., Visbeck, M., and Fischer, J.: Observations of vertical currents and convection in the Central Greenland Sea during the winter of 1988–1989, J. Geophys. Res., 98, 14401–14421, doi:10.1029/93JC00658, 1993.

Schlitzer, R.: Ocean Data View, http://odv.awi.de, last access: 3 December 2016, 2015.

Sieger, R. and Grobe, H.: Pan2Applic – a tool to convert and compile PANGAEA output files, doi:10.1594/PANGAEA.288115, 2005.

Stewart, R. H.: Introduction to Physical Oceanography, ORange: Grove Text Plus, 353 pp., 2009.

Warren, B. A.: Nansen-bottle stations at the Woods Hole Oceanographic Institution, Deep-Sea Res. Pt. I, 55, 379–395, doi:10.1016/j.dsr.2007.10.003, 2008.

14

The Ocean Carbon States Database: a proof-of-concept application of cluster analysis in the ocean carbon cycle

Rebecca Latto[1,2] **and Anastasia Romanou**[1,2]

[1] Applied Physics and Applied Math, Columbia University, New York, USA
[2] NASA-GISS, New York, NY, USA

Correspondence: Rebecca Latto (rl2797@columbia.edu)

Abstract. In this paper, we present a database of the basic regimes of the carbon cycle in the ocean, the "ocean carbon states", as obtained using a data mining/pattern recognition technique in observation-based as well as model data. The goal of this study is to establish a new data analysis methodology, test it and assess its utility in providing more insights into the regional and temporal variability of the marine carbon cycle. This is important as advanced data mining techniques are becoming widely used in climate and Earth sciences and in particular in studies of the global carbon cycle, where the interaction of physical and biogeochemical drivers confounds our ability to accurately describe, understand, and predict CO_2 concentrations and their changes in the major planetary carbon reservoirs. In this proof-of-concept study, we focus on using well-understood data that are based on observations, as well as model results from the NASA Goddard Institute for Space Studies (GISS) climate model. Our analysis shows that ocean carbon states are associated with the subtropical–subpolar gyre during the colder months of the year and the tropics during the warmer season in the North Atlantic basin. Conversely, in the Southern Ocean, the ocean carbon states can be associated with the subtropical and Antarctic convergence zones in the warmer season and the coastal Antarctic divergence zone in the colder season. With respect to model evaluation, we find that the GISS model reproduces the cold and warm season regimes more skillfully in the North Atlantic than in the Southern Ocean and matches the observed seasonality better than the spatial distribution of the regimes. Finally, the ocean carbon states provide useful information in the model error attribution. Model air–sea CO_2 flux biases in the North Atlantic stem from wind speed and salinity biases in the subpolar region and nutrient and wind speed biases in the subtropics and tropics. Nutrient biases are shown to be most important in the Southern Ocean flux bias.

1 Introduction

The ocean carbon cycle plays an important role in controlling the airborne fraction of CO_2 in the atmosphere, thereby regulating the rate of global warming, i.e., the rising temperatures in the Earth's troposphere. However, the ocean carbon cycle is controlled by a plethora of physical, biological and biogeochemical processes over a broad range of temporal and spatial scales. In this paper, we seek to present and assess a data mining/pattern recognition technique, namely cluster analysis, for the purpose of defining the basic regimes, or "ocean carbon states", that describe the oceanic carbon cycle variability. The goal is to increase our understanding of the marine carbon cycle by revealing patterns and information that other techniques do not provide.

For geophysical applications, climate datasets have inherent complexities that are not easily identifiable in the age of "big" data. Cluster analysis is a highly effective uni- or multi-variate classification method for large, high frequency datasets because it can find structure in a body of complex, geophysical data (Anderberg, 1973; Peron et al., 2014). Clustering seeks to identify the critical modes and natural patterns of a dataset without any training or predetermined spatial–

temporal guidelines; therefore, it is an "unsupervised" graph theory method. The merit of a novel, unsupervised method such as clustering is that it can recognize connectivity between multiple variables. This can be understood as connectivity in a temporal sense where cluster analysis can identify joint interannual or seasonal patterns and in a spatial sense where clustering has the power to identify patterns that relate different regions or basins (Jain, 2010; Phillips et al., 2015).

Traditional methods of univariate analysis, such as principal component analysis or spectral decomposition, cannot fully describe important physical states of the climate system or adequately detect change (Hoffman et al., 2011) because these methods neglect interactions between state variables as well as spatial and temporal co-variability. In contrast, cluster analysis has been successfully applied to various dynamical systems in order to extract the organized states and detect change as well as in novel applications of model–data intercomparison (Hoffman et al, 2008). For example, this technique has been used to define atmospheric weather states by identifying cloud regimes (Jakob and Tselioudis, 2003; Rossow et al., 2005; Williams and Webb, 2009; Tselioudis et al., 2013; Bodas-Salcedo et al., 2014; Oreopoulos et al., 2016). Bankert and Solbrig (2015) were able to extract a 3-D cloud representation using cluster analysis. This technique has also been used to characterize water types in lakes (Trochta et al., 2015), hydraulic habitat composition in rivers (Hugue et al., 2016), phenology patterns in forests (Trans Mills et al., 2011), solar variability (Zagouras et al., 2013), ENSO phenomena (Radebach et al., 2013), and regions with characteristic hydrological responses (Halverson and Fleming, 2015), among many other applications.

Beyond identifying regimes, cluster analysis can be useful in model assessment applications, like that of Wood et al. (2015), which used weather states derived from cluster analysis for process studies, satellite calibration, and model evaluation. Both regime identification and model evaluation are the focus of the cluster analysis presented in this paper as well.

Elsewhere in ocean carbon cycle science, clustering-type methods (self-organizing maps and neural networks) have been used to build reconstructions or as regression analysis alternatives for surface ocean pCO_2 (Lefèvre et al., 2005; Telszewski et al., 2009; Sasse et al., 2013; Landschützer et al., 2013, 2014; Nakaoka et al., 2013). Unlike these studies, here we seek to obtain the co-variability maps and conditions of different ocean-related variables and understand where, why and how they change.

Other non-statistical studies, but similar in concept to multivariate regime identification, have focused more on larger-scale geographic variations (Fay and McKinley, 2014; Trochta et al., 2015) than on the regional aspects of the ocean biogeochemistry and its interaction with physical circulation like in the western boundary current regions, in the upwelling zones on the eastern boundaries, and in the eddying field.

The structure of the paper is as follows. Section 2 describes the datasets used in this study, both the observation-based sources as well as the model experiments. Section 3 presents the k-means cluster analysis methodology and application, including discussion of the k-means clustering technique and sensitivity to number of clusters chosen, to binning, and to data normalization. The results of the methodology are provided in Sect. 4. Section 4.1 focuses on how the methodology is applied in observations from the North Atlantic basin. The observed ocean carbon states are then characterized temporally and spatially in order to reveal their physical meaning. Next, the model carbon states are computed and characterized in a similar way to the observations. Using the ocean carbon states, model biases are also discussed and evaluated. Section 4.2 repeats the analysis presented in Sect. 4.1, but now applied to the Southern Ocean. Finally, general discussion and conclusions are provided in Sect. 5. A note about the figures in the paper: some interesting but non-critical figures are offered in the Supplement and are denoted as Fig. S#. All data and analysis scripts are available at the https://data.giss.nasa.gov/oceans/carbonstates/ website (DOI: https://doi.org/10.5281/zenodo.996891).

2   Data

2.1   Choice of variables to represent ocean carbon regimes

One critical question to answer at the onset of any clustering analysis is what key geophysical variables should be used to base the analysis on. For the purposes of this study, we picked sea surface temperature (SST) and partial pressure of CO_2 in the ocean surface water ($pCO_{2\,sw}$). The rationale for this choice will be explained now. There are two main pathways that determine the ability of the ocean to take up CO_2 (Sarmiento and Gruber, 2006): the chemical disequilibrium, expressed by pCO_2, dissolved inorganic carbon (DIC is the sum of all inorganic carbon species) and nutrients, and the physical processes, such as air–sea interaction (expressed by the wind speed) and ocean circulation (expressed by sea surface temperature and salinity). Greater insight into the ocean's biogeochemical processes that control these pathways can inform the improved use of field measurements, the development of better metrics for model evaluation, and the selection of more suitable parameterizations in climate models in order to provide more accurate predictions. We select $pCO_{2\,sw}$ and SST because they are able to represent a broad range of biogeochemical and physical processes. We use them in cluster analysis to find temporal and spatial patterns in their joint parameter space that can be used to understand CO_2 flux distributions and its fluctuations. Other variable pairs can be alternatively used here; a comparison between choices is set aside for future work.

This study will focus on two oceanic basins, namely the North Atlantic (defined as 80° W to 45° E, 0 to 90° N) and the

Southern Ocean (defined as 180° W to 180° E, 90 to 40° S), because of their importance in the global carbon cycle (Takahashi et al., 2009).

2.2 Observation-based data

2.2.1 Air–sea flux of CO_2 and pCO_2, surface wind speed, sea surface temperature and salinity

The 12-month climatology of the air–sea flux is obtained from the Carbon Dioxide Information Analysis Center (LDEO database, NDP-088; Takahashi et al., 2009). It is derived from the difference between surface water pCO_2 ($pCO_{2\,sw}$), air pCO_2, and the air–sea gas transfer rate. Surface water pCO_2 climatological mean distribution was obtained from 3 million measurements from 1970 to 2007, and normalized to a reference year 2000. The pCO_2 of the air is computed from the GlobalView CO_2 concentration zonal mean, NCAR monthly mean barometric pressure, SST, and salinity. Other variables in the dataset pertinent to this analysis are wind speed (derived from the 1979–2005 climatological mean NCEP-DOE AMIP-II Reanalysis wind speed field), climatological sea surface temperature (from NOAA Climate Diagnostic Center Objective Interpolation), and salinity (from the NODC World Ocean Database 1998). All variables are available as a 12-month climatology at a $4° \times 5°$ resolution.

2.2.2 Nitrate

The nitrate monthly climatology at 1° horizontal resolution is obtained from the World Ocean Atlas 2013 version 2 (Boyer et al., 2013). It is collected from in situ measurements at standard depth levels and is available as annual, seasonal, and monthly climatologies. Nitrate is an essential nutrient that limits the growth of phytoplankton, which is responsible for fixating carbon dioxide from the atmosphere. Therefore, pCO_2 levels in the surface ocean depend partially on the abundance of nitrate.

2.3 Numerical simulations

The NASA-GISS modelE2.1 output used for this analysis comes from five ensemble coupled model simulations of the 20th century with realistic greenhouse gas, aerosol, land use and solar forcing, as used in CMIP5 experiments. The model physics is somewhat different than the modelE2 used in the CMIP5 experiments, mostly due to improved representation of the ocean mesoscale mixing. The physical ocean and the biogeochemistry modules are described in detail in Romanou et al. (2013, 2017). Briefly, here we note that the ocean model is a non-Boussinesq mass-conserving ocean model with 32 vertical levels and $1° \times 1.25°$ horizontal resolution. The vertical coordinate is a stretched z-level coordinate and has a free surface and natural surface boundary fluxes of freshwater and heat that are obtained by the atmospheric model. In addition

to advection and turbulent mixing, it also includes a scheme for isopycnal eddy fluxes and isopycnal thickness diffusion. The interactive ocean carbon cycle model consists of a biogeochemical model (NASA Ocean Biogeochemistry Model, NOBM; Gregg and Casey, 2007) and a gas exchange parameterization for the computation of the CO_2 flux between the ocean and the atmosphere (Romanou et al., 2013). Specifically, the air–sea exchange of CO_2 (Sarmiento and Gruber, 2006; Takahashi et al., 2009) is described by Eq. (1):

$$F = \mathrm{kw} K_0 \left(pCO_{2\,atm} - pCO_{2\,sw} \right), \tag{1}$$

where kw is the piston velocity for CO_2 (in $\mathrm{m\,s^{-1}}$) that depends on the wind speed, K_0 is the solubility coefficient – dependent on sea surface temperature (SST) and sea surface salinity (SSS) (expressed in mole, $CO_2\,\mathrm{kg^{-1}\,atm^{-1}}$) – and pCO_2 is the partial pressure of CO_2 (Wanninkhof et al., 2013) in the atmosphere (atm) and the surface ocean (sw). Equation (1) describes the chemical disequilibrium of CO_2 in the oceanic and atmospheric reservoirs due to the solubility and biological pumps. As discussed in Sarmiento and Gruber (2006), the $pCO_{2\,sw}$ in Eq. (1) is a function of temperature and salinity, wind speed, DIC, nutrients, and alkalinity (a measure of the excess of bases over acids) which can be expressed as follows:

$$pCO_{2\,sw} = f\,(\text{SST, SSS, DIC, windspeed, nutrients, alkalinity}). \tag{2}$$

NOBM utilizes ocean temperature and salinity, mixed layer depth and the ocean circulation fields, and the horizontal advection and vertical mixing schemes obtained from the host ocean model as well as shortwave radiation (direct and diffuse) and surface wind speed obtained from the atmospheric model to produce horizontal and vertical distributions of several biogeochemical constituents. The carbon submodel parameterizes the cycling of carbon through the phytoplankton, herbivore and detrital components, affecting the dissolved inorganic and organic carbon in the ocean and interacting with the atmosphere. Alkalinity is assumed analogous to surface salinity, which is an acceptable approximation for the sea surface but does not take into account changes in the carbonate pump. Temperature and salinity are affected only by physical processes such as circulation, advection, eddy mixing and stirring, and local upwelling/downwelling, while DIC distributions are influenced by all these physical processes and also several biogeochemical processes such as air–sea gas exchange, production by organisms, biological export to depth and remineralization there and nutrient availability in the water column. Atmospheric pCO_2 ($pCO_{2\,atm}$) is the saturation concentration of CO_2 in equilibrium with a water-vapor-saturated atmosphere at a total atmospheric pressure P and a given atmospheric pCO_2 level:

$$pCO_{2\,atm} = \frac{P}{P^0}\,CO_2^0, \tag{3}$$

where $P_0 = 1$ atm and $[CO_2]^0$ is the saturation concentration at 1 atm total pressure.

The gas transfer velocity is given by

$$\text{kw} = c \left(\frac{Sc}{660} \right)^{-1/2} u^2, \tag{4}$$

where u is the surface wind speed and c is the piston velocity coefficient taken here equal to $0.337/(3.6 \times 10^5)$. The value of c has been agreed upon by the Ocean Carbon Model Intercomparison Project, phase II (OCMIP-II) so that the global, annual mean gas transfer coefficient for carbon dioxide (kw, K_0) is equal to $0.061 \, \text{mol} \, \text{m}^{-2} \, \text{yr}^{-1} \, (\mu\text{atm})^{-1}$ for preindustrial times. Sc, the Schmidt number, is computed using the temperature of the host ocean model following Wanninkhof (1992). The gas transfer velocity kw is computed only over open water. The solubility of CO_2 in the water K_0 is also parameterized based on OCMIP using prognostic temperature, salinity and sea level pressure. In these model runs, the global average of the atmospheric concentration of CO_2 follows the Mauna Loa measurements (Dlugokencky and Tans, 2014), although regionally atmospheric CO_2 is allowed to vary due to the distributions of the ocean sources and sinks.

The five ensemble member runs were averaged into one ensemble mean to account for the intrinsic climate variability that is not adequately resolved in climate models of low spatial resolution. The model output for the years 1995–2005 was then averaged again to produce a 12-month climatology for the purpose of direct comparison with the observationally based data in the Takahashi database.

The model output and the observational data were interpolated onto the same grid, which is the Takahashi ocean grid at $4° \times 5°$ resolution, with no Arctic Ocean, and the ocean mask was conformed across all observational and model datasets.

In the rest of the paper, some conventions with regards to nomenclature should be noted. Firstly, the Takahashi carbon flux, pCO_2 and ancillary data as well as the nitrate climatology will be referred to as "observations", for brevity, keeping in mind that they are really observation-based estimates and not direct observations. Secondly, "model" will exclusively refer to the numerical simulations using the NASA-GISS climate model, and by "algorithm", "method" or "technique" we will refer to the clustering technique.

All data products are available in the Ocean Carbon States Database (https://data.giss.nasa.gov/oceans/carbonstates/).

3 Methodology

A schematic diagram of the methodology is presented in Fig. 1. First, the 2-D histograms pCO_2-SST are computed from the climatological data, then the histograms are clustered using a statistical method, the k-means clustering method, and finally the regimes or "ocean carbon states" are obtained. The methodology steps will be explained in detail below, using as an example the North Atlantic basin data.

3.1 pCO_2-SST 2-D histograms

$pCO_{2\,\text{sw}}$ values in the North Atlantic span the range 50–450 uatm, while sea surface temperatures range between -2 and $30\,°C$. The 2-D histograms (Fig. 2) show the highest frequency of occurrence for $pCO_{2\,\text{sw}}$ values in the range of 300–400 uatm and temperatures in the range of 10 to $30\,°C$. Certain months (December, January, February and March) show a higher frequency of occurrence of cold temperatures (-2 to $2\,°C$) and low $pCO_{2\,\text{sw}}$ (50–300 uatm) than others. Figure 2 also reveals that certain histograms appear similar in shape; for example, January–April exhibit an S-shaped curve and no tilt, while June–September exhibit a diagonal tilt that reflects a tendency for higher temperatures to co-locate with higher $pCO_{2\,\text{sw}}$ values. This being a small dataset of only 12 2-D histograms, one could easily sort them into groups of similar shape just by visual inspection only. The methodology presented in this paper seeks to more mechanistically identify these groups, so that it can be confidently applied to larger and more complex datasets. We will call those organized groups, clusters or regimes or "ocean carbon states".

It is noted here that despite the broad range of values for both variables, the 2-D histograms are very similar regardless of the number of bins chosen for each of the variables.

3.2 k-means clustering

The k-means clustering algorithm (Anderberg, 1973; Jakob and Tselioudis, 2003) partitions the 2-D histograms of pCO_2-SST shown in Fig. 2 into a predefined number k of groups, called clusters. In the first step of the algorithm, k histograms are randomly selected and are considered the centroid of each of the k clusters. Each other histogram in the input dataset is then assigned to its nearest centroid by computing the Euclidean distance of each bin of the 2-D histogram from the same bin of the centroid. The procedure is repeated an N number of iterations, each time the centroid of the resulting group is recalculated, if doing so reduces the sum of the distances of each histogram to the centroid. This iterative procedure stops when the squared distance between the mean of each cluster and all the 2-D histograms assigned per cluster is minimized (Jain, 2010). More than one iteration (N) is necessary to have convergent clustering results because each analysis initializes at a random cluster centroid. In this paper, convergence is reached after 10 iterations, if not fewer (Fig. S1 shows how this is determined for the example of the North Atlantic basin).

3.3 Sensitivity to predefined number of clusters

To ensure that the chosen number of clusters, k, is representative of the system, typically one needs to repeat the technique for various values of k, and, using visual inspection, select the optimal value for k when the resulting clusters become repetitive or contain no additional information. Objective methods have been proposed (e.g., Bankert and Solbrig,

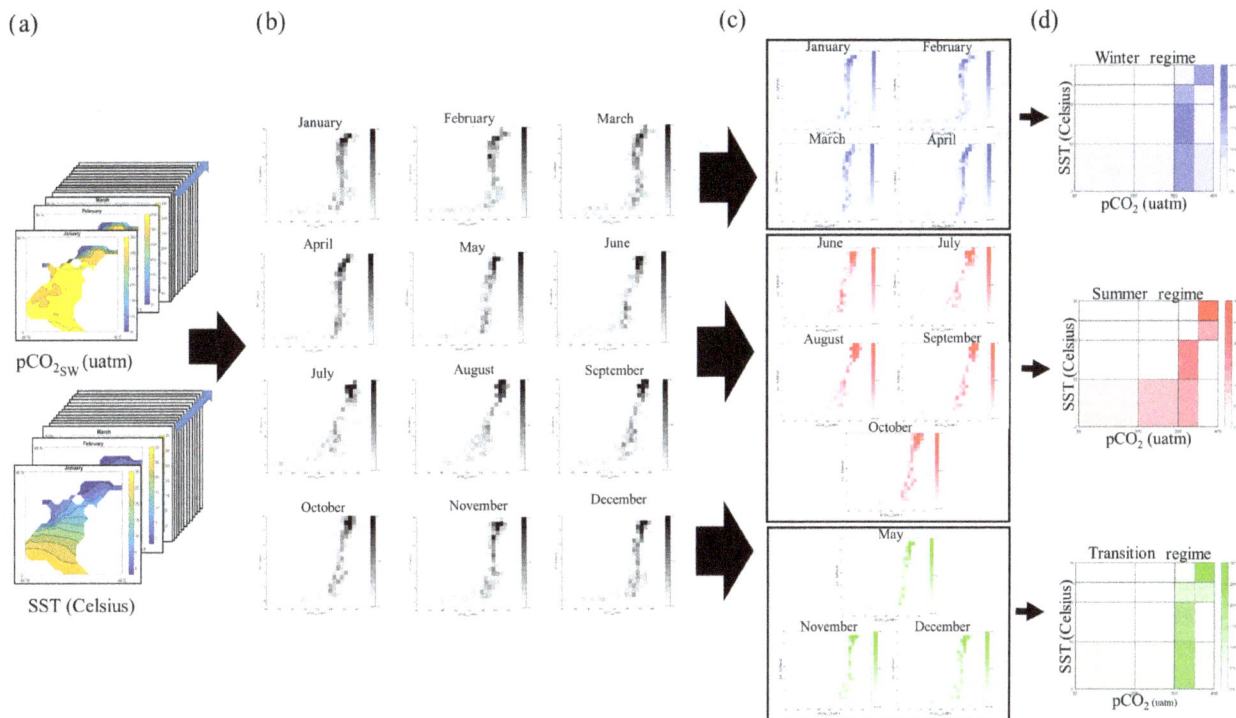

Figure 1. Schematic diagram of the clustering methodology used in this paper: **(a)** 12-monthly mean climatological year data of two variables, pCO_2 and SST, **(b)** monthly 2-D histograms, **(c)** clustering of the 2-D histograms into groups by similarity in the bivariate distributions, and **(d)** clusters resulting when $k = 3$ is assumed.

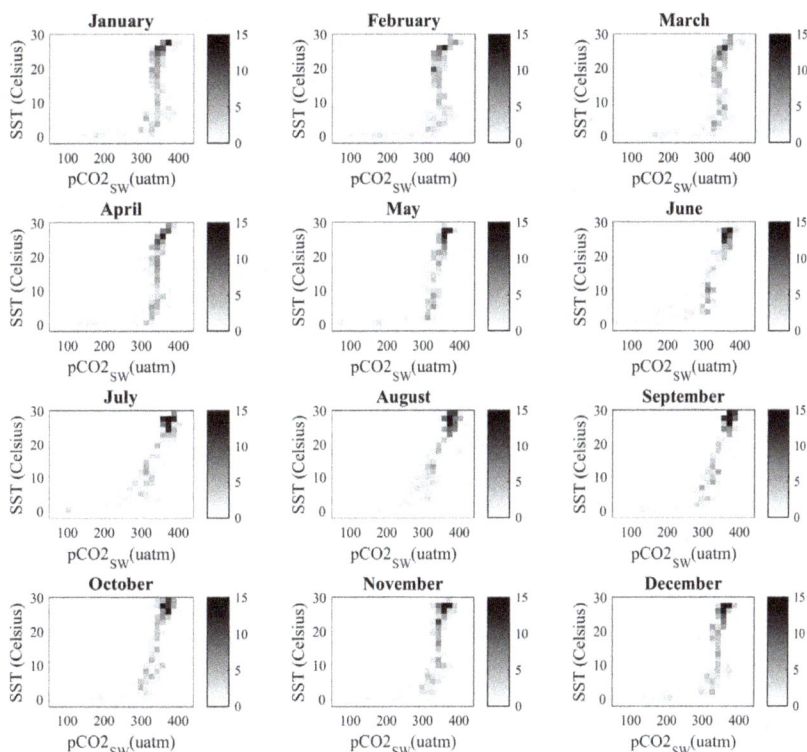

Figure 2. Monthly 2-D histograms of partial pressure of CO_2 in the surface water ($pCO_{2\,sw}$) and sea surface temperature (SST) in the North Atlantic (defined as $80°$ W to $45°$ E, 0 to $90°$ N) from the Takahashi observational dataset. The horizontal axis is $pCO_{2\,sw}$ (uatm) and the vertical axis is SST (°C). The bin interval is 15 uatm and 1.6 °C. The color bar describes the actual frequency of occurrence of each bin.

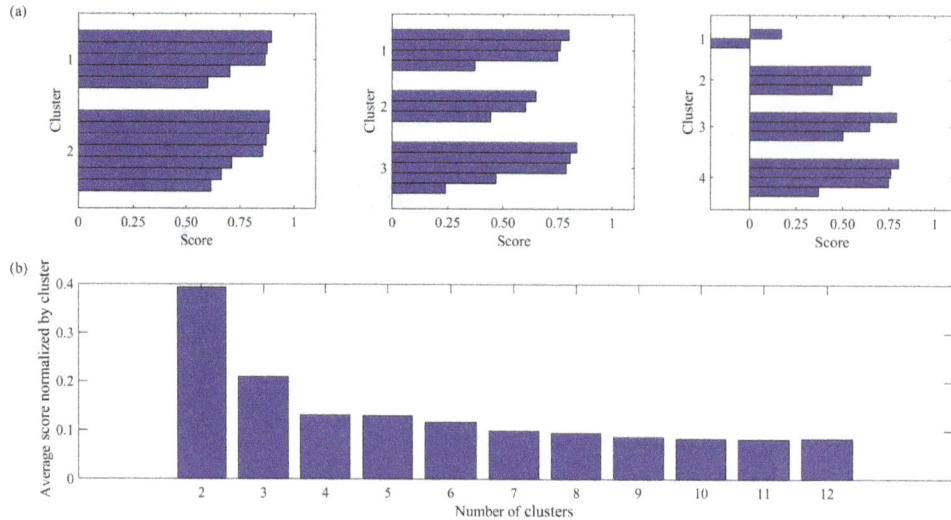

Figure 3. (a) Scores for each cluster analysis of observational data in the North Atlantic for $k = 2$, $k = 3$, $k = 4$, where cluster k is the predetermined number of clusters and each bar represents the score per 2-D histogram. **(b)** Average scores for each cluster analysis with increasing k, from $k = 2, \ldots, 12$, normalized by the number of clusters k.

2015), where the average radius of each cluster (the distance from the centroid to the most distant member within a cluster) is computed for decreasing k. Bankert and Solbrig (2015) found that when the number of clusters falls below the optimal k, the average radius grows rapidly. We employ a similar methodology here. First, we use a scoring algorithm that computes the distance of each 2-D histogram from the centroid of its cluster. The higher the score is, the closer the 2-D histograms are to the centroid. The maximum score is 1, which indicates a perfect match. Negative or low values indicate poorly matched histograms that are the farthest from the centroid in the cluster. We run the scoring algorithm for $k = 2$ through $k = 12$, since we have 12 2-D histograms, and therefore there can be up to 12 clusters. Figure 3a shows the scores only for $k = 2$, $k = 3$, and $k = 4$ as examples of the output of the scoring algorithm. We find that for $k = 2$ and $k = 3$ all 2-D histograms are well matched within a cluster (i.e., all scores are high), whereas for $k = 4$ there is 1 month with a negative score. To further summarize the scoring results, we introduce a "sensitivity criterion" to the predetermined number of clusters, k, in which we average all scores for each k and normalize by k. The results are shown in Fig. 3b, where we note that the averaged normalized score for $k = 2$ is 0.4, and it then quickly drops to 0.2 for $k = 3$ and to 0.1 for $k = 4$, and it plateaus after that. We choose as the optimal number of clusters the k with the highest score and no significant change in the normalized averaged score thereafter. This choice implies that any reorganization of the 2-D histograms within more than three clusters will not produce any "tighter" clusters, i.e., clusters where the members are closer together. We must note here that the method is not entirely objective, as one always needs to visually inspect the clusters themselves and ensure that the choice of k is indeed the best one.

3.4 Data normalization

As noted earlier, pCO_2 and SST have a broad range of values. Specifically, pCO_2 values vary by about 2 orders of magnitude between 50 and 450 uatm, and SST by 1 order of magnitude between -2 and 30 °C. It is customary in applications of statistical techniques such as clustering to normalize the data (subtract the mean and divide by the standard deviation) in order to force both datasets to be in the same range of values. However, this is not always necessary (Anderberg, 1973, p. 13; Kaufman and Rousseauw, 2005, p. 11). In our case, we are not clustering each variable separately in order to determine regression coefficients (as in Lefèvre et al., 2005). Rather, we are clustering the 2-D histograms and comparing them, in order to obtain groups of similar patterns. In addition, clusters represented in normalized data are not as easily understood physically and as well represented on geographical maps. Therefore we choose not to normalize the data for the purposes of this study.

4 Results

4.1 The North Atlantic Ocean carbon states

Figure 4 depicts the regimes for $k = 2$, $k = 3$, and $k = 4$. When only two clusters are predefined, i.e., for $k = 2$, the first cluster (Regime 2A, Fig. 4a) is dominated (30 % of the time) by pCO_2-SST pairs in the ranges of 350–400 uatm and 25–30 °C. The second cluster (Regime 2B) is dominated (20 %) by $pCO_{2\,sw}$ values within 300–350 uatm and SST values in the range -2 to 20 °C. When we choose more clusters initially, i.e., for $k = 3$, Regime 3B is very similar to Regime 2A and Regime 3C is analogous to Regime 2B, in the sense that the regimes have analogous bins of the highest

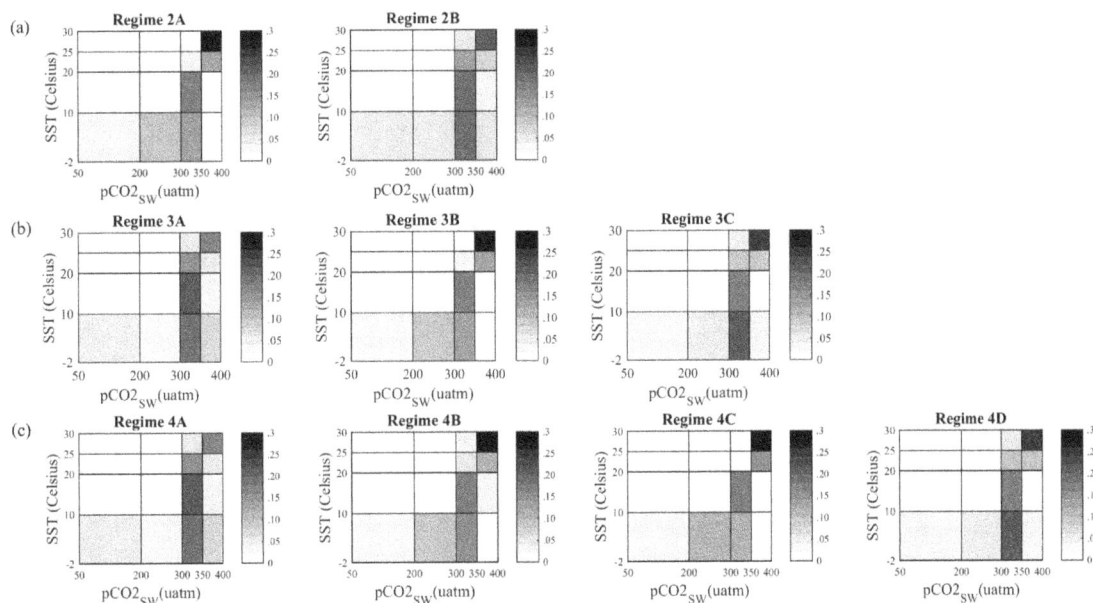

Figure 4. Cluster analysis output (regimes) for **(a)** $k = 2$, **(b)** $k = 3$, and **(c)** $k = 4$ for the North Atlantic, from the Takahashi observational dataset. The color bar represents the relative frequencies of occurrence of each value-pair interval; i.e., the frequencies are divided by the total number of frequencies per regime.

frequencies. Regime 3A is a new state that was unresolved in $k = 2$, but is not similar to 3B or 3C. For $k = 4$, Regimes 4B and 4C appear to be almost equivalent, and both derived from Regime 3B, which probably indicates that there is no new information gained by requiring four clusters. Similar visual inspection of the results for $k > 4$ confirms our more objective analysis result that $k = 3$ is the optimal number of clusters in the $p\mathrm{CO}_2$-SST space in the North Atlantic basin. It should be noted here that because we implement k clustering in the 2-D histograms and not the raw data, there is really no change in the results if we use a different number of bins in the histograms or if we use normalized data.

4.2   Temporal attribution for the North Atlantic carbon states

In order to characterize the ocean carbon states obtained in the previous section, we perform a temporal attribution analysis by determining when each cluster occurs. This is possible because for each cluster (regime), the k-means analysis routine computes the distance of every 2-D histogram in that cluster to the cluster's centroid. Since each 2-D histogram is associated with a certain month in the climatology, we are able to associate each cluster with certain months. Figure 5b shows that in the North Atlantic basin, regime 1 is represented by months January, February, March, and April and we call this the "winter regime"; regime 2 occurs during June, July, August, September and October and we will thus call it the "summer regime"; regime 3 occurs in May, November, and December and we will call this the "transi-

tion regime" because it reflects a mixed season in between the winter and summer regimes. Not surprisingly, these resulting regimes align themselves fairly well with the boreal winter and summer seasons in the North Atlantic. The cold season (winter regime) includes March and April but not November–December, which are included, rather, in the transition regime. The warm season (summer regime) includes the months between June and October, again broader than the typical boreal summer. It is not surprising that we recover the seasonal cycle from the 12-month climatology, and probably because our domain is the entire North Atlantic, from the Equator to the subpolar regions, these seasons are broader including months from the spring and fall, since the length of each season is different at different latitudes.

4.3   Spatial attribution of the North Atlantic carbon states

Next, we describe the geographical distribution of each regime (Fig. 5c). To do so, the frequencies of occurrence associated with each $p\mathrm{CO}_2$-SST bin in Fig. 5a are averaged over the months in each regime and mapped on the North Atlantic basin. We find that in the winter regime, the dominant value pairs (300–350 uatm and 10–20 °C) are found in the subtropical North Atlantic. In contrast, the dominant range of $p\mathrm{CO}_2$-SST pairs in the summer regime occurs in the tropics (values 350–400 uatm and 25–30 °C). The transition regime shows a mix of the winter and summer regimes.

We conclude that the ocean carbon states determined by a 12-month climatology of surface ocean $p\mathrm{CO}_2$ and SST are characterized by a cold season where most persistent value

(a)

(b)

(c)

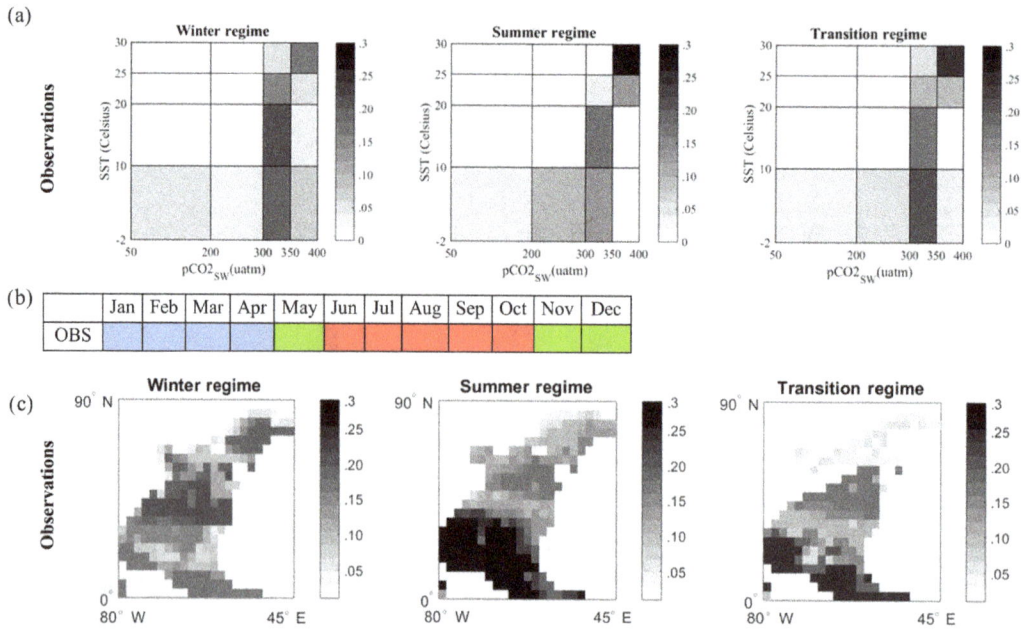

Figure 5. (a) Ocean carbon states (regimes) in observation-based data of the North Atlantic. (b) Monthly attribution of each ocean carbon regime in the Takahashi observational dataset. Temporal attribution is based on the distance of each monthly 2-D histogram to the centroid of each cluster. Each color represents a different regime: blue denotes the cold months' regime (winter regime), red the warm months (summer regime) and green the May/November–December transition regime, since this is more a mix of the other two. (c) Regional attribution of each regime depicted in (a). The colors in (c) correspond to the ones in (a), i.e., to the frequencies of occurrence of each bin (value pair) in the clusters of (a).

pairs occur in the subtropical North Atlantic and the subpolar region. In the warm season, however, the most persistent value pairs occur in the tropical Atlantic. For more complex datasets, e.g., when interannual variability is included, we expect to be able to detect regimes that correspond to processes controlled by the El Nino–Southern Oscillation (ENSO) or the North Atlantic Oscillation, for example.

4.4 The NASA-GISS climate model North Atlantic carbon states

Next we obtain the ocean carbon states from the GISS model simulations, following here the same methodology as for the observations. We are interested in understanding how similar the model regimes are to the observed ones we found earlier. As described in Sect. 2.3, we construct the ensemble mean climatologies for $pCO_{2\,sw}$ and SST for the period 1995–2005 from five simulations of Earth's historical climate of the 20th century performed with the NASA-GISS climate model. We then obtain the 12-monthly 2-D histograms from the model climatology and, using the same binning groups as in the observations, we obtain the model clusters. It should be noted here again that because we are actually clustering the 2-D histograms and not the raw data, our clusters are not sensitive to the number of bins or to normalization of the datasets prior to cluster analysis.

The sensitivity criterion (discussed in Sect. 3) for the model clusters is not as clear as in the case of the observations (Fig. 6). Note that there is a plateau after $k = 5$ and thus it appears that 5 would be a more suitable choice for k. However, as seen in Fig. 6a, some of the additional regimes include only one monthly 2-D histogram. We therefore chose here $k = 3$, recognizing that the model clusters have larger uncertainty. In a larger dataset that includes interannual variability, more than a single 2-D histogram would potentially be assigned to a regime, reducing that uncertainty.

Figure 7a shows the ocean carbon states for $k = 3$, while Fig. 7b characterizes their temporal occurrence: the model winter regime corresponds to the months December, January, February, March, April, and May; the summer regime corresponds to the months July, August, September, October, and November; the transition regime corresponds only to June. There is therefore good agreement with the regimes from observations (Fig. 5b). The model winter regime is somewhat broader than that observed by 2 months (December and May), while the model summer regime lags by 1 month (starts in July, while in the observations it starts in June).

The regimes (clusters) themselves are similar to the observations (comparing Figs. 5a and 7a) in that the same bins of most likely values are identified but with somewhat different frequencies of occurrence. As an example of comparison between the temporal regimes, for the winter regime both the model and the observations show that the dominant pairs are

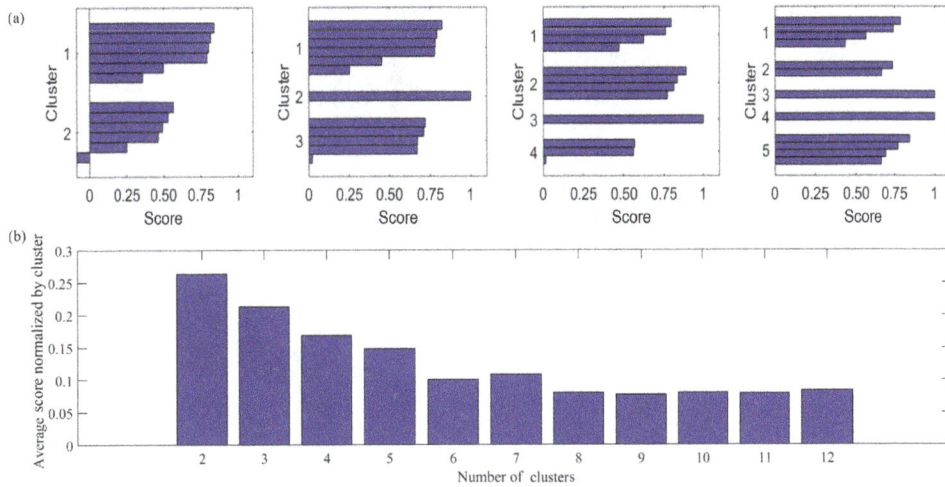

Figure 6. (a) Scores for each cluster analysis of the GISS model data in the North Atlantic for $k = 2$, $k = 3$, $k = 4$, $k = 5$, where cluster k is the predetermined number of clusters and each bar represents the score per 2-D histogram. **(b)** Average scores for each cluster analysis with increasing k, normalized by the number of clusters k.

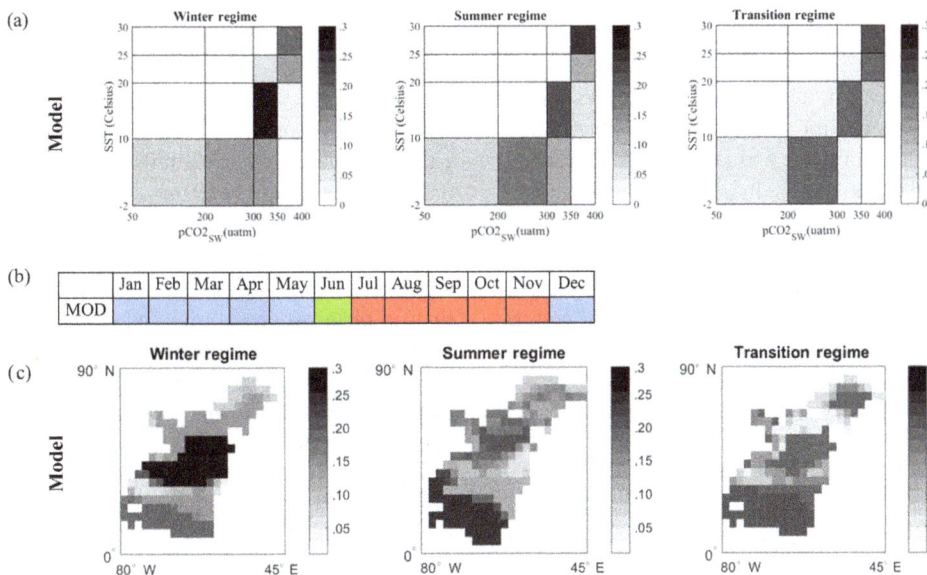

Figure 7. (a) Ocean carbon states (regimes) in the North Atlantic from the GISS model output. **(b)** Monthly attribution of each ocean carbon regime. Temporal attribution is based on the distance of each monthly 2-D histogram to the centroid of each cluster. Blue denotes the cold month regime (winter regime), red the warm months (summer regime) and green the transition regime (only June), since this is more a mix of the other two. **(c)** Regional attribution of each regime depicted in **(a)**. The frequencies of occurrence of each bin (value pair) in the clusters of **(a)** are mapped onto the North Atlantic grid.

in the range of 300–350 uatm and 20–25 °C, at 30 and 25% relative frequency. However, other weaker pairs are not well represented in the model, e.g., the range 50–200 uatm and -2 to 10 °C. During the winter regimes, the highest frequency of occurrence (25 %) is for the pair of values 300–350 uatm and 10–20 °C, whereas in the model the same pair of values is found 30 % of the time. Similarly, the summer and transition regime highest frequency pairs are well simulated. In contrast, the value pairs 200–350 uatm for very cold temperatures are not well represented in the model.

We also find (Fig. 7c) that in the winter regime, the dominant value pairs (300–350 uatm and 10–20°C, identified in Fig. S2) are found in the subtropical North Atlantic, with higher frequency in the model (darker shading) than in the observations. The GISS model however underestimates the frequency of occurrence of the value pairs in the subpolar region (values 50–350 uatm and -2 to 10 °C, identified in

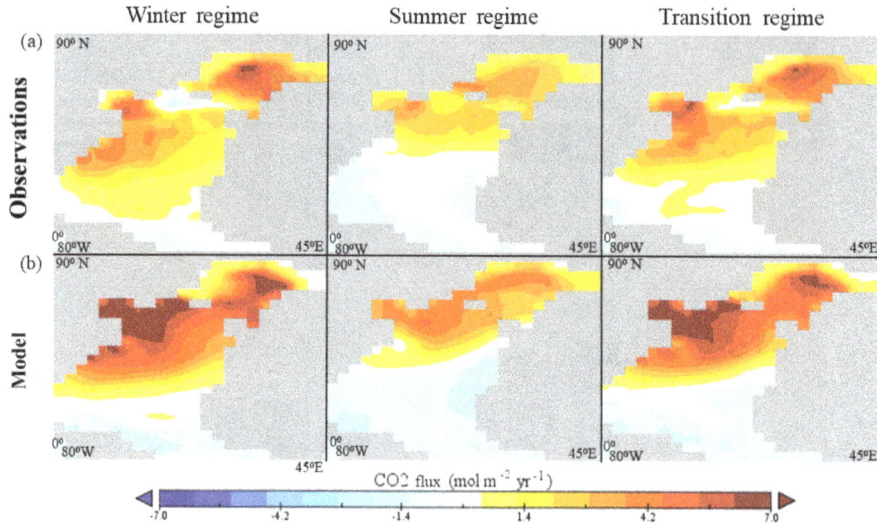

Figure 8. Composites of the CO_2 flux field over the observed regimes for **(a)** the observations and **(b)** the model. The composite fields are computed as averages of the field over the months included in each regime. Both the observations and the model data are composited over the same months as determined by the temporal attribution of the observations dataset, shown in Fig. 5b. Blue shades indicate outgassing, and red shades indicate uptake.

Fig. S2). In contrast, the dominant range of pCO_2-SST pairs in the summer regime occurs in the tropics (values 350–400 uatm and 25–30 °C; Fig. 7c) and is of higher frequency in the observations than the model. In other words, the model underestimates the extent of the tropical summer regime but reproduces well the other parts of the summer regime. The transition regime shows a mix of the winter and summer regimes for both observations and model. The model results in this regime indicate higher frequency in the subpolar region than in the observations.

4.5 Model North Atlantic air–sea flux of CO_2 error analysis and bias attribution

The ocean carbon states can provide a framework for model assessment against the observations. In this section we seek to identify biases in the simulated flux of CO_2 and attribute them to leading biases in physical and biogeochemical processes.

In the previous section we used the same methodology as in the observations to obtain independently the model regimes and assess how different they are. However, when we want to identify the causes of the model biases, we have to use the observed regimes as the basis of the comparison and assign the model data to the observed regimes, by averaging the model data over those months that the observed regimes occur (Fig. 5b). This approach was highlighted in Williams and Webb (2009) for general circulation model evaluations of the cloud regimes.

Figure 8 depicts the air–sea flux of CO_2 composited over the observed temporal regimes in both observations (Fig. 8a) and model output (Fig. 8b). In the winter regime, model out-

gassing (in shades of blue) is confined only to the tropics, whereas in the observations there is also a tongue of outgassing at about 60° N. Similarly, in the transition regime, the model has a more extended uptake region in the subpolar North Atlantic than in the observations. While the summer regime is better represented in the model than in the other two regimes, all three regimes show that model uptake is stronger than in observations at mid and high latitudes and that outgassing is also stronger in the model at mid to low latitudes.

To trace the source of model biases in the air–sea flux of CO_2 we need a better understanding of the physical or biogeochemical processes that control the air–sea flux of CO_2 in the model. Using the clustering analysis results from the previous section, we can investigate the underlying processes that might be responsible for the bias.

The process attribution is performed using a Taylor expansion of the model bias as shown in Eq. (5). The model flux bias, ΔF, depends on the biases of $pCO_{2\,sw}$, SST, salinity (SSS) and wind speed (wspd) such that

$$\Delta F \sim \frac{\partial F}{\partial pCO_{2\,sw}} \Delta pCO_{2\,sw} + \frac{\partial F}{\partial SST} \Delta SST$$
$$+ \frac{\partial F}{\partial SSS} \Delta SSS + \frac{\partial F}{\partial wspd} \Delta wspd, \qquad (5)$$

where Δq is the bias of the variable q, defined as the root mean squared error (RMSE) between the observations and the model, and q is any of the variables $\{pCO_{2\,sw}, SST, SSS, wspd\}$. $\frac{\partial F}{\partial q}$ is a weight term that represents dependence of the flux on that variable and is determined by the slope of a linear fit in the scatter plot of the flux F with each variable q for each carbon state. Since the North Atlantic basin is a very broad basin, both zonally and merid-

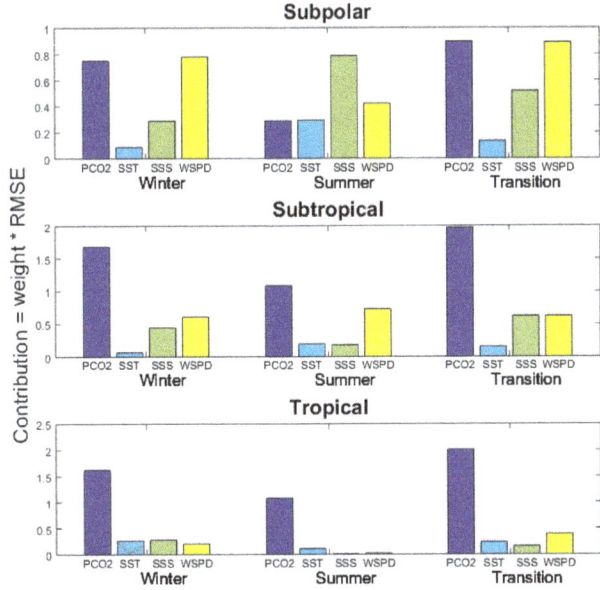

Figure 9. Contributions of each of the variables pCO$_2$, SST, SSS, and WSPD to the overall air–sea flux ΔF bias. The contributions are computed as the products of the weights and the RMSEs of each variable q as described in Eq. (5). See the text for a detailed explanation.

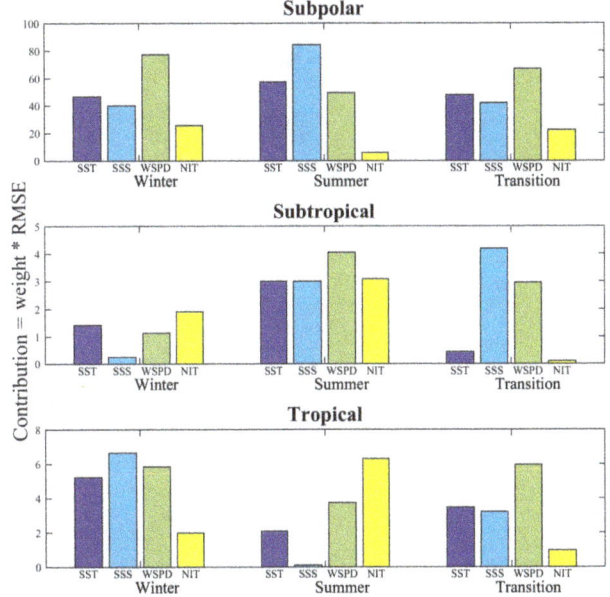

Figure 10. Contributions to the pCO$_{2\,\mathrm{sw}}$ bias in the model from SST, SSS, wind speed (WSPD) and nitrate (NIT) in the winter, summer, and transitional regimes. Contributions are computed as in Fig. 9 (see details in the text). The entire North Atlantic is differentiated into subpolar, subtropical, and tropical regions to better account for regional differences in the model biases and obtain a better linear fit for the computation of the weights in Eq. (6).

ionally, and because the carbon states' regional distribution (Fig. 5c) is quite complex, we identify areas where the linear fits will be more appropriate approximations of the $\{F, q\}$ relationships. The subpolar region where the value pairs are -2 to $10\,^{\circ}$C and 50 to 350 uatm, a subtropical region (10 to $20\,^{\circ}$C, 300 to 350 uatm), and a tropical region (20 to $30\,^{\circ}$C, 300 to 400 uatm) are demarcated in Fig. S2. Results of the regional scatter plots and the linear fit for each regime are shown in Fig. S3 and are synthesized in Fig. 9. Each contribution term (each term on the right-hand side of Eq. 5) is calculated from the multiplication of the weights and the RMSEs. Figure 9 then shows that over most of the North Atlantic, the flux biases are attributed mainly to errors in the pCO$_{2\,\mathrm{sw}}$, although in subpolar regions other terms such as salinity biases and wind speed biases become important. It therefore makes sense to further investigate biases in pCO$_{2\,\mathrm{sw}}$ and the processes these are attributed to, as presented in Eq. (2).

Similarly to Eq. (5),

$$\Delta p\mathrm{CO}_{2\,\mathrm{sw}} \sim \frac{\partial p\mathrm{CO}_{2\,\mathrm{sw}}}{\partial \mathrm{SST}} \Delta \mathrm{SST} + \frac{\partial p\mathrm{CO}_{2\,\mathrm{sw}}}{\partial \mathrm{SSS}} \Delta \mathrm{SSS}$$
$$+ \frac{\partial p\mathrm{CO}_{2\,\mathrm{sw}}}{\partial \mathrm{WSPD}} \Delta \mathrm{WSPD} + \frac{\partial p\mathrm{CO}_{2\,\mathrm{sw}}}{\partial \mathrm{NITRATE}} \Delta \mathrm{NIT}. \quad (6)$$

We perform the Taylor expansion of the bias for each of the regimes that we computed, calculating the weights and RMSEs in the same way as described for CO$_2$ flux biases. The estimates of the linear fit slopes of the scatter plots are shown in Fig. S4 and the composites of the contributions in Fig. 10.

Overall Fig. 10 shows that the biases in the subpolar region are larger than anywhere else in the North Atlantic basin, as the contributions to the bias in pCO$_{2\,\mathrm{sw}}$ are an order of magnitude larger there. Specifically, in the subpolar region, wind speed biases emerge as responsible for the winter and transition regime biases in pCO$_{2\,\mathrm{sw}}$, while salinity biases dominate the summer bias in pCO$_{2\,\mathrm{sw}}$. In the winter and transitional months, the quasi-cyclonic subpolar gyre, driven by energetic winds and wind outbreaks, leads to Ekman divergence in the surface layer that controls the pCO$_{2\,\mathrm{sw}}$ biases near the coast. At the same time, winter-time convective mixing is responsible for biases in the strength of the Meridional Overturning Circulation that are known to influence open ocean pCO$_{2\,\mathrm{sw}}$ (Romanou et al,. 2017). In the summer regime, GISS model sea-ice concentration is higher than observed; hence, melting will lead to significant surface salinity biases. Inaccurate model representation of the magnitude and fluctuations of the cyclonic wind stress curl as well as the sea-ice retreat and associated salinity changes are probably responsible for deficient physical characterization of the model ocean circulation, which would result in misrepresentations of the pCO$_{2\,\mathrm{sw}}$ and thus the CO$_2$ flux in the model.

In the subtropics, nitrate is found to be the largest contributor for the winter regime biases, wind speed is the main contributor in the summer, and salinity is the main contributor for the transition regime. The subtropics are characterized at

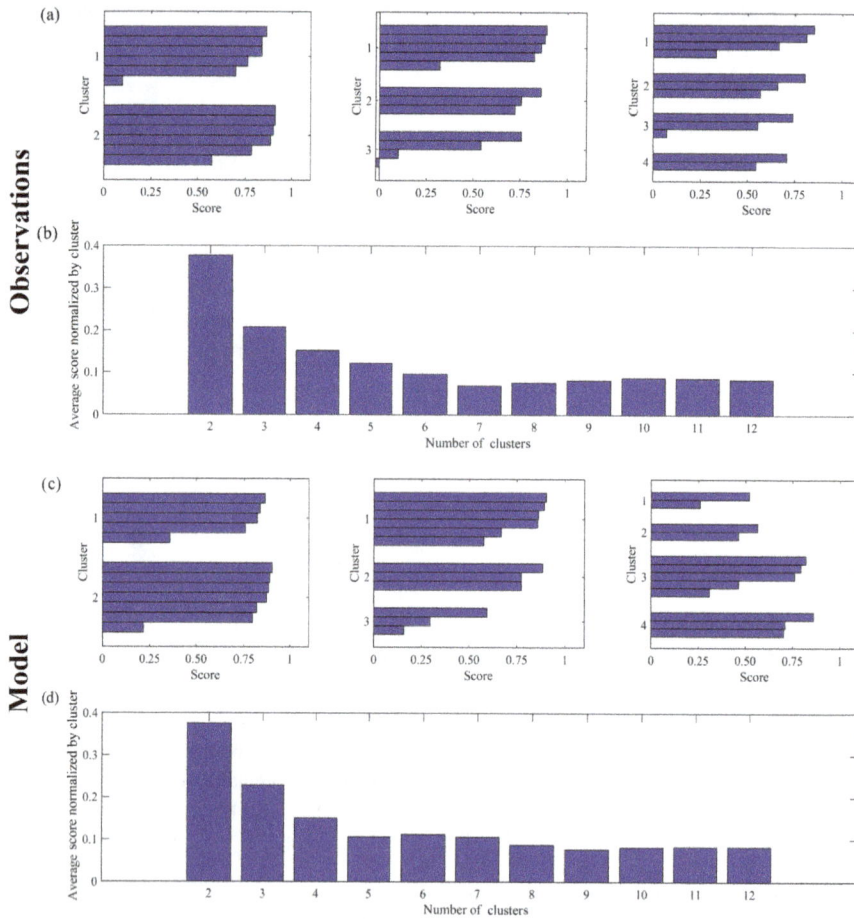

Figure 11. (a) Scores for each cluster analysis for $k = 2$, $k = 3$, $k = 4$ for the observational data in the Southern Ocean where cluster k is the predetermined number of clusters and each bar represents the score per 2-D histogram. **(b)** Average scores of each clustering analysis for increasing k, from $k = 1$ to $k = 12$. **(c)** Scores for each cluster analysis for $k = 2$, $k = 3$, $k = 4$ for the model data in the Southern Ocean. **(d)** Average scores of each clustering analysis for increasing k.

the surface by anticyclonic circulation and a strong western boundary current, the Gulf Stream. Gyre subduction supports downwelling which brings nutrients and pCO_2 to depth. Nitrate utilization by ocean biology during the winter regime is probably inaccurate in the model, while wind speed biases are known to be larger in the summer than the winter regime in the model.

In the tropics, biases in wind speed, nitrate and salinity are again found to be important. Here, nitrate biases, which are relatively higher in oligotrophic regions (Arteaga et al,. 2015), are probably due to misrepresentation of nitrogen fixation in the GISS climate model. Wind speed and salinity biases are associated with well-known biases in the intensity and position of the Inter Tropical Convergence Zone (ITCZ) that controls cloudiness, temperature gradient, and rainfall. The ITCZ moves north in the summer and south in the winter; therefore, a wind speed bias in the transition regime in the model could be explained by an inaccurate model reproduction of how the ITCZ affects the wind during its transitional

movement. The ITCZ increases precipitation, thus decreasing salinity; therefore, how salinity changes by season as a result of the shifting ITCZ could explain the winter regime bias.

4.6   The Southern Ocean carbon states

The application of cluster analysis in the Southern Ocean is presented here similarly to in the North Atlantic, with the purpose of examining whether the technique will also be able to identify some known aspects of the Southern Ocean carbon cycle. For brevity, though, the observation-based and model regimes will be presented alongside one another and will be followed by the model-error attribution analysis.

As done for the North Atlantic basin, we look at probability density distributions of observations in the Southern Ocean which show that both SST and $pCO_{2\,sw}$ exhibit a broad range of values. Temperatures range between -3 and $20\,^{\circ}C$, whereas $pCO_{2\,sw}$ values span 20 to 400 uatm. The 12-

monthly 2-D histograms of pCO_{2sw} and SST in the Takahashi dataset are shown in Fig. S5.

Figure 11a shows the scores of each regime for $k = 2, 3$, and 4 and Fig. 11b the results of the sensitivity criterion to the choice of k. Following the criterion established in Sect. 3, the largest score change is for $k = 3$ and $k = 4$. However, Fig. 11a shows that the overall score is better for $k = 3$, because more months have larger individual scores, with the exception of 1 month (November) which has a slightly negative score. Additionally, visual inspection of apparent patterns in the 12-monthly 2-D histograms (Fig. S5) also corroborates the choice of $k = 3$ as the optimal value for the number of clusters. Again, the small climatological dataset leads to some uncertainty in determining k.

A value of $k = 3$ is also chosen for the model analysis based on Fig. 11c and 11d, which show that this is also the optimal number of clusters. There is added ambiguity in the choice of k here, in addition to that due to the small dataset, which arises from the fact that the Southern Ocean is a very broad basin zonally and different processes become important in different regions more so than in the narrow North Atlantic basin, which make the choice of k not as clear as it was in the North Atlantic.

The observed and model ocean carbon states are shown in Fig. 12. In the summer regime, which includes January, February and March (see Fig. 13 and the explanation below), the highest frequency pair values (i.e., the most persistent pairs) are found around 20 % of the time in the observations and 25 % of the time in the model for the ranges of 250–350 uatm and 0–5 °C. Another range of pair values which also shows a high frequency of occurrence (25 %) is found for warmer temperatures (10–20 °C) and the same range of pCO_2, but the GISS model misplaces it towards higher values of pCO_2 (350–400 uatm). The winter regime comparison shows that the model captures the low pCO_2, low temperature state (20–150 uatm, −3 to 0 °C) well (30 % of the time in observations and in the model). The mid-range pCO_2, high temperature state (250–350 uatm, 10–20 °C) is not as well represented in the model. The observations there show the highest frequencies of occurrence for higher pCO_2 (350–400 uatm), in contrast to the model for lower pCO_2 (250–350 uatm). Comparison between the transition regimes reveals much less correspondence between observations and the GISS model, considering the high frequency states in observations are quite different than in the model (e.g., observations of high frequency states: 20–150 uatm, −3 to 0 °C; 250–350 uatm, 10–20 °C; 350–400 uatm, 0–5 °C; model high frequency state: 250–350 uatm, 0–10 °C).

4.6.1 Temporal attribution for the Southern Ocean carbon states

Temporal attribution, which is estimated using the method described in Sect. 4.1, is shown for both the model and the observations in Fig. 13. Note that all subsequent analysis

considers the austral seasons when referring to "winter" and "summer". The temporal attribution shows that the observations and model data are clustered in regimes that correspond to almost the same months. The only difference is that November is accounted for in the transition regime for the observations as opposed to the winter regime for the model. It is noted, however, that November is technically a "poorly matched" 2-D histogram in the observation cluster routine, as discussed earlier.

4.6.2 Spatial attribution of the Southern Ocean carbon states

In order to further explain the model and observed Southern Ocean regimes shown in Fig. 12, the frequencies of occurrence of each bin in the cluster are mapped onto the Southern Ocean regions (Fig. 14), where geographic nomenclature similar to that in Orsi et al. (1995) is used. In the coastal Antarctic divergence zone, SST varies within −3 to 3 °C and pCO_2 20–250 uatm, in the Antarctic convergence zone SST ranges within 3–10 °C and pCO_2 250–400 uatm, and in the subtropical convergence zone SST lies within 10–20 °C and pCO_2 250–400 uatm. Despite the strong temporal attribution agreement between the model and the observations, the regional attributions show much less correspondence. For example, in the summer regime, the observations show the highest frequency of occurrence between 250–350 uatm and 10–20 °C along the subtropical convergence zone (roughly along 40° S), while the highest frequency of occurrence for the model is for the pair 250–350 uatm and 0–5 °C, occurring in the coastal region (poleward of the divergence zone along 60° S). In the winter regime, both observations and the GISS model show the highest frequency of occurrence for the value pairs nearest to the coast where the pCO_2 is low and the temperatures coldest (20–150 uatm and −3 to 10 °C). Further offshore, the model highest frequency values occur for lower pCO_2 (250–350 uatm) than in the observations (350–400 uatm). Lastly, in the transition regime, the model shows a much higher persistency of values in the range (250–350 uatm and 0–10 °C) than in the observations.

It is therefore evident that the GISS model does not reproduce well the observed ocean carbon states in the Southern Ocean. There is a tendency to persistently underestimate pCO_2-SST values closer to the coast and overestimate it near the subtropical convergence zone, during the warm season, whereas, during the cold season, the model captures well the divergence zone regime but the errors further offshore switch: the model now overestimates pCO_2-SST value-pair frequency of occurrence in the Antarctic convergence zone but is performing better in the subtropical convergence zone. The transition regime is not well represented, indicating that not enough regimes are chosen to adequately describe ocean carbon states in this small dataset.

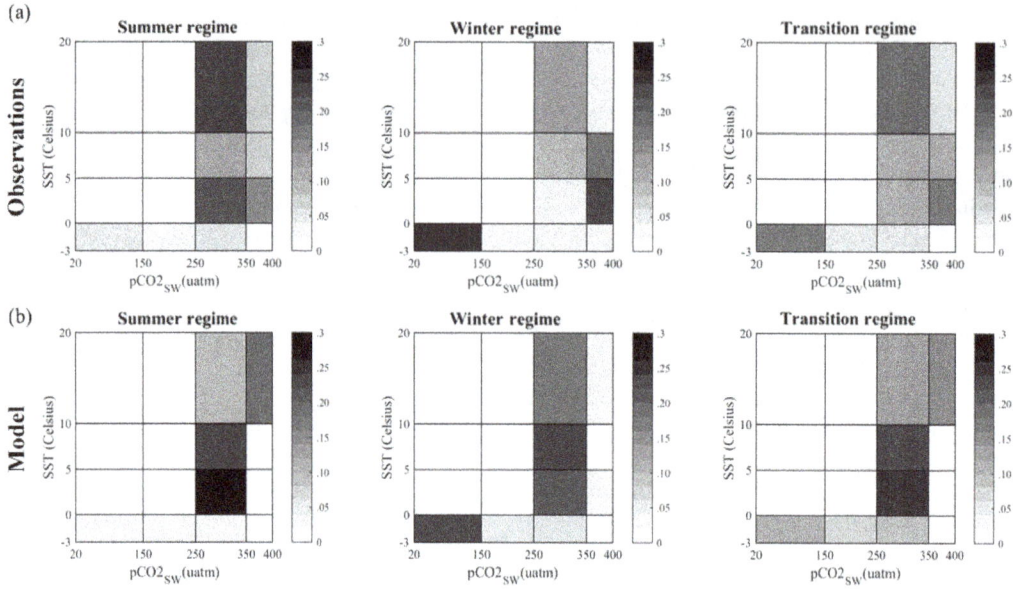

Figure 12. Comparison between the regimes in (**a**) the observations and (**b**) the model output.

Figure 13. Monthly attribution of each ocean carbon regime in observations and the GISS climate model. Temporal attribution is based on the distance of each monthly 2-D histogram to the centroid of each cluster. Referring to austral seasons, blue denotes the cold month regime (winter regime), red the warm months (summer regime) and green the transition regime, since this is more a mix of the other two.

4.6.3 Model Southern Ocean air–sea flux of CO_2 error analysis and bias attribution

Comparing the CO_2 flux composites on the observed regimes (Fig. 15) shows significant discrepancies between observations (Fig. 15a) and model (Fig. 15b). To better understand where these discrepancies come from, we perform an error attribution analysis, as in Sect. 4.1 above. In the summer regime (January through March), outgassing in the model is restricted to the subtropical convergence zone, whereas in the observations it is more localized and closer to Antarctica. At the same time, model uptake is stronger, confined to the coast and over a broader area than in the observations. This is consistent with the result from the previous section, where the model underestimates pCO_2 and SST in the Antarctic convergence zone. In the winter regime (June through October), the entire model basin is a sink for CO_2, whereas in the observations there is a zonally confined outgassing belt south of $50°$ S and an uptake belt north of it. Again, this result is consistent with the earlier finding that the model winter regime is closer to the observed near the Antarctic coast. The transi-

tion regime shares a mix of the same discrepancies as in the summer and winter regimes.

Bias attribution is computed for the three zonally defined regions indicated in Fig. S6. Based on the bias computations in Eq. (5) for $pCO_{2\,sw}$, SST, salinity, and wind speed with respect to CO_2 flux, $pCO_{2\,sw}$ is again shown to be the driving variable in most of the flux biases in the Southern Ocean (Fig. S7; Fig. S8 for scatter plots). We therefore seek to understand the processes that control the pCO_2 biases in the model, using the Taylor expansion in Eq. (6) (Fig. 16; Fig. S9 for scatter plots).

For almost all regimes and regions, biases in nitrate are large partly because of a lack of a closed, state-of-the-art nitrogen cycle representation in the climate model. On the other hand, observations are too scarce in the region, due to inclement weather and biases to specific seasons, so there is large observational uncertainty associated with the Takahashi climatology in the Southern Ocean. The model skill would be more adequately assessed as more in situ measurements are made (e.g., from the Southern Ocean Carbon and Climate Observations and Modeling, SOCCOM, experiment; Johnson et al., 2017). Nevertheless, the model underestimates surface nitrates in the Southern Ocean in particular because of a large nitrate deficit in the subsurface ocean which upwells in the subantarctic zone and flows into the Antarctic Circumpolar Current region. This is related to processes such as denitrification and accurate remineralization in the deep ocean. SST is the second-most dominating variable for biases in the coastal Antarctic. Inspection of the model biases shows that south of $70°$ S the model water column is colder than in observations; hence, upwelling there will bring colder waters near the surface. Interestingly, surface salinity biases are rel-

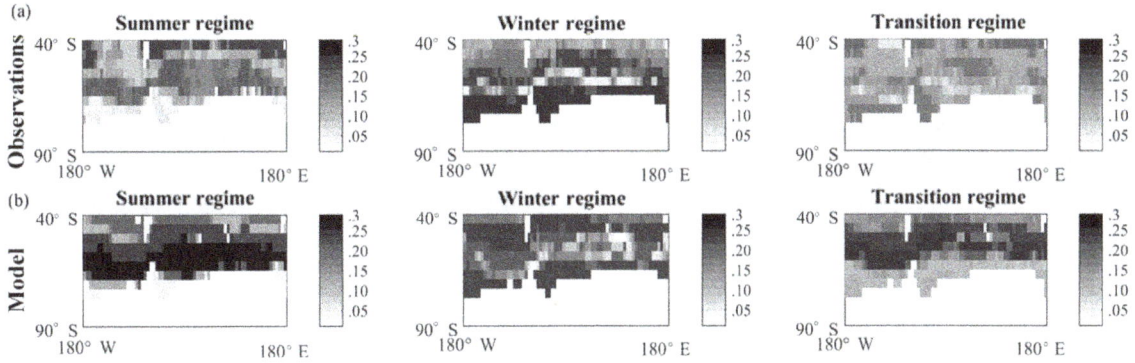

Figure 14. Regional attribution of each regime for $k = 3$ in the Southern Ocean in (a) the observations and (b) the GISS model simulations. Each spatial grid point for every month is associated with its relative frequency of occurrence in the cluster output, and then the months are averaged per regime to output the average frequency of occurrence in each regime. Model regimes are calculated using the monthly groups identified by the observations' temporal attribution.

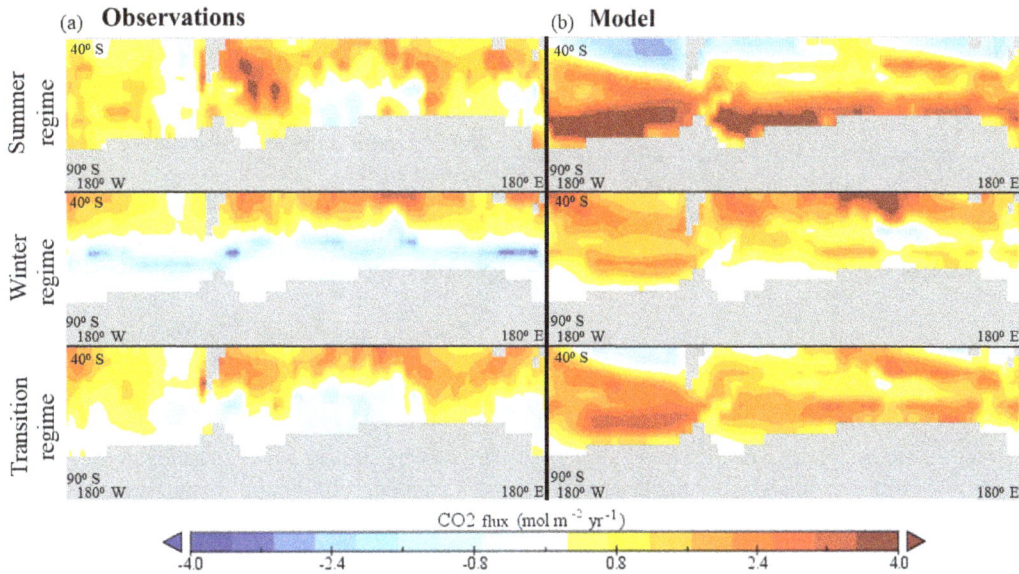

Figure 15. Composites of the CO_2 flux field over the regimes in the Southern Ocean for (a) the observations and (b) the GISS model. Both the observations and the model data are composited over the same months as determined by the temporal attribution of the observed regimes. Blue shades indicate outgassing, and red shades indicate uptake.

atively very important in the region south of the subtropical convergence zone, which suggests that a study of water mass formation in that region in the model and the observations would better explain the biases.

5 Data availability

Data and analysis scripts can be accessed at https://data.giss.nasa.gov/oceans/carbonstates.

6 Conclusions

This proof-of-concept study presents the k-means cluster analysis and the determination of the regimes called "ocean carbon states" in observation-based data of the ocean carbon cycle. A method is described here to determine the optimal number of clusters for the cluster analysis. The study also explores how to characterize the ocean carbon states temporally and spatially in order to determine the physical–biogeochemical processes related to each carbon state. Composites of the CO_2 flux and a quantitative exploration of the effect of each field on $pCO_{2\,sw}$ bias are also demonstrated.

In this study, pCO_2 and SST were chosen as the two variables that co-determine the carbon states, based on the fact that they both play critical roles in the biogeochemistry and the physics of the ocean system and control the flux of CO_2. One may choose different variables and it would be interest-

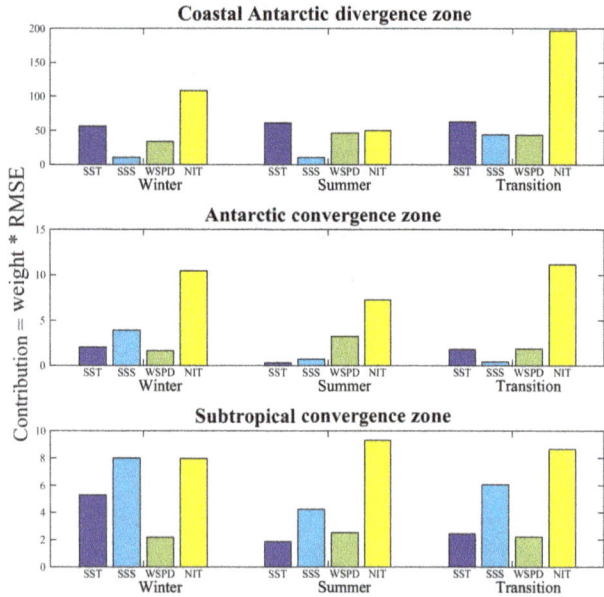

Figure 16. Contributions to the $pCO_{2\,sw}$ bias in the model from SST, SSS, wind speed (WSPD) and nitrate (NIT) in the winter, summer, and transitional regimes. The Southern Ocean is differentiated into the coastal Antarctic divergence zone (roughly polewards of 60° S), Antarctic convergence zone (roughly 60–50° S), and subtropical convergence zone (roughly 50–40° S) to better account for regional differences in the model biases and obtain better linear fit for the computation of the weights in Eq. (6) (see Fig. S9).

ing to see whether and how the regimes depend on the choice of variables.

We have also tested the importance of the choice of k. A main caveat of k-means cluster analysis is that k must be predetermined through reasoning that is subject to personal bias. However, we show that by assessing the clusters from multiple angles (i.e., the score plots, the sensitivity criterion visual inspection, analysing cluster outputs with increasing k, temporal attribution), it is possible to determine an optimal k that is semi-objective. Even so, in this proof of concept study, we acknowledge the uncertainty in our choice of k as a result of the small dataset of 12-monthly 2-D histograms, which occasionally results in there being only a small number of histograms per cluster.

The ocean carbon states we obtain from this climatological year dataset are interesting. We found that the subtropical North Atlantic is the dominant feature in the cold season regime (the months January through April). In the same regime, the subpolar North Atlantic also features prominently, which is associated with the high variability in this area due to sea ice retreat, the spring bloom and the winter–spring convection. In contrast, the tropical Atlantic dominates the warm months June through October, while the subtropical and subpolar regions play a smaller role. The transition regime, which is comprised of months that do not entirely fall into the winter or summer regimes, shows again

the lower tropics and the subtropical gyre to be more active. We would expect that a longer dataset that includes natural variability as well as the effects of longer-term climate and anthropogenic trends would result in more carbon states and hence that it would be an interesting extension of the present study.

The NASA-GISS model carbon states in the North Atlantic are similar, both in temporal as well as spatial characterization, to the observed ones, with better model skill in the summer than in the winter. Specifically, the model overestimates the importance of the subtropical gyre and underestimates the subpolar gyre during the cold months. During summer, the model underestimates the tropics, but not significantly. The transition months are found to behave differently in the model, although that might be a result of the small size of our input datasets.

In the Southern Ocean, during the warmer months (January–March), the observational states are more persistent along 40° S, the subtropical convergence zone, while the colder season has prominent states (higher persistency) mostly along the Antarctic coast. The transition regime shows a similar degree of variability across the entire Southern Ocean.

While the GISS model agrees in the temporal characterization of the ocean carbon states, it diverges from the observational spatial attribution, particularly in the summer and the transition regimes. It is of note here, however, that the Takahashi climatology is far more uncertain in the Southern Ocean (Takahashi et al., 2009) than it is in the North Atlantic, and therefore the model's lack of skill might not be as alarming. New observations in the area (e.g., from SOCCOM) will greatly benefit studies such as this.

Error analysis of the model response helps explain the GISS model biases. Applying k-means clustering analysis in the two main regions of the world that are known to be critical for the global ocean carbon cycle, namely the North Atlantic region and the Southern Ocean, defines the priorities for model improvement: in the North Atlantic biases in surface salinity, wind speed and surface temperature, whereas in the Southern Ocean priorities are nitrate and surface salinity. Clearly the GISS climate model would benefit from more realistic representation of the nitrogen cycle in the ocean as a whole.

The goal of this study is to enable us to apply this k-means clustering to "big" data, in order to find the interannual and regional patterns in larger, higher frequency climate datasets. This extended application will allow researchers to gain a much more comprehensive insight and intuition for physical systems by mechanistically and impartially grouping multiple variables that form the prominent features of these networks. Other variable pairs besides pCO_2 and SST will also be explored, such as CO_2 flux and chlorophyll, in order to assess other drivers in Eq. (1). Finally, higher order clustering and classification techniques will be analyzed in order to determine the most efficient and successful method for understanding the ocean carbon cycle.

All routines and datasets used in this study are freely available on the Ocean Carbon States page of the NASA Goddard Institute for Space Studies web portal (https://data.giss.nasa.gov/oceans/carbonstates).

Competing interests. The authors declare that they have no conflict of interest.

Acknowledgements. Resources supporting this work were provided by the NASA High-End Computing (HEC) program through the NASA Center for Climate Simulation (NCCS) at Goddard Space Flight Center.

Funding was provided by NASA-ROSES Modeling, Analysis and Prediction 2013 NNX14AB99A-MAP for GISS Model-E development and NNX15AJ05A NASA Cooperative Agreement 2015–2018.

Data used to generate figures, graphs, plots, as well as analysis were archived at the NCCS dirac repository, numerical codes are maintained and archived at GISS and all data and codes are available upon request from Anastasia Romanou. Clustering analysis was performed using the MATLAB ver 2015 computing environment. The authors wish to thank Robert Schmunk for his help in setting up the Zenodo page and the GISS portal.

Edited by: David Carlson

References

Anderberg, M. R.: Cluster analysis for applications, Academic Press, New York, 1973.

Arteaga, L., Pahlow, M., and Oschlies, A.: Global monthly sea surface nitrate fields estimated from remotely sensed sea surface temperature, chlorophyll, and modeled mixed layer depth, Geophys. Res. Lett, 42, 1130–1138, https://doi.org/10.1002/2014GL062937, 2015.

Bankert, R. L. and Solbrig, J. E.: Cluster Analysis of A-Train Data: Approximating the Vertical Cloud Structure of Oceanic Cloud Regimes, J. Appl. Meteorol. Clim., 54, 996–1008, https://doi.org/10.1175/JAMC-D-14-0227.1, 2015.

Bodas-Salcedo, A., Williams, K. D., Ringer, M. A., Beau, I., Cole, J. N., Dufresne, J., Koshiro, T., Stevens, B., Wang, Z., and Yokohata, T.: Origins of the Solar Radiation Biases over the Southern Ocean in CFMIP2 Models, J. Climate, 27, 41–56, https://doi.org/10.1175/JCLI-D-13-00169.1, 2014

Boyer, T. P., Antonov, J. I., Baranova, O. K., Coleman, C., Garcia, H. E., Grodsky, A., Johnson, D. R., Locarnini, R. A., Mishonov, A. V., O'Brien, T. D., Paver, C. R., Reagan, J. R., Seidov, D., Smolyar, I. V., and Zweng, M. M.: World Ocean Database 2013, in: NOAA Atlas NESDIS 72, edited by: Levitus, S. and Mishonov, A., Silver Spring, MD, 209 pp., https://doi.org/10.7289/V5NZ85MT, 2013.

Dlugokencky, E. and Tans, P.: Trends in atmospheric carbon dioxide, National Oceanic & Atmospheric Administration, Earth System Research Laboratory (NOAA/ESRL), available at: http://www.esrl.noaa.gov/gmd/ccgg/trends, last access: 8 August 2014.

Fay, A. R. and McKinley, G. A.: Global open-ocean biomes: mean and temporal variability, Earth Syst. Sci. Data, 6, 273–284, https://doi.org/10.5194/essd-6-273-2014, 2014.

Halverson, M. J. and Fleming, S. W.: Complex network theory, streamflow, and hydrometric monitoring system design, Hydrol. Earth Syst. Sci., 19, 3301–3318, https://doi.org/10.5194/hess-19-3301-2015, 2015.

Hoffman, F. M., Hargrove, W. W., Mills, R. T., Mahajan, S., Erickson, D. J., and Oglesby, R. J.: Multivariate Spatio-Temporal Clustering (MSTC) as a data mining tool for environmental applications, edited by: Sànchez-Marrè, M., Bèjar, J., Comas, J., Rizzoli, A. E., and Guariso, G., Proceedings of the iEMSs Fourth Biennial Meeting: International Congress on Environmental Modelling and Software Society (iEMSs 2008), Barcelona, Catalonia, Spain, July 2008.

Hoffman, F. M., Larson, J. W., Mills, R. T., Brooks, B. J., Ganguly, A. R., Hargrove, W. W., Huang, J., Kumar, J., and Vatsavai, R. R.: Data Mining in Earth System Science (DMESS 2011), Procedia Comput. Sci., 4, 1450–1455, https://doi.org/10.1016/j.procs.2011.04.157, 2011.

Hugue, F., Lapointe, M., Eaton, B. C., and Lepoutre, A.: Satellite-based remote sensing of running water habitats at large riverscape scales: Tools to analyze habitat heterogeneity for river ecosystem management, Geomorphology, 253, 353–369, https://doi.org/10.1016/j.geomorph.2015.10.025, 2016.

Jain, A. K.: Data clustering: 50 years beyond K-means, Pattern Recogn. Lett., 31, 651–666, https://doi.org/10.1016/j.patrec.2009.09.011, 2010.

Jakob, C. and Tselioudis, G.: Objective identification of cloud regimes in the Tropical Western Pacific, Geophys. Res. Lett., 30, 2082, https://doi.org/10.1029/2003GL018367, 2003.

Johnson, K. S., Plant, J. N., Coletti, L. J., Jannasch, H. W., Sakamoto, C. M., Riser, S. C., Swift, D. D., Williams, N. L., Boss, E., Haentjens, N., Talley, L. D., and Sarmiento, J. L.: Biogeochemical sensor performance in the SOCCOM profiling float array, J. Geophys. Res.-Oceans, 122, 6416–6436, https://doi.org/10.1002/2017JC012838, 2017

Kaufman, L. and Rousseauw, P.: Finding Groups in Data: An Introduction to Cluster Analysis, John Wiley & Sons, Inc., Hoboken, New Jersey, 2005.

Landschützer, P., Gruber, N., Bakker, D. C. E., Schuster, U., Nakaoka, S., Payne, M. R., Sasse, T. P., and Zeng, J.: A neural network-based estimate of the seasonal to inter-annual variability of the Atlantic Ocean carbon sink, Biogeosciences, 10, 7793–7815, https://doi.org/10.5194/bg-10-7793-2013, 2013.

Landschützer, P., Gruber, N., Bakker, D. C. E., and Schuster, U.: Recent variability of the global ocean carbon sink, Global Biogeochem. Cy., 28, 927–949, https://doi.org/10.1002/2014GB004853, 2014.

Lefèvre, N., Watson, A. J., and Watson, A. R.: A comparison of multiple regression and neural network techniques for mapping in situ pCO$_2$ data, Tellus B, 57, 375–384, https://doi.org/10.1111/j.1600-0889.2005.00164.x, 2005.

Nakaoka, S., Telszewski, M., Nojiri, Y., Yasunaka, S., Miyazaki, C., Mukai, H., and Usui, N.: Estimating temporal and spatial variation of ocean surface pCO$_2$ in the North Pacific using a self-organizing map neural network technique, Biogeosciences, 10, 6093–6106, https://doi.org/10.5194/bg-10-6093-2013, 2013.

Oreopoulos, L., Nayeong, C., Lee, D., and Kato, S.: Radiative effects of global MODIS cloud regimes, J. Geophys. Res., 121, 2299–2317, https://doi.org/10.1002/2015JD024502, 2016.

Orsi, A., Whitworth, T., and Nowlin, W. D.: On the meridional extent and fronts of the Antarctic Circumpolar Current, Deep-Sea Res. Pt. I, 42, 641–673, 1995.

Peron, T. K. D., Comin, C. H., Amancio, D. R., da F. Costa, L., Rodrigues, F. A., and Kurths, J.: Correlations between climate network and relief data, Nonlin. Processes Geophys., 21, 1127–1132, https://doi.org/10.5194/npg-21-1127-2014, 2014.

Phillips, J. D., Schwanghart, W., and Heckmann, T.: Graph theory in the geosciences, Earth-Sci. Rev., 143, 147–160, https://doi.org/10.1016/j.earscirev.2015.02.002, 2015.

Radebach, A., Donner, R. V., Runge, J., Donges, J. F., and Kurths, J.: Disentangling different types of El Niño episodes by evolving climate network analysis, Phys. Rev. E, 88, 052807, https://doi.org/10.1103/PhysRevE.88.052807, 2013.

Romanou, A., Gregg, W. W., Romanski, J., Kelley, M., Bleck, R., Healy, R., Nazarenko, L., Russell, G.. Schmidt, G. A., Sun, S., and Tausnev, N.: Natural air–sea flux of CO_2 in simulations of the NASA-GISS climate model: Sensitivity to the physical ocean model formulation, Ocean Model., 66, 26–44, https://doi.org/10.1016/j.ocemod.2013.01.008, 2013.

Romanou, A., Marshall, J., Kelley, M., and Scott, J.: Role of the ocean's AMOC in setting the uptake efficiency of transient tracers, Geophys. Res. Lett., 44, 5590–5598, https://doi.org/10.1002/2017gl072972, 2017.

Rossow, W. B., Zhang, Y.-C., and Wang, J.: A statistical model of cloud vertical structure based on reconciling cloud layer amounts inferred from satellites and radiosonde humidity profiles, J. Climate, 18, 3587–3605, https://doi.org/10.1175/JCLI3479.1, 2005.

Sarmiento, J. L. and Gruber, N.: Ocean Biogeochemical Dynamics, Princeton University Press, New Jersey, USA, 2006.

Sasse, T. P., McNeil, B. I., and Abramowitz, G.: A new constraint on global air–sea CO_2 fluxes using bottle carbon data, Geophys. Res. Lett., 40, 1594–1599, https://doi.org/10.1002/grl.50342, 2013.

Takahashi, T., Sutherland, S. C., Wanninkhof, R., Sweeney, C., Feely, R. A., Chipman, D. W., Hales, B., Friederich, G., Chavez, F., Watson, A., Bakker, D. C. E., Schuster, U., Metzl, N., Yoshikawa-Inoue, H., Ishii, M., Midorikawa, T., Nojiri, Y., Sabine, C., Olafsson, J., Arnarson, Th. S., Tilbrook, B., Johannessen, T., Olsen, A., Bellerby, R., Körtzinger, A., Steinhoff, T., Hoppema, M., de Baar, H. J. W., Wong, C. S., Delille B., and Bates, N. R.: Climatological mean and decadal changes in surface ocean pCO_2, and net sea-air CO_2 flux over the global oceans, Deep-Sea Res. Pt. II, 56, 554–577, https://doi.org/10.1016/j.dsr2.2008.12.009, 2009.

Telszewski, M., Chazottes, A., Schuster, U., Watson, A. J., Moulin, C., Bakker, D. C. E., González-Dávila, M., Johannessen, T., Körtzinger, A., Lüger, H., Olsen, A., Omar, A., Padin, X. A., Ríos, A. F., Steinhoff, T., Santana-Casiano, M., Wallace, D. W. R., and Wanninkhof, R.: Estimating the monthly pCO_2 distribution in the North Atlantic using a self-organizing neural network, Biogeosciences, 6, 1405–1421, https://doi.org/10.5194/bg-6-1405-2009, 2009.

Trans Mills, R., Hoffman, F. M., Kumar, J., and Hargrove, W. W.: Cluster Analysis-Based Approaches for Geospatiotemporal Data Mining of Massive Data Sets for Identification of Forest Threats, Procedia Comput. Sci., 4, 1612–1621, https://doi.org/10.1016/j.procs.2011.04.174, 2011.

Trochta, J. T., Mouw, C. B., and Moore, T. S.: Remote sensing of physical cycles in Lake Superior using a spatio-temporal analysis of optical water typologies, Remote Sens. Environ., 171, 149–161, https://doi.org/10.1016/j.rse.2015.10.008, 2015.

Tselioudis, G., Rossow, W., Zhang, Y., and Konsta, D.: Global Weather States and Their Properties from Passive and Active Satellite Cloud Retrievals, J. Climate, 26, 7734–7746, https://doi.org/10.1175/JCLI-D-13-00024.1, 2013.

Wanninkhof, R.: Relationship between wind speed and gas exchange over the ocean, J. Geophys. Res., 97, 7373–7382, https://doi.org/10.1029/92JC00188, 1992.

Wanninkhof, R., Park, G.-H., Takahashi, T., Sweeney, C., Feely, R., Nojiri, Y., Gruber, N., Doney, S. C., McKinley, G. A., Lenton, A., Le Quéré, C., Heinze, C., Schwinger, J., Graven, H., and Khatiwala, S.: Global ocean carbon uptake: magnitude, variability and trends, Biogeosciences, 10, 1983–2000, https://doi.org/10.5194/bg-10-1983-2013, 2013.

Williams, K. and Webb, M.: A quantitative performance assessment of cloud regimes in climate models, Clim. Dynam., 33, 141–157, https://doi.org/10.1007/s00382-008-0443-1, 2009.

Wood, R., Wyant, M., Bretherton, C. S., Rémillard, J., Kollias, P., Fletcher, J., Stemmler, J., de Szoeke, S., Yuter, S., Miller, M., Mechem, D., Tselioudis, G., Chiu, J. C., Mann, J. A., O'Connor, E. J., Hogan, R. J., Dong, X., Miller, M., Ghate, V., Jefferson, A., Min, Q., Minnis, P., Palikonda, R., Albrecht, B., Luke, E., Hannay, C., and Lin, Y.: Clouds, Aerosols, and Precipitation in the Marine Boundary Layer: An Arm Mobile Facility Deployment, B. Am. Meteorol. Soc., 96, 419–440, https://doi.org/10.1175/BAMS-D-13-00180.1, 2015.

Zagouras, A., Kazantzidis, A., Nikitidou, E., and Argiriou, A. A.: Determination of measuring sites for solar irradiance, based on cluster analysis of satellite-derived cloud estimations, Sol. Energy, 97, 1–11, https://doi.org/10.1016/j.solener.2013.08.005, 2013.

15

The Sub-Polar Gyre Index – a community data set for application in fisheries and environment research

Barbara Berx[1] **and Mark R. Payne**[2]

[1]Marine Scotland Science, 375 Victoria Road, Aberdeen, AB11 9DB, UK
[2]Centre for Ocean Life, Technical University of Denmark, National Institute of Aquatic Resources,
2920 Charlottenlund, Denmark

Correspondence to: Barbara Berx (b.berx@marlab.ac.uk)

Abstract. Scientific interest in the sub-polar gyre of the North Atlantic Ocean has increased in recent years. The sub-polar gyre has contracted and weakened, and changes in circulation pathways have been linked to changes in marine ecosystem productivity. To aid fisheries and environmental scientists, we present here a time series of the Sub-Polar Gyre Index (SPG-I) based on monthly mean maps of sea surface height. The established definition of the SPG-I is applied, and the first EOF (empirical orthogonal function) and PC (principal component) are presented. Sensitivity to the spatial domain and time series length are explored but found not to be important factors in terms of the SPG-I's interpretation. Our time series compares well with indices presented previously. The SPG-I time series is freely available online (http://dx.doi.org/10.7489/1806-1), and we invite the community to access, apply, and publish studies using this index time series.

1 Introduction

The sub-tropical and sub-polar gyres are the dominant features of the surface circulation of the North Atlantic Ocean (Fig. 1). Both are driven by the combination of permanent wind features, heat-input variation with latitude, and the global overturning circulation. The sub-tropical gyre is formed by the synthesis of the Gulf Stream, North Atlantic Current, Canary Current, and North Equatorial Current to yield a nearly continuous, anticyclonic circulation in the sub-tropical North Atlantic Ocean. Its equivalent in the sub-polar region can be considered as a cyclonic gyre encompassing the North Atlantic, East Greenland, and Labrador Currents (Fig. 1). Within the North Atlantic, changes in the strength and extent of the sub-polar gyre have been linked to changes in the advection of water masses (Häkkinen and Rhines, 2009) and changes in their properties (Holliday et al., 2008; Johnson et al., 2013). These changes in the strength and extent of the sub-polar gyre have been attributed to the strong overturning circulation observed in the preceding years (Häkkinen and Rhines, 2004; Bersch et al., 2007). More recently, marine ecologists have reported changes in

the ecosystem associated with changes in circulation in the North Atlantic and particularly the sub-polar gyre region (e.g. Hátún et al., 2009a, b; Payne et al., 2012).

Based on a survey of research scientists within the International Council on the Exploitation of the Seas (ICES) community, the Working Group on Operational Oceanographic products for Fisheries and Environment (WGOOFE; Berx et al., 2011) identified a need from fisheries and environmental scientists for freely accessible oceanographic data, in a suitable data format and with operational delivery. Within the climate community, the need for large volumes of data to be distilled into readily accessible, user-friendly data sets has recently driven the development of the Climate Data Guide (Schneider et al., 2013; Overpeck et al., 2011). This site provides a community-based overview of available data products and includes some expert guidance on the strengths and weaknesses of the products as well as information on their derivation. In its work as an interface between the ICES community and the operational oceanography community, WGOOFE found that index-based products – where a complex spatio-temporal data set, process, or state estimate may

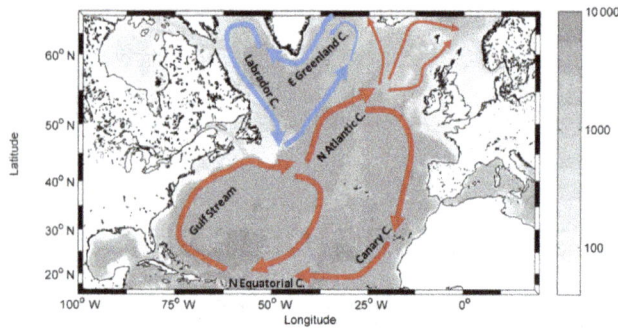

Figure 1. Map of the sub-tropical and sub-polar North Atlantic with a schematic representation of the ocean circulation.

be reduced to a single time series, such as the North Atlantic Oscillation Index (NAO; Hurrell and NCAR Staff, 2013) – remain a major gap in the available oceanographic data products (ICES, 2012a). Recently, Bessières et al. (2013) presented three index-based oceanographic data products developed by the MyOcean project: the El Niño indicator, the Kuroshio Extension indicator, and the Ionian Surface Circulation indicator. However, these index time series have not yet been made readily available to the wider community. To date, no readily accessible up-to-date data set exists summarizing the sub-polar gyre's dynamics. This limits some researchers in the field of ecosystem science in the search for drivers of ecosystem variability within the region and therefore also the development of improved fisheries management tools.

The sub-polar gyre index has been applied recently in a number of studies investigating ecosystem variability. Johnson et al. (2013) present a decline in nutrient concentrations (particularly nitrate and phosphate concentrations) in the Rockall Trough related to the strength of the sub-polar gyre. Recent work has even highlighted potential linkages between sub-polar gyre dynamics and higher trophic levels, including commercially important fish stocks (Hátún et al., 2009a, b, 2016). In 2012, the ICES Working Group on Widely Distributed Stocks (WGWIDE) emphasized the absence of such a data product as a key obstacle when studying distribution and abundance changes in economically important fish stocks, such as mackerel and blue whiting (ICES, 2012b).

The data set presented here aims to fill this gap in index-based operational oceanographic data products by presenting a time series of the Sub-Polar Gyre Index (SPG-I) extending from the start of satellite altimetry records (January 1993) to the present. The data set presented is freely available and easily citable. In Sect. 2, we outline the underlying data set and methodology for our SPG-I calculation; in Sect. 3 we present the data product and its sensitivities and compare our time series with other published results of SPG-I variability; and finally we present how to access the SPG-I data product and

acknowledge its use (Sect. 4), followed by a brief conclusion and outlook (Sect. 5).

2 Methodology

2.1 Sea surface height data

The altimeter products used to create the SPG-I were obtained through the Copernicus Marine Environment Monitoring Service (CMEMS; product identifier: SEALEVEL_GLO_SLA_MAP_L4_REP_OBSERVATIONS _008_027). For our analysis, we obtained the delayed time, global, daily Maps of Sea Level Anomaly (MSLAs) on a $1/4°$ by $1/4°$ grid. The product is the result of merging all available satellite missions at a given time, resulting in a better-quality product (particularly in recent years). Monthly mean maps were created by averaging the multimission daily maps by month, while seasonal climatology maps were calculated by averaging the monthly mean maps within the same month for all complete years (1993–2015). The climatological maps therefore represent the average conditions for each month of the year throughout the observation period. To avoid issues with observations in grid points on land, a land–sea mask was obtained by interpolating the $1/12°$ by $1/12°$ TerrainBase database (National Geophysical Data Center, 1995) on to the same grid as the altimeter data. The land–sea mask was also used to remove altimeter data in the Pacific Ocean, Great Lakes, and Mediterranean Sea. The altimeter data set starts in January 1993, and the latest update obtained from CMEMS extends to April 2016.

2.2 Calculation of SPG-I

We followed the method of Häkkinen and Rhines (2004) to calculate the SPG-I, which has been defined as the first principal component (PC1) of an empirical orthogonal function analysis (EOFA) of the sea level anomaly field in the North Atlantic. In Sect. 3, our results are compared to similarly defined gyre indices based on altimeter data, although alternative indices based on sea surface temperature and wind stress curl have also been defined by Hátún et al. (2009a) and Häkkinen et al. (2011), respectively.

In our analysis, we restricted the geographical extent to a rectangular area focused on the North Atlantic's sub-polar gyre (delimited by the 60° W and 10° E meridians and the 40 and 65° N parallels). The exact choice of spatial domain varies between authors – here we have chosen a region focused on the sub-polar gyre region itself. However, we also perform sensitivity analyses to examine the effect of this choice on the resulting index time series. Seasonality in the monthly mean observations of sea level anomaly was removed by subtracting the relevant climatological sea level anomaly map. We calculated the monthly SPG-I based on these deseasonalized maps of sea level anomaly. A yearly SPG-I is calculated based on the average of the deseasonal-

Table 1. Overview of defined regions to investigate sensitivity of SPG-I to the chosen spatial coverage (regions are also shown in Fig. 3). The abbreviations in the first column correspond to those used in figure legends. S: S. Häkkinen time series; B: time series presented here; R: time series based on different regions; 1X1: time series presented here based on a 1° by 1° grid; 2X2: time series presented here based on a 2° by 2° grid; all other time series presented here are based on a 1/4° by 1/4° grid.

Region	Longitude		Latitude	
	Left	Right	Bottom	Top
S	100° W	20° E	15° N	65° N
B, 1X1, 2X2	60° W	10° E	40° N	65° N
R1	100° W	30° E	0° N	66° N
R2	95° W	25° E	5° N	66° N
R3	90° W	20° E	10° N	66° N
R4	85° W	15° E	15° N	60° N
R5	80° W	10° E	15° N	60° N
R6	75° W	5° E	30° N	66° N
R7	65° W	0° E	45° N	66° N
R8	60° W	5° W	45° N	66° N

ized sea level anomalies by calendar year, with only complete years included in the analysis.

EOFA is a well-established analysis technique, but for completeness a short description follows. For more in-depth information, we refer the reader to Emery and Thomson (2001).

A major strength of EOFA is the reduction in data volume: a large data set can be reduced to a smaller one containing the most significant fraction of variability contained in the original data. In particular, it can often reduce large spatial data sets to a more manageable size. We can consider the altimeter data set as a time series of I time instances at J spatial locations, defined by the point's latitude and longitude on the grid. During EOFA, the data matrix is standardized (for each location, the mean is removed from the time series and the remaining anomalies then scaled by dividing by their standard deviation) and then decomposed into mutually uncorrelated (orthogonal) modes which have a spatial pattern (these are called the eigenvectors or empirical orthogonal functions) and a temporal amplitude (these are called the eigenfunctions or principal components). The first mode extracted using the EOFA technique explains the largest fraction of variability in the data set, and each subsequent mode explains the largest fraction of the remaining variability. By extracting the first mode, we obtain the time series of SPG-I based on the MSLA data. The sign of an EOF (empirical orthogonal function)/PC is not determined explicitly during the calculation process and may vary between machines and software versions. We therefore define the sign of the EOF and PC in 1993 to be positive and make adjustments as required.

The addition of data points either in space or time also changes the EOF/PC results, which is briefly explored here

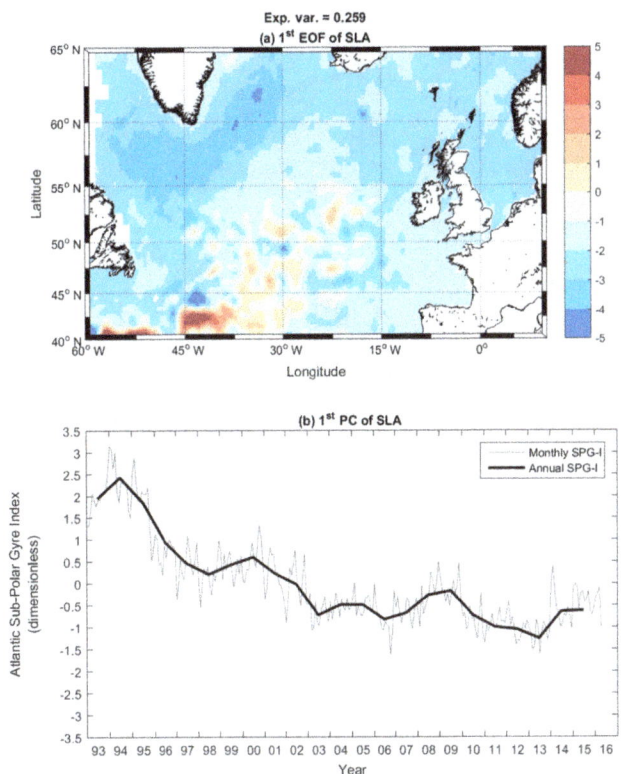

Figure 2. First mode of the EOFA of sea surface height: **(a)** empirical orthogonal function (spatial field, in centimetres); **(b)** principal component (temporal variability, dimensionless).

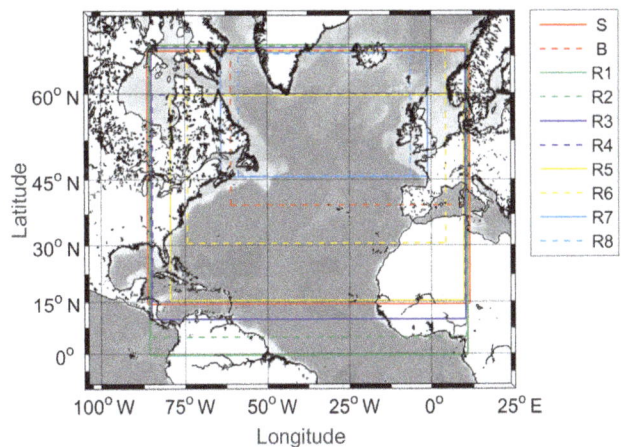

Figure 3. Map of defined regions to investigate sensitivity of SPG-I to the chosen spatial coverage. Details of boundaries are in Table 1. S: S. Häkkinen time series; B: time series presented here; R: time series based on different regions; 1X1: time series presented here based on a 1° by 1° grid; 2X2 grid: time series presented here based on a 2° by 2°; all other time series presented here are based on a 1/4° by 1/4° grid.

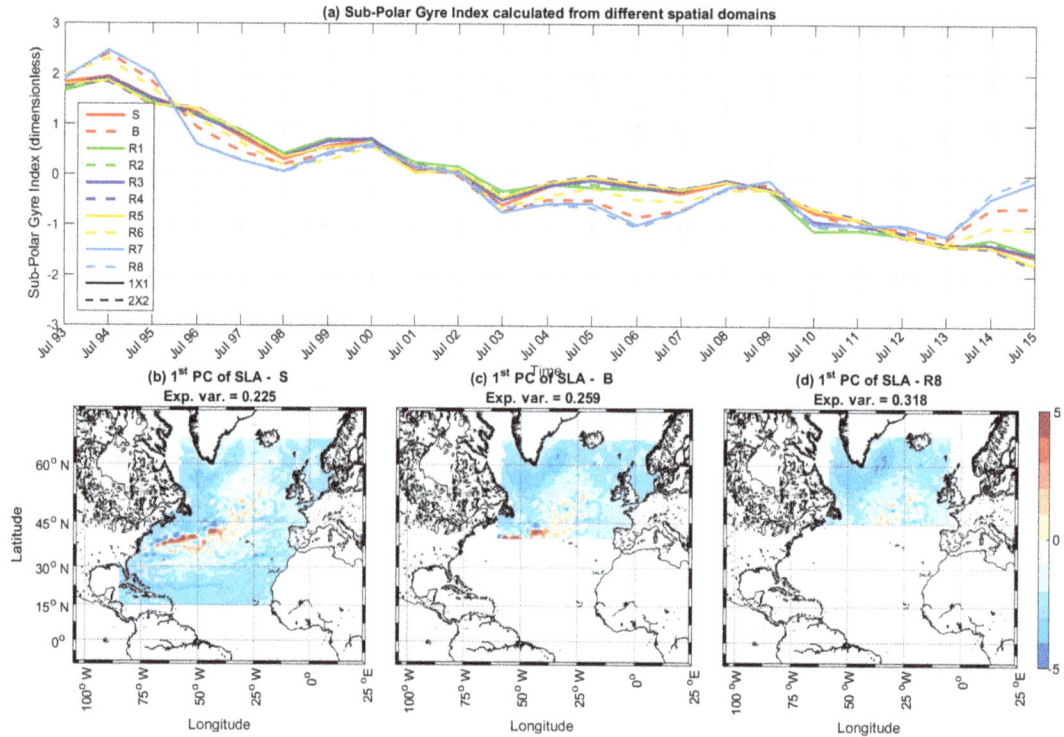

Figure 4. Comparison of yearly SPG-I for different domains within the North Atlantic: **(a)** yearly SPG-I time series; **(b–d)** corresponding empirical orthogonal function (spatial field, in centimetres) for three domains. Table 1 and Fig. 3 show the defined areas used in the EOFA. Colours in **(a)** correspond to those used in Fig. 3.

(Sect. 3). For the user, this means the entire time series needs to be updated when the temporal coverage of the data is updated. To investigate the impact of the chosen spatial extent of data included in the EOFA, we performed the analysis on a number of different regions centred on the sub-polar North Atlantic. These results were based on the $1/4°$ by $1/4°$ grid, although time series based on the $1°$ by $1°$ grid and $2°$ by $2°$ grids are also presented. The longitude and latitude limits of these regions are listed in Table 1 and shown in Fig. 3.

3 SPG-I time series

3.1 Interpretation

Figure 2 shows the first principal component (i.e. temporal variability) of the SPG-I and corresponding empirical orthogonal function (i.e. the spatial field). The spatial field shows the signature of the Gulf Stream–North Atlantic Current variability (eddy-like features in Fig. 2a) as well as the lower sea level in the sub-polar gyre region compared to the sub-tropical Atlantic. This first mode of the EOFA explains 25.9 % of the total variance. The second and third mode explain 8.6 and 6.4 % of the variance, respectively. Their interpretation is less studied, and as they are unrelated to published work on the sub-polar North Atlantic, they are not discussed further here.

In the SPG-I time series (Fig. 2b), positive values of the index are associated with a strong sub-polar gyre circulation with a wide spread. In comparison, negative values of SPG-I are associated with a weak sub-polar gyre and westward retraction. The commonly reported weakening and contraction of the sub-polar gyre can be seen in the mid to late 90s (Hátún et al., 2005; Häkkinen and Rhines, 2004). More recently the sub-polar gyre has been variable but remains weak (Fig. 2b).

The exact mechanism whereby changes in the sub-polar gyre state influence circulation across the North Atlantic basin continues to be an area of active research. Hátún et al. (2009b) present a schematic representation of how a weaker sub-polar gyre allows for greater advection of sub-tropical water masses (see their Fig. 2). The changes in the sub-polar gyre have recently been proposed to be a response to a prolonged positive NAO phase (Lohmann et al., 2009; Häkkinen et al., 2011; Robson et al., 2012). The atmospheric forcing and storm track variability strengthen the Atlantic Meridional Overturning Circulation, enhancing the northward transport of heat in the oceans and subsequent negative feedback on the sub-polar gyre circulation. Both Lohmann et al. (2009) and Robson et al. (2012) highlight the importance of the ocean's heat transport in the dynamics of the sub-polar gyre.

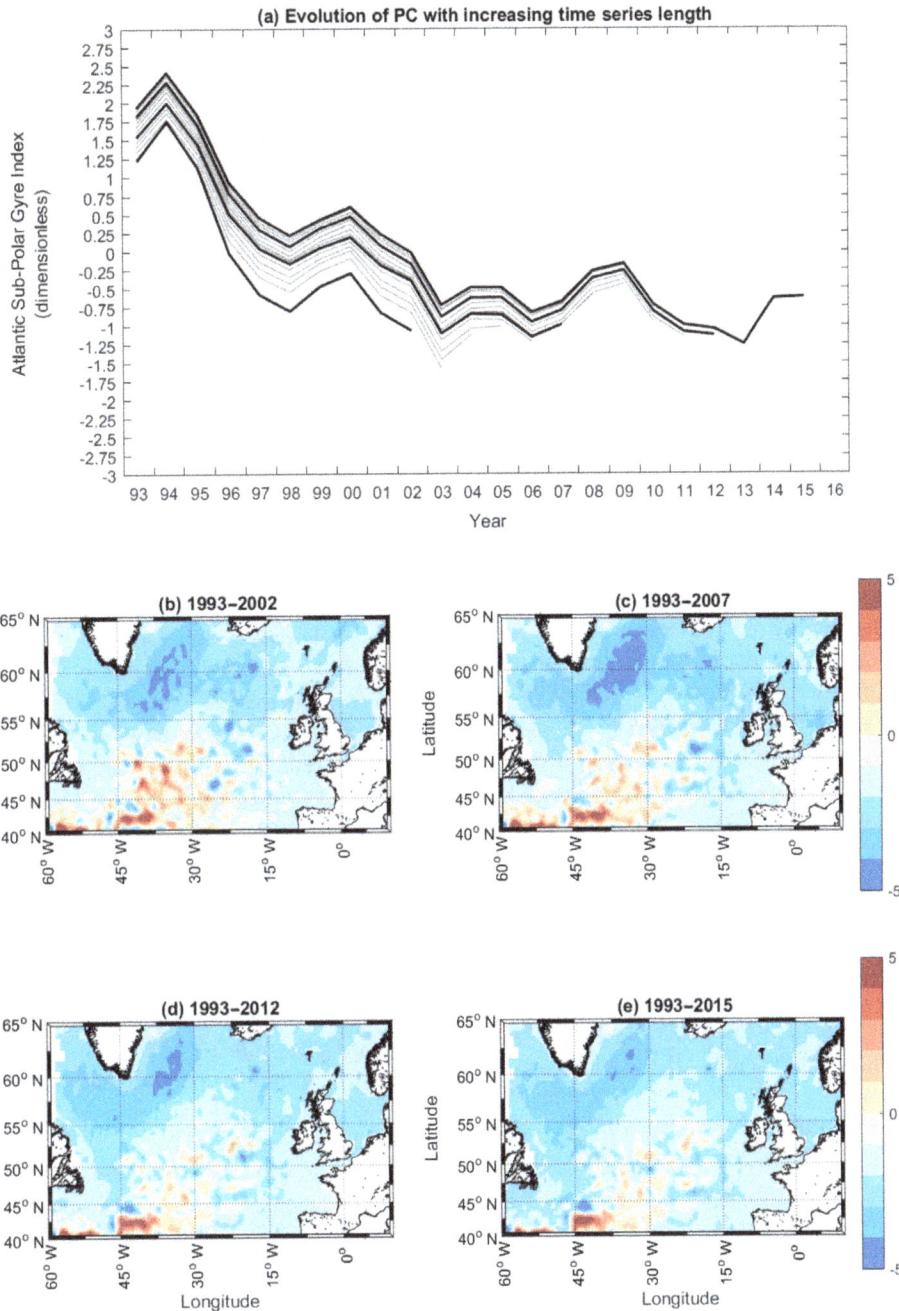

Figure 5. Sensitivity of yearly SPG-I to time series length: (**a**) principal component (temporal variability, dimensionless) for time periods from 1993 to 2002, increasing 1 year in length (grey lines) with periods shown in (**b–d**) in bold black lines; empirical orthogonal function (spatial field, in centimetres) for four time periods: (**b**) 1993 to 2002, (**c**) 1993 to 2007, (**d**) 1993 to 2012, and (**e**) 1993 to 2015.

3.2 Sensitivity to spatial extent

To investigate the sensitivity of the SPG-I to the chosen spatial domain for the EOFA, the analysis was repeated for a number of different regions (Fig. 3 and Table 1). The results for the annual SPG-I, shown in Fig. 4a, show that a separation does occur between indices focused solely on the sub-polar region (time series B, R6, R7, and R8 in Fig. 4a) and those incorporating a wider region of the North Atlantic. In particular, when including the region in the tropical North Atlantic Ocean, the SPG-I becomes strongly linear. The correlations between time series based on these various regions are high (all greater or equal to 0.90, $p < 0.001$). The correlation coefficients also highlight that indices calculated from regions restricted to the sub-polar North Atlantic correlate well to each other but less well to those calculated over the

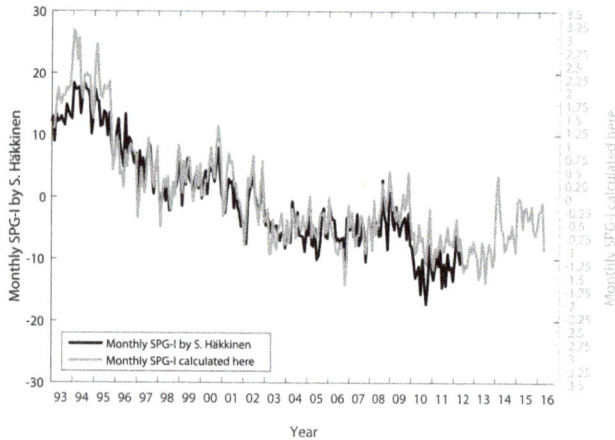

Figure 6. Comparison of SPG-I calculated here with that calculated by S. Häkkinen, based on monthly time series

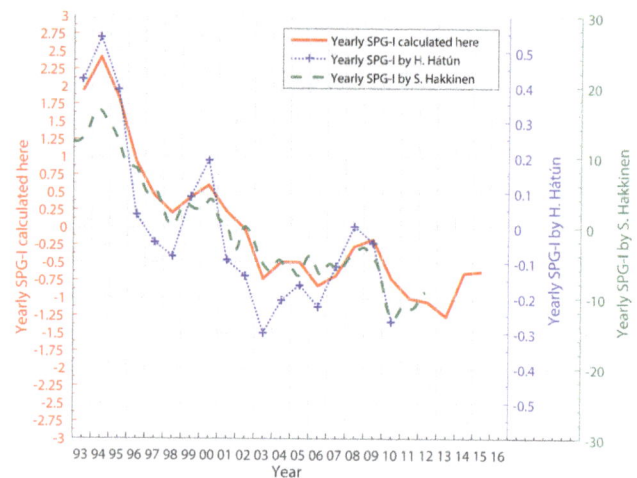

Figure 7. Comparison of yearly SPG-I with that calculated independently by S. Häkkinen and H. Hátún. See text for more details.

wider North Atlantic, with r values dropping from 0.99 to approx. 0.92. Linear detrending of the index time series prior to the correlation analysis does not influence this outcome: the lower correlation coefficients are between 0.53–0.98, but all remain statistically significant and exhibit the same pattern. This pattern is due to the fact that the SPG-I calculated for regions narrowly confined to the sub-polar North Atlantic has a higher level of inter-annual variability.

3.3 Sensitivity to time series length

To highlight the impact of increasing time series length on the SPG-I, the analysis was performed on the first 10 years of altimeter data (1993 to 2012), increasing time series length by time increments of 1 year (Fig. 5). The analysis highlights to the user the need to access the latest SPG-I time series: although the overall pattern of the SPG-I in preceding years remains unchanged, there are minor changes in the values of the time series. When updating the index, users should always download the entire SPG-I time series. The consistency of the index time series with increasing time series length, as seen in Fig. 5, confirms that the SPG-I can be defined robustly as the first PC of an EOFA of SLA (sea level anomaly) and that the statistical mode of the analysis has a dynamical meaning.

3.4 Comparison to previous results

The SPG-I presented here has been compared to similar indices previously presented by S. Häkkinen (Häkkinen and Rhines, 2004, 2009) and H. Hátún (Hátún et al., 2009a, b). A first comparison (Fig. 6) shows the comparison of the monthly resolution SPG-I time series to that previously presented by S. Häkkinen (S. Häkkinen, personal communication, 2013). These two time series show close agreement, and minor differences are likely due to different underlying altimeter data products and the change in time series extent.

There is also a difference in scaling between these two indices by 2 orders of magnitude, most likely due to differing units (centimetres vs. metres). However, as the typical usage of the data set is in terms of its correlations with other variables, rather than interpreting its absolute value, this discrepancy is not seen as important. In a second comparison (Fig. 7), the annual filtered time series by S. Häkkinen has been compared to the annual SPG-I calculated here and the time series estimated by H. Hátún based on annual mean SSH (sea surface height) data (H. Hátún, personal communication, 2012). Again, the overall pattern of these three indices shows good agreement, with all showing a clear reduction in the SPG-I in the mid to late nineties. The difference between this third index time series in comparison to the previous two is due to a different underlying altimeter product (a 1/3° by 1/3° Mercator grid, which placed additional emphasis on the sub-polar gyre region).

4 Data availability

Following the recommendations of Berx et al. (2011), we would like to ensure the SPG-I time series presented here is easily accessible and available to all. The data can therefore be downloaded from http://dx.doi.org/10.7489/1806-1 in ASCII format, and researchers can also access the supporting information there (including code to recreate the index). We are committed to updating the time series within 6 months from the time the data provider (CMEMS) publishes its updates. As the time series is based on the EOFA technique, we recommend users download the entire index time series when updating the time series they use. This will ensure the most relevant time series is used in any analyses. Version numbering will facilitate users identifying which version of the time series they are accessing. Finally, we would appreciate it if all those making use of this time series appropriately acknowledged its use by citing this paper and the digital object identifier (DOI) of the data set.

5 Summary

We have presented a time series of SPG-I based on monthly mean SLA maps obtained from CMEMS. Our time series compares well with indices presented previously. The variability in the index time series is influenced by the chosen spatial extent of the sea surface height data included in the EOFA, and inter-annual variability is suppressed when including the wider North Atlantic region. We also presented an indication of changes with temporal coverage of the data series, and users need to ensure downloading the entire time series when accessing future updates. The index data product we present is freely available from http://dx.doi.org/10.7489/1806-1, and we encourage all scientists interested in establishing linkages between North Atlantic climate variability and ecosystem function to access, apply, and publish.

Competing interests. The authors declare that they have no conflict of interest.

Acknowledgements. We would like to thank Sirpa Häkkinen and Hjalmar Hátún for providing their time series of SPG-I and ancillary information relating to their calculation; we are grateful to Hjalmar Hátún and Stuart Cunningham for useful discussions on sub-polar gyre dynamics. The research leading to these results has received funding from the European Union 7th Framework Programme (FP7 2007–2013) under grant agreement number 308299 (NACLIM).

Edited by: G. M. R. Manzella

References

Bersch, M., Yashayaev, I., and Koltermann, K. P.: Recent changes of the thermohaline circulation in the subpolar North Atlantic, Ocean Dynam., 57, 223–235, 2007.

Berx, B., Dickey-Collas, M., Skogen, M. D., De Roeck, Y.-H., Klein, H., Barciela, R., Forster, R., Dombrowsky, E., Huret, M., Payne, M., Sagarminaga, Y., and Schrum, C.: Does operational oceanography address the needs of fisheries and applied environmental scientists?, Oceanography, 24, 166–171, doi:10.5670/oceanog.2011.14, 2011.

Bessières, L., Rio, M. H., Dufau, C., Boone, C., and Pujol, M. I.: Ocean state indicators from MyOcean altimeter products, Ocean Sci., 9, 545–560, doi:10.5194/os-9-545-2013, 2013.

Emery, W. and Thomson, R.: Data Analysis Methods in Physical Oceanography, Elsevier, 2nd Edn., 2001.

Häkkinen, S. and Rhines, P. B.: Decline of Subpolar North Atlantic Circulation During the 1990s, Science, 304, 555–559, doi:10.1126/science.1094917, 2004.

Häkkinen, S. and Rhines, P. B.: Shifting surface currents in the northern North Atlantic Ocean, J. Geophys. Res., 114, C04005, doi:10.1029/2008JC004883, 2009.

Häkkinen, S., Rhines, P. B., and Worthen, D. L.: Atmospheric Blocking and Atlantic Multidecadal Ocean Variability, Science, 334, 655–659, doi:10.1126/science.1205683, 2011.

Hátún, H., Sandø, A., Drange, H., Hansen, B., and Valdimarsson, H.: Influence of the Atlantic subpolar gyre on the thermohaline circulation, Science, 309, 1841–1844, 2005.

Hátún, H., Payne, M., and Jacobsen, J. A.: The North Atlantic subpolar gyre regulates the spawning distribution of blue whiting (Micromesistius poutassou Risso), Can. J. Fish. Aquat. Sci., 66, 759–770, 2009a.

Hátún, H., Payne, M. R., Beaugrand, G., Reid, P. C., Sandø, A. B., Drange, H., Hansen, B., Jacobsen, J. A., and Bloch, D.: Large bio-geographical shifts in the north-eastern Atlantic Ocean: From the subpolar gyre, via plankton, to blue whiting and pilot whales, Prog. Oceanogr., 80, 149–162, doi:10.1016/j.pocean.2009.03.001, 2009b.

Hátún, H., Lohmann, K., Matei, D., Jungclaus, J., Pacariz, S., Bersch, M., Gislason, A., Ólafsson, J., and Reid, P.: An inflated subpolar gyre blows life toward the northeastern Atlantic, Prog. Oceanogr., 147, 49–66, doi:10.1016/j.pocean.2016.07.009, 2016.

Holliday, N. P., Hughes, S., Bacon, S., Beszczynska-Möller, A., Hansen, B., Lavín, A., Loeng, H., Mork, K., Østerhus, S., Sherwin, T., and Walczowski, W.: Reversal of the 1960s to 1990s freshening trend in the northeast North Atlantic and Nordic Seas, Geophys. Res. Lett., 35, L03614, doi:10.1029/2007GL032675, 2008.

Hurrell, J. and National Center for Atmospheric Research Staff: The Climate Data Guide: Hurrell North Atlantic Oscillation (NAO) Index (PC-based)., Tech. rep., last modified: 10 May 2013, available at: https://climatedataguide.ucar.edu/climate-data/hurrell-north-atlantic-oscillation-nao-index-pc-based (last access: 5 February 2016), 2013.

ICES: Report of the Working Group on Operational Oceanographic products for Fisheries and Environment (WGOOFE), 12–16 March 2012 and 6–8 November 2012, ICES HQ, Copenhagen and Brussels, Belgium., Tech. Rep. ICES CM 2012/SSGSUE:06, 2012a.

ICES: Report of the Working Group on Widely Distributed Stocks (WGWIDE), 21-27 August 2012, Lowestoft, United Kingdom, Tech. Rep. ICES CM 2012/ACOM:15, 2012b.

Johnson, C., Inall, M., and Häkkinen, S.: Declining nutrient concentrations in the northeast Atlantic as a result of a weakening Subpolar Gyre, Deep Sea Res. Pt. I, 82, 95–107, doi:10.1016/j.dsr.2013.08.007, 2013.

Lohmann, K., Drange, H., and Bentsen, M.: A possible mechanism for the strong weakening of the North Atlantic subpolar gyre in the mid-1990s, Geophys. Res. Lett., 36, 115602, doi:10.1029/2009GL039166, 2009.

National Geophysical Data Center, N. o. C.: TerrainBase, Global 5 Arc-minute Ocean Depth and Land Elevation from the US National Geophysical Data Center (NGDC), 1995.

Overpeck, J. T., Meehl, G. A., Bony, S., and Easterling, D. R.: Climate Data Challenges in the 21st Century, Science, 331, 700–702, doi:10.1126/science.1197869, 2011.

Payne, M. R., Egan, A., Fässler, S. M., Hátún, H., Holst, J. C., Jacobsen, J. A., Slotte, A., and Loeng, H.: The rise and fall of the NE Atlantic blue whiting (Micromesistius poutassou), Mar. Biol. Res., 8, 475–487, 2012.

Robson, J., Sutton, R., Lohmann, K., Smith, D., and Palmer, M. D.: Causes of the Rapid Warming of the North Atlantic Ocean in the Mid-1990s, J. Climate, 25, 4116–4134, doi:10.1175/JCLI-D-11-00443.1, 2012.

Schneider, D. P., Deser, C., Fasullo, J., and Trenberth, K. E.: Climate Data Guide Spurs Discovery and Understanding, Eos, Transactions American Geophysical Union, 94, 121–122, doi:10.1002/2013EO130001, 2013.

Permissions

The contributors of this book come from diverse backgrounds, making this book a truly international effort. This book will bring forth new frontiers with its revolutionizing research information and detailed analysis of the nascent developments around the world.

We would like to thank all the contributing authors for lending their expertise to make the book truly unique. They have played a crucial role in the development of this book. Without their invaluable contributions this book wouldn't have been possible. They have made vital efforts to compile up to date information on the varied aspects of this subject to make this book a valuable addition to the collection of many professionals and students.

This book was conceptualized with the vision of imparting up-to-date information and advanced data in this field. To ensure the same, a matchless editorial board was set up. Every individual on the board went through rigorous rounds of assessment to prove their worth. After which they invested a large part of their time researching and compiling the most relevant data for our readers.

The editorial board has been involved in producing this book since its inception. They have spent rigorous hours researching and exploring the diverse topics which have resulted in the successful publishing of this book. They have passed on their knowledge of decades through this book. To expedite this challenging task, the publisher supported the team at every step. A small team of assistant editors was also appointed to further simplify the editing procedure and attain best results for the readers.

Apart from the editorial board, the designing team has also invested a significant amount of their time in understanding the subject and creating the most relevant covers. They scrutinized every image to scout for the most suitable representation of the subject and create an appropriate cover for the book.

The publishing team has been an ardent support to the editorial, designing and production team. Their endless efforts to recruit the best for this project, has resulted in the accomplishment of this book. They are a veteran in the field of academics and their pool of knowledge is as vast as their experience in printing. Their expertise and guidance has proved useful at every step. Their uncompromising quality standards have made this book an exceptional effort. Their encouragement from time to time has been an inspiration for everyone.

The publisher and the editorial board hope that this book will prove to be a valuable piece of knowledge for researchers, students, practitioners and scholars across the globe.

List of Contributors

Matthew P. Humphreys, Florence M. Greatrix, Eithne Tynan and Claudia H. Fry
Ocean and Earth Science, University of Southampton, Southampton, UK

Eric P. Achterberg
Ocean and Earth Science, University of Southampton, Southampton, UK
GEOMAR Helmholtz Centre for Ocean Research, Kiel, Germany

Alex M. Griffiths
Department of Earth Science and Engineering, Imperial College London, London, UK

Rebecca Garley
Bermuda Institute of Ocean Sciences, St George's, Bermuda

Alison McDonald and Adrian J. Boyce
Scottish Universities Environmental Research Centre, East Kilbride, UK

Vincent Taillandier, Fabrizio D'Ortenzio, Joséphine Ras, Laurent Coppola, Emilie Diamond, Henry Bittig, Edouard Leymarie, Antoine Poteau and Louis Prieur
Sorbonne Universités, UPMC Université Paris 06, CNRS, LOV, Villefranche-sur-Mer, 06230, France

Nicolas Mayot
Sorbonne Universités, UPMC Université Paris 06, CNRS, LOV, Villefranche-sur-Mer, 06230, France
Bigelow Laboratory for Ocean Sciences, Maine, East Boothbay, USA

Orens Pasqueron de Fommervault
Sorbonne Universités, UPMC Université Paris 06, CNRS, LOV, Villefranche-sur-Mer, 06230, France
Laboratorio de Oceanografià Fisicà, CICESE, Ensenada, B.C., Mexico

Thibaut Wagener and Dominique Lefevre
Aix-Marseille Universiteì, CNRS/INSU, Universiteì de Toulon, IRD, Mediterranean Institute of Oceanography (MIO), UM 110, Marseille, 13288, France

Hervé Legoff
Sorbonne Universités, UPMC Univ Paris 06, CNRS, IRD, MNHN, LOCEAN, Paris, France

Catherine Schmechtig
Sorbonne Universités, UPMC Univ Paris 06, CNRS, UMS 3455, OSU Ecce-Terra, Paris CEDEX 5, France

Colleen B. Mouw and Audrey B. Ciochetto
University of Rhode Island, Graduate School of Oceanography, 215 South Ferry Road, Narragansett, RI 02882, USA

Brice Grunert and Angela Yu
Michigan Technological University, 1400 Townsend Drive, Houghton, MI 49931, USA

David J. Morris, John K. Pinnegar, David L. Maxwell, Stephen R. Dye, Liam J. Fernand, Stephen Flatman, Oliver J. Williams and Stuart I. Rogers
Cefas, Lowestoft Laboratory, Pakefield, Lowestoft, Suffolk, NR33 0HT, UK

Jean-François Legeais, Michaël Ablain and Lionel Zawadzki
Collecte Localisation Satellite (CLS), 31520 Ramonville-Saint-Agne, France

Hao Zuo
European Centre for Medium-Range Weather Forecasts, Reading, UK

Johnny A. Johannessen
Nansen Environmental and Remote Sensing Center (NERSC), Bergen, Norway

Martin G. Scharffenberg
University of Hamburg, Hamburg, Germany

Luciana Fenoglio-Marc
University of Bonn, Bonn, Germany

M. Joana Fernandes
Faculdade de Ciências, Universidade do Porto, 4169-007 Porto, Portugal
Centro Interdisciplinar de Investigação Marinha e Ambiental (CIIMAR), 4450-208 Matosinhos, Portugal

Ole Baltazar Andersen
DTU Space, 2800 Kongens Lyngby, Denmark

Marcello Passaro
Deutsches Geodätisches Forschungsinstitut, Technische Universität München, 80333 Munich, Germany

Sergei Rudenko
Deutsches Geodätisches Forschungsinstitut, Technische Universität München, 80333 Munich, Germany
Helmholtz Centre Potsdam – GFZ German Research Centre for Geosciences, 14473 Potsdam, Germany

Paolo Cipollini
National Oceanography Centre, Southampton, SO14 3ZH, UK

Graham D. Quartly
Plymouth Marine Laboratory, Plymouth, PL1 3DH, UK

Anny Cazenave
LEGOS, 31400 Toulouse, France
ISSI, Bern, Switzerland

Jérôme Benveniste
ESA/ESRIN, 00044 Frascati, Italy

Colleen B. Mouw and Audrey Barnett
Michigan Technological University, 1400 Townsend Drive, Houghton, MI 49931, USA
University of Rhode Island, Graduate School of Oceanography, 215 South Ferry Road, Narragansett, RI 02882, USA

Galen A. McKinley and Lucas Gloege
University of Wisconsin-Madison, 1225 W. Dayton Street, Madison, WI 53706, USA

Darren Pilcher
NOAA, Pacific Marine Environmental Laboratory, 7600 Sand Point Way NE, Seattle, WA 98115, USA

Giuseppe Aulicino
Department of Life and Environmental Sciences, Università Politecnica delle Marche, Ancona, 60131, Italy
Department of Science and Technologies, Università degli Studi di Napoli Parthenope, Naples, 80143, Italy

Yuri Cotroneo
Department of Science and Technologies, Università degli Studi di Napoli Parthenope, Naples, 80143, Italy

Isabelle Ansorge
Marine Research Institute, Oceanography Department, University of Cape Town, Rondebosch, Cape Town, 7701, South Africa

Marcel van den Berg
Department of Environmental Affairs, Cape Town, 8001, South Africa

Cinzia Cesarano
Progetto Terra, Gragnano, 80054, Italy

Estrella Olmedo Casal
Institute of Marine Sciences, ICM, Barcelona, 08003, Spain

Maria Belmonte Rivas
Institute of Marine Sciences, ICM, Barcelona, 08003, Spain
Royal Netherlands Meteorological Institute, KNMI, De Bilt, 3730, the Netherlands

Yao Luo, Dongxiao Wang, Fenghua Zhou and Charith Madusanka Widanage
State Key Laboratory of Tropical Oceanography, South China Sea Institute of Oceanology, Chinese Academy of Sciences, Guangzhou 510301, China

Tilak Priyadarshana Gamage
University of Ruhuna, Matara 810000, Sri Lanka

Taiwei Liu
China Harbour Engineering Company Ltd., Beijing 100027, China

Athanasia Iona
Hellenic Centre for Marine Research, Institute of Oceanography, Hellenic National Oceanographic Data Centre, 46,7 km Athens Sounio, Mavro Lithari Anavissos, Attica, Greece
University of Thessaly, Department of Ichthyology & Aquatic Environment, Laboratory of Oceanography, Fytoko Street, 38 445, Nea Ionia Magnesia, Greece

Athanasios Theodorou
University of Thessaly, Department of Ichthyology & Aquatic Environment, Laboratory of Oceanography, Fytoko Street, 38 445, Nea Ionia Magnesia, Greece

Sylvain Watelet, Charles Troupin and Jean-Marie Beckers
University of Liège, GeoHydrodynamics and Environment Research, Quartier Agora, Allée du 6-Août, 17, Sart Tilman, 4000 Liège 1, Belgium

Simona Simoncelli
Istituto Nazionale di Geofisica e Vulcanologia (INGV), Sezione di Bologna, Via Franceschini 31, 40128 Bologna, Italy

Graham D. Quartly
Plymouth Marine Laboratory, Plymouth, PL1 3DH, UK

Jean-François Legeais, Michaël Ablain, Lionel Zawadzki, Loren Carrère and Jean-Christophe Poisson
CLS, 31520 Ramonville-Saint-Agne, France

M. Joana Fernandes
Faculdade de Ciências, Universidade do Porto, 4169-007, Porto, Portugal
Centro Interdisciplinar de Investigação Marinha e Ambiental (CIIMAR), 4450-208 Matosinhos, Portugal

Sergei Rudenko
Deutsches Geodätisches Forschungsinstitut, Technische Universität München, 80333 Munich, Germany
Helmholtz Centre Potsdam GFZ German Research Centre for Geosciences, Telegrafenberg 14473 Potsdam, Germany

Pablo Nilo García
isardSAT, 08042 Barcelona, Spain

Paolo Cipollini
National Oceanography Centre, Southampton, SO14 3ZH, UK

Ole B. Andersen
DTU Space, 2800 Kongens Lyngby, Denmark

Sabrina Mbajon Njiche
CGI, Leatherhead, KT22 7LP, UK

Anny Cazenave
LEGOS, 31400 Toulouse, France
ISSI, 3912 Bern, Switzerland

Jérôme Benveniste
ESA/ESRIN, 00044 Frascati, Italy

André Valente and Vanda Brotas
Marine and Environmental Sciences Centre (MARE), University of Lisbon, Lisbon, Portugal

Shubha Sathyendranath, Steve Groom, Michael Grant and Timothy Smyth
Plymouth Marine Laboratory, Plymouth, PL1 3DH, UK

Malcolm Taberner
EUMETSAT, Eumetsat-Allee 1, 64295 Darmstadt, Germany

David Antoine
Sorbonne Universités, UPMC Univ. Paris 06, CNRS, Laboratoire d'Océanographie de Villefranche, Villefranche-sur-mer, 06238, France
Remote Sensing and Satellite Research Group, Department of Physics, Astronomy and Medical Radiation Sciences, Curtin University, Perth, WA 6845, Australia

Robert Arnone
University of Southern Mississippi, Stennis Space Center, Kiln, MS, USA

William M. Balch
Bigelow Laboratory for Ocean Sciences, East Boothbay, ME, USA

Kathryn Barker
ARGANS Ltd, Plymouth, UK

Ray Barlow
Bayworld Centre for Research and Education, Cape Town, South Africa

Simon Bélanger
Département de biologie, chimie et géographie, Université du Québec à Rimouski, Rimouski (Québec), Canada

Jean-François Berthon, Elisabetta Canuti and Giuseppe Zibordi
European Commission, Joint Research Centre, Ispra, Italy

Şükrü Beşiktepe
Institute of Marine Science and Technology, Dokuz Eylul University, Izmir, Turkey

Vittorio Brando
CSIRO Oceans and Atmosphere, Canberra, Australia
CNR IREA, Milan, Italy

Francisco Chavez
Monterey Bay Aquarium Research Institute, Moss Landing, CA, USA

Hervé Claustre
Laboratoire d'Océanographie de Villefranche (LOV), Sorbonne Universités, UPMC Univ Paris 06, INSU-CNRS, 181 Chemin du Lazaret, 06230 Villefranche-sur-Mer, France

Richard Crout and Richard Gould
Naval Research Laboratory, Stennis Space Center, Kiln, MS, USA

Robert Frouin, Mati Kahru and Brian G. Mitchell
Scripps Institution of Oceanography, University of California, San Diego, CA, USA

Carlos García-Soto
Spanish Institute of Oceanography (IEO), Corazón de María 8, 28002 Madrid, Spain
Plentziako Itsas Estazioa/Euskal Herriko Unibetsitatea (PIE/EHU), Areatza z/g, 48620 Plentzia, Spain

Stuart W. Gibb
Environmental Research Institute, North Highland College, University of the Highlands and Islands, Thurso, Scotland, UK

Stanford Hooker and Jeremy Werdell
NASA Goddard Space Flight Center, Greenbelt, Maryland, USA

Holger Klein
Operational Oceanography Group, Federal Maritime and Hydrographic Agency, Hamburg, Germany

Susanne Kratzer
Department of Ecology, Environment and Plant Sciences, Frescati Backe, Stockholm University, 106 91 Stockholm, Sweden

Hubert Loisel
Laboratoire d'Océanologie et de Géosciences, Université du Littoral – Côte d'Opale, Maison de la Recherche en Environnement Naturel, Wimereux, France

David McKee
Physics Department, University of Strathclyde, Glasgow G4 0NG, Scotland, UK

Tiffany Moisan
NASA Goddard Space Flight Center, Wallops Flight Facility, Wallops Island, VA, USA

Frank Muller-Karger
Institute for Marine Remote Sensing/ImaRS, College of Marine Science, University of South Florida, St. Petersburg, FL, USA

Leonie O'Dowd
Fisheries and Ecosystem Advisory Services, Marine Institute, Rinville, Oranmore, Galway, Ireland

Michael Ondrusek
NOAA/NESDIS/STAR/SOCD, College Park, MD, USA

Alex J. Poulton
Ocean Biogeochemistry and Ecosystems, National Oceanography Centre, Waterfront Campus, Southampton, UK

Michel Repecaud
IFREMER Centre de Brest, Plouzane, France

Heidi M. Sosik
Biology Department, Woods Hole Oceanographic Institution, Woods Hole, MA, USA

Michael Twardowski
Harbor Branch Oceanographic Institute, Fort Pierce, FL, USA

Kenneth Voss
Physics Department, University of Miami, Coral Gables, FL, USA

Marcel Wernand
Physical Oceanography, Marine Optics & Remote Sensing, Royal Netherlands Institute for Sea Research, Texel, Netherlands

Athanasia Iona
Hellenic Centre for Marine Research, Institute of Oceanography, Hellenic National Oceanographic Data Centre, 46,7 km Athens Sounio, Mavro Lithari Anavissos, Attica, Greece
University of Thessaly, Department of Ichthyology & Aquatic Environment, Laboratory of Oceanography, Fytoko Street, 38445, Nea Ionia Magnesia, Greece

Athanasios Theodorou
University of Thessaly, Department of Ichthyology & Aquatic Environment, Laboratory of Oceanography, Fytoko Street, 38445, Nea Ionia Magnesia, Greece

Sarantis Sofianos
Ocean Physics and Modelling Group, Division of Environmental Physics and Meteorology, University of Athens, University Campus, Phys–5, 15784 Athens, Greece

Sylvain Watelet, Charles Troupin and Jean-Marie Beckers
University of Liège, GeoHydrodynamics and Environment Research, Quartier Agora, Allée du 6-Août, 17, Sart Tilman, 4000 Liège 1, Belgium

Guðrún Nína Petersen
Icelandic Meteorological Office, Bústaðavegi 9, 108 Reykjavík, Iceland

Amelie Driemel, Eberhard Fahrbach, Gerd Rohardt, Antje Boetius, Gereon Budéus, Walter Geibert, Dieter Gerdes, Rainer Gersonde, Hannes Grobe, Hartmut H. Hellmer, Markus Janout, Wilfried Jokat, Gerhard Kuhn, Benjamin Rabe, Ursula Schauer, Michael Schröder, Stefanie Schumacher, Rainer Sieger, Thomas Soltwedel, Volker H Strass, Jörn Thiede, Sandra Tippenhauer,Wilken-Jon von Appen, Andreas Wisotzki, Dieter A. Wolf-Gladrow and Torsten Kanzow
Alfred-Wegener-Institut Helmholtz-Zentrum für Polar- und Meeresforschung, Bremerhaven, Germany

Agnieszka Beszczynska-Möller
Institute of Oceanography Polish Academy of Science, Sopot, Poland

Ralph Engbrodt, Steffen Gauger, Elena Stangeew and Marie-France Weirig
Independent researcher

Boris Cisewski and Manfred Stein
Thünen-Institut: Seefischerei, Hamburg, Germany

Patrizia Geprägs
MARUM – Zentrum für Marine Umweltwissenschaften, Bremen, Germany

Arnold L. Gordon and Stanley S. Jacobs
Lamont-Doherty Earth Observatory, Columbia University, New York, NY, USA

Enrique Isla
Institute of Marine Sciences-CSIC, Barcelona, Spain

Michael Klages
University of Gothenburg, Department of Marine Sciences, Gothenburg, Sweden

Jens Meincke
Institut für Meereskunde, Hamburg, Germany

Sven Ober and Cornelis Veth
Royal Netherlands Institute for Sea Research, 't Horntje, the Netherlands

Svein Østerhus
Uni Research Climate, Bergen, Norway

Ray G. Peterson
Scripps Institution of Oceanography, UC San Diego, USA

Bert Rudels
University of Helsinki, Helsinki, Finland

Jüri Sildam
Centre for Maritime Research and Experimentation, La Spezia, Italy

Rebecca Latto and Anastasia Romanou
Applied Physics and Applied Math, Columbia University, New York, USA
NASA-GISS, New York, NY, USA

Barbara Berx
Marine Scotland Science, 375 Victoria Road, Aberdeen, AB11 9DB, UK

Mark R. Payne
Centre for Ocean Life, Technical University of Denmark, National Institute of Aquatic Resources, 2920 Charlottenlund, Denmark

Index

www.ingramcontent.com/pod-product-compliance
Lightning Source LLC
Chambersburg PA
CBHW082100190326
41458CB00010B/3535